Shorebirds of the Pacific Northwest

Shorebirds
of the
Pacific Northwest

DENNIS PAULSON

Drawings by Jim Erckmann

UNIVERSITY OF WASHINGTON PRESS *Seattle and London*
SEATTLE AUDUBON SOCIETY

Copyright © 1993 by the University of Washington Press
First paperback edition, with corrections, 1998
Printed in Singapore by CS Graphics

Library of Congress Cataloging-in-Publication Data

Paulson, Dennis R.
Shorebirds of the Pacific Northwest / Dennis Paulson ; drawings by Jim Erckmann.
  p. cm.
Includes bibliographical references and index.
ISBN 0-295-97706-X (alk. paper)
 1. Shorebirds—Northwest, Pacific. 2. Birds—Northwest, Pacific. I. Title.
AL683.P36P38 1992   92-19050
598.3'3'09795—dc20   CIP

The paper used in this publication meets the minimum requirements of American National Standard for Information Sciences—Permanence of Paper for Printed Library Materials, ANSI Z39.48–1984. ⊗

Title-page illustration: Black-necked Stilt. Photo by Jo Anne Rosen

This book is dedicated to Ralph Hoffmann and Allan Brooks, who showed by dynamic prose and crisp illustrations in their inspiring *Birds of the Pacific States* that every bird species is memorable.

Contents

Illustrations

Figures

Color photographs

Distribution maps

Tables

Appendixes

Preface

THE PACIFIC NORTHWEST is a shorebird place. It is full of shorebird habitats, wet and dry. It lies far enough north for subarctic breeding species and far enough south for those of desert wetlands. Its temperate estuaries provide sustenance for great flocks of shorebirds through the winter. And it is positioned to capture migrants from all over the Northern Hemisphere, from the Pacific species that dominate it to the interior and Siberian species that come and go with the shifting winds.

The outer coast, a sharp border between land and sea, points a compass line for swift flocks of sandpipers and plovers that follow the sunlight to high latitudes every spring. Many of the same birds and many of their young return south almost before we have had time to miss them, to stay for our wet, gray winter or, continuing their journey, to seek a warmer, brighter one. The shorebirds that follow this course move by the dozens, hundreds, thousands, even millions up and down the Northwest coast, year after year. Our nearness to their northern breeding areas allows us to see them briefly in a different light, as many of them sing and display to one another on their northward rush, with a little of the same behavior—perhaps persistent hormones—left over in the earliest southbound birds each fall.

The coast's wave-washed rocks, sandy beaches, and protected estuaries are rest stops for the shorebird hordes. The rocks and beaches have their specialists, finely adapted to feed in these special habitats. But it is the estuaries, spaced along the ebb and flow of migratory movements and lined with mud so full of burrowing, crawling life it vibrates, that support an even greater variety of species. At the height of migration they arrive in flocks that at times seem to fill the sky. They spread out as they land, bills probing almost before their feet touch the ground. They stitch the mud again and again at each low tide, turning the seemingly inexhaustible invertebrate inventory into shorebird tissue. They refuel for some days and then, swaddled in sustaining fat, are on their way again to north or south. The concentrations of such birds in spring in Grays Harbor exceed those of shorebirds anywhere else on the American Pacific coast south of Alaska.

In protected straits and sounds away from the outer coast the spectacle is repeated on a smaller scale—in trickles rather than floods—with the flocks and the birds in them actually countable. Freshwater habitats are not prominent in the coastal lowlands of the Northwest, but each has its own shorebirds, a few breeding species and a potpourri of migrants in flooded pastures in spring and drying pond margins in fall.

In the Northwest interior, avian habitats succeed one another with each change of elevation and life zone. Far fewer shorebirds use the relatively arid country east of the Cascades, but they include species of marshy lakes and ponds and dry grasslands that breed only in this part of the region. Flooded in spring, lakes and ponds offer minimal habitat for migratory species, but, as they shrink during summer heat, their richly populated shores are exposed, and these scattered interior mud flats supplement the much more extensive coastal ones. In the interior it is the big humanmade reservoirs that furnish vital stopover spots, making up in part for the loss of natural wetlands.

The nutrients that supply the essential elements of the bed-and-breakfast basins of the shoreline are carried by lowland rivers from all the surrounding landscape, upstream almost to the mountain divides. It is clear that all parts of the region are important to shorebirds.

This book is timely because the estuaries and other wetlands that these beautiful and far-flying birds use are among the most endangered environments of the Northwest. As part of hemisphere-wide migration systems, they are important to birds occurring from the Alaskan and Canadian Arctic to the coasts of Peru and Chile.

With details on distribution and status, migration and molt patterns, and ways to distinguish not only different species but sex and age classes within each species, this book will contribute to our understanding of Northwest shorebirds. The more we know about the lives of these birds in this region, the better we will be able to manage their dwindling habitats and the more of them we will have to enjoy in the future. We hope the present volume will stimulate both the intellect and the emotions of its readers to work toward that goal.

Shorebirds of the Pacific Northwest is truly a joint venture. The author takes responsibility for the text, but it has been strengthened by the artist's knowledge of the birds. The artist takes responsibility for the illustrations, which in turn were strengthened by the critical eye of the author, in the midst of trying to determine exactly what each of these birds looks like. At times plates were being prepared and text written simultaneously, the division between the two blurring in our minds.

The separate sparks of interest in shorebird biology and in shorebird identification, shared strongly by writer and illustrator, come together here in what we hope will be a contribution to both of these fields.

Acknowledgments

THE QUALITY OF A BOOK such as this depends on many more people than those whose names are plainly visible on the cover. We are especially grateful to Jon Dunn, who offered many suggestions for fine-tuning the species accounts. His long field experience with most of the world's shorebirds, critical eye, and enthusiasm for the group enabled him to be a most constructive critic. Paula Thurman read the entire manuscript and greatly improved its organization and readability. Tom Schooley and Gary Mozel also criticized parts of the manuscript.

We were helped in many other ways, for which we wish to thank the following individuals. Phil Mattocks (Washington), Jeff Gilligan and Owen Schmidt (Oregon), and Dan Stephens (Idaho) checked status accounts for their areas of expertise. Phil Mattocks and Ellen Ratoosh (Washington), Charles Trost and Ian Paulsen (Idaho), David Beaudette, Winton Weydemeyer, and Philip Wright (Montana), and Dan Gibson (Alaska) submitted compilations of or information about shorebird records. Per Alström, Urban Olsson, and Claudia Wilds gave us tips on shorebird identification. Linda Feltner helped with graphic design. Wayne Campbell made analysis of shorebird occurrence in British Columbia much easier by providing us with the manuscript for a new book on birds of that province. Ted Miller donated a very helpful set of records of Russian shorebird vocalizations. Peter Connors furnished unpublished information about golden-plovers.

The following individuals facilitated examination of specimens in their care: Sievert Rohwer, Chris Wood, and Carol Spaw (Burke Museum, University of Washington); Gordon Alcorn and Terry Mace (Slater Museum, University of Puget Sound); Richard Johnson (Conner Museum, Washington State University); Richard Cannings (Cowan Vertebrate Museum, University of British Columbia); Robert Cannings and Wayne Campbell (Royal British Columbia Museum); Stephen Bailey (Museum of Vertebrate Zoology, University of California, and California Academy of Sciences); Robert Storer, Steven Hoffman, and Janet Hinshaw (Museum of Zoology, University of Michigan); Van Remsen, Steven Cardiff, and Donna Dittman (Museum of Zoology, Louisiana State University); Scott Wood and Stephen Rogers (Carnegie Museum of Natural History); and Philip Angle, Ralph Browning, and Roger Clapp (National Museum of Natural History).

Many photographers submitted material for our use, most of them for no gratification beyond public service. We greatly appreciate the efforts of Robert Armstrong, Robert Ashbaugh, Richard Chandler, Harold Christenson, Herbert Clarke, Tom Crabtree, Richard Droker, Linda Feltner, Jeff Gilligan, Edward Harper, Susan Hills, J. R. Jehl, Jr., Karl Kenyon, Jim Oakland, Urban Olsson, Wayne Petersen, Jan Pierson, Doug Plummer, Jo Anne Rosen, Richard Rowlett, Owen Schmidt, Takeshi Shiota, Robert Sundstrom, Joseph Van Os, Richard Veit, and Tim Zurowski to furnish us with beautiful as well as educational shorebird images.

Seattle Audubon Society played a very important part in this project from beginning to end, and we thank them for their continued enthusiasm and financial support. We would like to single out for special thanks Bob Grant, David Hutchinson, and Ann Musché, each of whom put in many hours of volunteer time to keep the book on track. Many other friends encouraged us throughout the incubation, among them Eileen Bryant, Wayne Campbell, Ben and Linda Feltner, Thero North, and Tom Schooley. We also thank the editors and staff of the University of Washington Press, with whom we have had a long and harmonious working relationship.

Above all we thank Netta Smith and Lynn Erckmann for their great patience as we spent weekend after weekend deliberating on shorebirds instead of working or playing at home.

Shorebirds of the Pacific Northwest

Fig. 1. Map of region

Introduction

THIS BOOK IS INTENDED to be a supplement to field guides, a "fact book" that increases the user's knowledge base rather than serving as a sole reference. Its purpose will have been fulfilled if it both increases the reader's interest in and appreciation of shorebirds and enables her or him to go into the field and correctly identify them.

GEOGRAPHIC COVERAGE

The geographic area covered by this book includes southern British Columbia from the latitude of the north tip of Vancouver Island, just including Cache Creek, Eagle Bay, and Mount Assiniboine on the mainland, south; Washington; Oregon; Idaho; and western Montana, east to the Continental Divide and the eastern boundaries of Silver Bow and Madison counties (Figure 1). This is not exactly the classical definition of the "Pacific Northwest," but it makes sense to us to include (a) southern British Columbia, the home of so many bird enthusiasts and so many bird records, and (b) western Montana, which relates biologically to the rest of the region and is appropriately included to fill out an approximate square.

The region is divided into two *subregions* for the purpose of discussing status of the species: COAST, which includes not only the coast itself but the entire area west of the Cascades crest; and INTERIOR, which includes that part of the region east of the Cascades crest. Within the coast subregion smaller areas are often distinguished: OUTER COAST, the shoreline of Vancouver Island, Washington, and Oregon immediately adjacent to the Pacific Ocean; PROTECTED WATERS, the marine environment away from the outer coast, i. e., the Strait of Georgia, Juan de Fuca Strait, Puget Sound, and the islands in these bodies of water; PUGET SOUND LOWLANDS, the lowland trough in Washington between the Olympic and Cascade mountains; and WILLAMETTE VALLEY, the lowland trough in Oregon between the Coast Range and Cascade Mountains. Within the interior subregion the following area names are used: OKANAGAN VALLEY, the valley of that river in south-central British Columbia; COLUMBIA RIVER BASIN, the central lowlands of Washington drained by the Columbia River; OREGON LAKES, the lakes from Klamath to Malheur on the southern plateau of Oregon; and SNAKE RIVER VALLEY, the southern lowlands of Idaho drained by the Snake River.

All localities cited in the text are included in a Gazetteer (Appendix 1), keyed to the map of the area (Figure 1).

SPECIES INCLUDED

This large group is represented by 42 regularly (annually) occurring species and an additional 20 species that vary from very rare (more than six records, not quite annual) to accidental (one record) in the region (Appendix 2). These 62 species, all documented from the region by specimens, photographs, or seemingly unequivocal sight records (American Oystercatcher and American Woodcock) are given full treatment in the text.

Shorebirds are highly migratory, and any of the Asian species known from Alaska or to the east or south of here might occur in the Northwest. Additionally, a few species originating to the east and south of this region might be expected to occur as vagrants. Thus we are including at least brief discussions of 16 "potential" species that could be expected to occur in the Northwest from their known distribution, although thus far unrecorded. There are sight records of a few of

Table 1. First Northwest Records of Rare Shorebirds

1887	White-rumped Sandpiper (October 18)	1977	Mongolian Plover (September 11)
1919	Hudsonian Godwit (July 7)	1978	Rufous-necked Stint (June 24)
1931	Bar-tailed Godwit (October 30)		Spoonbill Sandpiper (July 30)
1934	Eurasian Dotterel (September 3)		Great Knot (September 29)
1936	Mountain Plover (July 25)	1981	American Oystercatcher (April 19)
1960	American Woodcock (March 5)		Long-toed Stint (September 5)
1969	Bristle-thighed Curlew (May 31)	1982	Temminck's Stint (September 1)
1970	Spotted Redshank (October 17)	1984	Far Eastern Curlew (September 24)
1971	Ruff (August 7)	1985	Little Stint (September 7)
1972	Curlew Sandpiper (May 10)	1987	Terek Sandpiper (July 21)
1975	Gray-tailed Tattler (October 13)	1990	Piping Plover (July 13)

the potential species that I do not consider documented sufficiently for full inclusion, although some of them may be valid. Most of the last 15 species to be added to the Northwest avifauna (Table 1) would have been on a list of "expected" species, but two of them—American Oystercatcher and Spoonbill Sandpiper—would have been considered very unlikely to occur. Of the 14 species still considered potential, I consider the Common Greenshank, Wood Sandpiper, and Common Sandpiper the most likely to occur in this region.

Because of the wide distribution of so many species, the 78 species of shorebirds discussed herein include all the regularly occurring species of North America. We are not including two species that have occurred in North America only as visitors to the tropical border of the United States (Double-striped Thick-knee [*Burhinus bistriatus*] and Northern Jacana [*Jacana spinosa*]) and five species that have been recorded only in the eastern half of the continent as wanderers from western Europe (Northern Lapwing [*Vanellus vanellus*], Greater Golden-Plover [*Pluvialis apricaria*], Slender-billed Curlew [*Numenius tenuirostris*], Eurasian Curlew [*N. arquata*], and Eurasian Woodcock [*Scolopax rusticola*]).

All Eurasian shorebirds that breed in Siberia north of Sakhalin and east of 140° E. have now been recorded in North America, with the exception of the Eurasian Oystercatcher (*Haematopus ostralegus*), and that species is a candidate for occurrence on this continent. Less likely in Alaska and the Northwest, but by no means impossible based on the occurrence of equally unlikely species, are a group of additional migratory species that breed as far north as Korea and east of 120° E. (Table 2). All of these species are illustrated in Wild Bird Society of Japan (1982) and Hayman et al. (1986).

Common and scientific names of all species included in the text are from American Ornithologists' Union (1983), with the exception of the golden-plovers, for which we anticipate

Table 2. Asian Shorebirds of Potential Occurrence in Northwest

Northern Lapwing (*Vanellus vanellus*)	Eurasian Curlew (*Numenius arquata*)
Gray-headed Lapwing (*V. cinereus*)	Asiatic Dowitcher (*Limnodromus semipalmatus*)
Long-billed Plover (*Charadrius placidus*)	Pin-tailed Snipe (*Gallinago stenura*)
Oriental Plover (*C. veredus*)	Swinhoe's Snipe (*G. megala*)
Eurasian Oystercatcher (*Haematopus ostralegus*)	Latham's Snipe (*G. hardwickii*)
Nordmann's Greenshank (*Tringa guttifer*)	Solitary Snipe (*G. solitaria*)
Common Redshank (*T. totanus*)	Eurasian Woodcock (*Scolopax rusticola*)

an imminent change in taxonomic status. Subspecies names are from American Ornithologists' Union (1957), with some further modification (references under each species). For etymology of the Greek and Latin scientific names, see Terres (1980), Johnsgard (1981: 467-72), or Choate (1985).

REFERENCES

We are blessed with the recent publication of very high-quality books about shorebirds, and we are not faced, as we might have been only a decade ago, with the onerous task of attempting to correct published statements that range from slightly to grossly misleading.

Four recently published books can be considered standards for North American shorebird field identification. *Shorebirds of the World*, by Hayman, Marchant, and Prater (1986), is a profusely illustrated and information-packed compendium of worldwide scope. It is as fine a book treating a major taxonomic group of birds as I have seen. *The National Geographic Society Field Guide to North American Birds*, by The National Geographic Society (1987), and *The Audubon Society Master Guide to Birding*, by Farrand (1983), are the best available for North American shorebirds. These books complement each other because of their respective use of paintings and photographs to illustrate species. The former suffers because of the varied quality of its paintings, the latter because when photographs were unavailable, paintings were substituted for them, and the two illustration styles do not lend themselves well to direct comparison. *A Field Guide to Western Birds*, by Peterson (1990), will also be used by many birdwatchers. It is an excellent guide for beginners but does not go into as much detail as the other books. Furthermore, the shorebird illustrations, while aesthetically pleasing and fine for general identification, may be misleading in the critical area showing tertial patterns, primary projections, and wing projections.

Because these four books will be used as standards, I am taking the liberty of criticizing both their text statements and their illustrations whenever I consider criticism warranted; it is intended constructively. Criticisms will be levied against other published accounts only in cases in which misunderstanding might result from use of the information.

With the abundance of published photographs of shorebirds, it seems a valuable service to readers to point out their availability. I consider all the shorebird photographs in the following easily available books as resources and will cite them as examples of species, plumages, or particular individuals of interest: Armstrong (1983), Bull and Farrand (1977), Chandler (1989), Farrand (1983, 1988a, 1988b), Hammond and Everett (1980), Hosking and Hale (1983), Johnsgard (1981), Keith and Gooders (1980), Terres (1980), and Udvardy (1977). In addition, errors in species or plumage identification in photos in these publications will be corrected.

Distributional and seasonal information from the literature has come from many sources, with by far the most records from *American Birds* and its predecessor *Audubon Field Notes*. In addition, the following works have been of value: Bent (1927, 1929) for the entire region; Butler and Campbell (1987), Butler et al. (1986), Campbell et al. (1972), Campbell et al. (1973), Campbell et al. (1990), Cannings et al. (1987), Guiguet (1955), Hatler et al. (1978), Munro and Cowan (1947), and Taylor (1984) for southern British Columbia; Buchanan (1988a), Herman and Bulger (1981), Hoge and Hoge (1980), Hudson and Yocom (1954), Hunn (1982), Jewett et al. (1953), Lewis and Sharpe (1987), Weber and Larrison (1977), and Widrig (1979) for Washington; Browning (1975), Gabrielson and Jewett (1940), Gullion (1951), Littlefield (1990), Littlefield and Cornely (1984), and Littlefield and McLaury (1973) for Oregon; Burleigh (1972) and Taylor and Trost (1987) for Idaho; and Bonham and Cooper (1979), Skaar et al. (1985), and Weydemeyer (1973) for western Montana. Specific record sources are omitted with some regret, but to have included them would have compromised the readability of the book. All records included here are from the literature, with the exception of those that are asterisked (*), which are my own unpublished records (Washington) and those of David Beaudette and Winton Weydemeyer (Montana), and those with a number sign (#), which are specimens examined.

Good references on shorebird species include Bent (1927, 1929), Hayman et al. (1986), Johnsgard (1981), Stout (1967), and Terres (1980); on more general topics of shorebird biology, Burger and Olla (1984a, 1984b), Hale (1980), and Hosking and Hale (1983). Important regional references are Colston and Burton (1988), Cramp and Simmons (1983), and Nethersole-Thompson and Nethersole-Thompson (1986) for European (and many American) species; Dementiev et al. (1969) for Russian species; Lane (1987), Pringle (1987), and *Reader's Digest* (1986) for Australian species; and Urban et al. (1986) for African species. The *Wader Study Group Bulletin,* published in England, and *The Stilt,* published in Australia, are journals devoted specifically to shorebirds.

What Are Shorebirds?

The restlessness of shorebirds, their kinship with the distance and swift seasons, the wistful signal of their voices down the long coastlines of the world make them, for me, the most affecting of wild creatures.—Peter Matthiessen, 1973

SHOREBIRDS ARE the major group of birds that run, walk, and wade along the water's edge (they are called "waders" in Europe). All of our shorebirds are birds of the open, except the vagrant woodcock, although Solitary and Spotted sandpipers frequently occur at wooded ponds. None inhabits dense cattail or bulrush marshes as do the rails, although snipes and some small sandpipers such as Pectoral, Sharp-tailed, and Least commonly occur in dense sedge meadows or salt marshes, where they may disappear from sight in the inches-high vegetation.

Most shorebirds are gregarious by nature, and mixed flocks of several species are frequently seen; thus size comparisons will be easy if at least one species in the flock can be identified. Certain species (similar enough to be "birds of a feather") often flock together, and those associations will be mentioned in specific accounts. Typically, species of about the same size forage in a particular habitat, remaining near one another even though feeding in different ways or on different prey. Bear in mind that individuals of some species defend feeding territories in migration or winter, and these birds will be well-spaced apart from members of their own species. Nevertheless, they coalesce into flocks when a predator appears and causes them to take flight.

ANATOMY

Each shorebird is a beautifully functioning organism, the parts finely tuned by natural selection to work together to adapt the bird to its environment. The shorebird way of life is a special one, not paralleled by any other birds although approximated in some ways by the closely related gulls and the more distantly related rails.

By understanding the component parts of shorebirds, both in general and for each species, an observer will be better able to associate species and their field marks. Putting together the short legs and needlelike bill of a Red-necked Phalarope into a way of life provides a more memorable experience than the simple noting of their occurrence. And knowing that a short bill and brown, mottled back and wings are characteristics of upland species allows a Buff-breasted Sandpiper or golden-plover to fit more comfortably into pigeonholes of the mind.

Wings and Tail for Effective Flight

The typical shorebird breeds in a high-latitude interior marsh, tundra, or prairie and winters on a low-latitude coast. Long-distance flying is a part of its life, whether from winter to summer home and back or on a daily basis between feeding and roosting areas. Thus shorebirds have long, pointed wings, adaptations to rapid flight. This attribute characterizes virtually all shorebirds of our region, although the sedentary Black Oystercatcher has relatively rounded wings compared with most other species, and the round-winged, forest-dwelling woodcock shows that evolutionary history does not constrain totally. See under Shorebirds in Flight for additional information.

Shorebirds are rather conservative in tail shape, none of them featuring the variation found in many other groups of birds. Tail shape is often modified for maneuverability in flight, but shorebirds have evolved their surpassingly swift and agile flight with no such modifications. The fact that pratincoles, which feed in the air like swallows, do have forked tails, indicates that such modifications are important for birds that feed in the air, not just birds that fly from place to place or escape aerial predators with agility. The few shorebirds with tails slightly differently shaped from the mode perhaps are more effective at using their tails for display, particularly distraction display; the Killdeer is the best example. See under Coloration and Shorebirds in Flight for the importance of tail markings.

A Bill for Every Level

A bird's bill is its feeding appendage, and the tremendous variation in the diets of birds is mirrored by variation in bill shape and size. This is the case within the shorebird group, and a story is told by each species' bill, from the stubby forceps of a Semipalmated Plover to the incredibly long, arched probe of a female Long-billed Curlew.

Shorebirds feed in a variety of ways. Two basic methods are *picking* and *probing*. "Picking" refers to prey taken from the surface, either land or water, and presumably detected visually. "Probing" refers to immersion of the bill in the substrate—shallow or deep, slow or fast, regularly or intermittently—either to pursue prey already detected or to explore for prey tactilely. "Stitching" refers to a rapid series of probes made close together in a straight or zigzag line, clearly exploratory. When watching shorebirds it may be difficult to determine whether a probe was a response to evidence of prey or not, or whether it was successful or not.

It is usually safe to say that shorter bills are for picking, longer ones for probing, but remember this is a generality. A Black-bellied Plover may probe the superficial substrate to grab a worm, and a Long-billed Curlew may pick a crustacean from the surface. Short bills are generally the prerogative of plovers, which feed visually by spotting prey from a distance and taking it from the surface. But Black-bellied Plovers and golden-plovers regularly pull invertebrates out of shallow holes, coming as close to probing as a typical plover can.

Where probing is difficult if not impossible, as on rocky substrates, sandpipers evolve short, ploverlike bills, our examples including turnstones and Surfbird. Sandpipers dominate rocks and open water in the complete absence of plovers, probably because the plover run-and-stop mode does not work on such substrates. Interestingly, the Wandering Tattler and Rock and Purple sandpipers have become rock specialists with no apparent deviation from the normal bill types of their groups. Oystercatchers are specially adapted to capture bivalve mollusks exposed at low tide, which happens to occur primarily on rocky shores in our region; thus the Black Oyster-catcher is a rock species.

Upland foragers on dry substrates, which should be pickers rather than probers, tend to have short bills. This is an important habitat for plovers generally, and sandpipers such as Upland and Buff-breasted that utilize it are relatively short-billed. Of course the Long-billed Curlew, which not only breeds in upland habitats but also often winters in them, is a sensational exception, but its bill is primarily an adaptation for the coastal mud flats on which most curlews winter.

With longer bills come greater foraging depths, as can be seen where the big spring flocks of Western Sandpipers, Dunlins, and Short-billed Dowitchers are sorted into bands along the edge of the tide line. Even small differences in bill length are significant in the foraging behavior of these birds, for example, Semipalmated Sandpipers tending to forage higher and drier than Westerns. The Red Knot, Sanderling, and most stints may be considered generalists because of their relatively short and straight bills. These shorter-billed species freely alternate picking and shallow probing, but with longer-billed calidridines, with bills the length of a Dunlin's or longer, probing becomes the most effective feeding strategy. Slightly droopy bills characterize most of the larger *Calidris* sandpipers and are apparently just the right shape for probing to moderate depths.

Bill curvature varies in sandpipers from very slightly upcurved in godwits through just about straight in many species and slightly downcurved (or "drooped" at tip only) in some small sandpipers to spectacularly downcurved in big curlews. The straight, slender to stout bills of tringines are primarily adapted to picking, functionally equivalent to the slender, straight bills of insect-eating passerine birds. Many tringines feed in water, so a longer bill allows them to grab something as it is moving away. The tringine type of bill has been taken to its extreme as a needlelike forceps in the shorebirds that feed on tiny organisms on the water's surface—Black-necked Stilt and Wilson's and Red-necked phalaropes. The Red Phalarope feeds similarly but on slightly bulkier items, thus its heavier bill. The latter species may also filter-feed, as both bill and tongue seem slightly modified for straining.

Very long, straight or slightly upcurved bills are apparently adapted for continual probing in soft substrates, typically below water. Woodcocks feed in soft forest soil, and dowitcher, snipe, and godwit bills are so long that they can feed in an inch or more of water and still reach burrowing invertebrates. In fact they should prefer water, as their bills would probably get burdened with mud if they foraged for the same prey on the mud flat itself as it dried (look at the bill of a Sanderling foraging in wet but not inundated sand). As the water recedes, godwits and dowitchers leave the mud flats to Whimbrels, which probe selectively for large prey, and plovers, which capture invertebrates still visible on the moist surface.

The avocet bill is still another type, flattened at the base and upcurved, probably to pass through the water smoothly as the bird swings its head from side to side "scything" for planktonic crustaceans and other swimming animals. Some of the *Tringa* sandpipers and Wilson's Phalarope feed similarly at times without such modified bills. The bills of spoonbills are adapted for this same function, as apparently is the superficially similar bill of the Spoonbill Sandpiper, although its feeding habits are still imperfectly understood.

Legs Just Long Enough

Leg length varies about as much as bill length in shorebirds, and the extremes are found in birds that feed in water. A swimming leg should be short for best mechanical advantage in pushing through the water, and indeed phalaropes have the shortest shorebird legs. Wilson's, which feeds more on land than the other two, has longer legs for its size, and they are relatively shortest in the Red Phalarope, which is a more confirmed oceangoer than the other two and looks rather ludicrous on the infrequent occasions when it moves about on land. Shorebirds that usually feed on land also have short legs. Those that forage on rock substrates are among the shortest-legged, and these birds tend to have rather thick legs also, perhaps greater strengthening for constant movement on the hardest of shorebird substrates.

The longest legs, as might be expected, characterize those shorebirds that feed by wading. A series of related sandpipers from Baird's through Dunlin and Curlew to Stilt clearly shows the correlation between feeding substrate and leg length. Many tringine sandpipers wade, from the relatively short-legged Solitary Sandpiper to the yellowlegs and Spotted Redshank, among the longest-legged of sandpipers. Avocets and especially Black-necked Stilts represent the extreme.

The toes of shorebirds also vary for different functions. Rock shorebirds have rather short, thick toes, entirely adequate for irregular rock surfaces. Species feeding in soft mud typically have long, slender toes, surely for stability as well as support on a soft substrate. Small webs between the toes occur in some mud feeders (for example, Semipalmated Plovers and Semipalmated Sandpipers), larger ones in swimmers such as avocets. Phalaropes have rather short toes with lobed edges that increase their propulsive surface for swimming.

The Feather Coat

Except for their eyes, bills, and legs, birds are covered by feathers that provide insulation against temperature extremes. Much of the external anatomy of birds relates to the arrangement of their feathers, as can be seen in Figures 2-4. The feathers are arranged in tracts, and each of them overlaps with adjacent tracts to provide an aerodynamic surface as well as effective insulation. The feathers of the head and neck are short, allowing unobstructed movement for feeding. The body feathers are longer and overlap greatly, providing the bulk of the bird's insulation. The flight feathers (primaries and secondaries) and tail feathers (rectrices) are large and strong to furnish propulsion and lift in flight. The wing and tail coverts provide an aerodynamically smooth gradation between the body feathers and those of the wing and tail.

At rest and in flight the mantle feathers overlap the scapulars which in turn overlap the tertials, and the three tracts, originating on different parts of the bird's body, work as one smooth feather covering. The tertials cover and perhaps protect the primaries and at the same time tie the wings to the body and tail in a smooth, continuous line just like the rear edge of an airplane's wing.

The feathers also serve as the substrate for the bird's coloration. Many elements of the pattern are arranged along particular feather tracts, while others run across adjacent tracts. The tertials not only function aerodynamically but also provide part of the dorsal color pattern along with mantle, scapulars, and coverts. The flight feathers and tail, mostly hidden from view when the bird is at rest, have little to do with the color pattern until the bird takes flight.

PLUMAGES

Shorebirds exhibit three major post-fledging plumages and, in many species, two additional less distinctive ones (Table 3). "Breeding" and "nonbreeding" are used as plumage terms rather than "summer" and "winter" because the plumages are adaptations for breeding and nonbreeding life rather than any particular seasons. It seems misleading to call the gray plumage borne by many Sanderlings in migration in May and August a "winter" plumage. The terms "alternate"

Table 3. Shorebird Plumages

Plumage	Other Names	Usual Season	Duration
Downy		May-July	first month of life
Juvenal	juvenile	July-October	first autumn of life
First winter	first basic	October-April	first winter of life
First summer	first alternate	April-September	first full summer of life
Nonbreeding	basic winter	August/October-April	from first or second winter throughout life
Breeding	alternate summer nuptial	April-August/October	from first or second summer throughout life

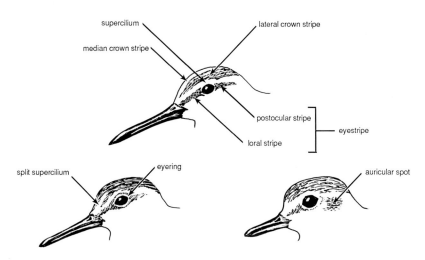

Fig. 2. Topography: head markings

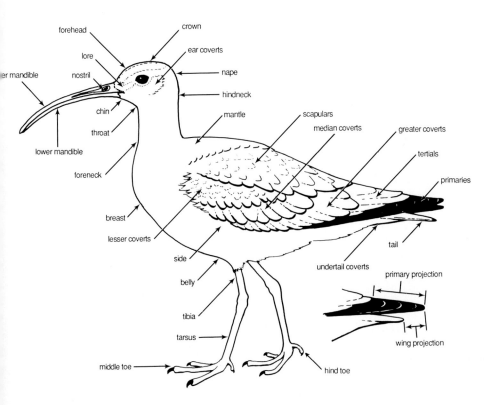

Fig. 3. Topography: standing

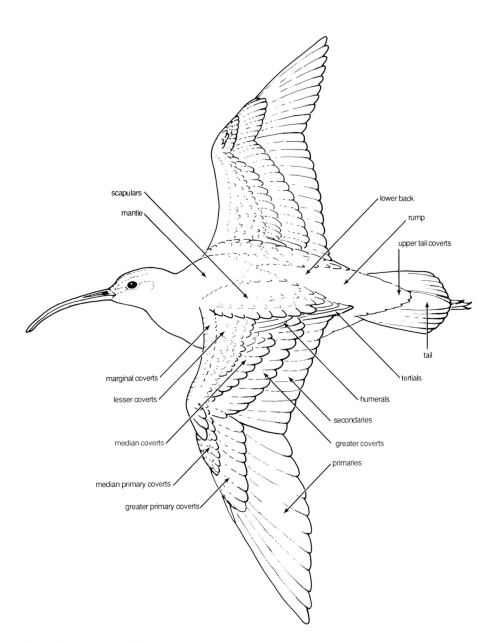

scapulars

mantle

lower back

rump

upper tail coverts

tail

tertials

humerals

marginal coverts

lesser coverts

secondaries

median coverts

greater coverts

median primary coverts

primaries

greater primary coverts

Fig. 4. Topography: in flight

and "basic" are used generally in the scientific literature but are not familiar to most users of bird field guides.

There has been much confusion between "juvenile" and "juvenal." In shorebirds "juvenal" describes the first post-downy plumage, a distinctive plumage in this group of birds. A "juvenile" is a bird in juvenal plumage; thus "juvenal" is an adjective, "juvenile" a noun. "Downy" is the accepted name for the chicks of precocial birds such as shorebirds. This plumage is seen in this region in species that breed here but is not described in the text, nor are the eggs of these species; eggs and chicks are almost invariably accompanied by adults. Descriptions and illustrations of these stages can be found in Harrison (1978).

Sources of Plumage Variation

Shorebird plumages vary greatly, and the timing and nature of this variation are of great consequence to anyone attempting to identify them. See under Coloration for further understanding of *why* shorebirds vary in color.

Plumage variation can be classified in five discrete categories:

Seasonal variation involves the difference between breeding and nonbreeding adults, marked in many species of shorebirds. Individuals in molt between the two plumages will look different from birds in either plumage, and an understanding of this will necessitate a certain amount of sophisticated interpolation. Times of molt differ among species *and* individuals, so one cannot depend on all shorebirds being in the same plumage or molt stage at once.

Age variation involves the differences between adult and juvenile shorebirds and, further, between these stages and immature stages that may be recognizable as such. Again, in the appropriate season many birds will be in molt between these stages. Knowledge of feather wear is extremely important to allow ageing of individual birds.

Sexual variation, the differences between males and females, has been inadequately emphasized in most field guides. Only in relatively few shorebird species are individuals easily identified by sex, but in others it can be done with careful scrutiny, especially if comparison is possible. Size and structure are often as important as plumage characteristics.

Geographic variation in shorebirds is fairly well understood at this time. Some species occur in distinct populations, many of them officially recognized as subspecies, and individuals of these populations are recognizably different in the hand and in some cases in the field. Usually not all plumages are equally distinctive of populations. In additional species there are slight geographic trends in size or coloration of populations, but the characteristics are not distinctive enough to provide sure knowledge of the origin of an individual bird.

Individual variation is surprisingly important in shorebirds. Besides the variation inherent in complexly patterned birds and that caused merely by feather wear and fading, unusual individuals also occur. With tens of thousands of Dunlins and Western Sandpipers to scrutinize, it should not be at all difficult to find one or more odd birds, and the only albino shorebirds I have ever seen were single individuals of these two species. Predators probably select out such birds from flocks, so the more unusual the bird the less likely it is to survive.

Molt

All shorebirds of the north temperate zone undergo a complete body molt in the fall, but its timing varies greatly in different species (and within some species as well), depending primarily on the timing and distance of the migration but also on factors such as the timing of breeding and the availability of prey concentrations. In some species, especially those that winter in the southern hemisphere, adults and juveniles pass through the Northwest in fall without molting and do so only on their wintering grounds. Individuals of any of these species, however, may begin molt soon after leaving their breeding grounds or even while still breeding and thus might occur here in mixed or nonbreeding plumage. In many species, particularly those that winter in the northern hemisphere, body molt takes place during migration, and adults change gradually from breeding to nonbreeding plumage and juveniles from juvenal to first-winter plumage while in the Northwest. In a few high-latitude breeders, definitely the exceptions, molt is completed before migration, and we see birds only in nonbreeding plumage in autumn. Finally, in some locally breeding species molt takes place in our area in late summer, again before migration.

In those species that change plumage seasonally, most if not all of the head and body feathers molt again in the spring. In addition, at least some of the tertials and wing coverts, which are exposed and furnish part of the bird's color pattern, also are replaced in spring. Even in species that do not change plumage perceptibly at that time, for example Killdeer, there is at least some spring molt. Most but not all shorebirds that winter to the south of us arrive here in spring already in full breeding plumage. Head and body molt may follow a sequence, often with the head feathers beginning the molt, then those of the underparts, followed by those of the upperparts, but there is much variation.

Wing and tail molt typically occur only once each year, in the period from about August (species breeding locally, with relatively short migrations) through September-October (many species wintering in this area or, in general, relatively closer to their breeding grounds) to November-January (species migrating long distances to the southern hemisphere). In a few species all or some of the rectrices are molted again in spring.

In some individuals of some species wing molt is suspended during migration, not to be completed until some time after arrival on the wintering ground; this appears to be rare in Northwest shorebirds. Most shorebirds accomplish their wing molt on their wintering grounds, so when one is seen in active wing molt it is probably a bird that will winter in our region. The gap in the series is perceptible on a bird at rest and easy to see on a bird in flight, and if it is on both sides it is presumably molt rather than replacement of otherwise lost feathers. Looking at a flock of adult Surfbirds and Black Turnstones in fall is always interesting, some birds in the midst of wing molt and others with no trace of it. These flocks probably consist of birds still on their way south and others that have stopped for the winter.

When a new feather grows out it is immediately affected by the environment, the effects more pronounced with time and rather substantial by the time of the next molt, whether in a year or a half year. Friction and airborne particles wear away at feathers, affecting their edges most. Those feathers that are most exposed wear the most rapidly, and wear is particularly evident on the scapulars, tertials, coverts, and primaries of shorebirds. Dark pigment protects feathers from wear, so lighter areas wear away most rapidly. Many shorebird feathers are margined with paler colors, and the pale margins wear away surprisingly quickly, leaving the bird darker overall and usually less patterned. The light fringes on darker juveniles produce cryptic (or adultlike) patterns on the breeding grounds that are then simplified by wear to bring them into a coloration more appropriate for their nonbreeding habitats as they migrate. Some lighter-colored juveniles have dark subterminal fringes on the feathers of the upperparts that may serve to arrest wear at that point.

Sunlight itself, perhaps its ultraviolet component, adds to the breaking down of feathers, and it certainly fades them. Paler colors—the brown, rufous, and buff so common in shorebirds— become increasingly pale over time, while black feathers do not seem to be much affected. A good example is furnished by the primaries of an early-fall adult, considerably paler than those of the same bird in winter after its wing molt, or those of a fresh-plumaged juvenile at the same time. This is evident in the hand and might be used to distinguish juveniles from adults in a flock in flight in early autumn.

Juveniles have relatively smaller and probably weaker body feathers than adults, and most of these feathers have to function only about three to four months until they molt into first-winter plumage in late fall. The feathers that are not molted then have to suffice for a considerably longer time, and immature birds in their first summer are maximally faded and worn, as they are carrying flight feathers, usually tail feathers, and some tertials and coverts that can be over a year old. Most of the other body feathers are molted in the first spring.

COLORATION

Shorebirds combine the best of all worlds of bird coloration. Some of them are as distinctly marked as any tanager or wood warbler, pleasing to the eye of anyone who is stimulated by swatches of pure, bright colors. Breeding-plumaged turnstones, stilts and avocets, oyster-catchers, a spring male Black-bellied Plover or Spotted Redshank at close range—any of them will elicit a gasp on sight. On the other hand, many shorebirds are marked with combinations of brown, gray, rufous, white, and buff, each bird drab at a distance but at close range a collection of complex and beautifully patterned feathers. These birds will warm the heart of the most discerning appreciator of subtlety, the person who prefers poorwill to hummingbird color.

Patterns

Shorebirds have complexly marked individual feathers that, grouped in feather tracts, produce particular patterns. At close range, the feather markings that are responsible for these patterns can be seen. Within each taxonomic group of shorebirds, different species typically share feather markings, but their color and the way they are arranged gives each of them its distinct appearance. The basic feather markings and the patterns they produce are illustrated and described in Figure 5.

Feather markings and patterns and their variation should be thoroughly understood to enhance identification skills, and the following guidelines should be helpful.

• Typically the upperparts of shorebirds are relatively dark with light markings on feather margins, the underparts light with dark markings associated with feather shafts, but there are many exceptions to this generalization.

• Most markings occur singly on feathers, but some (bars, dots, and notches) are repeated in series.

• Many of the feather-marking types grade into one another, as shown in Figure 5, even on the same bird. Dark and light patterns also blend with one another, for example a feather that is light with both dark bars and a central stripe grades into a notched feather when the dark markings exceed the light ones in extent.

• The gradations between marking types are often predictable. For example, the markings along a bird's side from neck to undertail may change gradually from stripes to spots to chevrons to bars, or the feathers of the upperparts change from pale, dark-striped mantle feathers to dark, pale-edged scapulars to dark, pale-margined tertials.

• Almost all markings vary in distinctness and in some cases in presence or absence. For example, stripes on the sides of the breast may vary from well-defined through weakly defined to absent in birds in the same plumage.

Fig. 5. Feather markings and patterns

DARK MARKINGS (USUALLY WITHIN FEATHERS)

MARKING	ILLUSTRATION	DESCRIPTION
STREAK		narrow longitudinal marking along and including shaft, visible only at close range
STRIPE		longitudinal marking along feather shaft
SPOT		round or oval marking, much smaller than feather and near its tip
CHEVRON		transverse V-shaped marking
BAR	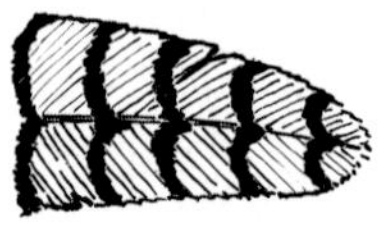	transverse marking, regular or irregular, at or near tip of feather when single
SUBTERMINAL FRINGE	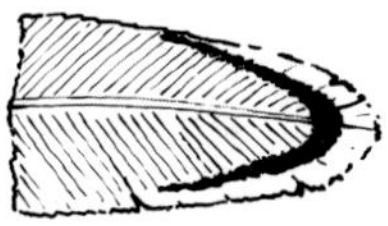	narrow marking just in from feather margin
SUBTERMINAL BLOTCH		large marking near but not at feather tip
DARK CENTER	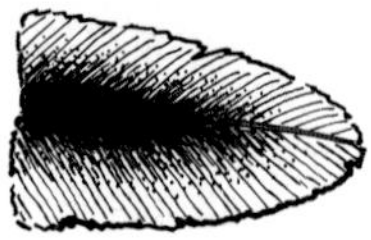	large oval or wedge-shaped marking filling much of feather center

PATTERN FORMED	EXAMPLES	GRADES INTO
streaked, visible only at close range	juv. yellowlegs necks juv. stint breasts	STRIPE

striped	curlew necks
	br. Western Sandpiper
	    mantle
	Pectoral Sandpiper
	    breast
		STREAK, SPOT

spotted	br. Spotted Sandpiper underparts nonbr. Surfbird sides	STRIPE, CHEVRON
between spotted and barred	br. Surfbird breast	SPOT, BAR

coarsely barred when single, finely barred when multiple	br. Wandering Tattler
	    underparts
	juv. Spotted Sandpiper
	    coverts
		SPOT, CHEVRON

finely scalloped, visible only at close range	juv. Surfbird upperparts juv. Red Knot upperparts	
irregularly blotched	br. Surfbird scapulars br. Western Sandpiper scapulars	DARK CENTER
blotched	nonbr. Least Sandpiper scapulars	BLOTCH

Fig. 5. **Continued**

MARKINGS THAT MAY BE DARK OR LIGHT

MARKING	ILLUSTRATION	DESCRIPTION
DOT	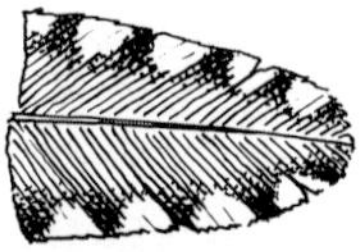	small spot, in series along feather edge
IRREGULAR MARKING		markings between feather margin and shaft, either longitudinal or transverse

LIGHT MARKINGS (USUALLY AT FEATHER EDGES)

MARKING	ILLUSTRATION	DESCRIPTION
TIP	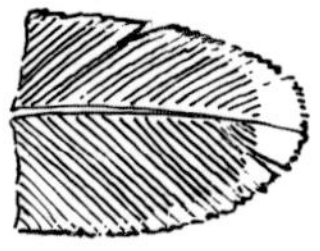	transverse marking at feather tip, following feather contour
EDGE	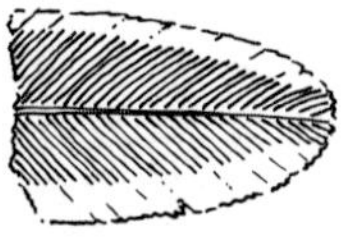	longitudinal marking on either inner or outer margin of feather, discontinuous at tip
FRINGE	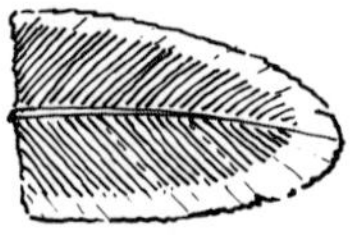	longitudinal marking all around feather margin, may be more prominent on outer side
NOTCH		marking extending perpendicularly from feather edge toward shaft as long wedge, in series

• Typically markings vary together in their brightness. A bird that is more vividly patterned above will be so below as well.

• Small markings on individual feathers can produce important differences in pattern, as for example the contrasting colors on the outer edges of calidridine mantle and scapular feathers that produce their characteristic mantle and scapular lines.

• Although feather edges and fringes are defined differently and generate different patterns on

PATTERN FORMED	EXAMPLES	GRADES INTO

dotted	juv. *Pluvialis* plover
	    upperparts (light)
	yellowlegs upperparts
	    (light and dark)
		NOTCH

	br. Spotted Sandpiper
	    upperparts (dark)
	juv. Short-billed Dow-
	    itcher tertials (light)	

PATTERN FORMED	EXAMPLES	GRADES INTO
scalloped	juv. Least Sandpiper scapulars	EDGE, FRINGE

often arranged into
long stripe
	Common Snipe
	juv. Pectoral Sandpiper
	    scapulars
	juv. Red-necked Phal-
	    arope scapulars
		TIP, FRINGE

scalloped
(can look striped when
on long, narrow feathers)
	juv. Ruddy Turnstone
	    upperparts
	juv. Baird's Sandpiper
	    upperparts
		TIP, EDGE, DOT

between barred and dotted, pattern finer if more notches	curlew tertials Marbled Godwit tertials	DOT

birds (edges make stripes while fringes make scallops), they are similar enough to grade into one another within species or individuals.

• The way a bird holds its feathers modifies its color and pattern as well as its shape. The best example is shown by the lower scapulars, which can be fluffed to cover many of the coverts or sleeked to expose them.

The patterns of the head are often simpler and bolder than those of the body and wings (Figure

2), and there are also important variations in head patterns. The pale supercilium can be simple or split, the upper fork extending upward as a lateral crown stripe. There can be dark stripes before or behind the eye or both, typically forming a dark eye stripe that contrasts with the pale supercilium above it. There is much variation in the distinctness of this stripe, even within species. In many species the feathers on the eyelids are pale, contrasting with the eye stripe and forming a distinct eye-ring. The feathers over the ear are often darker than those around it, forming an auricular spot (also called "ear coverts" or "ear patch").

Colors

Most of the colors on a typical shorebird are browns, along with black and white. Take the shorebird basic brown, make it darker or lighter, brighter or duller, add different proportions of gray, red, or yellow, and the result is a tremendous diversity of tones. Pure gray, untinged by brown, is not particularly common. Red, orange, and yellow are common on bare skin but not usually on plumage, although the American jacanas have bright yellow flight feathers and *Pluvialis* plovers yellow-spotted upperparts. Shorebirds appear to have little use for purples (except one courser), blues, or greens (except two lapwings). The brightest colors in the group are the almost-reds, the rich rufous or chestnut markings that characterize many of them, especially in breeding plumage.

For an educational exercise, catalogue the array of subtly different colors that are on an individual shorebird, in a species, or in shorebirds in general. See which species and plumages exhibit the most colors. The following section should make some of the variation understandable.

To Be Seen or Not—Shorebird Colors as Adaptations. Shorebird colors and patterns have evolved primarily for two diametrically opposed reasons: to make them less conspicuous in some circumstances (camouflage against nesting and feeding substrates) and more conspicuous in others (display markings for breeding/species recognition and flocking). A third reason, only recently considered, is that juveniles may mimic adults on the breeding grounds to reduce potential predation. In Table 4, it can be seen which of these factors affects each of the three shorebird plumages. With more major factors operating on them, differences between species should be greater in breeding and juvenal plumage than in nonbreeding plumage, which in fact describes the situation in many shorebirds. Species differences in breeding plumage are typically more pronounced than those in juvenal plumage, perhaps because species recognition is an especially important criterion. Little evidence is available, however, to show that these colors actually are used for species recognition!

Virtually all shorebirds are cryptically colored (camouflaged) above. The exceptions are a few big species—oystercatchers, avocets, and stilts—that by their size, habits, and nest sites cannot hide effectively and therefore might as well advertise their presence. They do so primarily from the ground, unlike the majority of sandpipers and plovers, which perform aerial displays and vocalizations. Flocking-display colors, which promote flocking when birds take flight, are probably limited to wings and tail and have little effect on head and body coloration.

Shorebirds spend a good part of their year in nonbreeding habitats, and it is primarily to these habitats that they are adapted in their overall size, shape, and feeding apparatus. This is of course the case for their nonbreeding plumage as well, and they are wonderfully camouflaged for their preferred substrates. We should be able to predict with ease the nonbreeding habitat of a newly discovered shorebird by the overall color and pattern on its upperparts in nonbreeding plumage.

Juvenal plumage may be a compromise between breeding-grounds camouflage (or mimicry) and nonbreeding-grounds camouflage, as it is often intermediate between the breeding and

Table 4. Factors Promoting Differences in Shorebird Plumages

	Plumage		
Selective Factor	Breeding	Nonbreeding	Juvenal
Breeding display (sex/species recognition)	x		
Nest-site camouflage	x		x
Feeding-substrate camouflage	x	x	x
Flocking display (species recognition)	x	x	x
Adult mimicry			x

nonbreeding plumages of adults. Interestingly, many of the markings that make juveniles relatively brightly colored either wear off or fade during their southward migration, and even before many species molt into their first adultlike plumage, their appearance becomes increasingly matched to their winter feeding substrates.

Breeding adults have the most complex plumages, usually variegated to camouflage them on nesting substrates and often with bright ventral colors for aerial or close-range ground displays. The larger species seem more likely to have darkly colored underparts during the breeding season, perhaps as they need to be a certain size for this rich color to be seen from a distance. They are anything but inconspicuous as they head north in full breeding plumage, still in the habitats for which their nonbreeding plumages evolved.

Shorebirds that feed on similar substrates are colored similarly, differing primarily in size and shape. Closely related species that feed on the same substrates may be extremely similar in all these qualities. Consider Little and Rufous-necked stints or Western and Semipalmated sandpipers. Not only are the members of either species pair not readily distinguishable on the basis of color, but all four are very similar. Common Ringed and Semipalmated plovers, Wandering and Gray-tailed tattlers, Short-billed and Long-billed dowitchers, Common and Spotted sandpipers, and Long-toed Stint and Least Sandpiper all represent species pairs that make this point. Members of each pair differ very little from the other member in size, shape, and color, and we can only wonder about the significance of the slight differences that do exist.

SHOREBIRDS IN FLIGHT

Flight Styles

Shorebirds are superlative flyers and are frequently in the air. We usually see them in flight as they move between feeding and roosting sites, or when they are flushed by potential predators or humans. A fortunate observer may see a flock passing in migration. With their long wings and rapid flight, they will be mistaken for few other types of birds, but among the shorebirds, flight patterns feature prominently in identification at the species level. At either close or distant range, most of them can be identified readily at this time, although quickness of response is important as they may flash past at high speeds. Shorebirds readily form mixed flocks in flight, which furnishes opportunities for comparison but may add confusion about just what species was present in a flock.

One simple generalization about shorebird flight is that it is swift, impressively so in many species. Migrating birds and especially birds attempting to escape aerial predators show the speed extremes, while birds in the course of a normal day's movements cruise considerably more slowly. Individual species of shorebirds have fairly characteristic flight styles, but there is also variation, especially between short-distance and long-distance flights (the Spotted Sandpiper is the best example).

Larger flocking species often fly in lines, while smaller ones flock in three dimensions. Lone shorebirds in flocks of other species are often at the edge or may drop out entirely, indicating incompatibility in flight styles. Conversely, shorebirds are so gregarious that they may try again and again to join such flocks. The choice may be between being falcon food and perhaps being injured in a midair collision.

In-flight characteristics include bird size, relative bill length, bill shape (few variations), and relative leg length, all visible on even a poorly lighted bird. If pattern (but not color) can be seen, look for presence of wing stripes, from obscure to vivid; markings of rump and tail; and markings of underparts. If lighting conditions are favorable, the colors themselves will be apparent, especially bright breeding-plumage ones like the reddish backs of Dunlins and Ruddy Turnstones or reddish bellies of Red Knots and dowitchers. See below for a more complete discussion of flight patterns.

Peculiar to shorebirds (and small gulls and terns) are their occasional erratic flight performances, which have been termed "crazy flights." A bird will fly at top speed, zigging and zagging in a manner that is breathtaking to watch, with ninety-degree course changes vertically as well as horizontally, and this may go on for several seconds and a hundred yards or more. Birds may do it when flushed or when already in flight. I have seen it in all sizes of shorebirds from stints to Long-billed Curlews, but it seems to occur in some species more than others. It has been considered an escape reaction, but I have never observed such a flight in a chase situation. It may instead be exhibited under a variety of circumstances as a signal to any predator that may be present that this bird is aware of its presence and essentially uncatchable.

Size

The Northwest species encompass just about the entire size range of shorebirds worldwide, from Long-billed Curlew to Least Sandpiper, and size must be considered an important attribute for flight identification. Size is difficult to estimate, increasingly so at greater distances, but with practice the ability to do so will improve. One fact to remember is that size is inversely proportionate to wingbeat frequency—*the slower the wingbeat the larger the bird*—in any group of birds, and with experience this is an important aid. Also, if any species can be identified out of a shorebird flock with certainty, it will provide an immediate size scale for others in the near vicinity. Obviously the strikingly marked species (for example Willets, turnstones, and Black-bellied Plovers) will be most readily used in this manner. A rough classification would allocate shorebird species in flight to *large* (Long-billed Curlew to Willet), *medium* (Black-bellied Plover and Greater Yellowlegs to dowitchers and Killdeer), and *small* (Pectoral Sandpiper and Lesser Yellowlegs to Least Sandpiper). Note that this does not correspond to the finer size categories applied to resting shorebirds. See Appendix 3 for sizes of all Northwest species and Figure 6 for size categories.

Proportions

Some of the differences in the relative proportions of shorebird body parts are readily seen in flight (Table 5). *Bill length* can be compared with head and neck length (that part of the bird in front of the wings), and species with bill about as long as the head and neck can be considered

Fig. 6. Representative sizes in flight

long-billed, those with bill distinctly longer than head and neck *very long-billed* (all others are considered *short-billed*). *Bill shape* is also important and can be seen at a considerable distance. The bird's bill can be categorized as *straight*, *upcurved*, or *downcurved*. *Leg length* is similarly important, and note should be taken of how much of the leg, if any, extends beyond the tail tip. The figures under Tarsus/Tail in Appendix 3 indicate the length of the legs in proportion to the tail. In most species with this number greater than 0.6 (and a few with it slightly less), the toe tips are visible beyond the tail tip, and the larger the number, the more of the leg projects beyond the tail. Species in which the toes and at least a trace of the tarsus extend beyond the tail (tarsus/tail about 0.85 or greater) are considered *long-legged*. Bear in mind that long-legged species on occasion fly with legs tucked forward in the body feathers rather than dangling behind, especially in cold weather.

Table 5. Flight Identification by Shape

This table categorizes shorebirds by attributes of their shapes, as defined in the text (p. 22–23). Large and very large shorebirds are indicated by an asterisk.

Regular Species (abundant to rare)	Irregular Species (casual/[potential])
Legs long	
Black-necked Stilt	[Black-winged Stilt]
*American Avocet	Spotted Redshank
Lesser Yellowlegs	
*Marbled Godwit	
Stilt Sandpiper	
Bill long, straight to slightly upcurved	
*Black Oystercatcher	*American Oystercatcher
Black-necked Stilt	[Black-winged Stilt]
*American Avocet	[Common Greenshank]
Greater Yellowlegs	Spotted Redshank
*Willet	Terek Sandpiper
	[Jack Snipe]
	American Woodcock
Bill long, downcurved	
*Whimbrel	*Bristle-thighed Curlew
Bill very long, straight to slightly upcurved	
*Hudsonian Godwit	*[Black-tailed Godwit]
*Bar-tailed Godwit	
*Marbled Godwit	
Short-billed Dowitcher	
Long-billed Dowitcher	
Common Snipe	
Bill very long, downcurved	
*Long-billed Curlew	*Far Eastern Curlew

Coloration

Besides size and proportions, *overall coloration* is important at some levels of identification, especially in species that are not conspicuously patterned. Table 6 lists species by most obvious coloration in flight. The majority of species are brown, varying from very dark brown (Solitary Sandpiper) to very pale gray-brown (Sanderling). The browns themselves vary from a "colder" gray-brown (as in Spotted Sandpiper) to a "warmer" reddish brown (as in Marbled Godwit). Some species are distinctly gray, from the dark Wandering Tattler to the pale nonbreeding Red Phalarope, others distinctly rufous, as in breeding Dunlin. Others appear black and white, like the turnstones, avocet, and stilt. Black Oystercatchers look entirely black, Spotted Redshanks in breeding plumage largely so.

Table 6. Flight Identification by Color and Pattern

This table categorizes shorebirds by their flight patterns, as defined in the text (p. 27–35). Large and very large shorebirds are indicated by an asterisk. Some species appear in two categories because they are somewhat intermediate between them.

Regular Species (abundant to rare)	Irregular Species (casual/[potential])

Plain-backed, black or very dark brown
*Black Oystercatcher

Plain-backed, gray

Wandering Tattler	Gray-tailed Tattler
	Terek Sandpiper

Plain-backed, brown

American Golden-Plover	[Oriental Pratincole]
Pacific Golden-Plover	[Little Curlew]
Upland Sandpiper	*Bristle-thighed Curlew
*Whimbrel	*Far Eastern Curlew
*Long-billed Curlew	[Jack Snipe]
*Marbled Godwit	American Woodcock
Buff-breasted Sandpiper	Eurasian Dotterel
Common Snipe	

Plain-winged, white-margined

Common Snipe	Eurasian Dotterel

Plain-winged, stripe-tailed

Solitary Sandpiper	[Little Ringed Plover]
Pectoral Sandpiper	
Sharp-tailed Sandpiper	
Ruff	

Plain-winged, white-backed

Short-billed Dowitcher	[Common Greenshank]
Long-billed Dowitcher	[Marsh Sandpiper]
	Spotted Redshank

Plain-winged, white-rumped

*Bar-tailed Godwit	*Whimbrel (Siberian)
	Great Knot

Plain-winged, white-tailed

Black-necked Stilt	[Black-winged Stilt]
Greater Yellowlegs	[Wood Sandpiper]
Lesser Yellowlegs	[Green Sandpiper]
Stilt Sandpiper	
Wilson's Phalarope	

Stripe-winged, plain-tailed

–	Terek Sandpiper

Table 6. Continued

Regular Species (abundant to rare)	Irregular Species (casual/[potential])
Stripe-winged, white-margined	
Semipalmated Plover	Mongolian Plover
Killdeer	[Wilson's Plover]
Spotted Sandpiper	[Common Ringed Plover]
	Piping Plover
	Mountain Plover
	[Common Sandpiper]
Stripe-winged, stripe-tailed	
Snowy Plover	Rufous-necked Stint
Sanderling	Little Stint
Semipalmated Sandpiper	Temminck's Stint
Western Sandpiper	Long-toed Stint
Least Sandpiper	[Purple Sandpiper]
Baird's Sandpiper	Spoonbill Sandpiper
Pectoral Sandpiper	[Broad-billed Sandpiper]
Sharp-tailed Sandpiper	
Rock Sandpiper	
Dunlin	
Ruff	
Red-necked Phalarope	
Red Phalarope	
Stripe-winged, white-rumped	
Curlew Sandpiper	Great Knot
	White-rumped Sandpiper
Stripe-winged, white-tailed	
Black-bellied Plover	
*Willet	
Red Knot	
Stripe-winged, band-tailed	
*Hudsonian Godwit	*American Oystercatcher
Surfbird	*[Black-tailed Godwit]
Calico-backed	
*American Avocet	
Ruddy Turnstone	
Black Turnstone	

Flight Patterns

Besides size and proportions, the flight patterns (Figures 7-12) of many species furnish field marks. These flight patterns are probably adaptations to promote instant and effective flocking by birds reacting to potential predators; when any bird takes flight it is immediately apparent not only what it is doing but what it is. Others of the same species (or similar-sized species) can join it and form one of the flocks that are often so effective in preventing predation. Watch a Merlin harrying Dunlins some time and see how this works; the Merlin will almost invariably attempt to isolate a single bird from the flock before it attempts capture. Alternatively, some of these flight patterns may function to show a predator that its intended prey is vigilant and unlikely to be caught. This may be more important in non-flocking species such as Solitary and Spotted sandpipers.

Wing Patterns

One component of the flight pattern is furnished by the wings. Many species show distinctly patterned wings, and those lacking such patterns are called *plain-winged*. Even "plain-winged" species may show some pattern, for example the contrast between flight feathers and coverts in curlews and godwits. Another subcategory is exemplified by dowitchers, which have a fairly conspicuous white stripe on the rear edge of the wing, and the Terek Sandpiper, with an even broader stripe in the same place.

Stripe-winged species typically show white stripes (consistently called "wing bars" by British authors) that run most of the length of each wing near its rear edge. The stripes are formed by white-tipped greater coverts (usually on the wing base) and white markings across the flight feathers (usually on the primaries). In inconspicuously stripe-winged species such as golden-plovers and Pectoral Sandpiper, the only white markings are the narrow tips to the coverts, with a faint indication of an extension of the wing stripe because of slightly paler parts of the primaries. In moderately conspicuously stripe-winged species such as Western and Baird's sandpipers, the bases of all the flight feathers are somewhat paler, and in flight the wing stripe is quite distinct and more obvious than it is in the preceding group, but still fairly narrow. In conspicuously stripe-winged species such as Semipalmated Plover and Dunlin, not only the covert tips are white, but most of the flight feathers show white markings that produce a conspicuous stripe running along much of the wing. In vividly stripe-winged species such as Willets, turnstones, and Sanderlings, the white stripes are wide and bordered with black to accentuate them.

Within each of these groups there is variation, with the extremes approaching species in the adjacent groups, but most species can be readily allocated to one of the categories.

Back and Tail Patterns

Back and tail patterns furnish another major component of flight identification. *Plain-backed* species are those in which the tail and back appear unmarked and similarly colored at a distance. A plain-winged, plain-backed species is plain indeed in flight, although, in the case of a Black Oystercatcher or Long-billed Curlew, certainly not inconspicuous.

White-margined species are those in which the tail is narrowly margined with white, always along its edges and in some species at its tip. The Spotted Sandpiper and the small plovers fall in this category. The Killdeer's rufous rump and back accentuate the length of its long, white-margined tail.

The *stripe-tailed* group includes a series of small species, primarily the stints and their relatives, with dark rumps and white- or gray-edged tails. Although they almost blend in one

Fig. 7. Flight-pattern groups

American Golden-Plover Pacific Golden-Plover Eurasian Dotterel Black Oystercatcher Wandering Tattler Terek Sandpiper Upland Sandpiper Whimbrel (*hudsonicus*) Bristle-thighed Curlew Far Eastern Curlew Long-billed Curlew Marbled Godwit Buff-breasted Sandpiper Common Snipe American Woodcock		Solitary Sandpiper	Spotted Redshank Whimbrel (*variegatus*) Short-billed Dowitcher Long-billed Dowitcher
	Mountain Plover	Pectoral Sandpiper Sharp-tailed Sandpiper Ruff	
	Mongolian Plover	Semipalmated Sandp. Western Sandpiper Rufous-necked Stint Little Stint Temminck's Stint Least Sandpiper Baird's Sandpiper Spoonbill Sandpiper	
Red Phalarope (breeding)	Semipalmated Plover Spotted Sandpiper	Snowy Plover Dunlin Red-necked Phalarope Red Phalarope	
	Killdeer	Sanderling Rock Sandpiper	
PLAIN	**WHITE-MARGINED**	**STRIPE-TAILED**	**WHITE-BACKED**

WHITE-RUMPED	WHITE-TAILED	BAND-TAILED
	Black-necked Stilt Greater Yellowlegs Lesser Yellowlegs Wilson's Phalarope	
Bar-tailed Godwit	Stilt Sandpiper	
Great Knot White-rumped Sandp.		
Curlew Sandpiper Red Knot	Black-bellied Plover Red Knot	Hudsonian Godwit
	American Avocet Willet	Amer. Oystercatcher Ruddy Turnstone Black Turnstone Surfbird

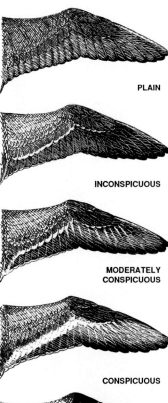

PLAIN

INCONSPICUOUS

MODERATELY CONSPICUOUS

CONSPICUOUS

VIVID

WHITE-RUMPED

WHITE-TAILED

BAND-TAILED

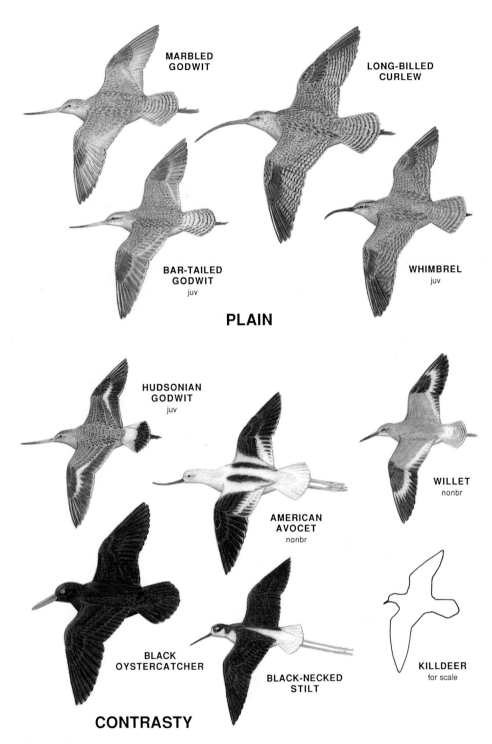

MARBLED GODWIT

LONG-BILLED CURLEW

BAR-TAILED GODWIT
juv

WHIMBREL
juv

PLAIN

HUDSONIAN GODWIT
juv

AMERICAN AVOCET
nonbr

WILLET
nonbr

BLACK OYSTERCATCHER

BLACK-NECKED STILT

KILLDEER
for scale

CONTRASTY

Fig. 8. Flight: Large species. Scale different from other flight figures; note Killdeer outline as reference. Plumages as indicated.

WANDERING TATTLER
juv

COMMON SNIPE

PACIFIC GOLDEN-PLOVER
juv

EURASIAN DOTTEREL
juv

BUFF-BREASTED SANDPIPER
juv

UPLAND SANDPIPER
juv

PLAIN-BACKED

Fig. 9. Flight: Plain-backed species. Plumages as indicated.

direction with the white-margined group, they are distinctive because the pale tail edges do not curve toward the tail tip as in the former. The Solitary Sandpiper has a dark tail, the outer feathers of which are conspicuously black-and-white barred, for a flashy flight pattern. The Ruff is a stripe-tailed shorebird that approaches the white-rumped group because its white outer uppertail coverts almost meet in the center.

The *white-backed* group includes dowitchers and a few Eurasian species, with the white on the lower back appearing as a white wedge up the back where bounded by the dark scapulars. The rump and tail may be pale like the lower back but are usually distinctly darker.

In *white-rumped* species, including the Curlew Sandpiper, the rump is white, set off from the distinctly darker back and tail.

White-tailed species, for example yellowlegs and Black-bellied Plover, have the rump (upper tail coverts) and tail (rectrices) paler than the back, usually appearing white or whitish. Often the upper tail coverts are white, the tail barred or dusky, but the translucency of the tail produces an

STRIPE-TAILED

SOLITARY SANDPIPER
juv

STILT SANDPIPER
juv

WILSON'S PHALAROPE
juv

LESSER YELLOWLEGS
juv

GREATER YELLOWLEGS
juv

WHITE-TAILED

SPOTTED REDSHANK
juv

LONG-BILLED DOWITCHER
juv

WHITE-BACKED

Fig. 10. Flight: Plain-winged species (stripe-tailed, white-tailed, & white-backed). Plumages as indicated.

Fig. 11. *(right)* **Flight: Stripe-winged species I (white-margined, white-tailed, & band-tailed).** Plumages as indicated.

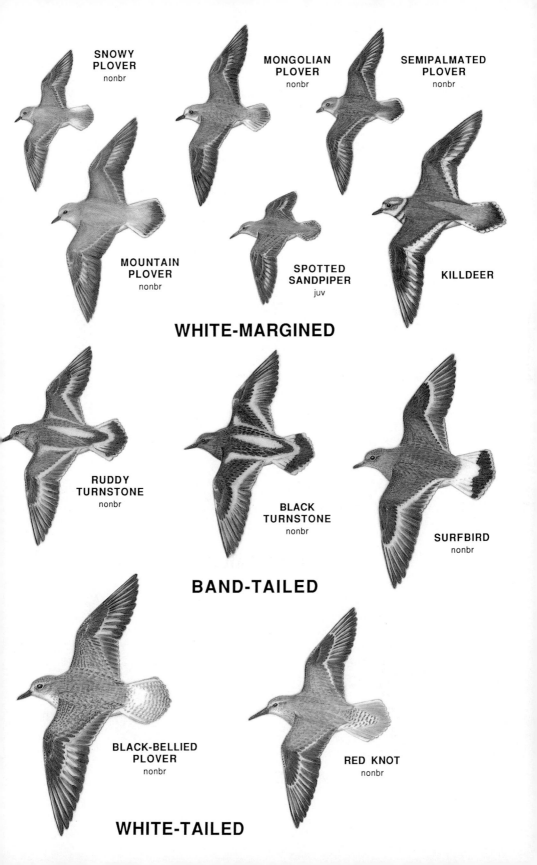

SNOWY
PLOVER
nonbr

MONGOLIAN
PLOVER
nonbr

SEMIPALMATED
PLOVER
nonbr

MOUNTAIN
PLOVER
nonbr

SPOTTED
SANDPIPER
juv

KILLDEER

WHITE-MARGINED

RUDDY
TURNSTONE
nonbr

BLACK
TURNSTONE
nonbr

SURFBIRD
nonbr

BAND-TAILED

BLACK-BELLIED
PLOVER
nonbr

RED KNOT
nonbr

WHITE-TAILED

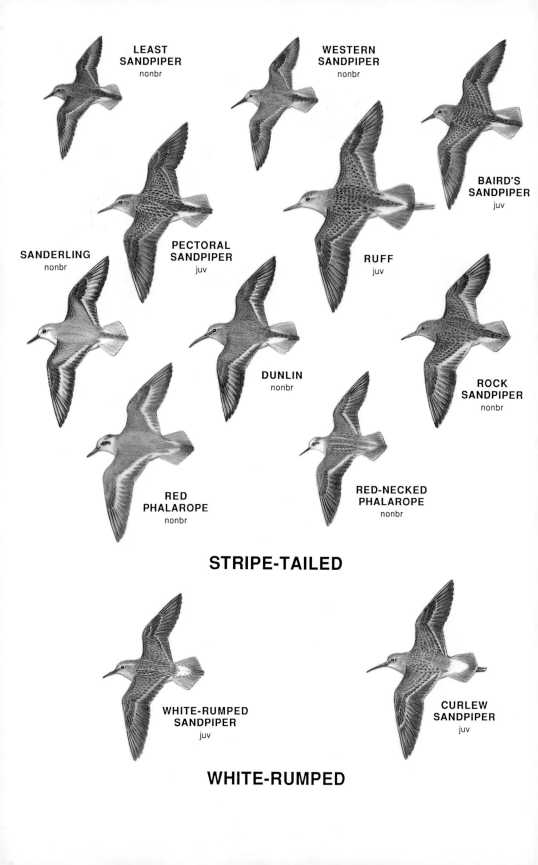

LEAST
SANDPIPER
nonbr

WESTERN
SANDPIPER
nonbr

BAIRD'S
SANDPIPER
juv

SANDERLING
nonbr

PECTORAL
SANDPIPER
juv

RUFF
juv

DUNLIN
nonbr

ROCK
SANDPIPER
nonbr

RED
PHALAROPE
nonbr

RED-NECKED
PHALAROPE
nonbr

STRIPE-TAILED

WHITE-RUMPED
SANDPIPER
juv

CURLEW
SANDPIPER
juv

WHITE-RUMPED

essentially white-tailed look at a distance. Generally the more a bird's tail is spread, the more light passes through it and the paler it looks.

The Surfbird, which has a white tail with a broad black band near the tip, is the only small *band-tailed* species, but the American Oystercatcher and Hudsonian Godwit are similarly marked.

In *calico-backed* species, the entire flight pattern is especially vivid. Turnstones have white lesser coverts that contrast with the dark greater coverts and, together with the vivid white wing stripe, give them a more complex wing pattern than other shorebirds. The American Avocet's wing pattern is also complex, with white outer scapulars and inner secondaries. Turnstones have a black band across the tail tip and another across the rump, both contrasting with the white back and tail. In all these species the black-and-white back adds to the conspicuousness of the flight pattern.

Thus, each shorebird species can be placed in a flight-pattern group (Figure 7), depending on the combination of wing and tail patterns. By being able to place a bird in question in one of these groups, the choices are fewer and the possibility of identification is much increased, especially when calls, location, and season of occurrence are taken into account.

Shorebirds Overhead

Some of the most difficult identifications are of shorebirds directly overhead, where the standard field marks of wing stripes and tail pattern are not visible. Nevertheless, even then many species are distinctive, and Table 7 lists readily identified species or groups of species. Points to look for include size, proportion, overall coloration of the underparts, contrast if any between breast and belly, distinctly paler or darker wing linings, and any special markings such as the black axillars of Black-bellied Plovers.

VOCALIZATIONS

The flight calls of shorebirds are species-specific and distinctive at surprising distances. The cultivation of an ear for shorebird calls will allow auditory surveillance of a diverse mud flat assemblage when a potential predator (falcon, harrier, or shorebirder) flushes the whole bunch and each gives its call. Whether these calls signify "watch out for the predator" or "come flock with me" or both, they impart an additional message to the knowledgeable birder—"this is who I am."

Some shorebird species are much less vocal than others, and it is of value to be aware that some are harder to detect in this way. In spring, as they are rushing toward their breeding grounds, many of our shorebirds begin to give their songs, completely different vocalizations that are longer and more complex than the familiar alarm/flocking calls. Early fall migrants may also give these songs, especially during territorial interactions.

Migrant species I have heard singing on more than one occasion in this region include Black-bellied Plover, Pacific Golden-Plover, Semipalmated Plover, both yellowlegs, Whimbrel, Western and Least sandpipers, Dunlin, and Short-billed Dowitcher. This is no more than a hint of the music of the arctic tundra and subarctic muskeg; the birds should be followed northward for the complete concert.

Fig. 12. Flight: Stripe-winged species II (stripe-tailed & white-rumped). Plumages as indicated.

Table 7. Shorebirds Overhead

This table categorizes shorebirds by their appearance from below. Large and very large species are indicated by an asterisk. The characteristic may refer to only one or two of the three plumages of a species (indicated by B - breeding, N - nonbreeding, and J - juvenal), and a given species can be in more than one category.

Regular Species (abundant to rare)	Irregular Species (casual/[potential])

Largely black below
Black-bellied Plover B   Spotted Redshank B
American Golden-Plover B
Pacific Golden-Plover B
*Black Oystercatcher

Black patch(es) on breast or belly
Black-bellied Plover B   Eurasian Dotterel B
American Golden-Plover B   Great Knot B
Pacific Golden-Plover B
Rock Sandpiper B
Dunlin B

Black patches at wing bases
Black-bellied Plover NJ

Largely reddish below
*Long-billed Curlew   Eurasian Dotterel B
*Hudsonian Godwit   *[Black-tailed Godwit B]
*Bar-tailed Godwit B   Curlew Sandpiper B
*Marbled Godwit
Red Knot B
Short-billed Dowitcher B
Long-billed Dowitcher B
Red Phalarope B

Light wing linings contrast with dark underparts
Black-bellied Plover B   Spotted Redshank B
Red Knot B
Buff-breasted Sandpiper
Red Phalarope B

Dark wing linings contrast with white underparts
Black-necked Stilt   [Black-winged Stilt]
Solitary Sandpiper   [Green Sandpiper]
Common Snipe   *Hudsonian Godwit NJ

Breast conspicuously dark against white belly
Ruddy Turnstone   Mongolian Plover B
Black Turnstone   *American Oystercatcher
Surfbird   Great Knot B
Least Sandpiper   Long-toed Stint
Pectoral Sandpiper   [Purple Sandpiper N]
Rock Sandpiper N
Dunlin N
Ruff B male
Red-necked Phalarope B

Dark band(s) on breast
Semipalmated Plover   Mongolian Plover N
Killdeer   [Wilson's Plover]
                                                [Common Ringed Plover]
                                                Piping Plover B
                                                [Little Ringed Plover]

Shorebirds in Time and Space

The economic, scientific, and esthetic values of these migratory species dictate they be permitted to continue their long-accustomed and to some extent still-mysterious habits of migration.—Frederick Lincoln and Steven Peterson, 1979

SHOREBIRDS MOVE rapidly over time and space, one of the attributes that makes them so exciting to experience. Their long-distance movements are such an important part of their story that any discussion of the annual cycle must emphasize migration.

THE SHOREBIRD'S YEAR

For a given species in the region, spring migration lasts a month or more, fall migration three to four months. When the diverse seasonal strategies of all our species are taken into account, however, migration takes place throughout much of the year. A typical shorebird of this region breeds in arctic or subarctic regions and migrates at least to the contiguous United States for the winter; many proceed farther, to the tropics or well into the southern hemisphere.

Because of their superb flight capabilities, many shorebirds migrate in long hops, even of up to thousands of miles between stops. The straightest lines between their landing points are great circle routes, curved lines when plotted on flat maps. In many cases spring and fall routes are different, doubtless to take advantage of optimal feeding opportunities provided by seasonal variation in prey abundance. Juveniles may follow adults in fall but almost invariably are much more widely distributed both geographically and ecologically. Populations with breeding distributions not far apart may have wintering grounds and migration routes very much removed from one another.

The migration strategies of shorebirds are among the more complicated and varied of those in the avian world, creating complex patterns of distribution and occurrence in a region such as the Northwest. The diversity of these patterns can be seen only by perusing all of the species accounts, although there are common features among some groups of species (see Distribution Groups). Figure 13 illustrates the more common patterns of seasonal occurrence in the region.

Many shorebird species have been recorded in the Northwest in every month of the year, while others are confined to particular seasons. Species that are present only seasonally typically have peak occurrence periods, with smaller numbers before and after these periods and stragglers still earlier and later. In many species that are normally present only seasonally, enough birds have wintered or summered so that extreme migration dates cannot be determined. Appendix 4 lists the extreme migration dates in both subregions for the species for which they can be determined. With continuing observations some of these dates will doubtless be altered.

Fall

The shorebird year can be examined by beginning in autumn, as the breeding season is coming to a close. Almost invariably adults appear before juveniles, in some species with little or no overlap between the age classes. Less well-known but probably prevalent, one sex may migrate somewhat before the other. This variation reflects the mating system of the species, in which one or the other sex may abandon parental care at any time from egg-laying through the fledging period. The pace of fall migration is leisurely in comparison with that of spring, in part because of these sex- and age-specific waves of birds but also probably because there is not the same push

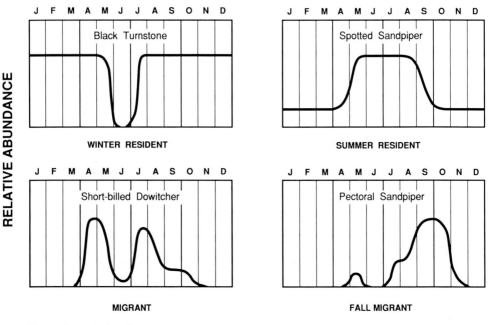

RELATIVE ABUNDANCE

Black Turnstone
WINTER RESIDENT

Spotted Sandpiper
SUMMER RESIDENT

Short-billed Dowitcher
MIGRANT

Pectoral Sandpiper
FALL MIGRANT

Fig. 13. Annual migration cycles

to get to wintering grounds as there is to get to arctic and subarctic breeding localities to produce another generation in the short time available.

Fall migration in the Northwest begins during the last week of June, when the first returning migrants of quite a few species appear; a few locally breeding species are on the move even earlier. These birds are in somewhat worn breeding plumage, and they arrive in increasing numbers through July and into August, when virtually all of our species are present. They are often darker at this time than in spring, as pale feather edgings on the back have worn off in part or entirely. Adults of some species have largely passed through the region by the end of August, but those of others remain common thereafter, whether migrants or winter visitors. By late July the first juveniles appear, and this age class becomes more common through August and September, in some species entirely replacing the adults. With the exception of the Rock Sandpiper and Dunlin, when a species does not arrive in the Northwest until August or September, it is represented here only by juveniles.

Juveniles of almost all species are clearly distinguishable from their respective adults, and because of this there may be two plumage types present for some time during the fall. The simultaneous occurrence of three plumages is possible in some species, as breeding-plumaged adults molt into nonbreeding plumage while juveniles are still present. Because of the complex mix of birds in different plumages as well as the greater diversity of species, it is in autumn that shorebird identification becomes a real challenge.

As September passes, adults of many species molt into their winter plumage, and juveniles begin to do the same during that month, so that by October most birds that are present are well along in body molt. Of course those individuals that are destined to winter in this area will complete that molt, and its entire progress can be witnessed during regular shorebird observations. Exceptions to this pattern are Dunlin and Rock Sandpiper, in which both adults and juveniles remain on or near their breeding grounds until they have entirely completed their autumn molt. These birds then make the trip south in nonbreeding plumage, and we do not normally see their juvenal plumages. These two species arrive in October, several months after most other species, and it is interesting that there are individuals of a number of other species with them that have apparently followed the same molting pattern, unlike the more typical members of their species. Migrants thought to have arrived with October Dunlin flocks include Black-bellied Plovers, Red Knots, Sanderlings, and Western Sandpipers.

Winter

Winter is a time of relative stability. Molt is essentially finished by November, individuals have settled on their wintering grounds, and not very much will change until March, when both spring molt and northbound migration begin to take place. Nevertheless, winter populations are not entirely stable, as there is evidence in some species of both continuing southbound retreat and early northbound advance during this period. In addition there is some movement of individuals (and doubtless flocks) from one habitat and area to another as conditions change. Both heavy rainfall and extreme cold affect freshwater and upland species such as snipes and Killdeers and cause populations of coastal species such as Dunlins to alter their foraging and roosting patterns.

Spring

The shorebirds that winter in our region begin to molt into breeding plumage in late March or April, and by mid April many birds are in full summer dress. Spring migrants from the south arrive in an avalanche at that time, many of them fully molted as well; but there are differences among species, Sanderlings, for example, molting very late in spring. The last week of April represents the peak of abundance for many species, with much larger concentrations than are ever seen in fall. By early May most shorebirds are in breeding plumage, and any that is not by the end of that month is probably a subadult bird that will not be breeding that year (however, some male phalaropes and female *Pluvialis* plovers may breed in a plumage not much different from nonbreeding).

By the end of May the beaches are relatively empty, but substantial northbound movement has been recorded even at the beginning of June, especially in high-arctic breeders such as Sanderlings and Red Phalaropes. Amazingly, there is only a short period in mid June at our latitudes between the last northbound migrants and the first southbound ones, and especially late or early migrant individuals sometimes cannot be placed in spring or fall categories. I have considered dates up to June 10 as spring migration dates, those after June 20 as fall migration dates, those in the period June 11-20 as equivocal unless some evidence indicated they represented migrating rather than summering individuals.

Summer

Summer for Northwest shorebirds lasts from late March, when Killdeers arrive in many areas and begin to set up their territories, to some time in August when the last locally produced juveniles lift off for points southward or coastward. Breeding is a fairly rapid process, even at these temperate latitudes, and only the Killdeer among our breeding species ever rears more than one brood. A complete breeding cycle includes territory establishment, courtship, nest building, laying, incubation, and care and fledging of young, a period typically lasting about three months in our species.

This discussion must include the phenomenon of oversummering (or merely "summering," as opposed to breeding), when birds spend the breeding season south of their breeding grounds, usually because they are not sexually mature but in some instances because something is wrong with them. Some small shorebirds breed at one year of age, but probably all larger ones wait at least another year, and a few may not mature until three or more years of age; this is still a poorly documented part of shorebird life. Oversummering is probably a normal part of the life cycle of all shorebirds with delayed maturation. Interestingly, little of it is done in the Pacific Northwest, although such birds are locally common south of this region. Summering individuals have been recorded for many but not all of the common Northwest shorebirds, in most cases in very small numbers. In a few species, flocks have been observed during more than one summer in outer-coastal estuaries, and these species can be considered regular summerers.

Summering birds may range from full nonbreeding to nearly full breeding plumage but are often intermediate. There may be a partial spring molt involving some of the body feathers. The unmolted body feathers and especially the tertials and primaries of one-year-old birds become extremely worn and faded during June and July, the most extreme case of this in a shorebird's life. Faded nonbreeding-plumaged birds during this period almost certainly belong to this age class. In late summer they begin their molt into a nonbreeding plumage indistinguishable from that of the adult.

It is peculiar that few of the shorebirds that winter here commonly, for example Black-bellied Plover, Black Turnstone, Surfbird, Dunlin, and Sanderling, also spend the summer in any numbers. Either individuals of Northwest populations all mature in one year, or first-year birds that winter here move elsewhere in summer. There is evidence that some nonbreeding subadult birds move through the Northwest at least in spring on their way to unknown destinations, for example Long-billed Curlew and Marbled Godwit. Without knowledge of the presence of individual birds over a period of weeks, it will be impossible to distinguish summering birds from migrants, although birds in nonbreeding plumage during the summer are more likely to be summering. Much more information is needed about this phenomenon, here and elsewhere.

Loftin (1962) is an important reference to oversummering shorebirds at lower latitudes in the western hemisphere.

DISTRIBUTION

Shorebirds are typically very widely distributed because of their long-distance migrations, which may take individual birds of some species from far northern to far southern latitudes and back in the course of a year. Not only are many of the species widely distributed latitudinally, but some of them occur virtually worldwide. Others are widespread in either Eastern or Western hemispheres. Some migratory or resident shorebirds occur only along Atlantic, Pacific, or Indian ocean shores or more locally in interior areas. Many tropical residents, particularly plovers, are limited to small parts of continents, and the most restricted ranges are found in species that occur only on single islands or island groups.

The Pacific Northwest is fairly typical of a temperate-latitude coastal region. It is far enough north to be within the breeding range of 14 species and far enough south for 21 species to winter regularly, 24 and 36 percent, respectively, of those species known to occur in the region. The remainder comprises regular migrants and species of irregular occurrence.

Subspecies

Of the 62 species given full treatment, 13 (21 percent) are represented by two or more subspecies in North America. Unlike passerine birds, in which distinctly different breeding populations very often intermingle in winter, shorebird subspecies usually have separate winter ranges and migration routes. They typically represent discrete populations that are of value to recognize for both their biology and their conservation; thus they are given some prominence here. However, in only four species and possibly a fifth (Ruddy Turnstone needs reassessment) do more than one subspecies certainly occur in the Northwest. In the Whimbrel, Ruddy Turnstone, Rock Sandpiper, and Short-billed Dowitcher, there is a common subspecies and another that is rare or irregular.

Only in the Solitary Sandpiper do two subspecies migrate through the region regularly, and in this species the pattern of subspecies distribution is very unusual for a shorebird, with two widespread forms replacing one another at different latitudes on a broad front. The usual pattern is for different subspecies to occur in different parts of the breeding range, often isolated from one another and typically at similar latitudes.

Distribution Groups

To understand the distribution of Northwest shorebirds, it is enlightening to know their overall distribution in North America and the rest of the world. North American shorebirds can be allocated to distribution groups based primarily on their breeding distribution and migration patterns. Questions of shorebird distribution may be approached by comparing distribution patterns as well as by comparing single species. For example, we may be able to understand the migration of some rare species in the Northwest, facilitating search and identification, from our knowledge of the migration of more common species within the same distribution group.

Although every shorebird species has its own unique distribution, the allocation of all North American species into discrete groups places the Northwest shorebird scene in perspective from the standpoint of the rest of the continent. In some species different *subspecies* fall readily into different distribution groups with different migration patterns, so the subspecies themselves are allocated to two or more of the groups.

The groupings stem from both breeding distribution and migration routes. *Northern breeders* are those that breed in arctic (tundra) and subarctic (taiga) habitats. *Continental breeders* are those that breed primarily south of those habitats, although some of them extend into northern regions. *Eurasian breeders* include both northern and continental breeders. *Coastal breeders* and *tropical residents* are obvious. Groups determined entirely by winter ranges would be similar but not always identical to those listed here.

Species or subspecies in brackets have not been officially recorded from the Northwest.

Eastern Eurasian/Alaskan Breeders, Asian/Pacific Migrants

[Oriental Pratincole]

Pacific Golden-Plover**

Mongolian Plover*

[Common Ringed Plover—*C. h. tundrae*]*

[Little Ringed Plover]

Eurasian Dotterel*

[Black-winged Stilt]

[Common Greenshank]

[Marsh Sandpiper]

Spotted Redshank

[Wood Sandpiper*]

[Green Sandpiper]

Gray-tailed Tattler

[Common Sandpiper]*

Terek Sandpiper

[Little Curlew]

Whimbrel—*N. p. variegatus*

Bristle-thighed Curlew**

Far Eastern Curlew

[Black-tailed Godwit—*L. l. melanuroides*]

Bar-tailed Godwit—*L. l. baueri***

Ruddy Turnstone—*A. i. interpres***

Great Knot

Red Knot—*C. c. rogersi***

Rufous-necked Stint*

Little Stint

Temminck's Stint

Long-toed Stint

Sharp-tailed Sandpiper

[Dunlin—*C. a. sakhalina*]**

Curlew Sandpiper*

Spoonbill Sandpiper

[Broad-billed Sandpiper]

Ruff*

[Jack Snipe]

This large group could be divided into a number of groups if the species were categorized from the standpoint of their own continental migration systems, rather than by the fact that birds that visit western North America come from Siberia.

Most shorebirds considered of Siberian origin are restricted to Eurasia as breeding species. Some of them, however, have extended their breeding range into Alaska, from sparingly to locally common. Asterisked species have bred at least once in Alaska, while those with two asterisks have well-established breeding populations in the state. All of these species migrate back into Asia during the nonbreeding season, and for the most part only individuals away from their usual migration route move down the east side of the Pacific to reach our region. The Bristle-thighed Curlew is an unusual member of this group, narrowly restricted to western Alaska as a breeder and the South Pacific as a nonbreeder.

Individuals of two subspecies on this list also occur regularly and in large numbers in migration on the American Pacific coast: Ruddy Turnstone and Red Knot. Both have breeding populations in Alaska, although most of our knots may come from Siberia, and both are more common in spring than fall.

With the exception of a very few birds that have wintered in the region, shorebirds from Siberia occur in the Northwest only during migration. The great majority of records are in autumn, the majority of individuals juveniles but with many adults represented as well (Table 8). The numbers of juvenile Ruffs and especially Sharp-tailed Sandpipers seen every autumn attest to major movements down the American Pacific coast of these species, an unusually great deviation from adult migration routes. Most of these birds must head out into the Pacific, as records of them decrease from north to south. Of great interest is the situation in the autumns of 1987 and 1988, when relatively few of these birds reached our region, almost surely correlated with the general scarcity of west winds.

I would have predicted that the Siberian-breeding shorebirds most likely to occur in North

Table 8. Northwest Occurrences of Siberian Shorebirds by Month, through 1990

Species	Spring						Fall						
	J	F	M	A	M	Jn	Jn	Jl	A	S	O	N	D
Mongolian Plover							I		I	I			
Eurasian Dotterel									2				
Spotted Redshank		I	I		I					2	I		
Gray-tailed Tattler										I			
Terek Sandpiper								I					
Far Eastern Curlew									I				
Whimbrel (*variegatus*)				I					I				
Bar-tailed Godwit				I	4	4			11	16	6		
Great Knot									2				
Rufous-necked Stint						I	3	3	2				
Little Stint										2			
Temminck's Stint									I				
Long-toed Stint									I				
Sharp-tailed Sandpiper				I				2	*	*	*	6	
Curlew Sandpiper					2			7	5	7	I		
Spoonbill Sandpiper									2				
Ruff	I			2	2		2	6	*	*	6		
Total	I	I	I	4	10	5	5	19	41+	54+	27+	7	0

Birds that were present in two or more months were recorded in that month in which they were present longest or, if present for equal time, that month in which they arrived. When more than one bird was present at one locality and date, only a single record is included. For the most commonly reported species, asterisks indicate more than 10 records. Numbers are conservative, not including sight records that are insufficiently documented.

America were those that wintered farthest east, thus with migration routes closer to this hemisphere. Table 9 lists the primary wintering grounds of the species breeding in eastern Siberia, and such a correlation is not evident. There are as many records of species such as Eurasian Dotterel and Little Stint that move southwest to their wintering grounds in Africa and India as there are of species such as Gray-tailed Tattler, Far Eastern Curlew, Great Knot, and Long-toed Stint that move directly south to southeast Asia and Australia. The Ruff is of surprisingly frequent occurrence considering its winter distribution, perhaps supporting the hypothesis that it breeds regularly in North America.

Some of these discrepancies could be because of the overall abundance of these species, with more numerous species more likely to occur as vagrants no matter what their migratory pathways, but there is no compelling evidence to indicate a direct correlation here.

On a more detailed level, it seems odd that there are five or six records of the Spotted Redshank from the region, while the unrecorded Common Greenshank has been observed in Alaska about twice as often, and the not-surely recorded Wood Sandpiper is a regular migrant in western Alaska and has even bred on Bering Sea islands.

The solution to this mystery may lie in the seasonality of Alaskan occurrence of these species (Table 10). Asiatic birds in general seem unlikely to stray into our region in spring, as there are virtually no records of any of them at that time. The few birds seen in spring, in fact, may be birds that have wintered in this hemisphere on their way back north. Both the greenshank and

Table 9. Primary Wintering Grounds of Eastern Siberian Breeding Shorebirds

Species	Africa	India	SE Asia	Australia
Mongolian Plover			x	x
Common Ringed Plover	x			
Eurasian Dotterel	x			
Common Greenshank	?	x	x	x
Spotted Redshank	?	x	x	
Green Sandpiper	?	x	x	
Wood Sandpiper	?	x	x	x
Gray-tailed Tattler				x
Common Sandpiper	?	x	x	x
Terek Sandpiper	?	x	x	x
Whimbrel			x	x
Far Eastern Curlew				x
Black-tailed Godwit			x	x
Bar-tailed Godwit			x	x
Great Knot			x	x
Rufous-necked Stint			x	x
Little Stint	x	x		
Temminck's Stint	x	x	x	
Long-toed Stint			x	
Sharp-tailed Sandpiper				x
Curlew Sandpiper	x	x	x	x
Spoonbill Sandpiper			x	
Broad-billed Sandpiper			x	x
Ruff	x	x		
Jack Snipe	x	x		

Distribution at the subspecies level wherever possible; question marks indicate wintering of Siberian population in region dubious.

Wood Sandpiper are more frequently reported in Alaska in spring migration than in fall, as is typical of many Siberian birds. Although storms come regularly across the Pacific from west to east during the spring, Asiatic migrants must be staying close to their coast at our latitudes.

On the contrary, the Spotted Redshank has been recorded from Alaska more frequently in fall, the season in which Siberian species have a higher probability of occurring in the Northwest, based on existing records. Of all the Siberian shorebirds recorded from Alaska in migration, only it and the Sharp-tailed Sandpiper are reported significantly more often in fall than spring. Nevertheless, even not counting the regularly reported Sharp-tailed Sandpiper and Ruff, fall records of Siberian shorebirds in the Northwest outnumber spring ones by about seven to one.

Finally, fall records of greenshank and Wood Sandpiper in Alaska are concentrated before mid September, those of Spotted Redshank and Sharp-tailed Sandpiper after that time, and wind patterns may be such that later-migrating birds are more likely to be swept across the north Pacific than those that precede them. Certainly more storm fronts move into the Northwest in October than in September.

Two Asian species recorded from North America, the Oriental Pratincole and Little Curlew, breed at some distance from the Bering Strait. They would have been considered unlikely to occur on this continent, yet have done so. Furthermore, a few Siberian species that are seldom

Table 10. Migratory Status of Siberian Shorebirds in Alaska

Species	Spring	Fall	Neither	NW records
	Species More Numerous in			
Mongolian Plover	x			x
Common Ringed Plover			x	
Eurasian Dotterel		x		x
Common Greenshank	x			
Spotted Redshank		x		x
Wood Sandpiper	x			
Green Sandpiper	x			
Gray-tailed Tattler			x	x
Common Sandpiper			x	
Terek Sandpiper	x			x
(Siberian) Whimbrel			x	x
Far Eastern Curlew	x			x
Black-tailed Godwit	x			
Great Knot	x			x
Rufous-necked Stint			x	x
Little Stint			x	x
Temminck's Stint			x	x
Long-toed Stint	x			x
Sharp-tailed Sandpiper		x		x
Curlew Sandpiper			x	x
Broad-billed Sandpiper		x		
Ruff		x		x

The list includes only species recorded at least three times in Alaska. Status information from Kessel and Gibson (1978), Byrd et al. (1978), Gibson (1981), and *American Birds*.

recorded even from Alaska—the Far Eastern Curlew, Little Stint, and Spoonbill Sandpiper—have been recorded in the Northwest, so other unexpected species might in fact be expected. They are mentioned under Species Included.

Northern Breeders, Widespread Migrants

Black-bellied Plover
Semipalmated Plover
Greater Yellowlegs
Whimbrel—*N. p. hudsonicus*

Ruddy Turnstone—*A. i. morinella*
Sanderling
Least Sandpiper

Generally shorebirds breed widely in the north and wander widely in migration, so it is not surprising that this is a sizable group. Red Knot, Dunlin, and Short-billed Dowitcher would be included in it if their subspecies were not allocated to more restricted categories. Whimbrel and Ruddy Turnstone could be separated into a subgroup of coastal migrants, as they are primarily coastal in the Northwest, while all other species of this group occur regularly if not commonly in the interior as well.

Most of the species in this group winter relatively far north, to at least California on the west coast and the Carolinas on the east coast, and many of them farther north. All of them winter at least occasionally in the Northwest. This is a primary distinction between members of this group and the group of northern breeders and interior/eastern migrants, which are heading for South America and thus must bear southeast from their northern breeding grounds.

Northern Breeders, Pacific Coast Migrants

Wandering Tattler	Rock Sandpiper
Black Turnstone	Dunlin—*C. a. pacifica*
Surfbird	Short-billed Dowitcher—*L. g. caurinus*

These forms have breeding ranges restricted to Alaska, and they are limited to the Pacific coast, from southern Alaska to northern South America, for their nonbreeding cycles. The Wandering Tattler winters widely in the South Pacific, but the other species or at least subspecies are entirely American.

All the members of this group are common on the Northwest coast in migration, all but the tattler and dowitcher also common in winter. The Dunlin and dowitcher are distinct from the other members of the group in not being rocky-shore specialists and in regularly occurring in the interior in very small numbers. Otherwise, no group is more closely restricted to the coast than this one, and there are only a few inland records of all the rock-inhabiting species put together.

Northern Breeders, Western Migrants

Western Sandpiper	Long-billed Dowitcher

This group includes species that breed primarily in Alaska and migrate widely through western and less commonly through eastern North America. They are placed in their own group because they are much more common on the Pacific than on the Atlantic coast, presumably because of their restricted breeding range. Both species occur in substantial numbers in the East, wintering fairly far north as in the widespread-migrating group. Wintering grounds range from southern United States to northern South America. The two species are similar in many aspects of their migration patterns, although the sandpiper is more abundant on the coast and the dowitcher in the interior.

Both of the species of this group are common migrants throughout the Northwest, and both winter in small numbers, especially to the south.

Northern Breeders, Interior/Eastern Migrants

American Golden-Plover	White-rumped Sandpiper
Lesser Yellowlegs	Baird's Sandpiper
Solitary Sandpiper—*T. s. solitaria*	Pectoral Sandpiper
and *T. s. cinnamomea*	[Dunlin—*C. a. hudsonia*]
[Eskimo Curlew]	Stilt Sandpiper
Hudsonian Godwit	Buff-breasted Sandpiper
[Red Knot—*C. c. rufa*]	Short-billed Dowitcher—*L. g.*
Semipalmated Sandpiper	*hendersoni*

These species migrate across the Great Plains, but not commonly farther west. Some of them are also common on the Atlantic coast, the Red Knot much more so than in the interior. Most of them are common on the plains both spring and fall, and the preponderance of records of this group from the Northwest in fall is because juveniles wander more widely than adults. Representatives of the eastern subspecies of Red Knot and Dunlin probably occur in the Northwest occasionally but have not been documented.

Lesser Yellowlegs and Solitary Sandpiper are a bit more difficult to place in this category, but both generally head away from the Pacific coast as they move southward and are more common in the Northwest in fall than in spring. Because of their more easterly migration routes, members of this group are usually observed in larger numbers in the protected waters of the straits of Georgia and Juan de Fuca than on the outer coast. Additionally, they are more common in the northern part of the region than the southern.

The American Golden-Plover, Hudsonian Godwit, and White-rumped Sandpiper have distinctive migration strategies, adults using the plains primarily in spring but migrating over the Atlantic Ocean directly to South America in fall. One or more populations of Semipalmated Sandpipers use similar routes, also once shared by Eskimo Curlews. In the godwit and White-rumped Sandpiper, juveniles apparently follow the adults across the Atlantic. As might be expected, there are more spring than fall records of the White-rumped Sandpiper in the Northwest, but the godwit is more often reported in fall. This could be explained if our records involve birds from the small southern Alaskan breeding population. Juvenile American Golden-Plovers and Buff-breasted Sandpipers use the plains route in large numbers in fall (some move out over the Atlantic also), and virtually all of the Northwest records of these birds pertain to juveniles.

Most of the members of this group winter in South America, many of them in the southern part of that continent. The yellowlegs, knot, and dowitcher are locally common in winter in the southern United States, the Dunlin farther north.

Winds could affect Northwest records of species that normally migrate to the east of our region, with increased records during periods of strong easterlies. Variation in these winds could determine whether an eastern species was more common in spring or fall and its status from year to year.

Water conditions on midcontinent could also affect the distribution of migrants through that area, with extremely wet years (shorelines not exposed) or extremely dry years (no water) causing some of them to move west toward the Pacific coast. In autumn 1985, a drought year on the Great Plains, five species of sandpipers—Lesser Yellowlegs, Semipalmated, Baird's, Stilt, and Buff-breasted—that are midcontinent migrants were especially common around well-surveyed Vancouver.

Northern Breeders, Atlantic Coast Migrants

[Purple Sandpiper]   [Short-billed Dowitcher—*L. g. griseus*]

The two species comprising this group are Atlantic coast equivalents to the rock shorebirds and the dowitcher subspecies in the northern-breeding, Pacific-coast-migrating group. The Purple Sandpiper is recorded sparingly but with some regularity west to the Great Lakes, and occasional individuals could reach the Northwest, especially to be watched for after strong northeasterlies or easterlies in October and November. The dowitcher is known only as an Atlantic coast migrant, with many individuals taking the offshore route to northern South America. It visits the Gulf coast also in small numbers but is quite unlikely to occur in the Northwest.

Northern Breeders, Oceanic Migrants

Red-necked Phalarope   Red Phalarope

The two seagoing phalaropes are included in this group, with the caveat that the Red-necked is also a widespread coastal and interior migrant in the West. They are most common in the Northwest in offshore waters or onshore during westerly winds.

Western Eurasian Breeders, Eurasian/African Migrants

[Northern Lapwing]
[Greater Golden-Plover]
[Common Ringed Plover—*C. h. hiaticula*]
[Common Greenshank]
Spotted Redshank
[Wood Sandpiper]
[Whimbrel—*N. p. phaeopus*]
[Slender-billed Curlew]

[Eurasian Curlew]
[Black-tailed Godwit—*L. l. limosa*]
[Bar-tailed Godwit—*L. l. lapponica*]
Little Stint
Curlew Sandpiper
Ruff
[Eurasian Woodcock]

This set of species and subspecies, recorded rarely or casually (regularly in the case of Curlew Sandpiper and Ruff) in northeastern North America, is unlikely to occur in the Northwest from European sources, although some of them have occurred or might occur in the region from Asiatic sources. The Greater Golden-Plover and Slender-billed Curlew breed only in western Eurasia, and records of them from our region are very unlikely. The Northern Lapwing, Eurasian Curlew, and Eurasian Woodcock, however, breed across Eurasia, as near as some species that have been recorded.

Continental Breeders, Widespread Migrants

Killdeer
Spotted Sandpiper

Common Snipe

To this group belong a few species that are widespread in temperate and subarctic North America, breeding throughout the Pacific Northwest. These species are associated with fresh water in both breeding and nonbreeding seasons but may move to the coast commonly in the winter. Variations in numbers among them appear to be primarily based on winter distribution and concentrations, in turn probably based on local and annual variations in winter weather.

Pacific Coast Breeders

Black Oystercatcher

This nonmigratory species occurs with the species that winter on rocky Pacific shores but is tied to its habitat throughout the year because of its foraging specializations.

Western Continental Breeders and Migrants

Snowy Plover—*C. a. tenuirostris*
Mountain Plover
Black-necked Stilt
American Avocet

Willet—*C. s. inornatus*
Long-billed Curlew
Marbled Godwit
Wilson's Phalarope

These species are temperate-zone rather than arctic breeders, reaching the northern limits of their breeding distribution in northern United States or southern Canada. One species, the Snowy

Plover, also breeds on the coast, but the other species are confined to the interior at these latitudes. The Black-necked Stilt, widespread in the Tropics, is included here because of its wide range in the West.

This group includes a subgroup of species—Black-necked Stilt, American Avocet, Willet, Long-billed Curlew, and Wilson's Phalarope—that are widespread in the West, breed widely if locally in the Northwest, and may be locally abundant in migration. Another subgroup includes Great Plains species—Mountain Plover and Marbled Godwit—that breed largely or entirely to the east and migrate primarily to the east and south of our region. All of these species but the Mountain Plover migrate in small numbers to the Atlantic and Gulf coasts, and the stilt breeds there.

Interior Continental Breeders and Migrants

Upland Sandpiper

This species is similar in many ways to the group of northern-breeding, eastern/interior-migrating species, except for its distinctly more southerly breeding distribution. It extends its breeding range sparingly into the Northwest but migrates out of the region toward the southeast. As in other interior species, it is the juveniles that furnish our records of birds in migration.

Eastern Continental Breeders and Migrants

Piping Plover   American Woodcock

This group includes two species restricted to eastern North America, the Piping Plover of inland and coastal beaches and the American Woodcock of deciduous woodlands. Although unlikely to occur in the region based on geography, both have done so.

Atlantic Coast Breeders

[Snowy Plover—*C. a. nivosus*]   [American Oystercatcher—*H. p. palliatus*]
[Wilson's Plover—*C. w. wilsonia*]   [Willet—*C. s. semipalmatus*]

Two species and two subspecies are largely restricted to the Atlantic and Gulf coasts within the United States. All are more widespread in the American Tropics, and all seem very unlikely to occur in the Northwest.

Tropical Pacific Coast Breeders

[Wilson's Plover—*C. w. beldingi*]   American Oystercatcher—*H. p. frazari*

The Pacific coast subspecies of Wilson's Plover and American Oystercatcher occur north rarely to southern California, and, unlikely as it would seem, a representative of one of them has been recorded from the Northwest.

Tropical Residents

[Double-striped Thick-knee]   [Northern Jacana]

Two species from the New World Tropics have occurred rarely or accidentally on the southern border of the continent. These species are the least likely of all North American shorebirds to be recorded from the Northwest.

Northwest Shorebird Distribution and Habitats

The Northwest supports a great diversity of habitats used by shorebirds, including virtually all shoreline habitats and most open upland ones. Substantially different constraints and opportunities are presented by breeding and nonbreeding habitats, however, as shorebirds are restricted in the breeding season not only by food supplies but by nest sites. In the breeding season they are spread out over two-dimensional breeding territories, while in the nonbreeding season they tend to be collected along shorelines at much higher densities. Thus the linear nature of coastal habitats virtually precludes breeding by the majority of shorebirds that use them for feeding, and only a few species nest on the coast.

Breeding Distribution and Habitats. Most of the breeding shorebirds of the Northwest breed in the interior of the region (Table 11). Only one species, the Black Oystercatcher, breeds only in the coast subregion, while five species breed only in the interior subregion and eight in both subregions. Of the eight breeding in both subregions, two (American Avocet and Wilson's Phalarope) are considerably more common in the interior. Even if the rare and local breeders (Semipalmated Plover, Greater Yellowlegs, Solitary Sandpiper, and Upland Sandpiper) are excluded, the pattern is the same.

Habitats used by breeding shorebirds include two (sand beaches and rocky shores) found only on the coast and two (dry grasslands and open ponds/lakes) found only in the interior. Three others (wet meadows/marshes, wooded ponds/lakes, and streams/rivers) occur widely in the region. Table 12 lists breeding habitats and their characteristic species. It is apparent that two freshwater environments (open ponds/lakes and fresh marshes) support the greatest diversity of shorebird species in the region, that shorebirds are much more likely to breed in freshwater than in marine habitats, and that each habitat has its characteristic and often unique species. The upland-breeding Killdeer shows the most varied breeding-habitat preferences.

Nonbreeding Distribution and Habitats. Shorebird species also show distinct habitat preferences in migration and winter, and Table 13 indicates the regularly occurring species and the habitats in which they are more likely to occur. As with breeding habitats, there are specialists in each of the nonbreeding habitats (except streams/rivers and flooded fields) and generalists that frequent a wider spectrum.

The following species are characteristic of these Northwest habitats: *open ocean*—Red-necked and Red phalaropes; *rocky shores*—Black Oystercatcher, Wandering Tattler, Black Turnstone, Surfbird, and Rock Sandpiper; *sand beaches*—Snowy and Semipalmated plovers and Sanderling; *mud flats*—Black-bellied and Semipalmated plovers, Whimbrel, Ruddy Turnstone, Red Knot, Western Sandpiper, Dunlin, and Short-billed and Long-billed dowitchers; *salt marshes*—Least and Pectoral sandpipers; *streams/rivers*—Spotted Sandpiper; *ponds/lakes*—Killdeer, American Avocet, Western and Baird's sandpipers, Long-billed Dowitcher, and Wilson's Phalarope; *fresh marshes*—Greater and Lesser yellowlegs, Solitary and Least sandpipers, Long-billed Dowitcher, and Common Snipe; and *flooded fields*—Killdeer, Dunlin, and Common Snipe.

SHOREBIRD CONSERVATION

Before the days of market hunting, when shorebirds were decimated throughout the continent, and before the days of widespread destruction of wetlands everywhere, relatively little was known about west-coast populations of these birds. Nevertheless, through some of the oldest records we have learned that at least some shorebird species were more common in our region two centuries ago than they are now. Marbled Godwits came in huge flocks to coastal estuaries,

Table 11. **Northwest Breeding Shorebirds**

	Coast	Interior
Snowy Plover	FC	FC
Semipalmated Plover	VU	cas
Killdeer	C	C
Black Oystercatcher	FC	
Black-necked Stilt		FC
American Avocet	cas	C
Greater Yellowlegs		VR
Solitary Sandpiper	VR	R
Willet		U
Spotted Sandpiper	U	U
Upland Sandpiper		VU
Long-billed Curlew		FC
Common Snipe	U	FC
Wilson's Phalarope	VU	FC

C - common; FC - fairly common; U - uncommon; VU - very uncommon; R - rare; VR - very rare; cas - casual; acc - accidental. See Appendix 2 for explanation of these status categories.
Italicized - marginal in region

Table 12. Northwest Shorebird Breeding Habitats

	RS	SB	SR	WPL	OPL	FM	DG
Snowy Plover		x			x		
Semipalmated Plover		o			o		
Killdeer		o	o		x	x	o
Black Oystercatcher	x						
Black-necked Stilt					x	x	
American Avocet					x	o	
Greater Yellowlegs				o			
Solitary Sandpiper				o			
Willet					x	x	
Spotted Sandpiper			x	x	x		
Upland Sandpiper							o
Long-billed Curlew							x
Common Snipe						x	
Wilson's Phalarope					x	x	

RS - rocky shores; SB - sand beaches; SR - streams/rivers; WPL - wooded ponds/lakes; OPL - open ponds/lakes; FM - fresh marshes; DG - dry grasslands
x - common to uncommon; o - rare or not typical

and Long-billed Curlews were harvested by the "wagon load" in the Columbia River basin. Perhaps only these large species suffered from hunting, common enough relative to human populations of the area to make golden-plover or Dunlin hunting unnecessary.

Habitat destruction has continued during the decades since all shorebirds of the region (except Common Snipe, a game species) were given total protection. Long-billed Curlews, unhunted, continued to decline in many parts of the region but are adapting locally to agricultural land. Marbled Godwits, rare through the early part of the century, may have increased in recent decades. Alternatively, the great increase in records of this species may be only a reflection of the increased concentration of observers in the region. This is probably the case with all of the

Table 13. Northwest Shorebird Nonbreeding Habitats; Regularly Occurring Species

	OO	RS	SB	MF	SM	SR	PL	FM	FF	DG
Black-bellied Plover			x	x			x		x	
American Golden-Plover		o		x	x		x		x	x
Pacific Golden-Plover		o		x	x					
Snowy Plover			x	o			x			
Semipalmated Plover			x	x			x			
Killdeer			x	x			x		x	x
Black Oystercatcher		x								
Black-necked Stilt				o			x	x		
American Avocet				o			x			
Greater Yellowlegs				x			x	x		
Lesser Yellowlegs				x			x	x		
Solitary Sandpiper							x	x		
Willet			o	x			o			
Wandering Tattler		x		o						
Spotted Sandpiper		o	o	o		x	x			
Upland Sandpiper										o
Whimbrel		o	o	x	x		o			o
Long-billed Curlew			o	x			o			
Hudsonian Godwit			o	o			o			
Bar-tailed Godwit			o	o						
Marbled Godwit			x	x			x			
Ruddy Turnstone		x	x	x			o			
Black Turnstone		x	o	o						
Surfbird		x								
Red Knot			x	x			o			
Sanderling		o	x	o			x			
Semipalmated Sandpiper			o	o			x			
Western Sandpiper			x	x			x		o	
Least Sandpiper			o	x	x		x	x	o	
Baird's Sandpiper			x	x			x			o
Pectoral Sandpiper				x	x		x	x	x	
Sharp-tailed Sandpiper				o	x			o		
Rock Sandpiper		x								
Dunlin		o	x	x			x		x	
Stilt Sandpiper				o			x	x		
Buff-breasted Sandpiper			o	o	o					o
Ruff			o	o			o			
Short-billed Dowitcher			o	x			o	o		
Long-billed Dowitcher			x	x			x,	x	x	
Common Snipe					o			x	x	
Wilson's Phalarope				o			x	x		
Red-necked Phalarope	x		o	o			x			
Red Phalarope	x		o				o			

OO - open ocean; RS - rocky shores; SB - sand beaches; MF - mud flats; SM - salt marshes; SR - streams/rivers; PL - ponds/lakes; FM - fresh marshes; FF - flooded fields; DG - dry grassland
x - common to uncommon; o - rare or not typical

rare species, although it remains possible that some shorebirds are still increasing from the market-hunting days, resulting in more peripheral records.

Only the annual Christmas Bird Counts may have been of sufficient extent in both space and time to document clearly any major changes in shorebird abundance in the region, and only in recent years have there been enough of them to furnish adequate information. I analyzed all Northwest coastal counts from 1974 to 1988, calculating five-year averages for each species of regular occurrence on each count. I compared the three time periods within each count and then assumed an *increase* within a species if there were substantially more *highest counts* in the third time period than in either of the earlier periods, and similarly a *decrease* if there were substantially more *lowest counts*. The results of this rather roundabout assessment can only be taken as indicative, but the next five years of Christmas Bird Counts can eventually be used to test hypotheses of increasing or decreasing populations.

Based on this analysis, I found little evidence of any major changes in abundance during that period, although some individual counts showed dramatic upswings or downswings. It is also apparent that numbers fluctuate considerably from year to year. The numbers do indicate increases in Black-bellied Plover and Sanderling and decreases in Black Turnstone, Rock Sandpiper, and Dunlin (see under those species). No obvious regionwide trends were observed in Killdeer, Black Oystercatcher, Surfbird, Western Sandpiper, Least Sandpiper, or Common Snipe. These species encompass almost the entire range of breeding and wintering distributions and habitats shown by Northwest shorebirds.

As habitats for shorebirds continue to decline, they will concentrate in those areas remaining intact, perhaps leading to brief, local increases. As we learn more and more about their movements, we realize that their well-being is tied to key locations throughout their range, very few such locations in species with conservative migration strategies. A given shorebird will have a breeding home, a winter home, and from one to a few stopover points on its migration, often different places in fall than in spring. Each place it visits could be considered a "vital spot" for this individual, and some vital spots on the migration route may support very large numbers of individuals of the species. An excellent account of this phenomenon is given by Myers (1983a).

The Northwest is an important part of the entire Pacific coast migration system and is involved to a lesser degree with migration systems through the North American interior and across the Pacific Ocean. To the south of us, estuaries in California, Mexico, Panama, and even Peru and Chile will have to be maintained for the well-being of our shorebirds, and of course, *our* estuaries will have to be maintained for *their* shorebirds. Shorebird populations anywhere from northern Alaska to southern Chile could be affected by substantial habitat losses in the Northwest. Relatively few of our birds end up on the Atlantic coasts of either North or South America, which are part of other major migration systems.

The vital spots for shorebirds in the Northwest, determined entirely by numerical concentrations, are some of the coastal estuaries and some of the Oregon lakes. These areas will be evident from the high counts of species in the text, but Grays Harbor, and in particular Bowerman Basin at Hoquiam, features the largest concentrations yet documented.

THE NEED FOR RECORDKEEPING

Anyone who wants to help the cause of shorebird conservation should keep records. With assiduous recordkeeping, we can continue to learn about populations of these birds in the region: where and when each species occurs, what the year-to-year variation in populations is, and what the long-term trends are. Only in about the last decade have there been enough records from a variety of Northwest sites to enable the writing of the species accounts in this book in any detail, and another ten years of increasingly complete recordkeeping will allow reassessment at intervals.

As an example of the need for information, there is almost nothing in the literature to indicate the status of the Western Sandpiper as a migrant through the interior of the Northwest, although it is probably one of the most common species. There are a few huge counts from Malheur Refuge, obviously numbers considered worthy of report. Otherwise, it is almost as if no one sees the species in the interior. Either observers do not submit interior records of this species or editors do not consider them worthy of publication in journals such as *American Birds*, just because they are common. Even on the coast, only unusual numbers or seasonally unusual observations are noted.

The Dunlin furnishes another example. If it were not for the annual Christmas Bird Counts, we would not have any documentation of its decline in the region. On the contrary, the status of relatively uncommon species such as the Baird's Sandpiper is well-known, as most birds observed are recorded and many find their way to publication. And every Hudsonian Godwit seen is reported, so it was possible to assess the relative numbers of this very rare species in spring versus fall and on protected waters versus the open coast, something that can be done for the Western Sandpiper only in very general terms.

A major gap in our knowledge is what species, sexes, and age classes of shorebirds oversummer in the Northwest. One approach to filling this gap would be an assault by shorebird counters on Northwest coastal localities on the two middle weekends of June over a period of several years. Are there substantial populations of summering shorebirds anywhere? Are they a regular feature where they occur? When there are only occasional individuals to be found, are they typically sick or crippled? In what plumages do they occur? For all the Christmas Bird Counts, we also have relatively little information about which sex and age classes spend the winter.

Although food habits of shorebirds are moderately well-known, records for each species are assembled from all over its range, and the details of what prey items are most important in the diets of shorebirds while in the Northwest are lacking for the most part. Any identification of prey items in the field is of value, therefore, although this is likely to be very difficult except with the largest species. Food-habit studies are of great importance, especially in major staging areas; knowing what a bird eats is an essential component of understanding its needs and making good management decisions.

What to Record

Birders who keep records typically tally species seen each day, by specific localities or the entire day's itinerary. Many estimate, at the end of the day, numbers of individuals of each species seen. All it takes to go beyond this to contribute something of great value to the ornithological record is to keep track of the numbers of birds in specific areas and record them separately, keeping track of both different habitats and different areas. On a trip to Ocean Shores, I usually record separately birds on the ocean beach (recording miles of beach covered), at the jetty at Point Brown, at the sewage pond, in the sand dunes, in the freshwater ponds and canals, in the mud flats and salt marshes of the "game range," and on the beaches of the spit that ends at Damon Point. Tide conditions must be recorded, as they play such a large part in the accessibility of shorebirds to the observer.

It is time to organize regular shorebird censuses in the Northwest to parallel those that have long been a feature of western Europe and the North American Atlantic coast. Regular censuses, *especially* of common species, in all parts of the Northwest would be of inestimable value now and in the future.

Accurate censuses, conducted over a period of years and sufficiently often to sample both resident and migratory populations effectively, would provide us with much information of interest and value, for example: (1) the use of juvenile-to-adult ratios to determine year-to-year variation in breeding success; (2) the use of temporal patterns in the migration of juveniles, as well as adults, to determine year-to-year variation in breeding seasons; (3) the determination of migration patterns, spatial as well as temporal, that differ between sex and age classes, and an attempt to integrate these patterns into evolutionary and ecological theory; (4) the monitoring of long-term population trends or cycles; and (5) effective management of shorebird populations based on a better understanding of their movements.

And beyond mere occurrence lies a whole universe of unknowns. Very few detailed studies of any sort have been done on shorebird biology in the Northwest, and the text references can be used as a quick guide to the paucity of studies about some species from anywhere in their range. There is so much to be learned about foraging behavior, movements between roosts and feeding areas, flocking tendencies, use of the environment over tidal cycles, molt schedules, aggression within and between species, and nocturnal feeding; the list goes on and on. The Dunlin is the most-referenced species in the Northwest, and there is much to be learned about it yet. Even the basic breeding biology of some species remains poorly known. Many potential projects for individual species are listed under Further Questions.

Computer Data Bases

In this era more and more data will be put on computers, and anyone intending to monitor populations of birds or compile records for geographic areas or taxonomic groups should consider putting their records into such data bases.

Shorebird Finding

With their intriguing displays and their haunting voices the waders have always attracted dedicated enthusiasts. Over the years they have drawn the watcher to many of the wild places of the world as well as to softer haunts close to towns and cities. —Desmond and Maimie Nethersole-Thompson, 1986

THERE ARE NO SECRETS to shorebird finding—unlike rails and owls, they are out there to be seen—but there are techniques that increase the probability of close encounters.

Broaden the Search

The simplest way to increase shorebird species found is to increase habitats scrutinized. From the list of nonbreeding habitat occurrence (Table 13), it is evident that there are enough habitat specialists so that observers must become generalists. Both salt water and fresh water must be visited, different coastal substrates have their own species assemblages, and even upland environments should not be ignored.

Narrow the Search

It is surprising how many shorebird species will turn up at one spot, vagile creatures that they are. Keeping close watch on a single patch of good shorebird habitat can be a rewarding endeavor, especially if accompanied by careful notes that can be compared over time. Probably the most thoroughly examined of such places in Washington is the complex of tiny ponds and grassland (Montlake Fill) on the edge of the University of Washington campus, where 23 species of shorebirds have been reported as of this writing. Much information about migration has come from daily records kept there over a period of years.

Prolong the Effort

Shorebird populations change almost continually through the year. Northward movements of birds have been detected in February, and spring migration of some species is in full swing by the beginning of April. Late spring migrants and early fall migrants overlap in mid June. Fall migrants are still moving south in November, perhaps December. Thus only during January is there little indication of movement, and this is a good month to check wintering populations (or sit out the rains studying shorebird books).

To encounter the greatest number of shorebird species in the year, intense field work should be carried out during the last half of April and first half of May for spring migrants and during July through October for fall migrants, thus for five months of the year. These periods encompass adult and juvenile migration, including the months during which most records of irregular species occur.

Pay Attention to Tides

The movements of coastal shorebirds are controlled by the tides, and encounters with them can be maximized by knowing the tidal cycles and what effects they will have on the birds' locations. Feeding and roosting birds offer different observation opportunities. Behavior can be observed in feeding birds, and bills are not frustratingly hidden. Birds that have been feeding in many areas, some of them inaccessible, come together at roosts. Comparisons are readily made, especially when a real or imagined alarm causes sleepy heads to be lifted from scapulars. Censuses of most of the shorebirds in the area are possible at this time, if roosts are limited.

Scan and Scan Again

Shorebirds are quite diverse in some habitats, with a dozen or more species using the same area simultaneously during migration. Sometimes a mud flat teeming with busy bodies seems overwhelming, especially if there are time constraints (the changing tide is often one). Scanning is most effectively done with a spotting scope, but binoculars may be even better for the quickest of scans (for example, if the birds are about to be disturbed). A *quick scan* provides an overall assessment of diversity, perhaps picking out the larger and more conspicuous species over the entire area. The observer may then return to particular species of interest or areas where there were too many birds to take them all in with a cursory scan. Alternatively, one can start at one end of the area and perform a *clean sweep*, attempting to identify every bird. This is the only way to discover uncommon species that are superficially similar to common ones, and observers with great patience are more likely to find these birds.

Contribute to the Cause

In all parts of the region there must be shorebird spots yet to be discovered, to be found by observers living in the interior or in coastal areas not mentioned in the text. Only a few places in the extensive estuarine complexes of Grays Harbor and Willapa Bay are regularly surveyed, and those areas and many of Oregon's productive bays are visited only on weekends. Sewage lagoons make excellent shorebird habitat, especially where aquatic habitats are otherwise scarce. Are all of them in this region now known to the ornithological community? Interior shorebird records come from only a handful of well-worked localities, but there are surely farm ponds and natural potholes that have not come within binocular range.

Shorebird Identification

The inscrutability of this group of birds is made even more painful by their enormous aesthetic appeal.—Jack Connor, 1988

ONE OF THE MOST typically human attributes is a persistent searching for truth, and in a small way the desire to know what species of plant or animal one is looking at is a manifestation of this attribute. The identification of birds in the field has become both a determined goal and a refined technique, perhaps both of these qualities stemming from the fact that bird watching is a game played with passion and devotion by many. It is no less important to thousands of amateur birders to peg a bird accurately for their pleasure than it is to the professional ornithologist conducting a census of a bird community to measure some environmental impact of human activities or to support a pet ecological hypothesis. In fact, birding for the fun of it prompts accurate identification even more than does science, as calling a sandpiper a sandpiper might be satisfactory for either an environmental impact statement or an ecological hypothesis, but such a bird will never make it onto a life list.

A point not sufficiently made in most guide books is that we do not have all the answers. There is much to learn on all frontiers of bird identification, vast gaps in our knowledge of the distribution, migration, feeding, and breeding biology of virtually all species. These points are made repeatedly in the species accounts below. No guide is an end-all to field identification; most active birders own at least several, and many own them all, gleaning important information from each new publication. I do not think that even the taxonomically specialized guides of today will "enable anybody who is interested in birds to identify any bird he or she is reasonably likely to see," as has at times been advertised; birds are just too variable.

STORING AND PROCESSING INFORMATION

Information, including how to identify birds, can be stored and processed best if it is *patterned*. Blocks of related information can be loosely or tightly tied together in larger blocks that are patterns. Knowing that all plovers run and stop and that all sandpipers keep moving as they forage is to know a pattern. Once this pattern is learned we do not have to remember specifically that a Dunlin should move continuously, a Killdeer stop and start. Taxonomic groupings of birds often represent patterns, like the very different coloration of *Pluvialis* and *Charadrius* plovers in any plumage. Knowing the general pattern should be followed by understanding the *variation* within it. All *Tringa* sandpipers have much in common, yet among them are species twice as large as others, species with red, green, or yellow legs, and species with medium or long bills. The group can still be characterized by relatively straight bills, relatively long legs, plain wings, foraging in water, typically an eye-ring, and a dotted back pattern in juveniles. *Exceptions* to the patterns are found in those species that fall conspicuously outside the variation in the group, species that make the group more difficult to define yet are clearly members of it, for example the Upland Sandpiper among the curlews.

A complete understanding of patterns, variation, and exceptions, even if not categorized specifically under these names, is of utmost value in learning to identify birds.

IDENTIFICATION TIPS

Most shorebirds will be viewed right out in the open on rocks, sand, or mud, where excellent, prolonged, and, because at least some of them are relatively tame, often close looks can be obtained. Therefore these birds, so often considered "impossible" to identify, are readily distinguishable by anyone provided with adequate literature, motivation, and patience. Would that warblers were so tractable!

Experienced observers use a great diversity of techniques to identify birds, more than they can usually name at short notice. A knowledge of plumage, voice, and behavior and a familiarity with geographic, ecological, and seasonal distribution of each species are only the first steps, although essential ones. Experience with the species and familiarity with the region can be augmented by insight into the methods that make identification both more simple and more certain. For those who are entranced by the thought of finding rarities, advanced techniques are beneficial. Kaufman (1990) presents many tips to enhance identification skills and treats a few shorebirds among other "difficult" groups and species pairs of North American birds. The following tips are provided for both beginning and intermediate observers but are appropriate for anyone who may have bypassed any of them along the way.

Give Yourself a Break

Take shorebirds a few species at a time. If possible begin in winter when, even at the best of Northwest localities, there will be no more than a dozen species, most of which are easily distinguished. When the new species arrive in April, they will be for the most part in distinctive breeding plumage. Don't start in August, a time when not only are most of the Northwest species present but also some of them are here in all three plumages! Or stick with freshwater localities for a while during the height of migration.

A second way to reduce the complexity of the group is to concentrate on easily identifiable species for a while. Be happy with Black Turnstones and Greater Yellowlegs and Killdeers and do not worry about the more obscure ones that at first elude integration into thought patterns.

The desire to see more and more new species is probably the biggest reason why many beginning birders do not learn identification skills well at first. They are exposed to too much too fast, jumping from area to area each weekend, seeing an array of new species before they have assimilated those from the weekend before. Classes that are taught in this manner serve more to familiarize students with good locations for birds and present them with some idea of the diversity and general groups of birds than they do to instill the specifics about the hundreds of species that may be encountered. Only with repeated visits and repeated encounters will some of these specifics emerge clearly from the multitude of quick impressions.

Use the Best Equipment Possible

Although as tame as birds go on the average, shorebirds are often seen at a distance just because of the nature of their habitat, from squishy mud flats to slippery jetties. Binoculars of as high a magnification as finances allow should be used. With their light weight, sharp optics, and excellent light-gathering abilities, the higher-priced 10x roof-prism binoculars are preferred by many shorebirders. Their only disadvantage is the inability of some of them to focus very close, which will rarely be a problem while shorebird watching.

For a shorebirder a spotting scope is an essential rather than a luxury. A zoom scope is a better value than a 20x fixed-power scope, as many shorebird identifications will depend on zooming to higher power. The type of tripod is as important as the type of scope, as it will be set up and taken down repeatedly along a shoreline. Important qualities include relatively light weight for lengthy estuarine treks, a head that allows simultaneous vertical and horizontal adjustment, smooth panning, quick and easy leg extension and retraction, and legs long enough for the tallest of your field companions.

Nevertheless, shorebird watching should not be given up by anyone not in possession of the highest technology. Instead choose a situation where a pair of battered 7x35 glasses are entirely adequate to observe Killdeers and Least and Western sandpipers at a pond near home. Many an unusual shorebird takes up brief residence at such ponds from time to time, and the common species can be learned well while saving for a spotting scope.

Do Your Homework

Learn the species to be expected in the area and season before going into the field. Use the information presented here to determine likely species in the habitats visited and their seasonal occurrence. Study only the plumages to be expected at that time of year; don't pay any attention to juveniles if it is April or to breeding adults if it is February. Mark field guides to facilitate appropriate scanning. It doesn't matter if the less likely species or plumages have not been committed to memory yet. They are after all less likely, and at the beginning virtually all birds encountered will be identifiable.

One of the most important steps on the pathway toward becoming proficient in bird identification is to learn all the common species of the area well, so they can be recognized wherever and whenever seen and so birds that are obviously something different can be interpreted as such. As there are 35 shorebird species that are at least fairly common at some place in the Northwest at some time of year, this is a substantial goal for a Northwest shorebirder.

Familiarize yourself with shorebird anatomy (Figures 2-4). Learn supercilium, mantle stripe, scapulars, and all other shorebird parts and markings. This is a worthwhile endeavor in itself and will prove invaluable when looking for tertial markings on dowitchers, mantle stripes on juvenile stints, and primary projections on golden-plovers. It is a heady experience just to learn that the knob on a bird's leg is not its knee!

As so many of our shorebirds come in three different-looking plumage classes (breeding, nonbreeding, and juvenile), establishing the plumage class of an unknown bird will decrease the problem of identifying it, especially if similar species are involved. See Figure 14 for a rough approximation of the seasons in which each plumage is to be expected, and the figures that express the same information under many individual species. The most likely complications occur in fall when adults and juveniles overlap, and juveniles are easily distinguished by their neatly patterned and unworn plumage. Almost half of the regularly occurring shorebirds are so distinctive they are unmistakable no matter the plumage, but when it comes to species pairs such as dowitchers and yellowlegs or similar groups such as stints and phalaropes, an effort should be made to determine the plumage type.

During both spring and fall many shorebirds are changing plumage, so construct mental pictures of birds in plumages intermediate between breeding and nonbreeding or between juvenal and first-winter. Bear in mind that individual birds may molt at unusual times; shorebirds in captivity often do so, probably indicating a disruption of the precise environmental control of their annual cycle.

Learn Birds in the Field

There is a tremendous inherent pleasure in seeing birds well and knowing them thoroughly. It should be an unnecessary reminder that there is much more to a bird than its name, while, even for the purest aficionado of identification, better-known birds are more easily identified birds.

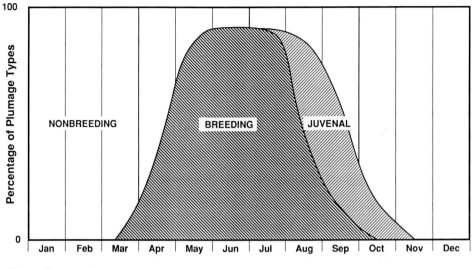

Fig. 14. Plumage-class occurrence

Sitting quietly at the edge of a mud flat or drying-up pond and thoroughly *seeing* the half-dozen species of shorebirds there can be a profound experience, a learning opportunity to combat the too-common attitude of "there's nothing interesting here, so let's go on to the next spot."

Shorebird calls can be learned with repeated encounters. Each species has its typical flight call; write their book descriptions down and commit them to memory. At some point it would be helpful to compile a list of them. Make a special point of paying attention to its vocalizations every time a known shorebird calls, easiest when only a single species is present. If you are fortunate enough to be able to whistle, respond to shorebirds you hear, fine-tuning imitations over time. At the least imitate them in your mind. It is more difficult for most people to identify birds by sound than by sight, and many birders ignore calls accordingly. It is clear that both patience and diligence are necessary to promote auditory sophistication.

Look at the bird, not the book. If you see an unknown bird and then immediately shift your attention to your guide, there is every chance you will lose track of the bird. Watch it as long as you can, noting everything relevant about its characteristics, even using a checklist such as this:

- size (very general unless nearby known species for reference)
- proportions (bill shape, relative bill and leg lengths, wing and primary projections)
- overall impression (all brown, brown above/white below, heavily spotted, rufous, etc.)
- colors (bill and leg)
- head markings (crown, supercilium, eye stripe, eye-ring, auricular patch, throat)
- back and wing markings (mantle, scapulars, tertials, coverts)
- breast and belly markings
- behavior (body movements and feeding behavior)
- calls (may have to wait until bird flushes)

Feel free to construct a better mnemonic device for yourself than the SPOCHBBBC acronym for this checklist.

Write down details about an unknown bird, even after it has departed. *As soon as you look in your book, your bird and the illustrations will begin to merge*, and the less time you have spent looking at the bird, the more likelihood of an inappropriate merging.

If the bird remains in sight, look in the field guide after absorbing as much as you can about it. Have someone else keep track of the bird if it is not likely to remain obvious. If this is not possible, make doubly sure you have observed everything about it before beginning your book work. Try scanning quickly to place the bird on the correct page or pages if there is a series of similar species. Quickly note range maps to eliminate very unlikely species. If the species seems unique in appearance, read about it to confirm field marks. If there are several similar species that might occur in the area, check them against one another.

Consider this method of committing birds to deeper levels of consciousness: pick individuals of known species at close range and write very careful descriptions of them. Not only will you know the species and plumage better when you are finished, but you will develop the skill of writing descriptions of unknown birds, for both personal edification and documentation of unusual records. Practice listing attributes of known species when they are in sight and—of great value—from memory. You will only rarely have to write down careful descriptions of *unknown* species, but of course these descriptions are of even greater value. Two of the region's four reported Great Knots were identified from their descriptions long after the birds were seen, as information available at the time of the sightings was not sufficient to allow identification.

Find a group of back-lighted shorebirds feeding and concentrate on their shapes, then move around to check them out for color and pattern. Shorebirds are often seen in poor lighting conditions, and anyone who can recognize silhouettes is ahead of the game.

Take Advantage of Flocks

Shorebirds are commonly seen in situations in which they are near other shorebirds, both because they are gregarious and because many species coexist in some habitats. Thus comparisons are generally possible, and one of the first things to note about an unidentified bird is its size and shape relative to one or more nearby, definitely identified birds.

Take careful note of whether an unfamiliar bird is consistently flocking with familiar ones. Aberrant individuals can occur in any species, and the bird itself, by being in a flock of its own species, might be sending a strong message. I have seen photographs of a "Ruff" and a "Curlew Sandpiper" that were respectively a short- and curve-billed dowitcher and a curve-billed Wilson's Phalarope; both were in flocks of the appropriate species.

Shorebirds at roosts present unparalleled gifts to the observer: they may be much closer than the same birds feeding out on the mud flats, and, especially, many or even all of the species in the area may be assembled together as if lined up on a field-guide page for comparison. Better yet, there are multiple individuals of each species, so adults and juveniles may be compared in season and individual variation is often evident. On the other hand, these same birds present perennial problems: bills are usually hidden, birds are close enough to one another to obscure other field marks, and the activity as birds enter or leave the roost is distracting. Figure 15 presents a set of birds that may be found sharing winter roosts on the Northwest coast; each has distinctive field marks to be used in lieu of its characteristic bill.

Recognize Nonshorebirds

Members of other groups—including herons, ibises, waterfowl, rails and coots, gulls, pigeons, larks, crows, pipits, starlings, blackbirds, buntings, and finches—may be seen moving along the same water-edge zone, and those that are brown, the prevailing shorebird color, may be mistaken

Fig. 15. Winter roost. Among small to medium-sized shorebirds in coastal winter roosts, Black-bellied Plovers (top center, bottom right) stand out by large size, short bill often visible. Long-billed Dowitchers (bottom, second and third from right) smaller, with distinctly gray sides. Dunlins (top, second from right; bottom, left two) smaller yet, most common species, with brown back and breast. Sanderlings (top row, two on left), of similar size, very pale with pure white underparts. Western Sandpipers (top right, bottom third from left) smallest, darker than Sanderling and paler than Dunlin; further distinguished from Dunlin by white breast.

at times for shorebirds in the same size category. Note that immature Mew Gulls, female Green-winged Teals, rails, and American Pipits are candidates for confusion, but virtually all of them are different in shape from the rather long-legged, long-necked, and short-tailed shorebirds. Perhaps the most similar in the Northwest are the Virginia Rail and Sora, which at times leave their dense marsh habitats and forage on freshwater mud flats like shorebirds; both of them have black and white barred flanks and heavier bills than sandpipers and plovers.

In flight, teals at a distance look like large sandpipers such as godwits and curlews, but something about the ducks, perhaps their larger heads and shorter wings, distinguishes them. Starlings and blackbirds perform aerial acrobatics in dense flocks much like small sandpipers, but they do not flash dark and light as do most shorebirds.

A Final Tip

The best-identified bird will always be the most completely seen bird. Look at the entire bird *and* all of its parts.

IDENTIFICATION PROBLEMS

Other than the variables that add difficulty to any identification—distance, poor lighting, partial views, and aberrant plumages, for example—there are some identification problems that are not so obvious.

Problems with Common Species

To the beginner, *shorebirds all look the same* (I have heard this phrase again and again, even at an estuary with everything from Whimbrels through Black-bellied Plovers and dowitchers to Semipalmated Plovers and Least Sandpipers). Earlier on, when everyone thought shorebirds had only summer and winter plumages, it was bad enough, but now that it is common knowledge that there is a third, somewhat different plumage in most species, the level of anxiety has gone up by a third. Some solutions to this problem are (1) begin in spring, (2) look only at species larger than stints, and (3) go out with a more experienced observer who brings to the endeavor both patience and a sense of humor.

With a different perspective, the identification of shorebirds can seem almost too simple. Authors of field guides, wanting to be definitive as well as helpful, often make it look easy because they show a plateful of birds, each of which looks different from all the rest, each with its own field marks indicated. Users of the book will tend to build their impression of the bird from the illustrations and text material in a typological fashion, as if the whole story about the bird is told there. But birds vary more than one would think from field guides, and only very recently has this issue been addressed by books showing examples of variation (Hayman et al. 1986, National Geographic Society 1987). Even these exemplary books fall short of discussing the degree of variation that is apparent in shorebirds on close scrutiny of a flock on a beach or a collection in a museum tray. Any graphic treatment of this variation would have to include more illustrations than is usually possible, for example showing all the variations to be expected in birds molting from one plumage to another.

Shorebird identification has been particularly short-shrifted because field-guide authors failed to acknowledge the substantial sexual dimorphism found in the group. "Typical" individuals were illustrated, usually male *Pluvialis* plovers and Ruddy Turnstones, female phalaropes and Long-billed Curlews. Thus half of all birds seen in spring deviated, in some cases substantially, from available illustrations. It is no wonder that shorebird identification has been considered difficult!

Problems with Perceived Rarities

Many birders appear poised to find rarities as a regular event, ready to turn a common species into a rare one. A point that must be made at least once is that *common species are more common than rare species*, and an unusual bird may as likely be an unfamiliar plumage or odd variant of a common species as a rare one. The most frequent manifestation of this problem occurs when a bird is seen that is not obviously a familiar species, and its identity in the mind of the beholder shifts to the next-commonest species that looks most like it. "A little knowledge is a dangerous thing" describes this phenomenon well and represents a significant problem in bird identi-fication.

Two occurrences that happened just as this section was being written illustrate this point. An especially bright Pectoral Sandpiper in Washington was called a Sharp-tailed Sandpiper by observers not familiar with the Pectoral's substantial variation. Perhaps even more seriously, a bird in Massachusetts that was clearly not a Pectoral Sandpiper was called a Sharp-tailed Sandpiper, very rare there but to the observers the only known similar species. This bird in fact turned out to be a much rarer hybrid "Cox's" or "Cooper's" sandpiper when finally looked at closely enough by observers who did understand the variation in both Pectoral and Sharp-tailed. "A lot of knowledge is a less dangerous thing" seems an adequate afterthought.

When an unknown shorebird appears, do not immediately assume it is something rare and exotic. Consider that it is a species to be expected at that place and time but perhaps in an unfamiliar plumage (for example, in molt between two distinctly different plumages) or even an aberrant individual. A Greater Yellowlegs with large white wing patches was correctly identified by an Oregon observer who saw more than just this obvious "field mark."

Many misidentifications are based entirely on expectations, when someone expects a particular bird to be in a particular place and then "sees" it. It is most likely to happen when one is looking for an uncommon species that is reported present, and the bright Pectoral Sandpiper mentioned above was probably called a Sharp-tailed because an individual of the rarer species had just been seen there. One can be easily tripped up while jumping to conclusions.

Problems with Real Rarities

Birds appear at times or in places at which they shouldn't, with the response "it couldn't be *that* species." There is then a chance that they will be considered the next-most-likely species that *should* be there. This problem is essentially the opposite of the previous problem. Location, habitat, and season are important clues to identity, but they are less important than the attributes of the bird itself. With experience should come greater confidence in one's own abilities, in this case a conscious suspension of disbelief.

Problems of Familiarity

A major problem in field identification of birds is caused by a bias in observer-detected variation based on familiarity: *well-known birds appear to vary more than poorly known ones*. As we learn more about any species, we begin to realize how variable it is. If a species is rarely seen, the less common individuals (those in the extremes of the normal curve of variation) are virtually never seen. To a European, Semipalmated Sandpipers may not seem particularly variable, while to an American the same is true with Little Stints.

Conversely, there may be times when an observer becomes more familiar with a rarer species. Most observers who have watched birds in a serious manner over long periods of time feel they know the common species very well, which in turn may lead to misidentifications when birds are seen that fall outside personal limits. Often unusual birds are studied with much more effort than is ever accorded common ones. The need to write feather-by-feather descriptions of rare birds to document their occurrence leads to a depth of knowledge about them that may be unnecessary in everyday identification of common ones. For example, the conspicuous white shaft of the outermost primary was mentioned as noteworthy in a Little Curlew in California, but many observers probably do not realize that all curlews and many other shorebirds exhibit the same characteristic.

Problems of Observation

Not only do individuals of a species vary among themselves, but also each individual varies, depending on what it is doing. It thus projects multiple images of itself, which complicates field identification.

Neck-length variation changes the appearance of a bird. Birds have long, sinuous necks that can be extended or contracted, and at the extremes they will obviously look quite different.

Apparent leg length varies with how a bird is standing. The same bird, standing up taller, looks longer-legged because more of its tibia is exposed below its belly feathering. Remember that absolute or relative leg length can be critically assessed only by using the tarsus, the lower segment of the leg that is always exposed.

Resting shorebirds look larger than feeding ones. Resting shorebirds fluff up to conserve heat, while feeding ones probably cannot so effectively do this as they run about. The apparent difference in body size can also make a difference in the apparent relative length of bill and legs, and of course fluffy feathers hide more of the legs, making them look shorter anyway.

There are also perspective problems associated with viewing birds. *The apparent bulk of a bird varies with the proportions of its appendages.* The relative bulkiness of birds is often noted by observers, who regularly describe birds by such terms as fat, slender, dumpy, or elegant, and the use of this attribute as an identification characteristic has been formalized by its mention in many field guides and identification articles. The terms imply differences in body bulk, yet they really refer to proportions of appendages. The proportions involved include the relative length of the bill, neck, wings, legs, and tail, and the longer these are (and the more of them that are long), the more likely a bird will be described as slender.

Thus, Semipalmated Sandpipers have been described as plump in several field guides and identification articles, when in fact virtually all of their body dimensions (as measured by skeletal elements, Cartar 1984) relate to one another just as do those of Western Sandpipers. This is particularly true when comparing female Semipalmated and male Western, which are essentially the same size except for bill length. Female Westerns are larger in body bulk than male Westerns as well as both sexes of Semipalmated, but their conspicuously longer bills and slightly longer legs make them look more extended, which registers as more slender. It may be entirely the shorter bill that makes the Semipalmated appear plump.

Dunlins are deemed dumpy, Curlew Sandpipers elegant, again because of differences in neck and leg length rather than any inherent difference in the shape of the birds' bodies, which are about the same size in these two species. Dumpy versus elegant would be an even more appropriate comparison between Willet and Greater Yellowlegs, as the former has considerably greater bulk but bill and legs about the same length as the latter.

Another term that has been used for shorebirds is "pot-bellied." This may refer to species in which the tail is held well above the belly level, producing an even curve from breast to tail and different from those species with tail held lower and a horizontal line from legs to tail. Looking at photographic collections of shorebird bellies, as in Farrand (1983) or Chandler (1989), is instructive and even amusing. This term has been applied to Upland Sandpiper and to distinguish Wilson's from the other phalaropes, but to me these two species are rather differently shaped. I wonder if their long necks contribute to such an impression.

Optical equipment confuses the issue of size. Estimating the sizes of birds at different distances from the observer but viewed in the same binoculars or spotting scope will be difficult. Contrary to common sense, birds at a greater distance may look larger than nearer ones. This also occurs in camera lenses and thus photographs, so caution is advised in judging size under high magnification.

Jizz

Field observers have always recognized birds at a distance by intangible qualities—by attributes that were more than merely the sum of their field-mark parts. These attributes have been termed *gestalt* (the German word for form or shape) or *jizz* (origins obscure). The latter word seems to have become a part of the language of British and to a lesser extent American birders. However, I believe that comments on size, shape, coloration, and behavior cover the attributes that jizz rather imprecisely describes.

This term is rarely used in field guides. Thus it is far from universal, yet many identification articles in *British Birds* utilize it, and it is much used by Harrison (1985) in his excellent seabird book. He calls it a "combination of ill-defined elements," although he allows that shape is a substantial component. Grant (1986) presented a tremendous amount of information on those difficult-to-distinguish birds, the gulls, without recourse to this term, although other discussions of gull identification make use of it.

Two recent identification articles on stints use the jizz concept, Veit and Jonsson (1984) more than Grant (1984). Both advise a certain amount of caution, but they nevertheless go on to point out subtle differences in species that are in some cases a consequence of clear morphological features and in others something not so easily assessed. As an example of the latter, "the horizontal posture of the Rufous-necked Stint differs from the more erect stance of the Little Stint" (Veit and Jonsson 1984: 856).

As the *concept* of jizz or gestalt is much in use, I feel a close look at how this concept is used is in order. Most bird watchers regularly identify birds at a glance just because of "something about them," which certainly encompasses the concept. On being queried we can often explain what it is that allowed instantaneous identification, and I believe that such an explanation could be forthcoming in all cases with further thought. Of course Dunlins look somewhat dumpy compared with Baird's Sandpipers; golden-plovers look fast and narrow-winged as they pass at a distance; and Least Sandpipers look different from Westerns when they take off. But I feel that these attributes can be simply described as consequences of real, nonoverlapping differences between the birds that can be seen consistently, say, if a set of photographs of them are examined.

In addition to these clear-cut differences, there may be *average* but overlapping differences between closely similar species, very useful for a quick call of a common species that can then be confirmed by further observation, but not complete field marks by themselves. In some cases I feel that these average differences, first described as "something different" about the birds, become by their use in the literature full-blown distinguishing characteristics, despite their not being definitive.

The concept of jizz is very useful as one gains familiarity with the common species of one's region. The general aspect, shape, and behavior of a bird furnish information that allows one to identify it at a great distance before details of coloration, size, or bill shape are seen. A probing group of dowitchers falls in its own gestalt pigeonhole, so to speak, and it could never inhabit the pigeonhole of a scattering of turnstones or march of curlews.

Observers at all levels of experience should develop the ability to identify birds at a distance; in fact this book is in part a response to our belief that it can be done. But the greatest caution should be exercised in identifying rare or unlikely species on the basis of jizz. *Characteristics* should be used to identify birds, and these should be objective ones that can be documented quantitatively or qualitatively and that can be perceived in the same way by two different observers. Anything that can be clearly described and compared with its alternative in writing or verbally (for example, a white wing stripe versus none or a bill longer or shorter than the head length) should be the basis for bird identification. These should be characteristics that do not change when the bird turns its head, fans its tail, stretches its neck, or fluffs its feathers against the cold.

More about the problems entailed here will be mentioned under individual species accounts, but the general caveat is applicable everywhere. As I have read more and more about the latest field marks, and talked with people at meetings and in the field, I have become convinced that the use of finer and finer distinctions between closely related species and the way these distinctions are perceived may be excessive in some cases. This is happening especially when observers try to pick out rare or very rare species from among the common ones. It is appropriate to *look for* unusual birds by general impressions, but it is not appropriate to *identify* them by these same impressions.

The Use of Single Field Marks

Finally, if possible all birds should be identified by more than one field mark. The curve-billed phalarope and dowitcher discussed above are examples of this phenomenon, as are breeding-plumaged Sanderlings and a Western Sandpiper with orange spots on its breast that were called Rufous-necked Stints. Probably all active field observers have seen examples of this; most have done it themselves. Field guides that emphasize single field marks contribute to the problem, which is easily avoided.

Birds are abundantly endowed with identification characteristics (field marks), from those that are absolutely definitive to those that are strongly or even weakly indicative. Many of the bird's attributes place it in a group of only a few species, eliminating almost all other possibilities. Definitive characteristics are of highest priority to note. Nevertheless, a suite of indicative characteristics might furnish a definitive identification if enough of them were noted, even though any single one would not be considered very important. A report of a juvenile Bar-tailed Godwit would be much better documented if all or most of its indicative characteristics were noted as well as its rather few definitive characteristics.

A Final Caveat

With the information explosion and subsequent appearance of many advanced field guides, a prevailing attitude is that enough information is now available and arranged and illustrated that we should have no more trouble identifying any bird of any species at any time. In this book we adhere to a more conservative attitude that we hope is enlightened rather than a throwback to the old days of really inadequate knowledge. We believe there are individual shorebirds that will fly or run past, even some that can be observed feather by feather, that may prove very difficult to identify; some may never give up their secret. At the same time these birds should be relatively few and far between, as we now have much more information available with which to make informed decisions about bird identification.

IDENTIFICATION OF SHOREBIRD GROUPS

A first step in the process of identification is to place a shorebird in its correct family. Species of two families—oystercatchers and avocets/stilts—are so distinctively shaped and colored that they furnish the joy of instant recognition (as would a pratincole). A slightly less instantaneous but still relatively easy step is to distinguish sandpipers from plovers. The sandpiper family contains 46 of the Northwest's 61 shorebird species (plus 13 of the 19 potentials), so restricting a bird in question to this family still leaves a lot of work. Nevertheless, the plovers, comprising one-fourth of the species, have been eliminated by this step.

Foraging behavior usually furnishes a simple dichotomy—*sandpipers keep moving, plovers run and stop* (Figure 16). Sandpipers forage by running or walking steadily, their bill pointed toward the substrate. They may or may not probe, but they keep moving, with the exception of some of the larger species such as curlews and yellowlegs that may stand still at times, but these long-necked, long-legged birds could never be mistaken for plovers. Plovers, on the other hand, are often still, alternating brief stationary periods with quick, short runs. At the end of a run they may peck at the substrate, presumably at a prey item seen from a distance.

Plovers are visual foragers, and their big eyes probably represent an adaptation to be able to do this even at night. Many sandpipers forage tactilely with a steady forward motion while probing in sand or mud substrates or among herbaceous plants. Even those that forage visually,

Fig. 16. Plover and sandpiper feeding modes. Plover stands, runs, and pecks or stops. Sandpiper moves forward continually, pecks or probes at intervals or probes continually.

which can include almost any species but is typically the case in tringines and a few other sandpipers of rocky and upland habitats, tend to hold to this steady forward progression. Variations on these themes are possible, however. On one cold winter day, probably when intertidal invertebrates were deep in their burrows, I watched Black-bellied Plovers foraging by moving steadily through shallow water just like the Dunlins with which they were feeding.

In addition to their primary behavioral difference, the two groups are generally distinguished by bill size and shape—shorter and thicker in plovers and longer and thinner in sandpipers. *All plover bills are much shorter than their head length, most sandpipers longer.* It is because of this that some short-billed sandpipers (Surfbird and turnstones) were long considered plovers, even though many other characteristics were clearly those of sandpipers. If these three species, with their sandpiper walk, can be recognized, then anything else with a short, thick bill is a plover. The other short-billed sandpipers (Buff-breasted the shortest) have quite slender bills.

Within each of the two larger families there are some clearly apparent subdivisions and some not so clear although qualifying for taxonomic status. Within the plovers the *Pluvialis* group and the typical *Charadrius* group stand apart, but the Eurasian Dotterel looks intermediate in some ways, although a member of the latter group.

Within the very diverse sandpiper family several groups stand out discretely, in particular the curlews (without the aberrant Upland Sandpiper), godwits, turnstones, dowitchers, snipes, woodcock, and phalaropes. The tringines and calidridines are just a bit too diverse to be easily characterized, but with practice members of these two groups can be allocated to the correct tribe. Get a feel for each of the groups from the text descriptions and illustrations.

FIELD-MARK GROUPS

Table 14 groups species by conspicuous single field marks or combinations of field marks. The observer may be able to narrow down the list of possible species by committing the categories, if not the species in them, to memory. Remember these are only clues to an identification that should be confirmed by further study.

Table 14. Quick Identification

Many shorebirds have special distinctive attributes—the classical Peterson field marks—by which they can immediately be recognized. The following lists will allow quick identification of these species. The characteristic may refer to only one or two of the three plumages of a species (indicated by B - breeding, N - nonbreeding, and J - juvenal), and a given species can be in more than one category. Note that irregular species are in the right-hand column, and that large and very large species are asterisked.

Regular Species (abundant to rare)	Irregular Species (casual/potential)

Tail length

Tail extends conspicuously beyond wing tips

Killdeer	[Common Sandpiper]
Spotted Sandpiper	Temminck's Stint
Upland Sandpiper	[Purple Sandpiper]
Rock Sandpiper	American Woodcock
Common Snipe	[Jack Snipe]

Leg length

Legs long to very long

Black-necked Stilt	[Black-winged Stilt]
*American Avocet	[Common Greenshank]
Greater Yellowlegs	[Marsh Sandpiper]
Lesser Yellowlegs	Spotted Redshank
*Marbled Godwit	
Stilt Sandpiper	

Leg color

Legs bright red, orange, or pink

Semipalmated Plover B	[Common Ringed Plover]
Black-necked Stilt	Piping Plover
Greater Yellowlegs (atypical)	[Black-winged Stilt]
Lesser Yellowlegs (atypical)	Spotted Redshank
Spotted Sandpiper B	
Ruddy Turnstone	
Ruff B	

Legs bright to dull yellow

Semipalmated Plover NJ	[Common Ringed Plover]
Greater Yellowlegs	Piping Plover
Lesser Yellowlegs	Eurasian Dotterel
Spotted Sandpiper	[Marsh Sandpiper B]
Wandering Tattler	[Wood Sandpiper]
Upland Sandpiper	Gray-tailed Tattler
Surfbird	Terek Sandpiper
Least Sandpiper	Long-toed Stint
Pectoral Sandpiper	[Purple Sandpiper]
Sharp-tailed Sandpiper	
Rock Sandpiper	
Stilt Sandpiper NJ	
Buff-breasted Sandpiper	
Ruff	
Wilson's Phalarope NJ	

Table 14. Quick Identification (continued)

Regular Species (abundant to rare)	Irregular Species (casual/potential)

Leg color (continued)

Legs green, greenish or olive

Solitary Sandpiper	[Common Greenshank]
Spotted Sandpiper NJ	[Marsh Sandpiper]
Red Knot	[Wood Sandpiper]
Pectoral Sandpiper	[Green Sandpiper]
Sharp-tailed Sandpiper	[Common Sandpiper]
Stilt Sandpiper	Great Knot
Ruff	Temminck's Stint
Short-billed Dowitcher	[Broad-billed Sandpiper]
Long-billed Dowitcher	
Common Snipe	
Wilson's Phalarope NJ	

Legs bright blue
*American Avocet

Legs gray to black

Black-bellied Plover	Mongolian Plover
American Golden-Plover	[Little Curlew]
Pacific Golden-Plover	[Eskimo Curlew]
Snowy Plover	*Bristle-thighed Curlew
*Willet	*Far Eastern Curlew
*Whimbrel	*[Black-tailed Godwit]
*Long-billed Curlew	Great Knot B
*Hudsonian Godwit	Rufous-necked Stint
*Bar-tailed Godwit	Little Stint
*Marbled Godwit	White-rumped Sandpiper
Red Knot B	Curlew Sandpiper
Sanderling	Spoonbill Sandpiper
Semipalmated Sandpiper	[Broad-billed Sandpiper]
Western Sandpiper	
Baird's Sandpiper	
Rock Sandpiper B	
Dunlin	
Wilson's Phalarope B	
Red-necked Phalarope	
Red Phalarope	

Bill shape and size

Bill spoon-shaped

–	Spoonbill Sandpiper

Bill obviously downcurved for much of its length

*Whimbrel	[Little Curlew]
*Long-billed Curlew	[Eskimo Curlew]
Curlew Sandpiper	*Bristle-thighed Curlew
	*Far Eastern Curlew

Table 14. Quick Identification (continued)

Regular Species (abundant to rare)	Irregular Species (casual/potential)

Bill shape and size (continued)

Bill slightly downcurved at tip
Western Sandpiper
Dunlin
Stilt Sandpiper

Bill slightly to obviously upcurved

*American Avocet	[Common Greenshank]
Greater Yellowlegs	Terek Sandpiper
*Hudsonian Godwit	
*Bar-tailed Godwit	
*Marbled godwit	

Bill very long, straight

*Hudsonian Godwit	*[Black-tailed Godwit]
*Bar-tailed Godwit	
*Marbled Godwit	
Short-billed Dowitcher	
Long-billed Dowitcher	
Common Snipe	

Bill color

Bill with conspicuous yellow or orange

Semipalmated Plover B	[Common Ringed Plover B]
Spotted Sandpiper	Piping Plover B
Red Phalarope B	[Purple Sandpiper]
Surfbird	
Rock Sandpiper NJ	

Bill with conspicuous red or pink

*Black oystercatcher	*American Oystercatcher
Spotted Sandpiper B	Spotted Redshank
*Whimbrel	[Little Curlew]
*Hudsonian Godwit	[Eskimo Curlew]
*Bar-tailed Godwit	*Bristle-thighed Curlew
*Marbled Godwit	*[Black-tailed Godwit]

Conspicuous black markings

Entirely dark at distance

*Black Oystercatcher	Spotted Redshank B

Upperparts black or very dark brown, underparts mostly white

Black-necked Stilt	[Black-winged Stilt]
Black Turnstone	

Upperparts contrasty black and white
*American Avocet
Surfbird B (worn)
Sanderling J

Table 14. Quick Identification (continued)

Regular Species (abundant to rare)	Irregular Species (casual/potential)

Conspicuous black markings (continued)

Most of underparts black, upperparts colored otherwise
Black-bellied Plover B
American Golden-Plover B
Pacific Golden-Plover B

Breast black, belly white
Ruddy Turnstone B
Black Turnstone
Ruff B (male)

Black patch on breast or belly

Rock Sandpiper B	Great Knot B
Dunlin B	
Ruff B (molting male)	

Black patch through eye and ear
Red-necked Phalarope NJ
Red Phalarope NJ

Two black bands across breast
Killdeer

Single black or brown band across breast

Semipalmated Plover	Mongolian Plover NJ
	[Wilson's Plover]
	[Common Ringed Plover]
	Piping Plover B
	[Little Ringed Plover]

Conspicuous reddish markings

Head orange
*American Avocet B

Rufous patches on neck
Wilson's Phalarope B
Red-necked Phalarope B

Upperparts mostly bright rufous
Ruddy Turnstone B
Dunlin B

Conspicuous bright rufous patches on scapulars, nowhere else

Surfbird B	Great Knot B
Rock Sandpiper B	[Purple Sandpiper B]

Entirely rich reddish or reddish brown

*Long-billed Curlew	*[Black-tailed Godwit B]
*Marbled Godwit	Curlew Sandpiper B

Table 14. Quick Identification (continued)

Regular Species (abundant to rare)	Irregular Species (casual/potential)

Conspicuous reddish markings (continued)

Underparts mostly bright reddish, brighter than upperparts

*Hudsonian Godwit B	*[Black-tailed Godwit B]
*Bar-tailed Godwit B	
Red Knot B	
Curlew Sandpiper B	
Short-billed Dowitcher B	
Long-billed Dowitcher B	
Red Phalarope B	

Breast rufous, belly white

Ruff B (male)	Mongolian Plover B

Uniform brown

Upperparts uniform medium to dark brown

Spotted Sandpiper	*American Oystercatcher
	[Oriental Pratincole]
	Mongolian Plover
	Mountain Plover
	[Common Sandpiper]
	Temminck's Stint N

Upperparts uniform brown with white collar

Snowy Plover	[Wilson's Plover]
Semipalmated Plover	[Common Ringed Plover]
Killdeer	[Little Ringed Plover]

Gray

Upperparts plain gray

Wandering Tattler	Gray-tailed Tattler
Surfbird N	[Purple Sandpiper N]
Rock Sandpiper N	

Spotted

Upperparts conspicuously spotted with white, yellow, or pale buff

Black-bellied Plover J	Spotted Redshank J
American Golden-Plover	[Wood Sandpiper BJ]
Pacific Golden-Plover	[Green Sandpiper BJ]
Greater Yellowlegs J	*Hudsonian Godwit B
Lesser Yellowlegs J	
Solitary Sandpiper BJ	
*Whimbrel J	

Underparts conspicuously spotted with black or dark gray

Spotted Sandpiper B	[Purple Sandpiper N]
Surfbird N	
Sharp-tailed Sandpiper B	
Rock Sandpiper N	

Table 14. Quick Identification (continued)

Regular Species (abundant to rare)	Irregular Species (casual/potential)

Striped

Upperparts with conspicuous white or buff stripes

Western Sandpiper J	Eurasian Dotterel J
Least Sandpiper J	Little Stint J
Pectoral Sandpiper J	Long-toed Stint J
Sharp-tailed Sandpiper J	White-rumped Sandpiper J
Rock Sandpiper J	[Purple Sandpiper J]
Dunlin J	Spoonbill Sandpiper J
Stilt Sandpiper J	[Broad-billed Sandpiper J]
Common Snipe	[Jack Snipe]
Red-necked Phalarope	
Red Phalarope B	

Barred

Underparts conspicuously crossbarred

Greater Yellowlegs B	Spotted Redshank J
Wandering Tattler B	Gray-tailed Tattler B
*Hudsonian Godwit B	*[Black-tailed Godwit B]
*Marbled Godwit B	
Stilt Sandpiper B	
Long-billed Dowitcher B	
Common Snipe	

Explanation of Species Accounts

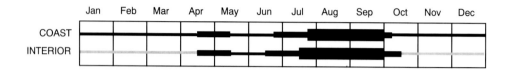

	Jan	Feb	Mar	Apr	May	Jun	Jul	Aug	Sep	Oct	Nov	Dec
COAST												
INTERIOR												

Distribution

The overall distribution and migration patterns of each species are briefly outlined here.

Northwest Status

The status in the Pacific Northwest is summarized by a phenological chart indicating the seasonal status of each species in the region by subregion. Each month is divided into four parts, and the status is thus assessed approximately by weeks. The widest lines indicate *common* during that week, the next widest *uncommon*, the narrowest *rare*. The gray line indicates *casual* (in some cases a single bird occurring over several weeks), and the individual points indicate *one or two records* during that week.

Appendix 2 lists the status for all shorebirds recorded from the region in simpler form. The seasons are "functional seasons" independent of the calendar year. *Winter* is defined as the period when the species is resident on its nonbreeding grounds, *spring* the period when it is moving through the area in northward migration, *summer* the period when it is resident on its breeding grounds (or summering grounds for those species that regularly summer south of their usual breeding range), *fall* the period when it is moving through the area in southward migration.

The regional status of each species is described in some detail, with records through 1990. As the status of so many shorebirds differs between coast and interior, it is described for both subregions. Additional attention is paid to local distribution. For example, some coastal species are also widespread away from the coast in the western lowlands, while others are largely confined to the coast. Most species that breed or winter in the area do so only locally. In most species the coastal status is described first, then the interior

status, but in species that breed only in the interior, their status in that subregion is described first. Seasonal status is usually described concurrently with geographic distribution.

Both general statements and many specific records have been included, the latter for extreme migration dates and largest recorded concentrations. From these records interested observers will have a guide to the optimal times and places to look for the species in particular parts of the region. These high counts are listed in the order Montana-Idaho-Oregon-Washington-British Columbia during spring and summer and in reverse during fall and winter. They are then ordered by date within each state or province. Imprecise localities and dates are listed as they were given in the literature. Both juvenile and adult extreme dates are given when they have been determined, although most literature records do not specify age class. Christmas Bird Count summaries are given for regularly wintering species.

For the rarest species (typically with fewer than 20 records for the region), all records accepted as valid are included. Furthermore, when seasonal or subregional records are few, all of them are usually listed. Specimens examined, some of them also cited in the literature, are indicated by number signs (#); unpublished sight records are indicated by asterisks (*). These records are from my own museum and field work, except those from Montana, which are from David Beaudette.

Habitat and Behavior

The preferred habitats and general behavior patterns of each species are discussed here. Emphasis is placed on the most common and typical features, but it should be emphasized that individuals of any species can behave in unusual ways or occur in unlikely habitats. A Red Phalarope is unlikely to feed on a rock jetty, and a dowitcher is unlikely to stride through an upland meadow picking insects off leaves, but almost any variation is possible in these highly motile and behaviorally varied birds.

Structure

The attributes of size and shape that are important in all plumages are described. The relative bill and leg lengths are important as identification characters of shorebirds, not to mention interesting features that define them biologically, and it seems useful to standardize the terms to describe these features.

All species are placed in size categories by weight, rather than by length, although length is typically used in field guides. These length measurements are unintentionally deceiving, as they measure from bill tip to tail tip. Because of this, different measurements are shown for two birds of the same weight if one has a longer bill or tail, and I have found that many users of these guides then expect the bird with the larger measurement to be a larger bird, which it is not. Throughout this text, when bird sizes are compared their *weights* are being compared, and these differences are apparent in the field as different body bulks. A Lesser Yellowlegs looks down on a Killdeer as it passes but is nevertheless a smaller bird. Perhaps the preoccupation of field guides with length rather than weight parallels our attitudes toward height and weight in our own species!

Appendix 3 lists weights and measurements for all species discussed herein, and Appendix 5 groups them into size categories by weight. All of these figures are averages, primarily taken from the literature (Cramp and Simmons 1983, Jehl and Murray 1986, Lane 1987, Prater et al. 1977, Ridgway 1919) but augmented in a few species by weights and measurements taken from specimens. A few weights are approximations. In species in which males and females are slightly different in size, the figures represent a midpoint between the two sexes. In species that are widely dimorphic in size (by more than 25 percent of the larger sex's weight), the sexes are listed separately. The relative size of the sexes and degree

of size dimorphism is determined to some degree by the mating system, with males distinctly larger in polygynous and promiscuous species and females distinctly larger in polyandrous species. Even in monogamous sandpipers, however, females are often slightly larger.

Weights and measurements vary substantially within each species. The larger the species the greater the variation, although usually comparable if expressed as a percent of the bird's size. Bill length appears to vary more in long-billed than in short-billed shorebirds, but this is primarily because of sex and age differences. Juveniles of longer-billed species have perceptibly shorter bills than adults, perhaps merely because it takes a long time to grow such a long bill. This is much more apparent in curlews, however, in which a shorter bill must function well in foraging, than in godwits or dowitchers, which must need their long bills right away for effective foraging in their nonbreeding habitats.

The significance of this variation is that the nearer two species are to one another in weight or any measurement, the more likely they overlap, and, even where there is no overlap, if differences are small relative to the size of the birds, they are essentially useless for distinction in the field. The ability to judge small differences seems to vary among observers and with experience, so it is difficult to assess exactly what degree of difference is necessary to be apparent in the field.

Proportions of tail, tarsus, and bill are also listed in Appendix 3, indicating the relative lengths of these appendages. They are thus useful for comparisons among species, especially related ones. The proportions are expressed relative to wing length, which is representative of bird size (weight), but bear in mind that some species, especially longer-distance migrants, have longer wings relative to their size than others, and this difference will affect the apparent proportions of tail, tarsus, and bill slightly. As with absolute size, the smaller differences in proportions will probably not be apparent in the field.

Bill Length and Shape. Shorebird bills are either straight, upcurved, or downcurved. They vary in length from very short to very long, a greater variance than in most other bird groups. Length is such an important variable because so many shorebird bills are used to capture prey at

Fig. 17. Relative bill lengths

different depths in soft substrates. Relative bill lengths are expressed for all species in Appendix 3, and these figures can be used to compare pairs or groups of species in a quantitative manner. Figure 17 shows representative bills for each length category.

Relative Bill Length Definitions.

Very short, no longer than distance from base of bill to back of eye

Short, distinctly shorter than head but longer than distance from bill to eye

Medium, about same length as head (from bill base to back of head)

Long, distinctly longer than head but less than twice head length

Very long, twice head length or longer

Leg Length. Shorebird feeding opportunities are both provided and constrained by leg length as well as bill length, so there is a substantial variation in shorebird leg length. Some species have relatively slender legs, others relatively stout, but variation in shape is considerably less than it is in the bill. The body diameter is an important dimension with which to compare relative leg lengths of different species. It is defined as the diameter of the bird at the widest part of its body (usually about over the legs) as seen from the side. Note that fluffed and sleeked shorebirds have bodies of different diameters, which complicates the assessment of this characteristic. Thus medium-legged species are those in which the legs look anywhere near the same length as

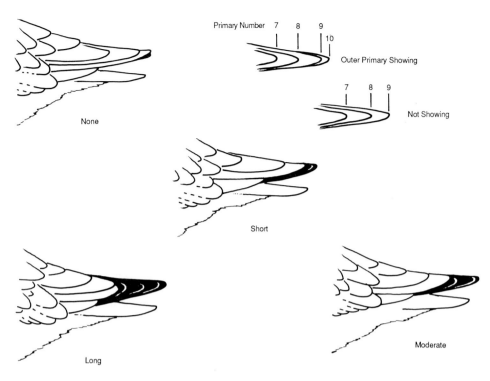

Primary Number 7 8 9

10

Outer Primary Showing

None

7 8 9

Not Showing

Short

Long

Moderate

Fig. 18. Primary projections

the diameter of the body. Relative leg lengths are expressed quantitatively for all species in Appendix 3, and these figures can be used for comparisons.

Relative Leg Length Definitions.

Short, exposed legs distinctly shorter than the diameter of the body

Medium, exposed legs about the same length as the body diameter

Long, exposed legs distinctly longer than the body diameter

Wing Length. As virtually all shorebirds have acutely pointed wings, no purpose is served by describing wing shapes. The few species that deviate from this mode are pointed out in the text. There are two characteristics apparent in nonflying birds that relate to wing pointedness, however. One involves the projection of the primaries past the tip of the longest tertial (*primary projection*). The primary projection is visible from the side, as the primaries are unpatterned and usually darker than the tertials (with entirely

dark wings, Black Oystercatcher and stilts are exceptions), and constitutes an important identification character in some groups.

In some species, in particular among the calidridines and phalaropes, breeding adults have distinctly longer tertials, and thus shorter primary projections, than do juveniles. This is noted wherever possible, and where the description refers to only one age stage, primary-projection length could not be determined for the other. Both age stages are mentioned when they are the same, and where neither is mentioned, they could not be differentiated.

The only complication in assessing the number of exposed primaries is that the two longest primaries may be about the same length, so the outermost one is not visible beyond the next-longest one (Figure 18).

The other characteristic, species-specific although somewhat variable, is the *wing projection*, the amount the wing tips project beyond the tail tip in a resting bird. This characteristic relates to both relative wing length and relative tail length and varies even within individuals, as birds hold their wings in different positions depending on what they are doing. For instance, a bird feeding with head down may pull its wings forward slightly, thus altering the relative positions of wing tips and tail tip. Note that the greater the relative tail length in Appendix 3, the more likelihood it will project beyond the wing tip. Because it is variable, this character is less diagnostic than some others, but nevertheless differences between species or among groups of species are pronounced enough to be significant in many cases. *Typically longer-winged species have both longer primary projections and longer wing projections*.

When birds are molting their primaries in autumn (rarely into winter in this region), the loss of the outermost primaries will compromise the assessment of both primary projection and wing projection. This should occur only in birds that intend to winter and can probably be disregarded in all those species and plumage stages that normally do not winter in the Northwest. Nevertheless, a bird that is unusual because out of place or out of season might also be unusual in its molt pattern.

Primary Projection Definitions.

None, longest tertial reaches wing tip, no primaries exposed

Short, 1-2 primary tips exposed beyond longest tertial (no long gaps—greater than eye diameter—between successive primary tips)

Moderate, at least 3 primary tips exposed (at least one long gap between primary tips)

Long, at least 4 primary tips exposed (at least two long gaps between primary tips)

Tail Shape. Shorebird tail shape varies over a fairly narrow range, with the exception of the long, forked or pointed tails of a few Old World species (including the potentially occurring Oriental Pratincole). The descriptions used here should be self-explanatory (Figure 19), but a few additional words are necessary. *Square* tails are those in which the outermost and central feathers are more or less the same length. The farther such a tail is spread, the more rounded it will look, however. In *rounded* tails the outer feathers are distinctly shorter than the central ones in the folded tail, and the tail looks more rounded than

in the square-tailed species when it is spread. A *long, graduated* tail is one in which the tail projects well beyond the wing tips at rest, and the outermost feathers are no more than three fourths the length of the central pair, thus still more rounded when spread.

A *wedge-shaped* tail is one that is approximately square or slightly rounded when folded, with a strongly projecting central pair of feathers that makes it look distinctly pointed when spread. Finally, a *double-notched* tail has a projecting central pair of feathers as in a wedge-shaped one, but the outermost feathers, although shorter than the central pair, are longer than some of the intermediate ones, so that each half of the tail is notched.

Plumage

A telegraphic style is used for some parts of the description, to save space and alleviate a certain amount of tedium from the sheer length of the plumage descriptions. Although illustrations are indeed worth thousands of words, these descriptions are included especially to show variation and to furnish information about parts of birds not shown in drawings or photographs.

Plumage Chart. A chart included for common species with distinguishable plumages shows approximately when birds in those plumages are likely to be present in the Northwest. Individual birds in a given plumage might be seen outside the periods indicated on the chart. Question marks indicate periods when so few birds of that age category are seen that the timing of molt is poorly understood.

Bare-part Colors. The colors of the bill and feet are described first. Almost all shorebirds of the region have dark brown eyes, and only in species that are exceptions to this (oystercatchers and stilt) has the eye color been described. It has not been generally recognized that *bill and leg colors change over the year in some shorebirds*, and these changes are noted. A simple generality is that many species have darker bills and legs

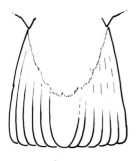

Square

Rounded

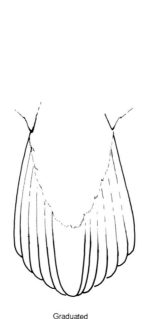

Graduated

Wedge-shaped

Double-notched

Fig. 19. Tail shapes

during the breeding season, paler during the non-breeding season, with juveniles colored like non-breeding adults. Birds in migration in breeding plumage may still show nonbreeding bill or leg coloration (or traces thereof), leading previous writers to conclude that these colors did not change. Paintings in recent field guides for the most part still fail to show these distinctions, except for the very obvious changes such as bill color in Red Phalarope and leg color in Wilson's Phalarope.

Plumage Distinction. Each of the distinct plumages (breeding, nonbreeding, and juvenal)

of each species is discussed in turn or combined if they are not distinct. Immature plumages (first winter and first summer) are often difficult to distinguish from those of adults, but they are discussed for species in which they might be recognizable in the field. Plumages beyond the first summer correspond to those of adults in all species, as far as we know, even if these birds are not yet sexually mature. Immature plumages are not discussed for potentially occurring species.

Molt. When "molt" is mentioned, it refers to molt of body feathers, which changes the appearance of the bird. Wing and tail molt, which do not do so, are specifically mentioned in some instances. Progression of molt is more easily seen in species with dark underparts such as Black-bellied Plovers, Red Knots, and Red Phalaropes. Adults of these species may be detected late in autumn when traces of their dark ventral plumage are still in evidence. Presumably white-bellied species also retain some breeding feathers late in fall that are not detectable in the field.

Subspecies

The status of and differences between subspecies are discussed for all species in which two or more subspecies are known to occur or might occur in the Northwest.

Identification

This section outlines the most important points of identification for the species and contrasts it with similar ones, both by plumage stages and by characteristics independent of sex, age, and season. Species names in this section and the following two are in small capitals for ease of comparisons. It is very important to determine the plumage stage (breeding, nonbreeding, or juvenal) of the bird in question, as many of the detailed comparisons are made between birds in the same plumage.

In Flight

Species distinctions in flight are enumerated here, in particular from above or behind but also from below. Style of flight is emphasized when distinctive, as well as field marks. In many cases species similar at rest are not similar in flight and vice versa, thus making in-flight observations of great significance.

Voice

Voice descriptions primarily refer to flight calls, those vocalizations made by birds when they are flushed or, less often, as they fly by. These descriptions have been kept simple, describing typical calls only, but there is some individual variation even in the typical calls, and occasional birds on their nonbreeding grounds give either atypical calls or calls from another context, in particular those that are used on the breeding grounds. Some descriptions of vocalizations in the literature combine breeding and nonbreeding calls, which complicates learning them, as vocalizations are considerably more diverse on the breeding grounds. For Northwest breeding species, some breeding vocalizations, in particular the display "songs," are described, but the variety of less distinctive calls between mates and between adults and young has not been included.

Further Questions

If one takes the time to look at birds, they will invariably be seen doing interesting things, and the questions asked in this book will allow observations to be directed in ways that will benefit the field of ornithology and the more special field of bird identification, as well as provide personal edification. Even though birds are without doubt the best-known animals, there are many unanswered questions about every species, even the most familiar ones. Amateurs in this field have already contributed immensely to the body of knowledge and can continue to do so more effectively with specific goals and directions.

Questions are posed under most species, and others may occur to the reader. The references, both under each species and to shorebirds in general, should provide resources to allow determination of questions that been answered satisfactorily or in part.

Notes

This section is reserved for comments on published statements about shorebirds that I consider dubious or erroneous (for the most part responding to statements in publications regularly used or considered particularly authoritative) and miscellaneous comments about some species that do not fit elsewhere in the text. Also, published records of occurrence in the region that I do not consider adequately documented are listed here.

Photos

The reader is referred to published photographs that would be of interest, in many cases supplementing the illustrative material presented herein. In addition, misidentifications of both species and plumage that have occurred in published guide books are pointed out.

References

All references about identification, distribution,

and general biology that are considered of potential interest to the field observer are included. These references are all published and all in English. Only the more significant and interesting references on general biology are included, and there may be other publications about some species, in some instances many of them. Many of the references involve distant populations in distant lands, in some cases because nothing has been written about our own populations but also because there is much evidence that shorebirds have similar habits in different parts of the world.

References to the general biology of potentially occurring species are given only if the species is not included in Cramp and Simmons (1983), the references were published subsequent to that work, or they are of particular importance for identification or biology. Old references as well as new are included, and perusal of some of those from the early part of the century as well as recent ones will show how scientific and popular writing have diverged over the years.

There are also many dissertations, theses, and unpublished reports in English and many publications in other languages on shorebirds, and references to some of them can be found in the following works: Blomqvist (1983a, 1983b), Burger and Olla (1984a, 1984b), Cramp and Simmons (1983), Hale (1980), Hayman et al. (1986), Johnsgard (1981), Lane (1987), and Nethersole-Thompson and Nethersole-Thompson (1986). The "recent publications on waders" section in the *Wader Study Group Bulletin* will provide continued up-to-date references.

Family and Species Accounts

PRATINCOLES Glareolidae

PRATINCOLES ARE MEMBERS of an Old World family that includes also the rather dissimilar coursers. Of the 17 species of the family, a single pratincole has been recorded twice in western Alaska, the only records of the family in the New World. Pratincoles are very distinctive shorebirds, much deserving their former name of "swallow-plovers" with their long wings, forked tails, and short legs. They feed by running along shorelines, picking their food in plover fashion, but they also forage in the air like swallows, the only shorebirds to do so, or fly up from the ground after insects they flush.

ORIENTAL PRATINCOLE *Glareola maldivarum*

Rarely are new families of birds added to the North American list, so the sighting of one of these birds at Attu in the Aleutian Islands on May 19-20, 1985, was accompanied by considerable excitement. Even though an unlikely visitor far to the northeast of its breeding range, another bird was seen still farther northeast at Gambell, June 5, 1986. It is unlikely to occur in the Northwest.

This bird is medium-small, about the size of a KILLDEER or LESSER YELLOWLEGS, with long wings projecting well to the rear that give it a distinctive swallowlike shape on the ground. The bill is short for a shorebird and also somewhat swallowlike, the legs distinctly short for a shorebird of its size. Breeding adults are medium brown on upperparts and breast, lighter and more buff-tinged on the belly, with a fine black line around the contrasty pale-buff throat. The upper belly and sides may be washed with rusty brown. Nonbreeding adults are duller, with more extensive white underneath and a broken black line around the pale throat patch. Juveniles have the crown, neck, and breast heavily marked with black spots and chevrons, and the mantle, scapulars, coverts, and tertials have pale buff fringes and dark subterminal bars. The breast is mottled rather than uniform.

An August specimen is apparently a year old and differs from any plumage described by Hayman et al. (1986) in having an entirely streaked throat, plain gray-brown breast, and whitish belly.

In the air its shape is somewhere between that of a tern and a swallow, with a prominently forked tail. In direct flight it flies much like a tern, with slower wingbeats than typical of shorebirds; because of its dark color and relatively short tail, it is most reminiscent of a BLACK TERN. In foraging flight it flies more like a swallow (BLACK TERNS forage aerially in similar fashion). From above, the white rump and blackish flight feathers contrast strongly with the entirely plain-brown upperparts. From beneath, the wings are dark—blackish with chestnut underwing coverts—and the tail is white with a black tip. The flight calls are ternlike but less harsh.

References

Lane 1987 (general habits), Pringle 1987 (description).

PLOVERS Charadriidae

Of the 65 species of plovers, 14 have been recorded in North America; all but the Greater Golden-Plover are discussed herein (Table 15).

Plovers are small to medium shorebirds (the largest species five times the weight of the smallest) with short, rather thick bills. They run and tip down to capture prey, then stand with head upright until they run again. This visual searching,

Table 15. Plover Distribution and Habitat Preference

Species	Hemisphere	Breeding Range	Breeding Habitat	Winter Range	Winter Habitat
Black-bellied	both	arctic	tundra	N/equatorial/S	beach/bay
American Golden	W	arctic	tundra	equatorial/S	grassland
Pacific Golden	E	arctic	tundra	equatorial	beach/grassland
Mongolian	E	arctic	tundra	equatorial/S	beach/bay
Snowy	both	temperate	beach/lake	N/equatorial	beach
Wilson's	W	tropical	beach	equatorial	beach
Common Ringed	E	arctic/subarctic	tundra/beach	N/equatorial	bay/beach/lake
Semipalmated	W	arctic/subarctic	tundra/beach	N/equatorial/S	bay/beach
Piping	W	temperate	beach/lake	N	beach
Little Ringed	E	temperate	lake/river	equatorial	bay/lake/river
Killdeer	W	temp./tropical	marsh/grassland	N/equatorial	marsh/grassland
Mountain	W	temperate	grassland	N	grassland
Eurasian Dotterel	E	arctic/subarctic	tundra/mountain	N	grassland

allowing them to detect prey at some distance, is an important aspect of their foraging behavior. Sandpipers keep their heads down, acquiring prey by close-range visual inspection and tactile probing. Both the plover tipping and the sandpiper probing are stereotyped enough to provide easy distinction between these two major groups of shorebirds.

With their conservative foraging behavior, plovers are all rather similarly shaped, with rounded heads, large eyes, short necks, and moderate-length legs. Appendix 3 shows the uniformity of bill and leg proportions, with less variation than comparably diverse groups of sandpipers. Wilson's and Black-bellied are relatively long-billed, prey size probably the important determinant. Leg length varies from short in the closely related Ringed, Semipalmated, and Piping to long in Mountain, Wilson's, and Pacific Golden-Plover.

Plovers vary greatly in habitat preference, encompassing all shorebird habitats but rocky shores. Many tropical species occur in inland habitats, including river and lake shores, marshes, dry plains, and savannas. Many species are beach dwellers in both breeding and non-breeding seasons. Only a few species breed in arctic and subarctic latitudes, and thus plovers as a group are not as highly migratory as are sand-

pipers. It is primarily the highly migratory species that undergo substantial habitat change in the course of the year.

Northwest plovers divide into two groups. Three large *Pluvialis* (Table 16) are sandpiperlike in plumage: strongly patterned above and largely black below in breeding plumage and finely patterned above and pale below in nonbreeding plumage. All are sexually dimorphic in breeding plumage. Note that these arctic-breeding species are colored like the tundra sandpipers among which they breed. Six smaller species of the genus *Charadrius* (Table 17) are plain-backed and white-bellied and undergo much less seasonal plumage change. Sexual dimorphism in breeding plumage varies from slight to absent. Only three *Charadrius* are of regular occurrence. The Eurasian Dotterel, also a *Charadrius*, is patterned above and dark below, with a juvenal plumage more like *Pluvialis*. The lapwings form a third large group of plovers, of which no members occur regularly in North America.

Killdeers are present on a year-round basis in the Northwest, while the other species visit the region seasonally. The Snowy Plover is a coastal breeder, while the Semipalmated Plover has bred only a few times in our region but is common during migration. Black-bellied Plovers are common throughout migration and winter. American

Table 16. Juvenile *Pluvialis* Plover Identification

Species	Weight	Wing	Tail	Tarsus	Bill	Back spots	Ear spot	Foreneck	Lower breast
Black-bellied	210	189	75	45	30	dull yellow to white	±	whitish, streaked	white, streaked
American Golden	145	178	67	42	22	yellow to white	±	whitish, mottled	gray, barred or mottled
Pacific Golden	130	166	63	43	23	bright to dull yellow	+	yellow, streaked	gray, streaked or mottled

Table 17. *Charadrius* Plover Identification

Species	Weight	Wing	Tail	Tarsus	Bill	Legs	Back	Collar	Breast	Wing stripe	Tail pattern
Snowy	38	105	44	23	14.5	gray	pale	±	plain	+	conspicuous white edges
Semi-palmated	42	119	55	22	12	orange	dark	+	1 band	++	narrow white edges and corners, dark subterminal spot
Piping	55	123	50	23	13	orange	pale	+	1 band	++	white rump and edges, dark tip
Mongolian	69	132	52	30	16	gray/green	dark	0	1 or partial band, rufous in breeding	+	like Semipalmated, dark tip less contrasty, less white on edges
Killdeer	91	160	95	35	20	pinkish	dark	+	2 bands	++	rufous base, white edges and tip, black subterminal band
Mountain	100	147	65	38	21	pinkish	medium	0	plain	±	narrow white tip, dark subterminal band
Dotterel	117	147	69	35	16	yellow	medium, striped	0	pale line	±	white corners

and Pacific golden-plovers are fairly common fall and rare spring migrants. All the migrant species are uncommon to rare anywhere in the interior.

Plovers regularly move from roosts to tidelands, and the Black-bellied and goldens are among the swiftest of shorebirds in flight. Killdeers prefer upland habitats and, with no de-manding tide-driven schedule, move languidly about the countryside. Members of this family tend to form small flocks, although Black-bellied Plovers may aggregate by the dozens or even hundreds.

Plovers are among the best examples of the problems entailed when illustrations are made using museum specimens as primary or sole references. The base of the bill in plovers is soft compared with the tip, and in dried specimens the skin at the base shrinks, contracting the bill base and causing it to look slender at the base and almost bulbous at the tip. Sandpiper bills are less distorted, but they also are often rendered too thin at the base. A comparison of photographs with the paintings in almost any field guide will show the problem clearly.

References

Bock 1958 (classification and relationships), Graul 1973a (head and breast markings), Nielsen 1975 (relationships among some *Charadrius*), Simmons 1961a, 1961b (foot-stirring behavior), Vaughan 1980 (general discussion of British species).

BLACK-BELLIED PLOVER *Pluvialis squatarola*

These largest plovers are among our wariest shorebirds, keeping their distance from the observer and one another as they run and stop on beaches and mud flats. In breeding plumage they look like well-spaced dominos, strikingly salted and peppered above and black below. They become Grey Plovers (their British name) in fall and winter, drab and gray-brown like the sandpipers that run among them, but they stand out by plover feeding mode and large size. Birds overhead are marked by mellifluous, melancholy whistles and black axillars.

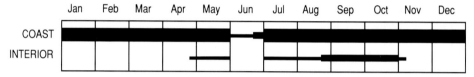

	Jan	Feb	Mar	Apr	May	Jun	Jul	Aug	Sep	Oct	Nov	Dec
COAST												
INTERIOR												

Distribution

The Black-bellied Plover breeds locally all across the arctic tundra and winters over a very broad latitudinal range on coasts from northern Europe and southern Canada to southern South America, Africa, and Australia. It is primarily a coastal migrant, but juveniles regularly move through the interior in small numbers.

Northwest Status

A year-round visitor to the outer coast, this species is abundant in migration and common in winter. Spring migrants, although difficult to distinguish from wintering birds, increase during April and peak near the end of that month. Males appear to migrate slightly earlier than females. Two-thirds of the birds in a flock of 50 at Grays Harbor on May 25, 1985*, were in nonbreeding plumage and were probably first-year birds. Most migrants are gone by late May, but small flocks of birds, virtually all immatures and of both sexes, frequent some coastal beaches and bays from one end of the region to the other through the summer.

On the outer coast the fall migration period usually begins in early July, exceptionally late

June. A peak in early August probably comprises adults, while another in early to mid September may comprise mainly juveniles, which first appear at the end of August. In October a third peak is probably caused by birds that have remained in Alaska to molt along with the flocks of Dunlins that arrive here during the same period.

Winter numbers are also substantial. Both adults and first-year birds winter commonly in the region, but there may be some segregation by flock. Twenty in one flock and six of seven in another flock seen at Ocean Shores on October 18, 1986*, were juveniles, while all of six birds collected at random there on December 4, 1984, were adult males. Many flocks, however, may be mixed. A feeding flock on the Skagit Flats on November 9, 1986*, included 30 adults and 20 immatures, and a roosting flock at Fort Flagler State Park on December 24, 1988*, consisted of 5 adults and 16 immatures.

In all parts of the interior, adult Black-bellied Plovers are rare migrants, with few records, usually of one to three birds, in late April and May; thus the flock of 100 in a flooded field at Merrill was unexpected and probably storm-displaced.

There are even fewer records of adults in fall. Juveniles are considerably more common, occurring widely from late August through October. Usually only a few birds are seen, but larger groups have been reported on occasion from most parts of the region.

Christmas Bird Count data indicate an increase in regional wintering populations in recent years. Seven of 12 possible lowest counts were obtained during the 1974-78 period and 7 highest counts during the 1984-88 period.

COAST SPRING. *High counts* 500 at Tillamook, OR, April 24, 1984; 900 at Willapa Bay, WA, April 22, 1983; 5,000 at Delta, BC, April 24, 1985.

COAST FALL. *Adult high counts* 1,000 at Boundary Bay, BC, June 22, 1974; 5,000 at Boundary Bay, BC, August 11, 1974. *Juvenile early dates* Ocean Shores, WA, August 26, 1979; Ocean Shores, WA, August 26, 1982*. *Juvenile high counts* 4,000 at Boundary Bay, BC, September 6, 1971; 1,400 at Leadbetter Point, WA, September 21, 1978; 1,500 at Leadbetter Point, WA, October 16-17, 1978; 1,300 at Grays Harbor, WA, October 20, 1979*.

COAST WINTER. *High counts* 1,000+ at Boundary Bay, BC, December 1, 1978; 815 at Ocean Shores, WA, December 19, 1976*; 1,500 at Padilla Bay, WA, December 18, 1982; 1,065 on southern coast of Washington, January 15-16, 1983; 607 at Coos Bay, OR, December 19, 1982.

INTERIOR SPRING. *Early dates* Minidoka Refuge, ID, April 9, 1964; Malheur Refuge, OR, April 9, 1972. *High counts* 18 at American Falls Reservoir, ID, spring 1987; 100 at Merrill, OR, May 3, 1981; 6 at Summer Lake, OR, May 1, 1987; 8 at Reardan, WA, May 14, 1960. *Late dates* Malheur Refuge, OR, May 24, 1970 and 1972; Reardan, WA, May 30, 1965; Okanagan Landing, BC, May 24, 1932.

INTERIOR FALL. *Early date* Malheur Refuge, OR, June 30, 1973. *Adult late date* Lind Coulee, WA, August 27, 1988. *Juvenile high counts* 22 at Nakusp, BC, September 22, 1985; 75 at Walla Walla River delta, WA, October 19, 1986; 25 at Hermiston, OR, fall 1982; 125 at American Falls Reservoir, ID, September 25, 1983; 52 at Springfield, ID, October 26, 1986; 12 at Pablo Refuge, MT, September 12, 1979*; 10 at Ninepipe Refuge, MT, September 26, 1980*.

Late dates Potholes, WA, November 5, 1950; 21 at Malheur Refuge, OR, November 7, 1987.

CHRISTMAS BIRD COUNTS	Five-Year Averages		
	74-78	79-83	84-88
Comox, BC	9	70	102
Ladner, BC	105	251	215
Vancouver, BC	61	16	61
Victoria, BC	65	112	128
White Rock, BC	16	9	39
Bellingham, WA	5	30	37
Grays Harbor, WA	691	457	249
Leadbetter Point, WA	114	275	309
Sequim-Dungeness, WA	220	214	377
Coos Bay, OR	150	289	192
Tillamook Bay, OR	61	248	231
Yaquina Bay, OR	29	45	16

Habitat and Behavior

Most frequently found on estuarine mud flats, Black-bellieds are also regularly seen on outermost sandy beaches, especially during migration. They are the second most common shorebird on Northwest mud flats in winter, when they feed in association with the hordes of Dunlins; often these two are the only shorebirds present at that time. In winter they also feed and roost with Dunlins and Killdeers in wet plowed fields and grassy meadows near the coast. In the interior they are most often seen at reservoirs.

In flight Black-bellieds coalesce into flocks of up to a few dozen individuals, sometimes to a hundred or more, but when feeding they spread out and avoid one another. They regularly fly with Red Knots and Short-billed Dowitchers during migration and at times with Dunlins during winter, although usually in their own flocks or at the edges of Dunlin clouds.

The Black-bellied feeds in typical plover fashion, but it also probes shallowly for partially, and perhaps even entirely, hidden prey. Polychaete worms and small bivalves are frequent prey, pulled from their burrows. Black-bellied Plovers and Whimbrels are among the few common shorebirds of the region that regularly take prey large enough to be attractive to gulls, and it is commonplace to see gulls chasing these shorebirds in attempts to appropriate the prey for themselves; they often succeed.

The Black-bellied is the only plover I have seen feeding in a decidedly unploverlike fashion. Some individuals foraging in shallow water walk along just like a sandpiper, picking at floating objects that are presumably small invertebrates.

Roosting is with other species, the Black-bellieds typically forming their own clumps and easily distinguishable from the next-larger Whimbrels and next-smaller dowitchers. Plovers usually roost with bill exposed rather than tucked under the scapulars as is typical of sandpipers, but Black-bellieds regularly roost in the hidden-bill sandpiper mode also (Figure 15).

Structure

Size on large end of medium. Bill short and thick, tapering evenly to bluntly pointed tip. Legs short. Only plover in region with hind toe, albeit small one, perhaps indicating relative primitiveness among plovers (some lapwings, but no other plovers, also with hind toes). Wings extend just beyond tail tip, primary projection moderate to long in adults and juveniles. Tail rounded.

Plumage

Breeding (Figure 20). Bill and legs black. Mated pairs conspicuously dimorphic, females duller but surprisingly variable. Males in full plumage mostly jet black below (the classical Black-bellied Plover of field guides), females with much white mixed with the black. Males look paler but more contrasty above, females duller and browner. At closer range differences between sexes more apparent, at least some differences caused by males having complete spring molt and females retaining many if not most nonbreeding feathers. Nevertheless, some of females' dull feathers fresh in spring.

Typical male with vivid white stripe on either side of neck running from front of crown to wing, this stripe more poorly defined in females. Crown in males pale gray, with or without blackish markings and usually conspicuously paler than mantle just because lightly marked; females show little crown-mantle contrast. A few males with

crowns about as dark as females, however. Mantle, scapulars, and tertials in males black with white tips and notches, looking vividly barred; coverts gray-brown with white fringes and notches. Thus upperparts in males vividly patterned, while crown, mantle, scapulars, tertials, and coverts in females all drab gray-brown, fringed with paler gray; some scapulars and/or tertials may be more heavily marked with dark brown to blackish. Age of females' tertials obvious by worn and pointed look.

Throat, foreneck, breast, and upper belly in males solid black; lower belly and undertail coverts white. Females black in same areas, but black usually mixed with white, often to extent of being thoroughly mottled or even largely white beneath. Finally, outermost undertail coverts in males with only a few well-defined black spots, same feathers in females often with more and browner spots. Some females almost as bright as males.

Lower back gray-brown, uppertail coverts white. Rectrices white, barred with brown to brownish black; bars quite variable in width from individual to individual. Wing brown above, prominent wing stripe formed from white tips of greater primary coverts and white markings on outer webs of inner six or seven primaries. Secondaries slightly paler than primaries. Greater primary coverts dark brown, darker than rest of wing. Underwing coverts light gray-brown to white; bases of inner webs of primaries and secondaries white, forming conspicuous pale underwings. Axillars black, blending with black underparts.

Breeding-plumaged birds first appear in early April (a few males in full plumage by beginning of month), become common plumage type by third week of that month and reappear southbound in July. By July white feather tips may have worn so much as to be almost absent, males looking much darker-backed accordingly. Females, with less contrasty plumage, not so much changed, but males have become more like them.

Nonbreeding (Figure 15). Legs black or gray (latter perhaps only in first-winter birds). Plain, entirely drab brownish gray above, with most feathers narrowly white-fringed. Tertials and coverts more patterned, with weakly defined alternating darker and lighter dots or notches along

Fig. 20. Breeding *Pluvialis* plovers. Flying (left to right): Pacific Golden (m, f), Black-bellied (m,f). On beach (left to right): Black-bellied (f), American Golden (f, f, m taking off), Black-bellied (f, m, m). Species differences evident in breeding plumage include conspicuous white wing stripes, white tail, white underwings, and black axillars of Black-bellied, plain back and gray underwings and axillars of golden-plovers. Male Pacific distinguished from male American Golden-Plover by white extending along sides to undertail; females not distinguishable. Males solid black beneath, females vary from largely black to largely white.

their edges. Markings obscure except at close range. Pale supercilium and dark auricular spots fairly well defined, more obscure dark smudge on lores. Breast varies from pure white through grayish to fairly heavily mottled with brown, neck with dark stripes. Black axillars conspicuous against white underwing and belly in nonbreeding plumage. Outermost undertail coverts spotted with brown. Some birds have molted largely into this plumage by mid August, while others retain remnants of black ventral breeding plumage well into September. Primary molt begins before migration in Europe, molt arrested in many birds and outer one or two primaries molted in late winter (or not at all?). Situation not described for American birds, but apparently normal primary molt finishing on a few October specimens from Northwest.

Juvenal (Figure 21). Legs gray. More heavily patterned than nonbreeding adults, with crown, mantle, scapulars, coverts, and tertials boldly fringed and dotted with white or pale yellow. Very young juveniles with these dots as bright yellow as any golden-plover. Breast striped with

gray-brown (also faint bars on some individuals); at least some also have stripes and bars on flanks, at the most extreme almost entire underparts marked. Axillars also black in this plumage.

Immature. From October well into winter, immatures distinguished from adults by pale tertial dots and striped breast. By early spring tertial markings mostly worn off, but tertials look ragged compared with still-intact tertials of adults. First-summer birds of both sexes patterned much like nonbreeding adults. At least some have complete molt into nonbreeding plumage by April, others do so during summer. Primary molt in May, June, and July.

Identification

The only other Northwest shorebirds likely to be mistaken for this species are the closely related GOLDEN-PLOVERS, and the comparisons below refer to those species unless otherwise indicated.

Breeding Plumage. Compared with GOLDEN-PLOVERS, BLACK-BELLIEDS are:
- *larger*, obvious if they are together;
- usually *paler above*, from whitish to light

EURASIAN DOTTEREL

AMERICAN GOLDEN-PLOVER

BLACK-BELLIED PLOVER

Fig. 21. Juvenile striped and spotted plovers. Dotterel distinguished from all other plovers by vivid white supercilium, narrow white breast band, and striped upperparts. Black-bellied distinguished from golden-plovers by larger size, proportionately larger head and bill, and more prominently streaked sides.

brown, while GOLDENS are golden to dark brown;

• *pure white under the tail*, while GOLDENS usually show much black under the tail.

From a distance spring BLACK-BELLIEDS, especially males, look *paler above than below*, while GOLDENS look uniformly dark except for the white lateral stripe. Note the possibility of worn, dark BLACK-BELLIEDS in early autumn and female or molting GOLDENS with varying amounts of white under the tail.

Nonbreeding Plumage. Nonbreeding adults can be distinguished from GOLDEN-PLOVERS by *larger size, larger head and bill, and paler coloration*, much more easily detected in comparison than when only one species is present. RED

KNOTS at a distance might be the other most likely candidate for confusion; they are smaller and longer-billed. Nonbreeding-plumaged adult GOLDEN-PLOVERS are rarely seen in our region, and those of the PACIFIC species are usually more marked with golden than are BLACK-BELLIED in the same plumage, while those of the AMERICAN species are considerably darker than BLACK-BELLIED.

Juvenal Plumage. Comparisons between BLACK-BELLIED and the two species of GOLDEN-PLOVERS will be most common in this plumage. Juvenile BLACK-BELLIEDS are light to medium brown or grayish brown, with slightly darker markings and white to yellow spots above; GOLDENS are medium to dark brown, with pale to bright yellow spots above. It is important to remember that juvenile BLACK-BELLIEDS can be almost as yellow-spotted above as GOLDENS, and almost as dark, so they are actually quite similar in this plumage. An early-season BLACK-BELLIED, in fact, can be much more golden above than a late-season GOLDEN, although such birds would not occur simultaneously. Older field guides portray "winter" BLACK-BELLIEDS with no indication of yellow and substantially paler than GOLDENS, and these are adults, but many such comparisons from September on will be of juveniles. See AMERICAN GOLDEN-PLOVER for additional details.

In Flight (Figure 11)

BLACK-BELLIED in flight is the most conspicuously patterned of the *stripe-winged and white-tailed* shorebirds. In all plumages there are *black axillars*, but in nonbreeding and juvenile birds they are conspicuous and distinctive. GOLDEN-PLOVERS look uniform in flight, all brown above and gray below, with a narrow, inconspicuous wing stripe. The only other shorebird near the size of a BLACK-BELLIED PLOVER with similar wing stripe and pale rump and tail is the RED KNOT, and in winter plumage the two look superficially similar. The plover is larger, with typical plover bill shorter than head (about length of head in knot), a more conspicuous wing stripe and pale tail, and, of course, black axillars.

Voice

BLACK-BELLIED PLOVERS are quite vocal, and their beautiful, plaintive *whee-er-eee* whistles can be heard at most times when birds are flying or alerted. No other bird in the Northwest sounds like this species, with the exception of the accidental BRISTLE-THIGHED CURLEW. Birds in spring and fall occasionally give the sharp whistles otherwise characteristic of the breeding grounds.

Further Questions

Although we do not understand the significance of the sexual plumage differences, we can use them to learn about sexual differences in behavior, habitat use, and migration patterns. Birds should be in full plumage by early May on the way north and when they return in early July. At both of these times, observers could record the percentage of individuals in full (presumably male) plumage to determine if the sexes have different schedules. The same records kept throughout spring and fall would furnish a general description of the timing of body molt. Also, the pattern of wing molt in this area is poorly known and should be watched for during July to October. In particular both the age and the degree of primary wear should be noted on birds in the big flocks that arrive in October; we do not know their age class or, if adults, whether they underwent wing molt before they migrated.

Plovers have been reported to be territorial when they feed, actually defending feeding territories by aggressive interactions, and Black-bellied Plovers are typical of the group in being well-spaced apart from one another while feeding. Different studies of this species have reported spacing by aggression and spacing by mutual avoidance, and it would be of interest to watch Black-bellied Plovers in the Northwest to see how they accomplish this. It may depend on the density of food resources and the amount of available habitat, but we need more information on when and where the plovers exhibit aggressive behavior to determine this.

Nocturnal feeding is common in British Black-bellied Plovers; is it in the Northwest? Does this allow them to winter here in large numbers, the largest shorebird to do so? Finally, we should document the conditions under which Black-bellied Plovers roost with bill exposed or hidden. Presumably it is related to wind speed, temperature, or both.

Photos

Fresh breeding-plumaged males are shown in Bull and Farrand (1977: Pl. 240), Udvardy (1977: Pl. 190), Keith and Gooders (1980: Pl. 145), Armstrong (1983: 115), and Farrand (1988a: 168; 1988b: 155); a worn breeding-plumaged male is in Farrand (1983: 319). Armstrong (1983: 115) and Farrand (1983: 319) illustrate the range in juveniles. The latter appears to be a very young bird, just off the breeding grounds and more brightly colored than will usually be the case in migration; note its pale legs. Nonbreeding adults are shown in Bull and Farrand (1977: Pl. 195), Hammond and Everett (1980: 110), Terres (1980:

717), and Farrand (1983: 317; 1988a: 168; 1988b: 155).

References

Baker 1974 (foraging behavior), Branson and Minton 1976 (molt and migration in Britain), Burger et al. 1979 (aggressive behavior in migration), Drury 1961 (breeding biology), Höhn 1957 (nesting behavior), Hussell and Page 1976 (breeding biology), Loftin 1962 (summering), Mayfield 1973 (nesting behavior), Paulson 1990 (sandpiperlike feeding), Pienkowski 1982 (winter diet and feeding behavior), Pienkowski 1983a (winter foraging patterns), Pienkowski 1983b (winter feeding behavior), Stinson 1977 (winter spacing behavior), Townshend et al. 1984 (winter territorial and feeding behavior), Wood 1986 (winter territoriality).

AMERICAN GOLDEN-PLOVER *Pluvialis dominica*

Golden-plovers have it all—swift flight, coloration both striking and subtle, melodious calls, long-distance travels and accompanying wide dispersal, even major questions of identification. Vividly marked breeding adults pass through the region rarely in spring and more often in fall, when they are accompanied by larger numbers of juveniles. They frequent upland sites but also cohabit mud flats with the closely related Black-bellied Plover, differing in darker plumage and smaller size, with relatively smaller bill. This continental species wanders in migration west to the Pacific coast where it meets and flocks with the Pacific Golden-Plover in coastal estuaries.

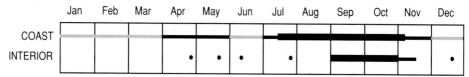

Distribution

The American Golden-Plover breeds throughout the North American Arctic, from western Alaska to the Atlantic coast of Canada, and winters in southern South America. Adults migrate north through the middle of the continent and south directly across the Atlantic to northern South America; they are rare anywhere in fall away from eastern Canada and northeastern United States. Juveniles range much more widely and are locally common in fall west to the Great Plains, in much smaller numbers to the Pacific coast.

Northwest Status

The accounts for the two species of golden-plovers are combined here, as most of the published record has not distinguished between them. See below for species-specific details.

These birds are rarer than Black-bellied Plovers anywhere in the region. A few are seen every spring, mostly on the outer coast, with records primarily in the last half of April and first half of May. At least six March records from Oregon, Washington, and British Columbia are anomalous but may indicate a small wave of early migrants. Birds have been recorded throughout June, with a half-dozen records between June 10 and 20. They were seen only briefly and were presumably in transit but were impossible to distinguish as "late spring" or "early fall." There is no evidence that individuals have summered in the region.

The fall migration period is lengthy, primarily August through October, with presumed southbound migrants regularly recorded as early as mid July. During fall, flocks of 50 to 150 birds are seen regularly in coastal localities such as Grays Harbor and Leadbetter Point, and sizable flocks occur even as far into protected waters as Vancouver. They are much less common on the coasts of Vancouver Island, Oregon, and the remainder of Washington, perhaps because of a scarcity of suitable habitat. Adults make up the majority of birds in the flocks through August and linger in small numbers thereafter, while juveniles dominate numerically in September and October. Numbers of birds are sufficient to cause speculation that they may represent the normal migration pathway of the westernmost breeding population of the species. But are they, too, heading for the interior of southern South America?

Although the majority of these birds depart by the end of October, there is a smattering of records thereafter that probably were wintering birds, including a few from the coasts of British Columbia, Washington, and Oregon from December to February. Occasional wintering is not surprising, as individuals of the Pacific Golden-Plover regularly winter in central California. March records are more difficult to allocate to season. American Golden-Plovers are quite early migrants, regularly arriving on the Texas coast by March, and could be responsible for early-spring records from the Northwest. Pacific Golden-Plovers do not usually migrate before April.

Very few adults have been seen in the interior of the region, with records totaling six in spring and three in fall. Records are much more numerous during the juvenile migration in September and October. Usually single birds or small flocks are seen, occasionally up to a few dozen. The December record from Kahlotus, a specimen, was presumably a very late migrant.

COAST SPRING. *High count* 11 at Tillamook Bay, OR, late April 1984.

COAST FALL. *Juvenile early date* Ocean Shores, WA, August 20, 1987#. *High counts* 208 at Iona Island, BC, September 16, 1972; 200 at Ocean Shores, WA, September 18, 1983; 30 at Tillamook, OR, September 26, 1987.

COAST WINTER. *High count* 19 at Comox, BC, December 22, 1974.

INTERIOR SPRING. Hauser, ID, May 14, 1950; Lewiston, ID, April 22-24, 1978; Malheur Refuge, OR, May 8 & 23, 1967; Moses Lake, WA, June 10, 1976; Shuswap Falls, BC, May 19, 1918.

INTERIOR FALL. *Adults* Okanagan Landing, BC, July 26, 1922; Klamath Falls, OR, July 29, 1978; Camas Refuge, ID, July 18, 1961 (4). *Juvenile early dates* Turnbull Refuge, WA, September 4, 1956#; Lewiston, ID, September 4, 1957#. *High counts* 20 at Nakusp, BC, September 15-20, 1965; 17 at Richland, WA, mid October 1975; 26 at Lewiston, ID, fall 1984; 30 at American Falls Reservoir, ID, October 12, 1985; 15 at Springfield, ID, November 2, 1986; 9 at Pablo Refuge, MT, September 12, 1979*; 9 at Ninepipe Refuge, MT, September 26, 1980*.

Late dates Okanagan Landing, BC, October 19, 1943; Kahlotus, WA, December 19, 1924; Walla Walla, WA, November 12, 1980#; Malheur Refuge, OR, November 4, 1987.

Status of the Two Species. The status of the two species is not well worked out in the Northwest because of the small number of specimens available. The only spring specimens from the Northwest that I examined are two American from British Columbia—Comox, May 21, 1937, and Courtenay, May 15, 1942—and one Pacific from Washington—Cohasset, May 14, 1938. A male photographed at Tillamook Bay on June 18, 1986, and two males photographed at Ocean Shores on May 11, 1987, also appear to be American. From the interior, a specimen from Hauser, May 14, 1950, was reported as American. Most males seen on the coast recently in spring were easily identified as Pacifics, which may be more likely on the coast. Although very rare, either species may be equally likely in the interior.

Nine adult specimens have been examined from fall, all from Grays Harbor. Five of them are Americans, on dates from August 7 to 20, and four are Pacifics, from August 7 to September 12. On two occasions, both species were collected from the same flock. As adult Americans are considered rare in autumn west of their offshore Atlantic migration route, this is surprising, yet these specimens match the characteristics of that species in size, proportions, and lack of any primary molt. Some Americans from western Alaska may get caught up in southbound flocks of Pacifics (surely the case with juveniles), although the adults should be orienting to the southeast toward South America. Much remains to be learned. A wintering bird on the Oregon coast was called a Pacific, as seems likely.

The specimen record from the Queen Charlotte Islands of British Columbia, north of our region, is interesting. First, it indicates that Pacifics may have different spring and fall routes. Of adult specimens from those islands, 18 were from spring and only two from fall, although adults were reported as fairly common there during August in one year. Second, of twelve juvenile specimens examined, two from August 23 and 24 are Americans, and ten from September 23 to November 9 are Pacifics. Again, this evidence supports the hypothesis of a regular Pacific coast migration for Americans, and the destination of these birds would be most interesting to learn. It is not surprising that there is a record of an American Golden-Plover from Australia.

To the south of the region, recent observations in California indicate the Pacific as the more common species on the coast, with interior records only from the Salton Sea. Americans decrease in proportion southward but are represented all the way to the southern end, and the few birds from east of the Sierra were all identified as Americans. Similarly, a few Pacifics wander off course into the interior of North America, whence I have examined specimens from interior British Columbia (north of our region, an October adult) and Alberta (three September juveniles).

Juveniles of both species occur commonly in the Northwest, at least at some times and places. Of 22 juvenile specimens examined from the interior of the region, collected September 4 to November 12, all are American. Of 53 from coastal and near-coastal localities, 33 are American and 20 Pacific. On the coast juvenile Americans were collected from August 20 to October 30, juvenile Pacifics from September 6 to November 24. Although the two overlap greatly, it is evident that the Pacific is a later migrant on the average (Table 18).

Pacifics may be more common, however, than is indicated by these specimen records. A flock of golden-plovers, many members of which were photographed at Ocean Shores on September 15, 1985*, included seven adults, all probably Pacifics, and 23 juveniles, of which more were Pacific than American. Four juveniles collected from a flock at Ocean Shores on October 1-2, 1986, included two of each species.

Habitat and Behavior

Golden-plovers are upland plovers in many parts of their range, occurring on plowed fields and among sand dunes, but on our coast they also move with the falling tide onto the mud flats, where they associate freely with Black-bellieds. Roosting birds are commonly seen in adjacent salt marshes, fields, and even on the open ocean beach.

The two species typically use different habitats on their winter ranges, American grasslands and

Table 18. Juvenile Golden-plover Specimens Examined from Northwest Coast Subregion

	Aug	Sep	Oct	Nov
American	2	21	10	0
Pacific	0	3	12	5

Pacific beaches. Nevertheless, Pacifics also feed in uplands throughout their range, especially with the clearing that has taken place on Pacific islands, and Americans readily forage on mud flats during migration. Birds on Northwest oceanfront beaches may more likely be Pacific.

These birds join flocks of their own or other species in flight but space out across a salt marsh or mud flat while foraging. They run, stop, and peck in typical plover fashion, often standing erect with extended neck. Individuals at times feed by poking their bills shallowly into the mud, possibly tactile feeding. A typical golden-plover roost consists of loosely associated individuals, either in *Salicornia* or on vegetated but semiopen sand dunes.

Golden-plovers must be the high-speed champions among shorebirds, as I have seen individuals both on breeding grounds and in migration hurtle past at speeds I estimated as approaching 100 miles per hour—unequalled by any other shorebird I have seen. As Dunlins have been clocked at 110 miles per hour from an airplane, I suspect these plovers exceed even that speed, an understandable adaptation in birds that migrate nonstop across vast ocean reaches.

Structure

Size on small end of medium. Bill very short, fairly slender for a plover and tapering smoothly to moderately pointed tip. Legs short, although fairly lengthy for a plover. Wings projecting conspicuously beyond tail tip, primary projection long in adults and juveniles. Tail rounded.

Plumage

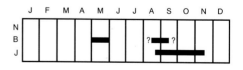

Breeding (Figure 20). Bill black, legs gray to black. Breeding plumage illustrated in most field guides is that of male—solid black from chin to undertail coverts. Upperparts dark brown, all feather groups heavily dotted and/or margined with bright golden to pale yellow or whitish. Conspicuous white stripe from forehead over eyes and down each side of neck, where becoming wider, to wings. Retained tertials (few) and coverts (many) gray-brown, with slightly darker brown bars and whitish notches.

Rectrices gray-brown, barred with blackish. Wing brown above; narrow grayish white tips on all greater coverts form faint wing stripe, accentuated by slightly paler bases of flight feathers. Underwing and axillars brownish gray.

Female averages duller, about like male above but with less well-defined head and neck stripe and a considerable admixture of white in the black underparts. Those with least black much like nonbreeding adult below. As in Black-bellied, female variability and sexual difference caused by partial spring molt in many females.

Spring birds in Northwest usually in breeding plumage, although at least one nonbreeding-plumaged golden-plover seen in April, possibly immature or very dull female. Virtually all fall adults heavily mixed with white below; full-plumaged males rarely seen, probably because autumn molt begins on breeding grounds.

Breeding plumage acquired during northbound migration, in April; retained at least as late as end of September (in East). Most primary molt on wintering grounds in October to December, while body molt finishing.

Nonbreeding. Legs gray. Mantle feathers, scapulars, and tertials gray-brown, usually with darker centers and bars; mantle may be sparsely dotted or fringed with whitish or golden markings. Upperparts look irregularly mottled or virtually plain, depending upon stage of wear. Prominent white supercilium, dark auricular spot, and dark smudge on lores. Throat white, breast gray-brown, and belly white. Breast and sides vary from profusely striped with brown (only in first-winter birds?) to unmarked. This plumage rarely if ever seen in Northwest, and autumn adults that are largely pale beneath still have gold-dotted upperparts of breeding plumage. In fall, likely to be mistaken for juveniles at

a distance; but at close range, worn state of plumage should be apparent, and almost always some black feathers on underparts.

Juvenal (Figure 21). Legs gray. Basically similar to nonbreeding adult, but with both upperparts and underparts somewhat more heavily marked. Crown, mantle, scapulars, and tertials dark brown, marked with bright to pale yellow: crown feathers edged, scapulars and tertials dotted, mantle feathers with paired spots at tips. Supercilium white to pale yellowish, prominent but usually interrupted by dark smudge on lores. Face otherwise pale except for narrow, dark postocular stripe and auricular spot. Underparts dull whitish to very pale sandy brown, lower neck and upper breast striped with gray-brown to brown. Breast pattern complex: many feathers gray-brown with pair of sandy brown spots on either side of tip, producing both barred and striped effect. Lower breast, sides, and sometimes belly faintly barred with gray-brown.

Rectrices plain compared with those of adults, with dark bars only obscurely indicated and pale yellow or whitish notching on outer edges. Juveniles first appear in late August and largely replace adults by mid September. Molt into first-winter plumage on wintering grounds from October to December but retain juvenal plumage without molting while in our region, even well into November.

Typical juveniles distinguished from nonbreeding adults by spotted rather than striped upperparts. Some very heavily marked juveniles with yellow-edged feathers look more like adults, but profuse ventral markings, especially bars, should distinguish them.

Immature. Distinguished from nonbreeding adult by dark, worn tertials and, at least in early winter, mottled breast. Most if not all young birds molt primaries in first winter, attain breeding plumage (inseparable from adults), and migrate north to breed.

Identification

This section deals with differences between golden-plovers and other shorebirds. For differences between the two species of golden-plovers, see under Pacific Golden-Plover.

All Plumages. GOLDEN-PLOVERS are distinctly *smaller* than BLACK-BELLIED, actually

Breeding female Black-bellied Plover. As male, but some brown on upperparts and white on underparts. Cambridge Bay, NT, 16 June 1975, Dennis Paulson

Breeding male Black-bellied Plover. Large plover with black underparts in breeding plumage; differs from other *Pluvialis* in larger size, heavier bill, and pure white undertail; crisp black and white pattern diagnostic of male. Cambridge Bay, NT, 18 June 1975, Dennis Paulson

Breeding male American Golden-Plover. Breeding golden-plovers have gold-dotted upperparts and at least partially black undertails; male of this species with broad white neck stripe ending at breast and entirely black undertail. Cambridge Bay, NT, 12 June 1975, Dennis Paulson

Breeding male Pacific Golden-Plover. Black underparts with narrow white neck stripe continuous onto flanks diagnostic of species and sex. Honolulu, HI, 17 April 1987, Dennis Paulson

Breeding female American Golden-Plover. Typical female golden-plover, with less yellow above and black below than male; species difficult to distinguish, but long primary projections characteristic of this species. Cambridge Bay, NT, 25 June 1975, Dennis Paulson

halfway between the size of that species and a KILLDEER. The *relatively smaller bill* is a helpful point, especially in comparison, although the difference is not striking.

Breeding Plumage. In breeding plumage GOLDEN-PLOVERS are distinguished by *dark brown and golden upperparts* (gray, blackish, or light brown in BLACK-BELLIED) and *black or mottled undertail coverts*. The undertail coverts of most GOLDENS in this region will be mottled with white, but none should have the extensive pure white undertail of a BLACK-BELLIED.

Nonbreeding and Juvenal Plumages. Nonbreeding adults and juveniles are most like juvenile BLACK-BELLIED but differ in the following ways:

• GOLDENS are *darker above*, with both darker ground color and darker spots, but juvenile BLACK-BELLIEDS can have spots that are as yellow as those of many GOLDENS.

• GOLDENS have a *slightly more contrasty head pattern*, the cap darker and thus the pale supercilium more distinct, and a better defined dark auricular spot, but there is overlap.

• Juvenile AMERICAN GOLDENS have a *gray belly* that contrasts little with the breast color. The belly of BLACK-BELLIED, by virtue of being white, contrasts more with the patterned breast. PACIFIC GOLDEN-PLOVERS have whiter bellies, and some individuals are almost as contrasty as BLACK-BELLIEDS.

• GOLDENS are generally *more extensively but less distinctly marked below*, with relatively inconspicuous *dots or bars* on the lower breast and sides, while BLACK-BELLIEDS have sharp striping on the breast and often the sides. GOLDENS are clearly striped, if at all, only on the lower throat and upper breast (more conspicuously so in the PACIFIC, as stated above).

The only other plover with conspicuously marked upperparts is the juvenile EURASIAN DOTTEREL, striped rather than spotted above and with a more conspicuous supercilium and light line across the breast. The BUFF-BREASTED SANDPIPER is vaguely similar but forages like a sandpiper and has an almost unpatterned head and yellow legs.

In Flight

Entirely brown above in flight, with a narrow pale wing stripe, this bird is unlike any other common shorebird in the Northwest. The underparts are primarily pale gray in fall, all or partly black in spring and late summer, in both cases not duplicating any other local species that is basically brown above. BLACK-BELLIED PLOVERS are extensively black beneath but have vivid white wing linings and undertail coverts in breeding plumage, white bellies and black axillars in nonbreeding plumage, and from above they are conspicuously white-tailed with a much more obvious wing stripe than GOLDENS.

Similar-sized brown shorebirds include PECTORAL and SHARP-TAILED SANDPIPERS and RUFFS, all of which share the salt-marsh and mud-flat habitat of GOLDEN-PLOVERS, and all of which have conspicuous white sides to the rump, especially the RUFF. The BUFF-BREASTED SANDPIPER lacks white rump markings but is considerably smaller than the plover, with wing linings white against a buffy breast; the plover shows little contrast between breast and gray wing linings. See PACIFIC GOLDEN-PLOVER for minor differences and MOUNTAIN PLOVER and EURASIAN DOTTEREL for two irregular species that might be mistaken for this one in flight.

Voice

Both GOLDEN-PLOVERS are quite vocal in flight. The typical call of this species is a short *queedle* that breaks in the middle, not like any other shorebird call except the *chuweedle* of the PACIFIC GOLDEN-PLOVER. I do not know if the AMERICAN ever gives that call or the *chuwi* call typical of the PACIFIC, although most descriptions of GOLDEN-PLOVER vocalizations in regions where only *dominica* occurs make no mention of such calls. As some of the breeding-ground vocalizations of the two species are completely different, it would not be surprising if there were flight-call differences between them as well.

Further Questions

Much is still to be known about golden-plovers in the Northwest, particularly the exact status of the two species in the region. Interior birds should be scrutinized, the different migration schedules of the juveniles should be confirmed, and much attention should be paid to possible differences in habitat preferences and vocalizations. The calls of birds carefully determined to species should be described and tape-recorded if possible; we still do not know if the flight calls of the two species are different. There are records in the literature of the Pacific species from the Northwest interior, but these have not been confirmed.

Some of the field marks to distinguish the species need to be further elucidated by repeatedly checking each mark on juveniles. A valuable exercise would be to photograph, preferably from the side, a series of individuals in autumn flocks. These photographs would be of tremendous value to augment existing museum specimens. Good photographs in flight would furnish information about leg length and primary molt of those individuals.

Notes

The two golden-plovers, long considered subspecies of the same species, overlap in western Alaska with little if any interbreeding and will certainly be considered distinct species by the American Ornithologists' Union in the near future. I am so considering them in anticipation of that decision.

Some nonbreeding adults (quite rare here, only Pacific for sure), breeding males (few in spring, Pacific more likely) and many juveniles (both fairly common in fall) of these two species can be distinguished in the field by color pattern, but close study and photographic documentation are advised. It must be emphasized that some individuals are difficult to identify even in the hand, and, at our present level of knowledge, field identification to species will not always be possible. In fact, these two species represent the only Northwest shorebirds that frequently cannot be distinguished in the field.

The statement made by Hayman et al. (1986)

that first-year American show little or no black on their underparts is contradicted by evidence presented by Johnson (1985). I consider slips of the pen the statements by Hayman et al. (1986: 100) that the American is the largest-billed golden-plover and by Chandler (1989: 76) that the toes of the American project farther beyond the tail tip in flight than do those of the Pacific. The plates in Peterson (1990: 123) confuse the distinction between juvenile and nonbreeding individuals in this species.

Photos

Older field-guide photographs of American Golden-Plovers in "winter plumage" tend to be of juveniles, as in Udvardy (1977: Pl. 235) and Terres (1980: 717), while those of Black-bellied in the same books are of winter adults, thus encouraging the idea that distinction is easy. Armstrong (1983: 114) and Farrand (1983: 321; 1988a: 169; 1988b: 154) provide photos of juveniles, Bull and Farrand (1977: Pl. 194) a fresh-plumaged nonbreeding adult. Breeding adults in books are males (Bull and Farrand 1977: Pl. 239; Udvardy 1977: Pl. 189; Johnsgard 1981: color Pl. 17; Armstrong 1983: 114; Farrand 1983: 319, 1988a: 169, 1988b: 154; Pringle 1987: 78, 83). The "Lesser golden plover" in Johnsgard (1981: black-and-white Pl. 6) is a Black-bellied.

References

Baker 1977 (summer diet), Byrkjedal 1989a (choice of nesting habitat), Byrkjedal 1989b (nest defense behavior), Connors 1983 (discussion of breeding overlap and species status of *dominica* and *fulva*, with differences between them), Conover 1945 (identification and Alaskan distribution), Drury 1961 (breeding biology), Jackson 1918 (molt), Johnson 1985 (timing of primary molt in immatures), Myers and Myers 1979 (wintering behavior), Pym 1982 (identification), Vaurie 1964 (differences between *dominica* and *fulva*).

PACIFIC GOLDEN-PLOVER *Pluvialis fulva*

Both species of golden-plovers—plains-migrating *dominica* and Pacific-migrating *fulva*—are locally common in the Northwest, but only recently have we attempted to differentiate them, and there is still much to be learned. The Pacific species, just brushing the Northwest region at the eastern edge of its wide range, is largely restricted to the coast.

Nonbreeding Black-bellied Plover. Rather plain breast and upperparts and lack of yellow diagnostic of plumage and species. North Padre Island, TX, 24 January 1991, Dennis Paulson

Juvenile Black-bellied Plover. Pale-dotted upperparts and heavily marked underparts diagnostic of plumage; distinguished from golden-plovers by larger head and bill and short-necked look, as well as pattern details on the underparts. NJ, late September 1985, R. J. Chandler

Nonbreeding American Golden-Plover. Plain upperparts and breast diagnostic of plumage; lack of yellow distinguishes from Pacific Golden; black of breeding plumage appearing on underparts (most migrate north in this plumage). Southeastern TX, April 1982, Linda M. Feltner

Nonbreeding Pacific Golden-Plover. Yellow-edged mantle and scapulars and yellowish head and neck distinguishes from Black-bellied and American Golden in any plumage; primary projection unusually long. Honolulu, HI, 22 February 1983, Dennis Paulson

Juvenile American Golden-Plover. Dotted upperparts and heavily marked underparts diagnostic of plumage; virtual lack of yellow, barred belly, and long primary projection distinguishes from Pacific Golden. Monomoy, MA, 6 October 1984, Wayne Petersen

Juvenile Pacific Golden-Plover. Dotted upperparts and heavily marked underparts diagnostic of plumage; golden cast to head and neck, prominently streaked neck, mostly whitish belly, and short primary projection distinguishes from American Golden. Attu, AK, 13 September 1983, Edward Harper

Distribution

The Pacific Golden-Plover breeds in western Alaska and all across Siberia and winters in southeast Asia and Australasia and on islands across much of the tropical Pacific. Adults and juveniles appear to have similar migration routes through eastern Asia and across the Pacific.

Northwest Status

See under American Golden-Plover.

Habitat and Behavior

The Pacific Golden-Plover is generally similar to the American in habitat choice and behavior, as far as we know. However, on their wintering grounds Pacifics occur much more often in coastal habitats. In Hawaii, birds often probe vigorously in grass lawns.

Structure

Size on small end of medium. Bill very short, fairly slender for a plover and tapering smoothly to moderately pointed tip. Legs short, although fairly lengthy for a plover. Wings projecting to or slightly beyond tail tip, primary projection moderate in adults and juveniles. Tail rounded.

Plumage

Breeding (Figure 20). Bill black, legs gray. Male black from chin to belly, undertail coverts and sides mostly white but may be mixed with some black. Upperparts dark brown, all feather groups heavily dotted and/or margined with bright golden to pale yellow or whitish. Conspicuous white stripe from forehead over eyes and down each side of neck continuous with side stripe. Retained tertials (few) and coverts (many) gray-brown, with slightly darker brown bars and whitish notches.

Rectrices gray-brown, barred with blackish. Wing brown above; narrow grayish white tips on all greater coverts form faint wing stripe, accentuated by slightly paler bases of flight feathers.

Underwing and axillars brownish gray.

Female averages duller, about like male above but with less well-defined head and neck stripe and even more white in the underparts than females of the preceding species. Those with least black much like nonbreeding adult below. Female variability and sexual difference caused in part by less complete spring molt in females.

Spring birds in Northwest usually in breeding plumage, although at least one nonbreeding-plumaged golden-plover seen in April, possibly immature or very dull female. Virtually all fall adults heavily mixed with white below; full-plumaged males rarely seen, probably because autumn molt begins on breeding grounds.

Breeding plumage detected in adults as long as present in region, generally into early November (as late as mid December in one on Oregon coast). Most primary molt on wintering grounds in October to December, while body molt finishing.

Nonbreeding. Legs gray. Mantle feathers, scapulars, and tertials gray-brown, with or without darker centers and bars, typically fringed with whitish or golden; fringes may be broken. Upperparts typically look striped but may be rather plain, depending upon stage of wear. Prominent white or yellowish supercilium, dark auricular spot, and dark smudge on lores. Throat white to yellowish, breast gray-brown to yellowish brown, and belly white. Breast and sides vary from profusely striped with brown (only in first-winter birds?) to unmarked. This plumage rarely seen in Northwest, and autumn adults that are largely pale beneath still have gold-dotted upperparts of breeding plumage. In fall, likely to be mistaken for juveniles at a distance, but at close range worn state of plumage should be apparent, and almost always some black feathers on underparts.

Juvenal (Figure 21). Legs gray. Basically similar to nonbreeding adult, but with both upperparts and underparts somewhat more heavily marked. Crown, mantle, scapulars, and tertials dark brown, marked with bright to pale yellow: crown feathers edged, scapulars and tertials dotted, mantle feathers with paired spots at tips. Supercilium typically yellowish, prominent but usually interrupted by dark smudge on lores.

Face otherwise pale except for narrow, dark postocular stripe and auricular spot. Underparts dull whitish to very pale sandy brown, lower neck and upper breast striped with gray-brown to brown. Breast pattern complex: many feathers gray-brown with pair of sandy brown spots on either side of tip, producing both barred and striped effect. Lower breast, sides, and sometimes belly faintly barred with gray-brown.

Rectrices plain compared with those of adults, with dark bars only obscurely indicated and pale yellow or whitish notching on outer edges. Juveniles first appear in late August and largely replace adults by mid September. Molt into first-winter plumage on wintering grounds from October to December but retain juvenal plumage without molting while in our region, even well into November.

Typical juveniles distinguished from non-breeding adults by spotted rather than striped upperparts. Some with yellow-edged feathers look more like adults, but abundant ventral markings, especially bars, should distinguish them.

Immature. Distinguished from nonbreeding adult by dark, worn tertials and, at least in early winter, mottled breast. Some birds attain adult breeding plumage by first spring and leave wintering grounds, presumably to migrate to breeding localities. These birds recognizable as first-year birds only at close range by very worn primaries. Most first-year birds remain in nonbreeding plumage on wintering grounds, molting primaries in midsummer or fall. Individuals of both sexes may become more brightly marked in each succeeding year until they attain full breeding plumage.

Identification

The following material pertains only to the difference between the two species of golden-plovers. See under American Golden-Plover for differences from the Black-bellied Plover and other shorebirds.

All Plumages. It appears that the AMERICAN GOLDEN-PLOVER is a slightly larger bird, but separation by size would be impossible in the field. The PACIFIC has bill and tarsal measurements very slightly larger than the AMERICAN, and, with its smaller size, it should look slightly longer-billed and longer-legged; these differences are minor at best. The entire head may look relatively larger in the PACIFIC, as, from the skull configuration, it must have slightly larger salt glands than the AMERICAN. The most obvious differences are in wing proportions.

• Specimens can be distinguished by wing length, most *fulva* falling below 170 mm and most *dominica* falling above that measurement; adults have slightly longer wings than juveniles. Nevertheless there is overlap, and some birds will not be identifiable by this characteristic.

• Relative wing length may prove to be of great value as a field mark: typically in PACIFIC the *wings project to or slightly beyond the tail tip*, while in AMERICAN this projection is distinctly longer, at least as much as the bill-to-eye distance.

The *primary projection is typically shorter* in PACIFIC; only three primaries (or three plus part of the fourth) are visible, a distance just longer than the bill length, while in AMERICAN it is longer, substantially longer than the bill length and with three and a half or four primaries visible. Again, there may be overlap, and this character will not work for birds with worn tertials (especially breeding-plumaged adults in autumn).

Breeding Plumage.

• In breeding plumage the male PACIFIC tends to have *much white under the tail and on the sides* (but not the extensive pure white under the tail of the Black-bellied Plover), while the male AMERICAN is usually entirely black in these areas. Probably very few male AMERICANS in molt would ever look just like PACIFICS. At their most heavily pigmented, females of both species might look something like the male PACIFIC.

• In the PACIFIC, the *white neck stripe is slightly narrower* than that of AMERICAN, in which the width is emphasized because it stops suddenly at the wing and widens there. Almost all photographs of breeding-plumaged adults in standard reference works are AMERICANS.

• One average difference is evident from examination of museum specimens. In AMERICAN the mantle feathers typically have paired apical spots of yellow or whitish, and only in some feathers of some individuals is there another pair of lateral spots. In PACIFIC there are typically two

pairs of spots, so the mantle averages somewhat more heavily marked; this is why PACIFIC may appear more golden above than AMERICAN, rather than because of differences in the color or size of the spots.

The timing of the molt into breeding plumage differs in adults (and immatures?) of the two species. In spring many AMERICANS reach North America in nonbreeding plumage, molting as they proceed north, and many birds retain that plumage well into April. In PACIFICS, contrarily, molt of the upperparts begins early on the wintering grounds, and by the end of March all appear to be fully spangled with gold above even if showing much white below. By late April, when they migrate, most birds of both sexes are in full breeding plumage.

In autumn, adult PACIFIC apparently begin molting the underparts before the upperparts, as in September many individuals with largely white bellies still retain breeding plumage above. By October the underparts are usually entirely light, the upperparts in molt, and by November full nonbreeding plumage has been attained. Adult AMERICAN retain remnants of breeding plumage into mid October in northern South America, but by December they are normally in full nonbreeding plumage.

Nonbreeding Plumage.
Nonbreeding adult AMERICAN GOLDEN-PLOVERS are typically dull and may lack any trace of yellow above or below, looking like dark versions of Black-bellied Plovers, although with distinctive plain gray breast; PACIFIC GOLDEN-PLOVERS are more typically heavily marked with yellow above. The most brightly marked AMERICAN show yellow-fringed mantle feathers and black-dotted scapulars and tertials, while the most brightly marked PACIFIC have yellow dots on the mantle and yellow notches or fringes on the tertials. Individuals of both species can be quite dull above by late winter, however. Most or perhaps all AMERICAN are entirely gray-brown and whitish below, while at least some PACIFIC are buffy-brown to yellowish around the head and breast. Some individual birds could not be distinguished.

Juvenal Plumage.
There are a number of average differences in color and pattern between juveniles of the two species, but there is much

overlap, and perhaps only the more extreme individuals can be definitely identified.

• Some PACIFICS have *especially conspicuous yellow markings on the crown, mantle, scapulars, and tertials*, producing a strikingly golden-backed bird, but others look essentially like AMERICANS, with smaller spots and narrower fringes on the upperparts. The more brightly marked AMERICANS are quite golden above and have probably been called PACIFICS in the past. The golden spots appear to fade fairly quickly, so fresh-plumaged juveniles are more brightly colored above than older birds. Thus birds seen late in fall that are conspicuously golden are more likely to be PACIFICS.

• The supercilium tends to be narrower and, at least in some individuals, tinged with yellow in PACIFIC, a bit whiter and broader in AMERICAN.

• The auricular spot is more sharply focused (darker and smaller) in PACIFIC, a bit more diffuse and extensive in AMERICAN.

• Juvenile PACIFICS usually have the *ground color of the sides of the head and neck, throat, and even the breast yellowish*, while AMERICANS usually have the same area gray to whitish.

• On the average, the *stripes on the nape and side of the neck and breast are dark and well-defined* in PACIFIC, on a paler ground color, while they are blurry in AMERICAN, the feathers as likely to look barred as striped, and on a darker gray ground color. There is considerable overlap in this character, however.

• The lower breast and sides are more likely to be barred in AMERICAN, with stripes and bars intermingled in PACIFIC. The lower belly is darker gray and often barred in AMERICAN, more likely to be paler and unmarked in PACIFIC.

Immature Plumage. At least some if not all first-winter birds of the two species look very different because of a different molting schedule. Those of the AMERICAN do not molt the feathers of the upperparts, and the dark brown crown, mantle, and upper scapulars contrast strongly with gray-brown lower scapulars, coverts, and tertials at least into late winter, while those of the PACIFIC replace many of the juvenile upperpart feathers with gold-fringed adult-type feathers.

Some PACIFICS in autumn migration (probably only birds at least two years of age) have one or more innermost primaries darker in color and less worn than the feathers on either side of them, presumably molted on the breeding grounds. Specimens of AMERICAN do not show this peculiar characteristic, which may be visible in photographs of flying birds.

In Flight (Figure 9)

The two species of GOLDEN-PLOVERS look essentially the same in flight, although at close range breeding-plumaged males could be distinguished from below. The longer legs of the PACIFIC may provide a good field mark in flight: in that species, about *half the length of the toes project beyond the tail tip*, while in the AMERICAN they barely project if at all. This characteristic needs to be confirmed by more observations in eastern North America, where only AMERICANS occur.

Voice

The call of this species is typically a two-noted whistled *chuwi* somewhat like that of a SEMIPALMATED PLOVER or a three-noted *chuwheedle*. I have not heard the *queedle* that is typical of the AMERICAN GOLDEN-PLOVER. Adults in spring in Hawaii also give a sharp, whistled *wheet*, possibly a territorial call.

Further Questions

See under American Golden-Plover.

Notes

Frequently cited records of Pacific Golden-Plovers from the interior of the region rest on a paper by Sloanaker (1925). He listed specimens from Lake Chactolet (=Chatcolet), October 1, 1923, and Kahlotus, December 19, 1924, under this name. I suspect he called all these birds *fulva* merely because they were taken in the western part of the range of the composite species, as he apparently did not examine the specimens himself. Both Burleigh (1972) and Jewett et al. (1953) appear to have accepted this identification uncritically, but I have not been able to examine these specimens. Because of its late date, the Kahlotus specimen might more likely be a Pacific, which migrates late and occasionally winters on the Pacific coast, but there are late records of Americans from the East, even a very few wintering birds on the Atlantic and Gulf coasts.

The statement by Peterson (1990: 122) that this is the golden-plover more likely to be seen in the Pacific states is contradicted by the evidence. See also under American Golden-Plover.

Photos

See under American Golden-Plover. Few photos of this species are published, but they include nonbreeding adults in Keith and Gooders (1980: Pl. 102) and Pringle (1987: 80-81).

References

Connors 1983 (discussion of breeding overlap and species status of *dominica* and *fulva*, with differences between them), Conover 1945 (identification and Alaskan distribution), Henshaw 1910 (Pacific migration), Hindwood and Hoskin 1954 (winter ecology), Jackson 1918 (molt), Johnson 1973 (reproductive condition of summering birds), Johnson 1977 (plumage and molt of summering birds), Johnson 1979 (summary of 1973/1977 papers), Johnson 1985 (timing of primary molt in immatures), Johnson and Johnson 1983 (molt, sexing, and ageing), Johnson and Nakamura 1981 (roof roosting), Johnson et al. 1981 (winter site tenacity and behavior), Johnson et al. 1989 (fat cycles and wintering behavior), Johnston and McFarlane 1967 (Pacific migration), Kinsky and Yaldwyn 1981 (molt and migration in southwest Pacific), Pym 1982 (identification), Sauer 1962 (breeding biology), Vaurie 1964 (differences between *dominica* and *fulva*).

MONGOLIAN PLOVER *Charadrius mongolus*

Slightly larger than the Semipalmated Plover, the longer bill and lack of a white collar in this very rare visitor provide distinguishing features in any plumage. Rufous-breasted breeding adults are striking and unmistakable.

	Jan	Feb	Mar	Apr	May	Jun	Jul	Aug	Sep	Oct	Nov	Dec
COAST							—		—	—		

Distribution

The Mongolian Plover breeds in scattered populations from the Siberian tundra west to the Central Asian plateaus. It winters on coasts all around the Indian Ocean and from the Philippines to southern Australia in the southwest Pacific.

Northwest Status

Three coastal-Oregon records of this species between July 11 and October 21 and four California records in the period from July 12 to October 3 (some of them perhaps involving the same bird) provide a clear pattern of occurrence. Most of these birds stayed where first discovered for one to three weeks, a good indication of the length of stay of migrant shorebirds at stopover points. As the species has nested in western Alaska and is a rare but regular spring and fall migrant on Bering Sea islands, it is as likely to occur in our region as any of the "accidental" Siberian shorebirds. Of the Oregon birds, those in July and October were adults, the September one apparently a juvenile.

COAST FALL. Tillamook Bay, OR, September 11-17, 1977; south jetty of Columbia River, OR, October 16-21, 1979; Bandon, OR, July 11-29, 1986.

Habitat and Behavior

The Oregon birds were seen on sand and mud flats, typical habitat for this species in Asia and Australia where it is a coastal bird in both migration and winter. Where common it forms large flocks, but it will likely be solitary in our region. Individuals in Oregon were loosely associated with Semipalmated Plovers and small sandpipers. Foraging is in typical plover fashion by running and picking from the surface.

Structure

Size small. Bill very short and fairly heavy, legs short. Wings usually extend just beyond tail, primary projection short in adults and juveniles. Tail shallowly double-notched.

Fig. 22. Breeding Mongolian Plover. Dark rufous breast and contrasting white throat distinctive.

Plumage

Breeding (Figure 22). Bill black, legs gray to blackish. Plumage bright and contrasty, males with black mask and white head markings much like those of Semipalmated Plover. Narrow loral stripe, ear coverts, and bar across forecrown black. Supercilium white, only behind eye. Forehead white, often bisected by black vertical line; collar rufous, often some rufous on cap. Upperparts uniform medium brown. Breast rufous, usually bordered in front by narrow black line. Lower cheeks and throat white, contrasting strongly with dark surrounding areas; lower breast and belly white. Females average duller, black head markings mixed with or replaced by brown.

Tail brown, darker brown on terminal third (except outermost feathers), narrowly tipped with white; outermost pair of rectrices white-edged. Wing brown above with white stripe formed from narrowly white-tipped secondary coverts and white patches on about three innermost primaries. Underwing coverts white, some tinged brownish gray. Axillars white.

Molt into breeding plumage typically takes place in March and April, into nonbreeding plumage in August and September; some adults with reddish on breast into mid September.

Nonbreeding. All black and rufous markings lost, head pattern brown and white. Fairly well-defined breast band brown and much narrower than that of breeding plumage; usually widely interrupted in middle, leaving no more than smudge at either side of breast.

Juvenal (Figure 23). Legs gray-green to greenish yellow. Much like nonbreeding adult

but with conspicuous pale buff feather edges on mantle, scapulars, and coverts. Underparts may look more like breeding adult, with breast and side of head tinged with bright buff, or like nonbreeding adult, with interrupted wide brown band contrasting strongly with white throat. Buff edges usually worn off by October.

Immature. First-winter birds probably indistinguishable from adults. One-year olds commonly oversummer in southern hemisphere, usually in worn nonbreeding plumage.

Subspecies

The subspecies occurring in the Northwest is presumably *C. m. stegmanni*, which breeds in eastern Siberia and is documented by specimens from Alaska.

Identification

Breeding Plumage These birds, with rufous breast and white throat, are unmistakable.

Nonbreeding and Juvenal Plumages. Fall individuals with full breast bands may be confused with SEMIPALMATED PLOVERS but are distinctive in the following characteristics:

• MONGOLIAN has *no white collar* on the hindneck (because of this the white throat is bet-

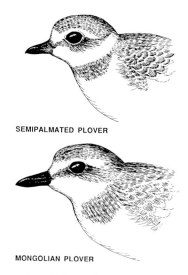

SEMIPALMATED PLOVER

MONGOLIAN PLOVER

Fig. 23. Juvenile Semipalmated and Mongolian plovers. Mongolian distinguished from Semipalmated by larger size, distinctly longer and heavier bill, more obscure head pattern, and lack of white collar.

cially to buff-breasted juveniles, but MONGOLIAN is *smaller* and averages *darker* than MOUNTAIN (which is closer to KILLDEER in bulk). MONGOLIAN also has a *shorter and thicker bill and a narrower ear patch* (dark area below and behind eye is separated from eye, paler, and extends farther onto throat in MOUNTAIN, which thus has less contrast between face pattern and throat).

Some fall MONGOLIANS have virtually no white on the forehead and a well-defined facial mask and are thus distinct from any other regional plover.

In Flight (Figure 11)
The flight pattern is obscurely *stripe-winged* and distinctly *stripe-tailed*, like a drab Dunlin although with narrower white tail edges. The MONGOLIAN has a *less conspicuous wing stripe and less white in the tail* than either SEMIPALMATED or SNOWY and shows no white collar. The *tail tip is not so much darker than the tail base* as it is in the similarly colored SEMIPALMATED and MOUNTAIN PLOVERS.

Voice
MONGOLIAN PLOVERS are not especially vocal, but the flight call is distinctive, a short, rapid trill on one level. It is the only plover in the region with such a call, reminiscent of the call of a PECTORAL SANDPIPER.

Juvenile Mongolian Plover. Small *Charadrius* plover lacking white collar and with relatively large bill and gray legs; pale fringes on upperparts diagnostic of plumage. Kyushu, Japan, mid September 1984, Urban Olsson

ter defined and more restricted than in those species such as SEMIPALMATED in which the white continues around as a collar).

• MONGOLIAN has a *longer bill*, as long as the distance from its base to the rear of the eye (SEMIPALMATED's bill about the same length as distance from base to front of eye).

• MONGOLIAN has relatively longer *gray to greenish-yellow legs* (yellow to orange in SEMIPALMATED).

MONGOLIANS are really more like SNOWY PLOVERS, with their longer bills and subdued markings, than they are like SEMIPALMATEDS, and birds with very restricted breast bands could be confused with SNOWIES but for being both *larger and considerably darker*. See Snowy Plover for discussion of an Asian subspecies of that species that might occur in North America and is more similar to MONGOLIAN. MOUNTAIN PLOVER, also very rare, is superficially similar, espe-

Photos
See photographs in Roberson (1980: 123 and Pl. 11) and Sinclair and Nicholls (1980) for the substantial variation in this species, but note that in the latter article, birds called "Lesser Sand-Plover" (=Mongolian Plover) in Figs. 121 and 126 were subsequently identified as Greater Sand-Plovers. The flight photograph in Roberson (1980: 123) appears too pale, with too much white on wings and tail, doubtless an artifact of lighting. The "juvenile" in Farrand (1983: 321) is in nonbreeding plumage.

References
Hindwood and Hoskin 1954 (winter ecology), Sinclair and Nicholls 1980 (distinction from similar Greater Sand-Plover, only remotely likely to occur in the Northwest; see *Brit. Birds* 75: 94-96, 1982, for additional information).

SNOWY PLOVER *Charadrius alexandrinus*

These little beach ghosts are scattered in small populations on upper coastal beaches and interior alkaline flats. To see them is an accomplishment, as they whisk across the sand, stopping abruptly as one's eyes travel on. Plover shape, pale back, and lack of breast band indicate this species. Coastal breeding populations from Grays Harbor south are threatened by development and off-road vehicles, those in the southern Oregon interior only by droughts and floods.

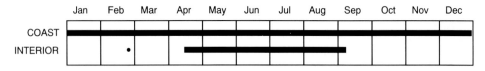

Distribution

The Snowy Plover breeds locally on most continents. In North America it breeds on the Pacific and Gulf coasts and very locally in the western interior, and in South America on the Humboldt Current coast. In Eurasia it is much more widespread in the interior and occurs also on European, Asian, and north African coasts. It is absent from much of the tropics as a breeding bird, although northern populations move to tropical coasts in the nonbreeding season.

Northwest Status

This plover is an uncommon breeder on the outer coast, north to the north side of the mouth of Grays Harbor, where a few pairs are present every year at Damon Point. More are to be found at Leadbetter Point and still more scattered along many appropriate beaches on Oregon's coast. Recent population estimates had 32 breeding on the Washington coast and 84 on the Oregon coast. In July 1986 a small population was found on Dungeness Spit, but no evidence of breeding is yet at hand.

Most birds disappear from Washington in winter, but a few remain as far north as Leadbetter Point in at least some years. Many birds apparently pull out of Oregon also, but wintering is regular on the Oregon coast, with most birds reported on Tillamook, Lane, and Coos county beaches, and a total of 50 to 100 wintering. Surprisingly, birds banded on the central California coast as both breeders and nestlings have turned up on the Oregon and Washington coasts subsequently, and a bird banded in the Oregon interior turned up the next year on the Oregon coast.

The species has declined in numbers considerably, probably because of constant human disturbance to outer beaches, and Washington populations in particular are vulnerable. Recent protection of its nesting areas at both Damon and Leadbetter points may have been successful in stopping, if not reversing, the population decline in Washington.

There are also a few spring and summer records farther north on the coasts of Washington (May 14 to September 14) and southern British Columbia (April 29 to July 17), including both the west coast of Vancouver Island (Tofino and Long Beach) and protected waters (Iona Island and Dungeness). These are probably birds that migrated too far and failed to find a mate or appropriate breeding habitat. There is a single record from the Willamette Valley, at White City, May 6, 1984.

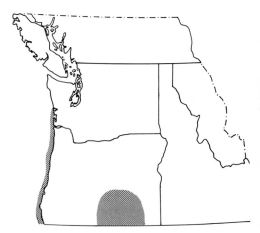

Snowy Plover breeding distribution

There is a more substantial breeding population at the Oregon lakes, north to Malheur Refuge and west to Summer Lake, with the largest populations at Harney Lake and Lake Abert. They are typically present from late April through August. The breeding population estimate for this region is on the order of a thousand birds. Elsewhere in the interior there are more than a dozen records from eastern Washington (Potholes Reservoir, Reardan, and Wallula) and southern Idaho (the length of the Snake River valley from Nampa to Rigby) from April 11 to May 28, presumably northbound birds that had overshot their breeding grounds. The few records from Montana are from east of this region, but one of them involved a breeding pair.

INTERIOR SPRING. *Early date* Harney Lake, OR, February 27, 1968 (3).

INTERIOR FALL. *Late date* Alvord Lake, OR, September 9, 1979.

Habitat and Behavior

Snowy Plovers are birds of the dry sand beaches and open dunes fronting the open ocean or alkaline lake shores in the interior. They are well-camouflaged, turning away from an intruder or squatting on the sand to hide their more conspicuous dark head and breast markings and white underparts. Typical of their family, they forage with quick runs, and, not moving very far in the course of a day, they fly relatively seldom. Although primarily found on the outer sand beaches, they also forage on nearby mud flats, especially after the breeding season. Many of these birds may be juveniles, their habitat-choice mechanisms not yet finely tuned to their adaptations. Foraging is always by picking, and beach crustaceans and insects are primary foods. Interior birds take many ephydrid flies, an especially abundant food source for shorebirds at alkaline lakes.

Unlike the other *Charadrius* plovers of the region, there is no flight display, males patrolling their territories on foot. The nest is a scrape in a sandy area, often near driftwood or vegetation and sparsely lined with small objects. The clutch size is three, one fewer egg than is typical of other plovers (Semipalmated and Killdeer) potentially breeding in the same area. Both sexes incubate and care for the young.

Structure

Size small. Bill very short, legs short. Wings short for a plover, extending about to tail tip, with scarcely any primary projection in adults and juveniles. Tail shallowly double-notched, central rectrices projecting slightly as in a *Calidris* sandpiper.

Plumage

Breeding. Bill black, legs gray (typical) to blue-gray to greenish. Upperparts light sandy brown; forehead, narrow supercilium, narrow collar, and underparts white. Breeding males with black bar across forecrown, dark brown postocular stripe (and loral stripes in some individuals), and black patch on side of upper breast (forming half-ring). Birds may be in this plumage from January to November. Breeding females variable, some looking virtually like males and others considerably duller, with brown markings where male has black. Sexes often distinguishable in pairs. Some males show warm buff to reddish cast to crown, usually in spring.

Central rectrices light brown with slightly darker tips, outer two pairs white. Wing brown above, broad white tips of greater secondary coverts and white patches on outer webs of about five inner primaries form conspicuous wing stripe. Underside of flight feathers light sandy brown, coverts and axillars white.

Upperparts fade substantially between one autumn molt and next.

Nonbreeding (Figure 24). Dark head markings dull brown, otherwise similar to breeding adult.

Juvenal. Legs often paler than in adult, pale gray or even dull yellowish. Most lightly marked plumage, with no contrasty head or breast markings and distinct pale buffy fringes to all feathers of upperparts. Molts into adult plumage during first winter, may mature in one year.

Subspecies

Only one subspecies—*C. a. nivosus*—of this wide-ranging and variable species has to date been recorded from the region. Individuals of the Gulf coast subspecies *tenuirostris* are considerably paler than those of interior and west-coast *nivosus* and would look distinctive in direct comparison with Northwest birds. Worn summer in-

PIPING PLOVER

SNOWY PLOVER

Fig. 24. Nonbreeding Piping and Snowy Plovers.
Piping distinguished from Snowy by slightly larger size
and shorter, thicker bill.

dividuals of *tenuirostris* are almost pale enough
to be called whitish above. Hayman et al. (1986)
merge *tenuirostris* with *nivosus*, but the two are
quite distinct. Many *nivosus*, presumably from
interior populations, migrate to the Gulf coast in
winter, and this may have caused other authors to
consider the Gulf coast breeding birds more variable than they are. This resident Gulf Coast subspecies is quite unlikely to reach our region.
However, the following subspecies, although not
recorded from North America, might be expected
on this side of the Pacific, as it is migratory and
occurs at similar latitudes.

East Asian birds (*C. a. dealbatus*) are distinctly darker than Northwest populations, almost
as dark as Mongolian Plovers. In addition they
are larger and relatively longer-billed than American Snowy Plovers, with brighter buff on the
crown of males and almost invariably a dark loral
stripe. In nonbreeding and juvenal plumages they
could as easily be mistaken for Mongolian as for
American Snowy plovers, but they can be distinguished from Mongolian by smaller size, white
collar, and less distinct supercilium.

Although American Snowy Plovers are supposed to differ from those of Eurasia by the lack
of brown or black in front of the bill (Prater et al.
1977, Cramp and Simmons 1983), at least some
birds have distinct dark lores (see photos in Page
and Stenzel 1981), so observers should not be
misled into thinking they have found an Asian
vagrant because of this character. The same is
true of buffy or reddish coloration on the head, a
trait also exhibited by some American birds. The
darkness of the upperparts and bill size remain
the best distinction between the two subspecies.

Identification

All Plumages. This is a small plover, with less
bulk than the much darker SEMIPALMATED, but it
stands as high because of longer legs. It differs in
the following ways:

• The *pale upperparts and lack of breast
band* quickly distinguish SNOWY from SEMI-
PALMATED. Even juveniles of SEMIPALMATED
have a complete breast band, usually narrower in
the middle.

• SNOWY has *grayish legs and black bill*. SEMI-
PALMATED has yellow to orange legs, and adults
have orange-based bills, but those of juveniles
are black as in the SNOWY.

• SNOWY has a *longer, more slender bill*,
longer than eye-to-bill distance but about the
same as that distance or very slightly shorter in
SEMIPALMATED. The SEMIPALMATED's bill is
relatively thicker as well as shorter than that of
the SNOWY.

The very rare MONGOLIAN PLOVER is distinctly larger and darker than SNOWY and lacks
the white hindneck collar of that species (but see
above under Subspecies). Also see PIPING PLO-
VER. Nonbreeding-plumaged SANDERLINGS are
the only other shorebirds of the region as pale as
SNOWY PLOVERS, but they are typical sandpipers
that run or walk continually while foraging. The
SANDERLING's bill is almost the length of its
head, the plover's much shorter. Both species
may be found together when SANDERLINGS move
up from the water's edge to forage or roost on the
upper beach.

In Flight (Figure 11)

In flight SNOWY PLOVERS can be distinguished
by *short plover bill, pale coloration, and moder-*

ately conspicuous wing stripe. Winter SANDER-LINGS and RED PHALAROPES have more conspicuous wing stripes and are larger. The plover has a *partial dark bar at its tail tip*, differing in this way from the rather similar SANDERLING. The similarly shaped SEMIPALMATED PLOVER is considerably darker than the SNOWY with tail less edged with white and with a conspicuously darker tip.

Voice

SNOWY PLOVERS are not highly vocal, but their flight calls include a short, low-pitched note that may be extended into a trill—*prit* or *prit-it*. Males give a two- or three-noted whistle, the second note highest and loudest, to advertise their presence on territory.

Further Questions

Breeding pairs of Snowy Plovers should be checked to see if the sexes can be distinguished, and if there are birds with dark loral stripes and reddish head markings in the Northwest. Observers should monitor the success of known nesting populations of this threatened species and search for additional ones. All birds should be checked for color bands (see under Sanderling).

Photos

The "juvenile" in Farrand (1983: 323) is in non-breeding plumage; note its pale legs, comparable in color to a Wilson's Plover or Killdeer. The adult male in Farrand (1988a: 174) looks much darker than the usual Snowy Plover.

References

Herman et al. 1988 (populations in southeastern Oregon), Lessels 1984 (mating system), Marshall 1989 (status in Oregon), Page and Stenzel 1981 (breeding habitat use), Page et al. 1983 (breeding success and predation), Page et al. 1985 (nest sites and predation), Page et al. 1986 (wintering in Pacific states and information about breeding populations), Page et al. 1991 (distribution of western breeding populations), Purdue 1976a (parental response to high temperature at nest), Purdue 1976b (adaptations for breeding on salt flats), Purdue and Haines 1977 (saltwater tolerance), Warriner et al. 1986 (breeding biology and polygamy), Wilson-Jacobs and Dorsey 1985 (occurrence at Coos Bay, Oregon), Wilson-Jacobs and Meslow 1984 (breeding on Oregon coast).

WILSON'S PLOVER *Charadrius wilsonia*

The Wilson's Plover is perhaps the least likely North American plover to occur in the Northwest, as it normally occurs only on the Gulf and Atlantic coasts north to New Jersey and the Pacific coast north to central Baja California. Nevertheless, it has occurred in spring and summer in southern California several times (presumably Pacific birds) and in the continent's interior, north to Minnesota, Ontario, and Oklahoma, several times (presumably Gulf birds). It appears about as likely to occur in the Northwest as the American Oystercatcher, a species that has done so.

WILSON'S PLOVERS are closer in size to the larger MONGOLIAN than to the smaller SEMIPALMATED, but their bill is even longer and heavier than that of MONGOLIAN, looking disproportionately large. The legs are also intermediate in length. There is a distinct white collar around the hind neck; thus the bird looks like a big-billed, long-legged SEMIPALMATED, or a big-billed, collared MONGOLIAN. Adults have a complete breast band, black in adult males and brown in adult females. Juveniles have a complete or incomplete brown breast band and the pale-scaled back pattern typical of the group. The bill is always black, the legs pinkish to brownish olive.

The direction of origin of an adult in breeding plumage that turned up in the Northwest would be determinable with close study. Birds from the Pacific coast population (*C. w. beldingi*) typically show some rufous, males on the crown and females on the breast band. Atlantic and Gulf coast birds (*C. w. wilsonia*) show no rufous.

In flight WILSON'S looks much like SEMIPALMATED, but the bill is noticeably large. The call is a piercing, whistled *whit* or *wheet*, sometimes repeated in a series of several notes. Birds of this species are often long-distance runners when approached.

References

Bergstrom 1988a (breeding biology), Bergstrom 1988b (breeding behavior), Morrier and McNeil 1991 (activity budget).

COMMON RINGED PLOVER *Charadrius hiaticula*

Not recorded from our region, the Common Ringed Plover is a possible vagrant, based on its occasional occurrence in spring and fall on Bering Sea islands and possible breeding on St. Lawrence Island. Because it is not easily distinguished from the Semipalmated, it may be of more regular occurrence in Alaska than is now thought to be the case. However, Siberian birds normally orient in the opposite direction from North America, wintering in the Near East and Africa.

Differences between the RINGED and the slightly smaller SEMIPALMATED in breeding plumage include (Figure 25):
• RINGED often has a distinctly wider black breast band, probably the best distant field mark, often as wide as or wider than the white area in front of it; that of the SEMIPALMATED is often no more than half that width. The band in the RINGED is often wider at the sides of the breast and slightly constricted in the middle, while that of the SEMIPALMATED tends to be of even width. Remember that extreme individuals overlap and especially that band thickness changes with posture.
• RINGED has a more extensive white supercilium, conspicuous and longer than the eye diameter, whereas that of SEMIPALMATED varies from a moderately conspicuous spot smaller than the eye to entirely absent. This is most distinct in males (sex differences are comparable in the two species) but may approach overlap in females. A cause for confusion would be furnished by molting or first-summer SEMIPALMATED that combined the black breast band and head markings of breeding plumage with the well-developed white supercilium of nonbreeding plumage.
• RINGED has more extensive white on the forehead, usually more pointed at the rear and approaching the eye more closely.
• RINGED has somewhat more black on the head.
• RINGED has almost no indication of naked yellow skin around the eye, this eye-ring showing up moderately well at close range in SEMIPALMATED. This may be one of the best close-range field marks.

There may be overlap in all these characters, but a bird that shows a wide breast band, extensive white supercilium and generally more contrasty head markings might be a COMMON RINGED PLOVER.

Juveniles and nonbreeding adults of these two species are much more difficult to distinguish, but on the average RINGED has a wider breast band. Most RINGED I saw in early February had wider bands than any nonbreeding SEMIPALMATED I have seen. Birds in any plumage differ in additional ways:
• RINGED has a relatively longer bill, the culmen straight, whereas the culmen looks slightly concave in the shorter-billed SEMIPALMATED.

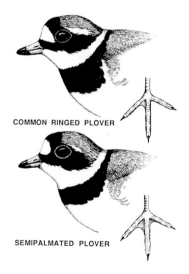

COMMON RINGED PLOVER

SEMIPALMATED PLOVER

Fig. 25. Breeding Common Ringed and Semipalmated Plovers. Ringed distinguished from Semipalmated by broader breast band, more black on head, better developed white postocular stripe, and white forehead patch usually pointed behind. Toe webbing diagnostic but extremely difficult to see in field.

- RINGED has less-webbed toes, the webbing less than that of SEMIPALMATED between outer and middle toes and entirely lacking between middle and inner toes (Figure 25); this character is useful only under the most ideal viewing conditions.

- The wing stripe in RINGED is more conspicuous than in SEMIPALMATED, probably evident only in comparison.

- RINGED has a distinctive flight call, similar to the SEMIPALMATED'S sharp *chuwee* but more plaintive and slightly lower-pitched with less force and less sharply accented second syllable (somewhat more like a GRAY-TAILED TATTLER). The calls are different enough so that the only

RINGED I have heard in North America was immediately recognizable as something different from the familiar SEMIPALMATED.

References

Bateson and Barth 1957 (geographic variation), Chandler 1987a (identification), Dukes 1980 (identification), Lifjeld 1984 (feeding ecology), Pienkowski 1982 (winter diet and feeding behavior), Pienkowski 1983a (winter foraging patterns), Pienkowski 1983b (winter feeding behavior), Pienkowski 1984 (breeding biology and population dynamics), Taylor 1980 (migration system), Vaurie 1964 (geographic variation).

SEMIPALMATED PLOVER *Charadrius semipalmatus*

Like a miniature one-banded Killdeer, this perky little plover can be found on beaches and mud flats during migration, stopping and starting in the midst of a flow of sandpipers. At closer range its short bill and bright yellow-orange legs are distinctive. Normally an arctic and subarctic breeder, it causes excitement when an occasional pair decides to nest in the region.

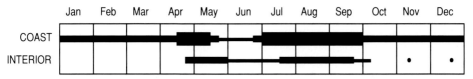

Distribution

The breeding range of the Semipalmated Plover encompasses the North American Arctic and Subarctic, from tundra to cobble and sand beaches of both fresh and salt waters. Migration is continentwide, wintering on coasts from southern United States south through much of South America.

Northwest Status

This species is abundant on the outer coast and its estuaries and less common in protected waters. It occurs in our area primarily as a migrant from mid April to mid May and from July through September; peaks are evident at the end of April and the end of July. Adults are present into at least early September but are exceeded in number by juveniles as August progresses. Juveniles do not show a separate peak, and numbers in September are much lower than earlier in the fall.

Birds regularly winter on the outer coast in small numbers, with January and February re-

cords north even to southern British Columbia. On the Washington coast there are more records in December and late February than in January, perhaps indicating some withdrawal of individuals (or withdrawal of observers?) during the coldest part of the winter.

Breeding has occurred during recent years at Iona Island (annually, up to three pairs) and at Damon Point on the north shore of Grays Harbor (sporadically, one pair), but these are quite peripheral records for this species. Even more extralimital records were furnished by pairs that bred successfully in the Oregon interior at Stinking Lake in 1987 and Harney Lake in 1989. Occasional nonbreeding birds, even small flocks, summer on the coast; the largest counts have come from Leadbetter Point.

In the interior the species is generally an uncommon migrant, more common in spring, when records come primarily from late April to late May with a few in June. The spring high count at Summer Lake is extraordinary, and it will be

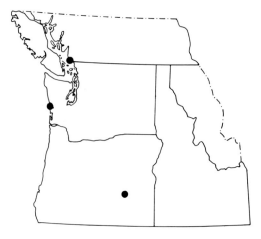

Semipalmated Plover breeding distribution

interesting to determine whether such numbers are typical of this locality. In fall it is more regular, although not seen annually even at well-studied localities such as Malheur Refuge. Records typically extend from mid July to late September, with many more juveniles than adults.

COAST SPRING. *High counts* 400 at New River, OR, late April 1984; 830 at Grays Harbor, WA, April 25, 1981; 2,000 at Ocean Shores, WA, May 3, 1986; 400+at Long Beach, BC, May 8, 1974.

COAST FALL. *High counts* 250+at Delta, BC, July 29, 1984; 250+at Boundary Bay, BC, August 10, 1986; 1,700 between Ocean Park and Leadbetter Point, WA, July 26, 1978; 500 at Ocean Shores, WA, July 26 and August 4, 1979*; 600 at Tillamook, OR, July 24, 1985; 400 at Tillamook, OR, August 6, 1986; 400 at Waldport, OR, August 22, 1986. *Juvenile early date* Ocean Shores, WA, August 8, 1981*.

COAST WINTER. *High count* 40 at Bandon, OR, February 23, 1987.

INTERIOR SPRING. *Early dates* Boise, ID, April 29, 1978; Nyssa, OR, April 29, 1981; Okanagan Landing, BC, April 28, 1941. *High counts* 10 at Boise, ID, May 7, 1978; 10 at Nyssa, OR, April 29, 1981; 12 at Bend, OR, May 3, 1986; 1,029 at Summer Lake, OR, May 1, 1987; 30 at Banks Lake, WA, spring 1963; 15 at Yakima River delta, WA, spring 1977. *Late dates* Boise, ID, June 6, 1982; White Lake, BC, June 10, 1973.

INTERIOR SUMMER. Nampa, ID, June 16, 1980.

INTERIOR FALL. *Early dates* Yakima River delta, WA, July 4, 1979; Malheur Refuge, OR, July 10, 1982. *High counts* 20 at Salmon Arm, BC, September 6, 1971; 12 at Twelve Mile Slough, WA, September 3, 1951; 17 at Malheur Refuge, OR, July 26, 1983; 15 at Medicine Lake, ID, August 28, 1951; 15 at Kootenai Refuge, ID, September 15, 1986; 10 at Pablo Refuge, MT, August 22, 1979*. *Late dates* Central Ferry, WA, November 12, 1985; Malheur Refuge, OR, October 14, 1985, December 16, 1989 (7); Potlatch, ID, October 6, 1953.

Habitat and Behavior

Although typical inhabitants of mud flats, these plovers are at times even more common on exposed sandy beaches where they are often the only shorebirds to associate with Sanderlings in numbers. An open, even substrate for running and an unimpeded view for searching seem essential, the birds staying entirely out of marsh vegetation. They spread out when feeding, often with considerable aggressive territorial behavior.

It may seem strange to characterize a species by its absence, but because it is a plover and spends much of its time standing still, is well camouflaged, and is small, this species easily escapes detection on a mud flat covered with other shorebirds. It roosts as scattered individuals or in small groups, not packing into dense flocks as do many other shorebirds and usually separated from the flocks of sandpipers with which it shares the beach.

Semipalmated Plovers forage in typical plover fashion by running, stopping, and running again. They very commonly "foot-stir," with one foot extended forward at about 45 degrees and vibrated on the substrate, presumably causing some invertebrates in their visual field to move and be detected. Polychaete worms, making up in length what they lack in substance (they look like the thinnest worm possible), seem to be major food items on ocean beaches.

Birds on their breeding territories fly around and around in a "butterfly flight" with slowly beating wings and sharp, liquid, whistled calls that accelerate to a chatter or trill. This behavior is an unmistakable sign on the part of the male of

Breeding Semipalmated Plover. Brown, single-banded plover with orange bill base and legs; black forehead, lores, and breast band characteristic of breeding male. Ocean Shores, WA, 3 May 1986, Dennis Paulson

intended breeding. The nest is a slight depression in a sandy or gravelly area, often with some detritus as lining material. The clutch size is usually four, and a plover nest with four small (33x23 mm) eggs in it on the coast would likely belong to a Semipalmated (Snowy lays three eggs; Killdeer's four are larger [36x27 mm]). Both sexes incubate and care for the young.

Structure

Size small. Bill very short and stubby, legs short. Small webs between outer and middle toes and even smaller ones between middle and inner toes. Wings extend to tail tip, primary projection short in adults and juveniles. Tail rounded.

Plumage

Breeding (Figure 25). Bill black with yellow or orange base, legs dull to bright yellow-orange. Eyelid skin yellow, producing very narrow but evident eye-ring. Simply plumaged, medium brown above and white below. Uniformity of upperparts broken by white collar, of underparts by dark breast band (narrowly continued around behind collar). Breast band, sides of head, and forecrown black in both sexes or mixed with brown in females. Sex distinction possible in mated pairs and at close range in migration. Forehead white, may be small white dot above and behind each eye. Breast band of even width or moderately incised below, almost divided at most extreme.

Rectrices brown, blacker toward their ends and with white tips; outermost feathers white. Flight feathers and primary coverts distinctly darker than secondary coverts. Wing brown above, broad white tips of greater secondary coverts and white patches on outer webs of inner five or six primaries form conspicuous wing stripe. Underwing coverts white, anterior ones tinged light brownish; axillars white. Adults appear to pass through the region in fall without molting.

Nonbreeding. Bill black, legs dull yellow or yellow-orange. Breast band, sides of head, and forecrown always brown like back. Well-defined white supercilium extends behind eye.

Juvenal (Figure 23). Like nonbreeding adult but may have narrower breast band, almost interrupted in middle. Feathers of upperparts tipped to variable extent with pale buff, creating scaled effect. White supercilium allows easy distinction from breeding adults when both age stages occur together in August.

Immature. At least some birds take two years to mature. Summering birds vary from full breeding to full nonbreeding plumage, presumably indicating range of variation in first-summer plumage. In individual birds attributes of nonbreeding plumage such as white supercilium can be combined with attributes of breeding plumage such as black breast band.

Identification

All Plumages. A *small brown plover with a single breast band* will be this species, save for a very unlikely encounter with one of the eastern or far eastern species that may drift our way (see WILSON'S PLOVER, COMMON RINGED PLOVER, LITTLE RINGED PLOVER, and MONGOLIAN PLOVER). Also see discussion of downy KILLDEER. The PIPING PLOVER is very similar to the SEMIPALMATED but is much paler, and the SNOWY PLOVER is similarly pale but has dark markings only at the sides of the breast in lieu of a breast band.

In Flight (Figure 11)

This species is a *small brown shorebird with moderately distinct wing stripe* and a tail with dark tip and narrow white edges and corners. The tail is often fanned, especially as a bird lands, and it is the only common shorebird of its size and color with a *distinctly darker tail tip.* Additionally, it is the only common and widespread shorebird with a *very short bill,* scarcely visible at a distance. The considerably larger KILLDEER, with somewhat similar flight pattern, has a longer tail, bright rufous rump from above, and two breast bands from below. See MONGOLIAN PLOVER.

Voice

The SEMIPALMATED calls frequently in flight, and its clear double whistle (*chuwee*), the second note emphatically pronounced and higher-pitched, is often the first indication of its presence. Only the PACIFIC (and perhaps AMERICAN) GOLDEN-PLOVER among regular Northwest species has a call similar enough to cause confusion.

Further Questions

Photos of roosting groups would be of great value to show variation in head and breast patterns, especially of breeding-plumaged adults, for more effective comparison with Common Ringed Plover. Also, information is needed about the age of both summering and wintering Semipalmateds; look for midsummer wing molt, indicating a one-year-old bird. This species is as good a species as any in our region to watch for territorial feeding behavior. Displays are obvious, and the birds are spread out enough so that aggressive runs are obvious. Watch in the fall when they are together to see if adults dominate juveniles. Observers who spend considerable time in both hemispheres should determine whether Semipalmated Plovers, with their slightly greater toe webbing, typically feed on softer substrates than Common Ringed Plovers.

The status of breeding birds in the region should be assessed on a yearly basis, with attempts to protect known sites from disturbance.

Notes

The legs of juveniles are usually duller and paler than those of adults, contrary to National Geographic Society (1987: 104-5).

References

Baker 1977 (summer diet), Baker and Baker 1973 (summer and winter foraging behavior), Bock 1959 (relationship to Common Ringed Plover), Burger et al. 1979 (aggressive behavior in migration), Campbell and Luscher 1972 (breeding in British Columbia), Chandler 1987 (identification), Ivey and Baars 1990 (breeding in Oregon), Ivey et al. 1988 (breeding in Oregon), Loftin 1962 (summering), Morrier and McNeil 1991 (winter activity budget), Smith 1969 (relationship to Common Ringed Plover on Baffin Island), Strauch and Abele 1979 (winter feeding ecology), Sutton and Parmelee 1955 (breeding biology), Vaurie 1964 (distribution and relationship to Common Ringed Plover), Weber et al. 1976 (breeding in British Columbia).

PIPING PLOVER *Charadrius melodus*

Imagine a Semipalmated Plover bleached to Snowy Plover hues and you envision the Piping. Although colored for the light sand beaches of the Atlantic and Gulf coasts, our only visitor turned up on dark mud flats at an interior slough. Its high whistle is ever less likely to be heard, however, as it is threatened throughout its range by development.

	Jan	Feb	Mar	Apr	May	Jun	Jul	Aug	Sep	Oct	Nov	Dec
INTERIOR							•					

Distribution

The Piping Plover breeds sparingly on river and lake shores from the northern Great Plains to the Great Lakes and on Atlantic coastal beaches from the Maritimes to North Carolina. Wintering is on the south Atlantic and Gulf coasts.

Northwest Status

An adult that appeared briefly early in fall migration in eastern Washington furnished the only photographically documented record from the region. The species has occurred once as a spring transient just east of this region in Montana and has bred in the northeastern part of that state. Additionally, at least three different birds have wintered to the south of the region, on the coast of southern California. Piping Plovers could be expected as occasional stragglers during migration, although this has become steadily less likely with the recent substantial decrease over much of its range. The closest breeding populations at this time are in south-central Alberta, the closest wintering populations on the Texas coast.

INTERIOR FALL. Reardan, WA, July 13-16, 1990.

Habitat and Behavior

This species is everywhere characteristic of sand beaches, whether coastal or at inland lakes and rivers. It forages by running and stopping like other small plovers, with individuals usually well-spaced along beaches. I have never seen them densely packed, as is often the case with Semipalmated on Northwest coasts. The Reardan bird foraged—apparently successfully in this atypical habitat—on mud flats around an alkaline slough, at times associating loosely with feeding sandpiper flocks but flying by itself when flushed.

Structure

Size small. Bill very short and stubby, legs short. Wings extend to tail tip, primary projection short in adults and juveniles. Tail rounded.

Plumage

Breeding. Bill black with orange base, legs bright yellow-orange. Overall plumage light sandy brown above, including cheeks, and white below. Uniformity of upperparts broken by white collar, of underparts by black breast band (narrowly continued around behind collar but not always visible). Some individuals, especially in eastern populations, with breast band interrupted in center. Forehead and broad supercilium white, latter interrupted by black band across forecrown from eye to eye. Breast band and forecrown black in both sexes or mixed with brown in females. Sex distinction possible in mated pairs and at close range in migration.

Rump white, contrasting with back in flight. Rectrices light sandy brown, darker brown toward their ends and with white tips; outermost two pairs white. Flight feathers and primary coverts distinctly darker than secondary coverts. Wing light brown above, broad white tips of greater secondary coverts and white patches on outer webs of inner six primaries form conspicuous wing stripe. Underwing coverts and axillars white.

Nonbreeding (Figure 24). Bill black, legs dull yellow or yellow-orange. Breast band (complete or incomplete), sides of head, and forecrown always light sandy brown like back. Well-defined white supercilium extends behind eye.

Juvenal. Like nonbreeding adult but feathers of upperparts tipped to variable extent with whitish, creating scaled effect (less obvious than in darker plovers).

Identification

All Plumages. The PIPING PLOVER is slightly larger than the SEMIPALMATED, and its legs average brighter orange. In all plumages the two species look almost identical except for the PIPING'S *considerably paler coloration* above, so much paler that there is normally little likelihood of confusion. The upperparts are so pale that the paler edges of the scapulars and coverts, already poorly developed in juveniles of this species, scarcely contrast with the ground color. A faded autumn adult SEMIPALMATED might cause confusion but is really much darker. From such a bird, PIPING would still be differentiated by flight pattern and call.

The PIPING is similar in coloration to the SNOWY PLOVER, but there are easily seen differences. PIPING in all plumages has *orange legs* (brightest in breeding plumage), while those of SNOWY are grayish. The leg color should always be diagnostic, although juvenile SNOWIES may have pinkish-tinged legs. PIPING has a slightly shorter and *considerably thicker bill* than SNOWY. Its overall coloration is slightly paler than that of western SNOWY PLOVERS in comparable plumages, but direct comparison would be necessary to detect this.

Breeding Plumage. Adults in breeding plumage differ from SNOWY PLOVERS in additional ways. PIPING has a yellow or orange bill base, a black crossband behind the white forehead that extends from eye to eye (interrupted by white supercilium in SNOWY), and a complete (or, less commonly, incomplete) black breast band. The complete breast band is typical of western populations, those more likely to occur in this region.

Nonbreeding and Juvenal Plumages. In nonbreeding adult and juvenile PIPING, which are like SNOWY in having entirely black bills, the area between eye and bill is always white (sometimes with dark line in SNOWY), and there is usually more white over the eye. In these birds the contrast between forehead and crown is less obvious in PIPING than in SNOWY (Figure 24).

In Flight

In flight PIPING looks extremely pale—whitish in bright sun—with a *white tail base*, the tail differing from both SNOWY (white edges prominent) and SEMIPALMATED (tail base same color as back).

Voice

The flight call is a sharp, whistled *peep-u*, quite distinct from the SNOWY'S ascending whistle or short trill and the SEMIPALMATED'S ascending double whistle.

Notes

There is a well-described but not further documented sight record of one at Manzanita, Oregon, on September 8, 1986.

References

Cairns 1982 (breeding behavior), Cairns and McLaren 1980 (distribution), Haig and Oring 1985 (distribution), Haig and Oring 1988a (site tenacity and mate fidelity), Haig and Oring 1988b (distribution and dispersal), Johnson and Baldassarre 1988 (winter ecology), Nicholls and Baldassarre 1990a (winter distribution), Nicholls and Baldassarre 1990b (winter habitat associations), Niemi and Davis 1979 (breeding biology), Russell 1983 (distribution), Wiens and Cuthbert 1988 (site tenacity and mate fidelity).

LITTLE RINGED PLOVER *Charadrius dubius*

This small plover, known from Pacific North America by only two late spring records from the western Aleutians, is unlikely to occur in the region. If seen it would be noteworthy in comparison with the rather similar SEMIPALMATED PLOVER by its more slender appearance, with relatively longer tail, longer legs, and slimmer bill. The tail projects distinctly beyond the wing tips, shaping this bird somewhat like a KILLDEER.

On closer scrutiny the observer should look for duller legs (yellowish or pale pinkish, not as bright as in SEMIPALMATED), an entirely black bill, although some males have a touch of yellow or orange at the base of the lower mandible (SEMI-PALMATED bill is orange-based in breeding plumage, all black in fall and winter), a conspicuous yellow eye-ring (the SEMIPALMATED'S eye-ring is visible at closer range), and, in breeding-plumaged birds, a fine white line behind the black bar crossing the forecrown (lacking in SEMIPALMATED). Juvenile LITTLE RINGED PLOVERS are similar to juvenile SEMIPALMATED but have a more obscure head pattern, the white supercilium scarcely developed.

In flight the LITTLE RINGED is at its most distinctive, as it lacks a wing stripe entirely. The tail is patterned like that of the SEMIPALMATED. The flight call is totally different from that of SEMIPALMATED or COMMON RINGED, a plaintive whistled *peeeo*. This species usually prefers freshwater habitats, both ponds and streams, but visits coastal mud flats as well.

KILLDEER *Charadrius vociferus*

The ubiquitous, onomatopoetic Killdeer is the beginner's shorebird. Give it a vacant lot of sufficient size and there it is, with instant vocal response to any disturbance. Learn the call of this natural sentinel and realize how widespread it is, present throughout the year in many parts of the Northwest. Males in butterfly flight hint of arctic aerial shorebird shows, and parents in distraction display reveal exciting behavior if not eggs or young. A distinctive pair of breast bands and long, mostly rufous tail ensure identification.

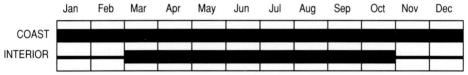

Distribution

The Killdeer breeds widely in open country throughout subarctic and temperate North America, south through Central America and the West Indies and along the northwest coast of South America. Populations of southern Canada and northern United States vacate their breeding range during winter.

Northwest Status

The most ubiquitous shorebird in North America, this species breeds throughout the lowlands of the Northwest. Breeding birds arrive early, mid February in the southern part of the region and March in the northern part. The average arrival date over 46 years at Fortine was March 9, the range February 18 to March 24. Similarly, they stay late (through October and often into November) in those regions where they are not present through the winter. After breeding they may collect into large flocks, but primarily they remain in small, scattered groups.

Most Killdeers leave the intermountain interior during the winter, although small numbers may be found in snow-free river valleys and agricul-

tural land there. Every one of 34 Christmas Bird Counts in the interior reported Killdeers at least once during the five-year period 1979-83, and ten of the counts reported them every year. Thus wintering is a normal phenomenon, and the scattered individuals found even in snow-covered and frozen countryside attest to their hardiness. West of the Cascades the species is widespread and fairly common throughout the year. Greatest winter abundance is reached in the Willamette Valley, where hundreds to thousands are regularly tallied on Christmas Bird Counts, and the high counts listed below were the highest for North America in those years.

COAST FALL. *High count* 250 at Boundary Bay, BC, September 17, 1922.

COAST WINTER. *High counts* 6,464 at Eugene, OR, December 28, 1969; 6,921 at Corvallis, OR, December 23, 1980; 7,307 at Roseburg, OR, December 18, 1982.

INTERIOR FALL. *High counts* 300 at Salmon Arm, BC, August 26, 1973; 2,000 at Turnbull Refuge, WA, early September 1969; nearly 1,000 at American Falls Reservoir, ID, August 1987.

CHRISTMAS BIRD COUNTS	Five-Year Averages		
	74-78	79-83	84-88
Campbell River, BC	21	8	14
Comox, BC	32	52	23
Deep Bay, BC	31	35	17
Duncan, BC	123	25	33
Ladner, BC	140	110	47
Nanaimo, BC	51	27	9
Pitt Meadows, BC	29	·12	26
Vancouver, BC	46	95	59
Victoria, BC	168	113	123
White Rock, BC	73	25	33
Bellingham, WA	128	83	111
Everett, WA	24	48	78
Grays Harbor, WA	69	93	44
Kitsap County, WA	57	63	76
Leadbetter Point, WA	8	16	19
Seattle, WA	76	114	68
Sequim-Dungeness, WA	54	26	58
Tacoma, WA	218	132	64
Coos Bay, OR	234	111	173
Corvallis, OR	1,741	4,946	5,414
Dallas, OR	258	824	424
Eugene, OR	1,591	2,039	2,369
Medford, OR	314	466	385
Portland, OR	153	139	32
Roseburg-Sutherlin, OR	387	2,119	1,235
Salem, OR	383	924	266
Sauvie Island, OR	958	532	87
Tillamook Bay, OR	199	35	177
Yaquina Bay, OR	127	58	50

Habitat and Behavior

This is a species of open habitats—fields, meadows, gravel bars of rivers, and muddy or grassy shores of ponds and lakes, probably on any substrate with an open view and adequate food supply. Breeding is common from sea level to the subalpine zone. Climate matters not, although the presence of nearby water may be important, as most nesting pairs in drier regions are near bodies of water or irrigated fields. Even if the nest is placed in the most arid of sites, the young have a productive habitat in which to forage.

After breeding, the birds remain common in many breeding habitats but also spread into additional waterside ones. At this time they are also frequent on freshwater beaches and in marine habitats such as estuaries, where they feed with their more water-oriented relatives. Killdeers of-

ten move onto mud flats well after they are left high and relatively dry, while other shorebirds follow the water's edge.

Killdeers forage in typical plover fashion, running at intervals and just as suddenly stopping to pick up a prey item or watch for another. Insects are primary year-round foods. On mud flats (and probably elsewhere) they practice the plover trick of "foot-stirring," rapidly vibrating a foot in front of them that presumably stimulates movement by potential prey. They spread out as they feed and tend to roost only in loose assemblages if together at all.

On open lawns and mud flats, Killdeers are as conspicuous as any other shorebird, but a bird standing still over a visually complex substrate can be perfectly camouflaged, and one may have to walk into its habitat to find it, which will be no problem when this reactive bird begins to call. It is well known as one of the most persistent alarm-callers of our avifauna, perhaps because breeding Killdeers and wandering humans come into contact so frequently. It is one of the characteristic "sentinel" species of open country, much more vocal than most of our shorebirds. This is at least in part because it is a resident, adults on their territories trying to deter potential nest predators, but it is quite vocal during the nonbreeding season as well.

Nests are placed on almost any open natural substrate, from short grass at the waterside to entirely unvegetated gravelly and stony soil on coastal beaches and river banks. In addition, Killdeers have taken advantage of all the open situations provided by humans, including the disturbed soil of new construction sites, the lawns of golf courses, cemeteries, and parks, and especially gravelly roadbeds, a favored site in the interior marshlands. A Killdeer apparently incubating right at the edge of a busy road is probably doing so! The four eggs are surprisingly cryptic in almost any of these situations. Nests on uniform-looking substrates, for example smooth green lawns, are usually placed near objects that break up the uniformity, and I am convinced one could preordain the location of such nests just by placing clusters of sticks or stones in such an area.

Both sexes incubate eggs and care for young, and some pairs raise two broods in a season. The

Killdeer is the virtual paradigm for the distraction displaying that occurs in so many ground-nesting birds, as it flops about with seemingly broken wings outspread and distressed-sounding cries. There is no better time to see the beautiful rump and tail pattern of one of these birds than when you are near its chicks and one or both adults trail along the ground in this apparent state of dysfunction. Watch your step when confronted by this impressive display; the chicks are too well camouflaged for their own good (as are the eggs, laid right out in the open) in a land of heavy-footed humans.

In winter and migration Killdeers often form small flocks, rarely up to dozens of individuals. Although not typically a highly social species, at times a hundred or more are present in a field, and thousands have been recorded in late-summer flocks in Oregon. Even when flying in flocks, Killdeers are usually more dispersed than many other small or medium species, but I have on rare occasions seen tight flocks, presumably migrants, pass by with rapid wingbeats. Killdeers commonly fly around at night over suburbs and countryside, their white bellies conspicuous in the glow of lights and their presence usually first detected by ear.

Structure

Size medium-small. Bill very short and fairly slender for a plover, legs short. Wings pointed but fall far short of tail tip, virtually no primary projection. Tail long, graduated.

Plumage

Adult (Figure 26). Bill black, legs pinkish. Eyelid skin reddish, producing pronounced eye-ring. Adults look about the same year round. Upperparts brown, underparts white. Crown and ear coverts brown, band between forehead and crown blackish, stripe from bill to below eye blackish to brown. Forehead white, extending just below eye; conspicuous white supercilium behind eye. Throat and broad collar white. Two conspicuous black bands that cross upper breast (upper one encircles neck) variable in width, also look broader or narrower depending on bird's posture; width not sex-related.

Lower back and uppertail coverts rufous. Central rectrices brown, darker toward ends and

Fig. 26. Killdeer. Distinguished by two conspicuous breast bands.

tipped with white. Next two pairs on either side rufous-tinged, outermost feathers white with blackish bars. Flight feathers and primary coverts distinctly darker than secondary coverts. Wing brown above, broad white tips of greater secondary coverts and white patches at mid-length on all flight feathers (much reduced on outermost four primaries) form conspicuous wide wing stripe. Underwing and axillars white, greater primary coverts light gray-brown except for tips. Tips of flight feathers gray-brown, forming dark band on rear edge of underwing. Wing molt begins in July on breeding grounds.

Fresh autumn adults show rufous fringes on tertials and scapulars that wear off during winter.

Juvenal. Buff fringes on mantle feathers, coverts, scapulars, and tertials evident at close range, but these markings wear off quickly during first autumn. Otherwise this plumage exactly like that of adult. No further plumage change, but age at maturity not established. Virtual lack of plumage variation correlates with minimal change between breeding and nonbreeding habitats.

Identification

All Plumages. A plover with *two prominent black breast bands* must be a KILLDEER in this part of the world. Only the single-banded SEMI-PALMATED PLOVER is at all similar, and the semipalmated is considerably smaller, about half the bulk of a KILLDEER and two-thirds the length. The double breast band and much longer bill of the KILLDEER are obvious field marks in addition to size, and the KILLDEER's *long tail* projects beyond the wing tips and gives it a more

Downy Killdeer. Fuzzy look and long down feathers on tail tip distinguishes from Semipalmated when in one-banded stage. Near Potholes Reservoir, WA, 29 May 1976, Dennis Paulson

pointed look toward the rear than other plovers.

Note also the KILLDEER's long, all-black bill (also all dark in juvenile SEMIPALMATED but orange-based in breeding adults), its red, fleshy eye-ring (yellowish or not visible in SEMIPALMATED), and longer supercilium. KILLDEER has pale pinkish legs, SEMIPALMATED yellow to orange. Downy young KILLDEER furnish possible confusion, as their breast band is originally single and still looks more or less so when they are about the size of a SEMIPALMATED. They have the downy, gawky look of a chick, however, and should be easily distinguishable at second glance. The stubby wings will do it if nothing else. See also the casual MOUNTAIN and MONGOLIAN PLOVERS, both of which lack the KILLDEER's white collar.

In Flight (Figure 11)

The *long, rufous-based tail* with white borders distinguishes this species from any other brown shorebird with *conspicuous white wing stripes*. Its flight is more leisurely than that of other shorebirds, as it moves slowly over the landscape instead of rapidly between feeding and roosting areas. Rapidly migrating KILLDEERS are rarely seen exceptions to this.

In open country the KILLDEER is not the only common bird with pointed wings and long tail. Two other fliers-by of similar size that coexist with it must be borne in mind: the MOURNING DOVE is all brown and has short wings and rapid wingbeat, and the AMERICAN KESTREL is bulkier with longer tail. Of the three, the KILLDEER has the most conspicuous contrast between brown upperparts and white underparts and usually the slowest flight, with long and narrow wings.

Voice

The noisy calls we often hear (a musical *tir-eeee*, a harder *dee*, and others) are predator-alarm calls, brought forth to signal our presence, or flocking calls. The *kill-deer, kill-deer* calls are given primarily by displaying males (occasionally by females) as they circle overhead with slowly beating wings in a "butterfly flight" that is characteristic of many plovers.

Further Questions

Killdeers, along with a variety of other open-country birds, have been reported as nesting on gravel roofs, and roof-nesting should be watched for in our region. Resident Killdeers in one eastern population were reported to maintain pair bonds and site fidelity through the winter, dominating birds that migrated into their territories. We should watch for winter territoriality in this area and attempt to monitor what our local nesting pairs do after they have bred. More information on second broods would be of interest, in particular a comparison between coastal and interior breeding populations.

Notes

Statements in the literature that Killdeers can be sexed by plumage characteristics are not borne out by either museum specimens or studies that have involved looking at many pairs.

References

Baldassarre and Fischer 1984 (diet and foraging behavior in migration), Brunton 1988a (summer time-activity budgets), Brunton 1988b (reproductive effort), Brunton 1988c (polyandry), Deane 1944 (distraction displays), Demaree 1975 (roof nesting), Furniss 1933 (nesting), Kull 1978 (nest material), Lenington 1980 (mating system), Lenington and Mace 1975 (fidelity and site tenacity), Mace 1978 (breeding densities), Mundahl 1982 (territoriality and mating system), Nickell 1943 (nesting), Nol and Lambert 1984 (breeding biology), Phillips 1972 (breeding behavior), Pickwell 1925 (nesting).

MOUNTAIN PLOVER *Charadrius montanus*

The Mountain Plover has little to do with mountains, except as they loom in the distance over its plains habitats. A casual visitor to the region, this open-country distance runner can be distinguished by Killdeer size and essential lack of conspicuous markings.

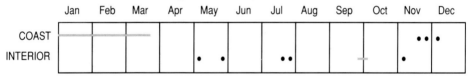

Distribution

The Mountain Plover occupies a restricted breeding distribution on the short-grass prairies of the North American Great Plains, wintering in a band of open grasslands from central California to southern Texas. It has decreased substantially because of habitat destruction in historic times.

Northwest Status

This species may have once bred in this region, as two young juveniles were collected in central Idaho in 1936, but there is no definite breeding record from west of the Continental Divide. As recently as 1973, breeding birds were reported from the Helena area of Montana, just east of our region. Otherwise the species is casual in the Northwest, with five records from the coast subregion (a winter visitor, November 19 to March 10) and six from the interior (a migrant, May 6 to 29 and July 19 to November 3). The Montana record involved a flock of 25 birds that frequented a field for a week; the identification was confirmed by examination of a specimen.

Records from Oregon and southern Idaho are more to be expected than from elsewhere in the region, given the interior breeding and winter ranges of the species. With much diminished numbers, it may be even less likely to be reported in the future, but there were four records in the most recent decade.

COAST FALL/WINTER. North Cove, WA, November 28, 1964; Corvallis, OR, January 2 to March 19, 1967 (2); Tillamook, OR, November 19-20, 1977; Siletz Bay, OR, February 3-26, 1983; 1 mi. N Tahkenitch Creek, OR, January 23, 1988; Bandon, OR, December 6, 1989 (2).

INTERIOR SPRING. 35 mi. N Arco, ID, May 29, 1977; Turnbull Refuge, WA, May 6, 1968.

INTERIOR FALL. Pahsimeroi Valley, ID, July 25, 1936#; Thorn Creek Reservoir, ID, July 19, 1979; Springfield, ID, November 3, 1984; Fortine, MT, September 27 to October 3, 1965 (25).

Habitat and Behavior

This is an open-country bird, breeding in short-grass prairies and wintering in low valleys with open soil or scattered grasses, now primarily open agricultural land. It is to be expected in sites

Juvenile Mountain Plover. Large upland *Charadrius* with plain face and long, grayish legs; scaly upperparts diagnostic of plumage. Pawnee Grasslands, CO, 9 August 1986, Wayne Petersen

similar to those inhabited by Killdeers and golden-plovers, often far from any wetlands. However, two November birds were found on ocean beaches foraging by themselves, and birds are occasionally seen at freshwater wetlands in California. Northwest interior birds have been in expected habitats, one with a flock of Killdeers in a field. Like many birds of the plains, it is a long-legged runner, and it is likely to move away on foot, whereas waterside plovers are more likely to stand their ground when approached and then fly away.

Structure

Size medium-small. Bill very short and fairly slender, legs short. Wings extend about to tail tip, primary projection moderate in adults and juveniles. Tail roughly square or with central pair of rectrices projecting slightly.

Plumage

Breeding. Bill black, legs grayish to pale pinkish. Upperparts sandy brown; underparts white, breast suffused with pinkish buff. Forehead white, extending back as short supercilium; band across forecrown and loral stripe black.

Rectrices brown, darker toward tips; outermost feathers slightly lighter and white-edged. Wing brown above, marked only by conspicuous wing spot produced by white outer webs at mid length of third to sixth primaries. Underwing entirely pale, coverts and axillars white.

Breeding plumage assumed by mid March.

Nonbreeding. Loral stripe and forecrown brown like crown in this plumage, pale supercilium slightly better defined. Tertials and coverts with rusty fringes that wear off later in fall. Molt into nonbreeding plumage begins to occur on breeding grounds, well along by August. Birds later in winter become duller and faded.

Juvenal. Like nonbreeding adult, but with pale buff fringes to all feathers of upperparts, presenting faintly scalloped effect as in juvenile Red Knot; these markings paler than rufous fringes of adults. Molts into adult plumage in first winter, may mature in one year (no summer records on wintering grounds).

Identification

All Plumages. About the bulk of a KILLDEER but longer-legged and shorter-tailed, this bird *lacks a breast band* in any plumage, and only in summer is its sandy color relieved by a few contrasty black head markings. Even from behind it is distinctively *unpatterned*, with no indication of the white collar present in our other plain-backed plovers (but see MONGOLIAN PLOVER). Nonbreeding-plumaged GOLDEN-PLOVERS and BLACK-BELLIED PLOVERS are speckled above and distinctly larger, and the very rare EURASIAN DOTTEREL is striped above in juvenal plumage. The plain-backed adult DOTTEREL is dark-bellied, with conspicuous white supercilium and pale crossband on the breast.

In Flight (Figure 11)

MOUNTAIN PLOVERS are *very plain* in flight, paler and with less conspicuous wing stripes than KILLDEERS or SEMIPALMATED PLOVERS. The wing pattern is unique, with *white primary patches*. The tail pattern is much like that of the SEMIPALMATED, a smaller species, but with less white around the edge and tip. The tail is very slightly paler than the back, so the *dark tail tip* stands out. The flight is usually low and rapid with *downcurved wings*, in contrast to the high, rapid flight of the SEMIPALMATED and the much more leisurely flight of the KILLDEER. Plain like an autumn GOLDEN-PLOVER, the MOUNTAIN is smaller, with more conspicuous wing pattern and white rather than gray underparts and underwings.

Voice

The flight call is a low, rolled whistle, almost a croak, but this species is generally less vocal than other plovers.

Photos

The "winter plumage" bird in Udvardy (1977: Pl. 233) is a juvenile; Farrand (1983: 329) shows another one.

References

Graul 1973b (mating system), Graul 1974 (vocalizations), Graul 1975 (breeding biology), Graul and Webster 1976 (breeding distribution), Johnson and Spicer 1981 (nesting), Knowles et al. 1982 (use of prairie dog towns), McCaffery et al. 1984 (behavioral ecology), Wallis and Wershler 1981 (status in Canada).

EURASIAN DOTTEREL *Charadrius morinellus*

Sandy in color and habitat, this upland species has been reported in fall on our coast and can be expected again. In any plumage it has a distinct long white supercilium, and the fine "smile" across its breast will certainly be duplicated by anyone with the good fortune to find one.

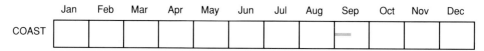

	Jan	Feb	Mar	Apr	May	Jun	Jul	Aug	Sep	Oct	Nov	Dec
COAST									▬			

Distribution

The Eurasian Dotterel breeds on arctic and alpine tundra all across northern Eurasia and winters in arid open country in north Africa and the Middle East.

Northwest Status

The two Northwest records of this species are from the mouth of Grays Harbor in early September. Four occurrences on the coast of central California from 6 to 20 September further indicate that month as the time to be on the lookout for this vagrant from Siberia. The Westport bird was an adult in advanced primary molt, the other five juveniles. This species has bred in far western Alaska but is casual anywhere south of there in this hemisphere, with most records in autumn. The normal migration route takes Siberian birds directly away from North America.

COAST FALL. Westport, WA, September 3, 1934#; Ocean Shores, WA, September 8, 1979.

Habitat and Behavior

This is an upland species, occurring on grasslands during migration; wintering birds in north Africa occur in habitats as arid as any occupied by shorebirds. The more recent Washington bird roosted on open, sparsely vegetated sand not far from the ocean beach. Not usually seen at water, it might associate with Killdeers and golden-plovers.

Shorebirds of grassy areas more often stand with neck extended and head held higher than similar-sized species of open habitats without vegetation, presumably for the augmented visibility. Thus photographs of dotterels often show them in this posture, which is characteristic, yet any plover can do this. Nevertheless, in grassland plovers such as Eurasian Dotterel, Mountain Plover, and golden-plovers, the line of the back makes a greater angle relative to the ground than in mud-flat species, raising the head slightly higher even without the neck extended.

Structure

Size medium-small. Bill very short and slightly more slender than in other plovers, legs short. Wings fall short of tail tip, primary projection short to nonexistent in adults and juveniles. Tail rounded.

Juvenile Eurasian Dotterel. Large, somewhat aberrant *Charadrius* with vivid supercilium and pale line across breast; heavily striped and scalloped upperparts and mottled breast diagnostic of plumage. Lapland, Sweden, early August 1984, Urban Olsson

Plumage

Breeding. Bill black, legs dull yellow to yellow-green. Breeding plumage vivid, with blackish crown and lores, sandy brown upperparts and upper breast, blackish belly, chestnut lower breast and sides, and white undertail coverts. Throat finely streaked with brown. Supercilium strikingly white, bordered above by black and long enough almost to meet its opposite on rear of head, where darkened to buff. White curved line across breast bordered in front by black. Scapulars and tertials narrowly to broadly fringed buff or white.

Rectrices brown, darker toward ends. Outer ones white-tipped, outermost pair white-edged as well. Wing brown above, paler gray-brown below, with no conspicuous markings other than vivid white shaft of outermost primary. Axillars light gray-brown.

Both sexes may be equally bright, but males, the incubating sex, often duller than females. Dullest males with crown finely brown-and-white streaked, less contrasty head and breast pattern and paler belly; lighter rufous and brown replace dark rufous and black of typical females.

Nonbreeding. Upperparts as in breeding adults but head and underparts duller and paler. Belly pale gray to whitish, thus paler than breast. Markings on head and underparts less distinct than in breeding adult, some faint streaking on neck and breast. Body molt begins on breeding grounds, half finished by early September. Primary molt begins before migration, some birds migrate with growing primaries.

Juvenal (Figure 21). More striped-looking than adult. Fringes on upperparts paler, usually interrupted at tips, producing more broken pattern. White crossband on breast less distinct than in adult and bordered by short streaks fore and aft, belly somewhat buffier.

Immature. First-winter birds recognizable by retention of some worn juvenile coverts and tertials in contrast with darker, complete margins of new adultlike feathers. First-summer birds mostly in worn nonbreeding plumage; most if not all migrate north in spring and presumably breed.

Identification

All Plumages. With typical plover shape and actions, this bird is likely to be encountered away from water in places frequented by KILLDEERS, GOLDEN-PLOVERS, and BUFF-BREASTED SANDPIPERS. Although superficially similar to the last two species and between them in size, it is immediately recognizable by its vivid *white supercilium.* From the front the *pale, slightly curved line on the lower breast* is apparent in any plumage.

Juvenal Plumage. This is the only plover with *strongly striped upperparts* in this plumage. Juvenile BLACK-BELLIED PLOVERS and GOLDEN-PLOVERS are more diagonally marked above with small spots.

In Flight (Figure 9)

This species has a rapid and straight flight, although, like the MOUNTAIN PLOVER, it is an excellent runner as well. In flight it is entirely *plain-backed and plain-winged* and could be mistaken for the similar and equally rare MOUNTAIN PLOVER, but the latter has distinct white primary patches. DOTTERELS are smaller than GOLDEN-PLOVERS, and the supercilium should be more distinct in flight, but these two species could be mistaken for one another. The DOTTEREL has a tail pattern much like that of the smaller plovers, with slightly darker tip and white edges, but the outer feathers are extensively white-tipped, so the *tail shows conspicuous white corners.*

Voice

The usual call is a soft trill or twittering, but the only bird I have encountered uttered a soft, repeated single note—*put, put*.

Notes

Contrary to Farrand (1983: 330), the juvenile's tail is virtually identical to that of the adult.

Photos

The "winter" bird in Hammond and Everett (1980: 111) is not only a juvenile but looks suspiciously like a museum specimen.

References

Byrkjedal 1987 (antipredator behavior), Kålås 1986 (incubation schedules), Kålås and Byrkjedal 1984 (mating system), Nethersole-Thompson 1973 (detailed biology), Nielsen 1975 (relationships to other plovers), Paulson 1979 (Washington record), Pulliainen 1970 (breeding biology).

OYSTERCATCHERS Haematopodidae

Of the 11 species of oystercatchers, there are two in North America (Table 19). One is a coastal resident in the Northwest, and the other has occurred once as a vagrant.

Oystercatchers are very large, short-legged shorebirds, all species surprisingly similar in structure. Some species are entirely dark ("black"), others extensively marked with white ("pied"). The great similarities among related species within both black and pied groups and polymorphism (both black and pied individuals) in at least one species still confuses classification. The knifelike red bill is specialized for bivalve-eating, although the birds take other marine invertebrate prey also. An oystercatcher well seen is not likely to be mistaken for any other bird.

This small family occurs throughout the world, especially along temperate coasts. In some regions both a black and a pied species occur. On the average, black oystercatchers occur on rocky shores, pied ones on sandy or muddy shores, but there is overlap. Breeding territories are linear, spread along appropriate coastlines, although a few pied species also nest along interior shores and on upland meadows. There are no territorial flight displays, but pairs and small groups engage in conspicuous, low pursuit flights. In most species individuals and family parties assemble into flocks after breeding, the flocks larger in the pied group.

References

Hockey 1983 (two-part bibliography on oystercatchers).

Table 19. Oystercatcher Distribution and Habitat Preference

Species	Hemisphere	Breeding Range	Breeding Habitat	Winter Range	Winter Habitat
American	W	temp./tropical	beach	N/equatorial	beach
Black	W	temperate	rocks	N	rocks

AMERICAN OYSTERCATCHER *Haematopus palliatus*

With its contrasty plumage and long, red bill, this unlikely visitor would be unmistakable should it ever reappear in the Northwest. Look for it well to the south and east on sand beaches in warmer climates.

	Jan	Feb	Mar	Apr	May	Jun	Jul	Aug	Sep	Oct	Nov	Dec
INTERIOR				•								

Distribution

The American Oystercatcher is resident on both coasts of North and South America from the Middle Atlantic states and northern Mexico south through the entire Neotropical region to southern South America.

Northwest Status

Unexpected in the Northwest, an individual of this species was observed in southwestern Idaho in April, from its description a first-year bird of the Mexican Pacific-coast population. The species is very rarely reported from the interior of the continent (with the closest record to the east from southern Ontario, to the south from southern New Mexico and the Salton Sea in southern California) and occurs on the Pacific coast only as a casual visitor to southern California. It seems highly unlikely to occur in the region again.

INTERIOR SPRING. Fruitland, ID, April 19, 1981.

Habitat and Behavior

This is a bird of sandy to muddy saltwater beaches and might be expected on large bodies of water on its rare visits to the interior. Its beach-bivalve-opening behavior is much like that of its dark relative, which would poorly prepare it for any sojourn away from the coast. The Idaho bird was seen at an alkaline marshy pond in association with Black-necked Stilts, and one can only wonder about its fate.

Structure

Size very large. Females slightly larger than males, bill-length difference averaging about six millimeters. Conversely, male bill slightly thicker than that of female; this difference in proportions may allow separation of sexes in pairs. Bill long and laterally compressed, legs short and thick. Wings relatively broader than in most shorebirds, about reaching tail tip, primary projection moderate. Tail square.

Plumage

Adult. Bill bright orange-red; iris yellow, with red eyelids forming narrow eye-ring; legs pale pink. Dark brown above and white below, with distinct black head and neck, appearing black and white at a distance. Uppertail coverts white; rec-trices brown with white bases, white more extensive on two outermost pairs. Wing dark brown above. Greater secondary coverts and bases of secondaries white, forming wide, conspicuous wing stripe on wing base, extended partway onto wing tip by white patches on outer webs of second to fourth or fifth primaries. From below axillars, wing coverts, and bases of flight feathers white, so underwing largely pale with darker rear border.

Juvenal. Basically similar to adult but duller and scalloped above with buff. Iris brown, eye-ring duller than in adult, bill dull orange with brown tip, legs gray.

Immature. Similar to juvenile, adult plumage acquired during first summer. Bill becoming brighter during first winter, not fully adult-colored until about one year old. Probably matures at two-three years.

Subspecies

Geographic variation in the species might make it possible to identify the origin of any future American Oystercatchers in the Northwest. Those from the Atlantic coast (*H. p. palliatus*) usually have an entirely white breast, the basal half of the tail white, a wide white band on the inner wing (entirely white greater coverts), and some white in the inner primaries. Those from Baja California (*H. p. frazari*) usually have some spotting on the lower breast, less than half the tail white (because of dark markings on upper tail coverts), a narrower band on the inner wing (greater coverts dark-based), and no white in the primaries.

Many birds observed in California have shown no indication of spotting on the lower breast, although they are presumed to come from the Pacific coast population.

Identification

All Plumages. The AMERICAN OYSTER-CATCHER is unmistakable, with its *large size, black-and-white pattern, and big red bill.* Even a partial-albino BLACK OYSTERCATCHER would not be patterned just like this species. Black-and-white AVOCETS and STILTS are very differently shaped, with slender, black bills.

In Flight

The *bold, white inner wing stripes* (with white

spots on the wing tip in Atlantic and Gulf coast birds) and *white tail base against an overall dark back* produce a very conspicuous flight pattern, and the *long, red bill* should be visible except in birds going directly away from the observer. Vaguely similar species include BLACK-NECKED STILT, AMERICAN AVOCET, HUDSONIAN GODWIT, and WILLET, each of which differs in several characteristics from this species.

Voice

The calls are loud, clear whistles much like those of BLACK OYSTERCATCHERS.

Photos

Farrand (1983: 333) shows an individual of *H. p. palliatus* in flight.

References

Cadman 1979 (territoriality), Jehl 1985 (relationship between American and Black oystercatchers in western Mexico), Kenyon 1949 (breeding behavior), Lauro and Burger 1989 (nest-site selection), Nol 1989 (food supply and reproduction), Stephens and Stephens 1987 (record from Idaho).

BLACK OYSTERCATCHER *Haematopus bachmani*

Foraging with heads down or roosting with bills tucked away, these blackish birds are easily overlooked on dark coastal rocks. Then, as one approaches, outrageous scarlet bills appear, and the birds take flight with their loud "wheep, wheep" ringing over the sounds of the surf. A pair on the wing leaves an indelible image, as if a couple of crows had decided to race over the waves, each carrying a firecracker.

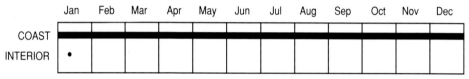

Distribution

The Black Oystercatcher is resident on rocky coasts from southern Alaska to northern Baja California.

Northwest Status

The linear range of this species extends from one end to the other of our outer coast, where it is fairly common on rocky shores. As this habitat is not ubiquitous, neither are the birds, which are normally absent from the Washington coast south of Point Grenville and from some of the sandy coastline of Oregon. Even in these regions birds turn up occasionally on jetties far from native rock, apparently wanderers rather than residents. The rocky coast bordering the straits of Georgia and Juan de Fuca and all the associated islands also support oystercatchers, south to Admiralty Inlet in Washington.

Oystercatchers are scarce to absent from the sandy and muddy mainland from Vancouver south, and there are few records south of Admiralty Inlet, where rocky shores are scattered.

They are essentially resident, although there is some movement of birds in autumn to more sheltered areas. Vancouver Island supports the largest populations, Washington fewer; 200 to 400 have been estimated as the total population in Oregon.

Black Oystercatcher breeding distribution

CHRISTMAS BIRD COUNTS	Five-Year Averages		
	74-78	79-83	84-88
Deep Bay, BC	10	17	6
Nanaimo, BC	19	21	14
Pender Islands, BC	4	13	28
Victoria, BC	36	44	37
Coos Bay, OR	15	24	17
Tillamook Bay, OR	15	15	11
Yaquina Bay, OR	21	12	31

The dispersal that produces occasional extra-limital records, for example at Seattle on January 3, 1981, and November 25, 1986, may be primarily by juveniles. However, the few records from the Ocean Shores jetty, some distance from the nearest resident population, have occurred in both spring and fall, perhaps indicating limited migration, and two birds were continually present at Mukilteo during 1985-86.

An amazing record of this coastal resident was provided by a bird "in distress" captured at 3,435 feet in the Washington Cascades in mid winter.

COAST. *High counts* 65 at Tofino Inlet, BC, May 31, 1931; 96 at Victoria, BC, November 7, 1962; 60 at Sidney, BC, October 16, 1965; 106 (nonbreeders) at Cleland Island, BC, July 17, 1970; 74 at Pulteney Point, BC, December 18, 1976; 70 at Pulteney Point, BC, March 12, 1977.

INTERIOR. Bumping Lake, WA, January 8, 1947.

Habitat and Behavior

Black Oystercatchers are birds of rocky shores, occurring less frequently on cobble beaches and occasionally appearing on rock jetties. Usually pairs or family groups are seen, but flocks may be seen at any time of year, apparently composed of nonbreeding immatures during the breeding season. They roost in loose groups near their feeding grounds, amazingly hard to see on the dark rocks. They forage in sandpiper mode, moving slowly and methodically over waterside, even wave-washed, rocks in search of invertebrates, primarily limpets and mussels. Watch with what finesse they open mussels and chip limpets from rocks with their combination knife and chisel.

The diagnostic calls can be heard from a long distance, a necessity to a bird in such a noisy habitat, and they are given throughout the year, probably to facilitate paired individuals keeping in touch with one another. The spectacular displays of pairs and small groups, both in flight and on the ground, make oystercatcher-watching especially rewarding in spring, and at that time the typical piping calls are often accelerated and multiplied for an incomparable auditory experience as well.

Nesting is largely, if not entirely, on islands, from small, rocky islets to considerably larger, forested ones. The nest is located on exposed beach substrates near the water, a slight depression without lining or with beach detritus, sometimes with considerable plant material. The clutch size is typically two or three, and both sexes incubate and care for the young. The Black Oystercatcher is one of two regional breeding shorebirds (Common Snipe the other) in which the young are fed by the adults from hatching nearly until fledging, and a half-grown young in hunched posture begging for food reminds one immediately of the same behavior in gulls.

Structure

Size very large. Females average slightly larger than males, and individuals in pairs might be sexed by female's slightly longer bill. Bill long and laterally compressed, legs short and thick. Wings relatively broader than in most shorebirds and fall short of tail tip, primary projection short. Tail square.

Plumage

Adult. Bill bright orange-red; iris yellow, with red eyelids forming narrow eye-ring; legs pale pink. In all plumages dark brown with black head and neck, looking black at a distance. From below flight feathers slightly paler than coverts.

Adult birds after autumn molt with narrow whitish tips to belly feathers that wear off during winter, scarcely changing their appearance; summer birds appear slightly lighter than winter ones.

Juvenal. Distinctively colored, with buff tips to mantle feathers, scapulars, coverts, and tertials. Bill dull orange with brown tip, iris brownish with indistinct eye-ring, legs dull brownish or yellowish. Bill distinctly smaller at fledging than that of adult, grows to adult size during first winter, by which time buff feather tips wear off and bill and iris become brighter.

Immature. First-winter bird may be slightly paler than juvenile, about color of summer adult, after midwinter molt of body feathers. Attains full adult coloration just after one year of age, after complete midsummer molt which is a month or two in advance of that of adults. In first spring and early summer distinguished from adult by duller, dark-tipped bill and more worn primaries. Probably does not breed before third summer.

Identification

All Plumages. Absolutely unmistakable, these *chunky, dark birds with bright-red bills* will be identifiable at great distance if seen at all well. Only CROWS, SCOTERS, and PELAGIC CORMO-RANTS perch on rocks, are black, and are about the same body bulk as an oystercatcher, but the four birds have such different shapes and habits that this confusion should not often arise. OYSTERCATCHERS roost when they are not feeding, and at that time they tuck their bills away; the pale legs, longer than those of the other black rock birds, are distinctive. CROWS are unlikely to roost where an OYSTERCATCHER does, CORMO-RANTS don't tuck their long necks away, and they and SCOTERS incline their bodies upward to perch because of their rearward-located legs.

In Flight (Figure 8)

This is a *large, all-dark* shorebird with a *long, red bill*. If the bill is seen there is no problem, but I once saw a CROW at the ocean carrying a red object, and my first thought was "OYSTER-CATCHER." The flight of an OYSTERCATCH-ER is distinctive, rapid and straight with *bowed wings*, whereas a CROW flies with flat wings and slower wingbeat, the wing-flexion difference comparable to that between a heron and an eagle. OYSTERCATCHERS typically fly low over the water, although they ascend to cross over low headlands.

Voice

The flight calls are loud, piping whistles— *wheeep, wheeep*—often given as the birds move about among feeding and roosting areas. Displaying birds, which may be seen at any time of year, seem demented as they run and fly about with long series of piping and rolling calls.

Photos

Farrand (1983: 335) shows a good comparison of adult and juvenile.

Further Questions

Careful observers could determine whether Black Oystercatchers exhibit the two different feeding modes of Eurasian Oystercatchers (Norton-Griffiths 1967). Our birds open mussels by "stabbing" them when the shell is partly opened or penetrating through the opening for the threads that attach them to the rock when that side is uppermost. Do any birds crush these bivalves by a blow of the bill ("hammering") as some Eurasian Oystercatchers do? The question is of interest, as individual Eurasian Oystercatchers usually use only one of these techniques, learned from the parents.

Notes

Contrary to the illustrations in Peterson (1990: 121), the short legs of oystercatchers do not project beyond the tail tip in flight.

References

Butler and Kirbyson 1979 (oyster predation), Drent et al. 1964 (breeding biology), Frank 1982 (effects of feeding on limpets), Groves 1984 (chick biology), Hartwick 1974 (breeding biology), Hartwick 1976 (foraging behavior), Hartwick 1978a (foraging behavior), Hartwick 1978b (foraging outside territory), Hartwick and Blaylock 1979 (winter ecology), Legg 1954 (nesting and feeding behavior), L'Hyver and Miller 1991 (nesting phenology and clutch size), Marsh 1986 (predation on mussels), Webster 1941a (breeding biology), Webster 1941b (feeding habits), Webster 1942 (growth and plumages).

AVOCETS AND STILTS Recurvirostridae

Like the oystercatchers, this is a small but world-wide family. There are nine species, although the list would fall to seven if all the black-and-white stilts were combined into one species as numerous authors suggest. Some regions have both an avocet and a stilt, although one or both types are missing from other regions, while Australia has three recurvirostrids. North America also has three species, of which only two are resident (Table 20); both occur in the Northwest.

All members of this family are medium to large shorebirds with long bills, small heads and long necks, and long legs. Their colors are contrasty black and white, augmented by reddish head and neck in some avocets. Seasonal plumage change is minor, best developed in the American Avocet. The family members are unmistakable, and there is no other bird similar enough to any of them to justify mentioning here.

Recurvirostrids are characteristic of shallow, marshy pond and lake edges and interior alkaline or coastal salt flats, where their long legs allow them to wade in water deeper than that frequented by most other shorebirds. The very long-legged stilts pick small invertebrate prey from the water's surface with their pair of fine, straight forceps, while avocets capture similar or larger prey (even fish) from below the surface by sweeping the recurved bill back and forth while moving rapidly through the water. Avocets swim frequently and effectively with their partially webbed feet.

Most species are common where they occur, breeding semicolonially. They lack aerial displays but make up for them with their fancy pre- and postcopulatory behavior, which can be readily observed in early spring in their breeding areas. Avocets assemble in large, sometimes huge, flocks after breeding, while stilts remain more scattered.

Table 20. Recurvirostrid Distribution and Habitat Preference

Species	Hemisphere	Breeding Range	Breeding Habitat	Winter Range	Winter Habitat
Black-winged Stilt	E	temp./tropical	marsh	equatorial	marsh/bay
Black-necked Stilt	W	temp./tropical	marsh	equatorial	marsh/bay
American Avocet	W	temperate	marsh	N	bay

BLACK-WINGED STILT *Himantopus himantopus*

The Black-winged Stilt breeds no nearer to this region than central China (rarely in Japan), and it winters from the Philippines south. There are only a few records from Siberia, so a single bird observed in the western Aleutians in spring 1983 was quite unexpected. It is unlikely to occur in the Northwest, but it might be kept in mind, especially when a stilt is seen in a peripheral area such as the north coast of the region.

The BLACK-WINGED is essentially identical to the BLACK-NECKED STILT in habitat preference and behavior and very similar in general appearance. Although slightly larger, this medium-sized shorebird could probably not be distinguished in the field from the BLACK-NECKED except by the color of its head and neck. In females this area is entirely white, while in males it is marked variably with dark gray to blackish above in breeding plumage and paler in nonbreeding plumage, never so sharply black as in the BLACK-NECKED. There is no white mark above the eye, and the dark area is not continuous with the black back as in the BLACK-NECKED. Juveniles should be distinguishable from juvenile BLACK-NECKED STILTS by the absence of the white supraocular spot. The Aleutian Island bird was apparently a first-summer individual, with dark-smudged head, a wash of dusky on the neck, and white-tipped secondaries. In flight, the mostly pale head and neck of the BLACK-WINGED is distinctive.

References

Espin et al. 1983 (age and foraging success), Goriup 1982 (behavior), Zeillemaker et al. 1985 (Aleutian Islands record).

BLACK-NECKED STILT *Himantopus mexicanus*

This striking wader trips about shallow interior marshes, looking exceptionally delicate with its very long, pointed wings, long neck, and very long legs. However, its strident calls belie its delicacy and, together with vivid black-and-white coloration and shocking pink legs, make it about as overt as a shorebird can be. Like a barking dog on a farm road, a stilt colony compromises the peacefulness of nature.

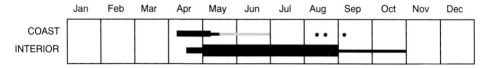

	Jan	Feb	Mar	Apr	May	Jun	Jul	Aug	Sep	Oct	Nov	Dec
COAST								• •	•			
INTERIOR												

Distribution

The Black-necked Stilt is a tropical marsh bird, distributed throughout most of South and Middle America and migrating into southern United States, farther north in the West than in the East. Breeding also occurs in coastal salt marshes, and many wintering birds move onto saltwater habitats.

Northwest Status

Breeding locally in the region's interior, the stilt's center of abundance is at the Oregon lakes and in the Snake River valley, primarily in the wetlands of the national wildlife refuges. At none of them is it abundant, but pairs are scattered widely and can be seen at any time from April to August, occasionally later. Daily counts at Malheur Refuge have varied from 15 to 105 birds in recent years, and 1987 counts at Summer Lake were the

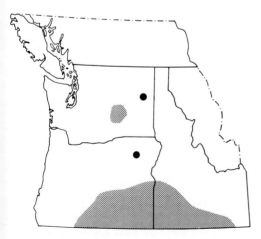

Black-necked Stilt breeding distribution

largest numbers anywhere in the region. Peripheral to these areas, stilts occur in much more scattered small colonies as far north as the vicinity of the Potholes Reservoir in central Washington, rarely to Reardan farther northeast. Drought in Nevada and California in 1977 caused stilts to increase in Oregon and move farther north into Washington, where they became established and have nested annually since then. Numbers continue to fluctuate from year to year and place to place but with an overall upward trend, and a pair spent the summer of 1988 at the Nisqually Refuge, in the coastal subregion, although there was no definite evidence of nesting.

Stragglers turn up rarely but regularly outside the breeding range, particularly in the Willamette Valley in spring. Virtual invasions occurred there in 1981; in 1984, when at least 130 birds were recorded in western Oregon from April 14 to May 4; and in 1985, when 70 were recorded in western Oregon from April 6 to May 3. During the invasion years these flights also carried a few birds to western Washington (each year) and the Vancouver area (1981), where stilts have been recorded only a few times.

This invasion pattern may have heralded a general increase in the region, as birds were common again in 1987, reported from numerous coastal and Willamette Valley locations from April 11 to May 5. More than ever before appeared in both Washington and British Columbia. Oddly, there is virtually no westward drift in fall, with only three coastal records from that season.

The coastal invasions parallel the increase of the species as a breeding bird in the interior, although some of them may be correlated with excessive flooding rather than drought at the Oregon refuges. In the interior a few birds have wandered as far north and east as southern British Columbia, northern Idaho, and western Montana, again with records only in spring and primarily in the invasion years. There seems little doubt that, at least at present, stilts are increasing in the region and will probably continue to extend their breeding range.

COAST SPRING. *Early dates* Medford, OR, April 6, 1985; Brownsville, OR, April 6, 1985. *High counts* 11 at Finley Refuge, OR, April 18-23, 1981; 40 at Fern Ridge Reservoir, OR, April 17, 1984; 38 at Medford, OR, April 17, 1984; 21 at Waldport, OR, April 21, 1985; 17 at Forest Grove, OR, April 16, 1987; 8 at Hoquiam, WA, April 21, 1985; 13 at Reifel Island, BC, April 19, 1987. *Late dates* Seattle, WA, May 12, 1988; Hansen Lagoon, BC, May 17, 1974.

COAST FALL. San Juan Island, WA, August 7, 1977; Woodburn, OR, August 17, 1979; 3 at Roseburg, OR, September 6, 1979.

INTERIOR SPRING. *Early date* Malheur Refuge, OR, March 20, 1984. *High counts* 12 at Stevensville, MT, spring 1979; 20 at Kootenai Refuge, ID, April 19, 1987; 263 at Summer Lake, OR, April 21-22, 1987; 45 at Dodson Road, WA, April 16, 1989*; 14 at Vaseux Lake, BC, April 27, 1987.

INTERIOR FALL. *High counts* 200 at Malheur Lake, OR, August 8, 1970; 563 at Summer Lake, OR, August 3, 1987. *Late date* Malheur Refuge, OR, October 26.

Habitat and Behavior

Stilts breed most commonly at marshy lakes and ponds, very often associated with avocets. They typically breed in shallow marshes that are likely to dry up in some years, and they are thus adapted to move about the countryside in search of new breeding sites. In fact they may move when breeding sites are either too dry *or* too wet for nesting, and both drying and flooding of Malheur Refuge in recent years has probably contributed to their vagrant tendencies. During migration they usually occur in habitats similar to those in which they breed, but they have also been seen on coastal mud flats. Only where they are common do stilts assemble in roosting flocks, and they spread out when feeding.

Stilts do most of their foraging in the water, their long legs allowing them access to deeper water than other shorebirds. Males feed in deeper water than females on the average. They feed much more delicately than do the dashing avocets and yellowlegs, striding gracefully through the water and picking tiny arthropods from the surface (often brine flies) or just below it (often brine shrimp) like a phalarope. Also unlike avocets, they rarely swim. The length of their legs can be appreciated when they are roosting on shore or as they flex at the ankle joint to reach the ground while feeding on shore as they often do. More than other shorebirds, they often rest flat on their tarsi on the ground. Like avocets they are very vocal on their breeding grounds, and their loud, yipping calls often mark their presence before they are seen as one approaches a marsh.

Stilts typically nest on small islands in their marshland, at times in surprisingly dense populations. They must depend on vigilance and aggressive antipredator displays rather than any use of camouflage, the more typical shorebird mode. The nest is a shallow depression with little lining or, especially in wetter spots, with a considerable amount of plant material built up. The clutch is typically four, but some nests, doubtless laid in by more than one female, contain up to eight eggs. The eggs are distinguished from those of avocets by smaller size (44x31 mm in stilt, 50x35 mm in avocet). Both sexes incubate and care for the young.

Structure

Size medium (looks larger because of long neck and legs), head relatively small. Male larger than female, difference greater than in avocet. Bill long, straight and very slender; legs very long, distinctly longer in male than female. Toes surprisingly short for such long legs, with small web between outer and middle ones. Wings long and pointed, projecting well beyond tail tip, with long, although scarcely visible, primary projection. Tail square.

Plumage

Adult. Bill black; iris red, although eye color apparent only at close range; legs bright pink. Glossy black above and white below with white forehead and spot over eye. Lower back and uppertail coverts white, tail light brownish gray. Wing black above and below, except for white basal lesser underwing coverts and axillars. Sexes easily distinguished in good light, male with entirely black upperparts, female with mantle and scapulars duller dark brown (faded in summer). Legs often brighter in breeding birds than in nonbreeding birds, but no seasonal plumage changes. Wing molt on breeding grounds.

Juvenal. Brown above with buff scallops all over back and wings, legs grayish pink. Inner primaries and secondaries white-tipped, white line along rear edge of wing conspicuous in flight. Apparently move out of region in this plumage.

Immature. Many juvenile coverts and tertials as well as distinctive flight feathers retained into first winter, easily distinguishable from adults. Molt into more adultlike plumage during first summer, still with heavily worn primaries. Most breed in second summer.

Identification

All Plumages. This is the only shorebird, and the only bird, of our region that is vividly *black above and white below, with long, bright pink legs.* Its long neck and legs suggest a bird much larger than it is, but it weighs only half as much as the AVOCETS with which it stands eye to eye. AVOCETS are much bulkier, with buffy or gray head and neck, much white on back and wings, and shorter, blue-gray legs. A GREATER YELLOWLEGS in poor light would look something like a STILT, but its legs and neck are much shorter relative to its body bulk.

In Flight (Figure 8)

Again, the *black-and-white coloration and very long, pink legs* are diagnostic. The rump and tail are white, and the white of the lower back extends forward between the black wings in a point, much as in a DOWITCHER. The *wings are black below*, dramatic against the white underparts. The flight is often slow and halting, with wing strokes almost ternlike in their deepness and angling toward the rear; the long legs extend far beyond the tail tip and often dangle, perhaps as a display. Wing beats are slower than in the AVO-

CET, unusual considering the STILT is a much smaller bird. Migrating flocks fly more swiftly, like other shorebirds.

Voice

This species is noisy on the breeding grounds, much less vocal in migration and winter. Birds call *kek, kek, kek, kek* loudly and persistently when disturbed, the individual notes rail-like in their volume and stridence. The calls of large juveniles on the breeding grounds are quite different, sounding virtually identical to the *peep* calls of Long-billed Dowitchers.

Further Questions

See under American Avocet for comments about "dump nests."

Notes

This account follows the AOU Check-list in recognizing several species of stilts, but Hayman et al. (1986) and numerous other authors combine Black-necked and Black-winged. The undersides of the wings are almost entirely black, contrary to Farrand (1983: 336).

Photos

A male and female are shown together in Hosking and Hale (1983: 156-57).

References

Burger 1980 (winter foraging behavior), Chapman et al. 1985 (British Columbia records), Hamilton 1975 (behavior), Rohwer et al. 1979 (early Northwest records), Sordahl 1982 (chick behavior), Sordahl 1984 (breeding site fidelity), Sordahl 1990 (antipredator behavior), Wetmore 1925 (food habits).

AMERICAN AVOCET *Recurvirostra americana*

In our interior alkaline marshes each summer, avocets strain crustacean soup with scythelike sweeps of their upcurved bills or noisily crisscross the airways as their nests are approached. Often breeding in colonies, and gathering in huge flocks in autumn, they are truly spectacular shorebirds. Their flashy pattern and ducklike flight distinguish them even at a distance, and at close range the beautiful orange hoods and blue legs add counterpoint to basic black and white.

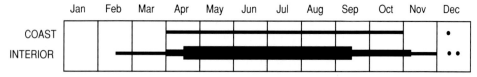

	Jan	Feb	Mar	Apr	May	Jun	Jul	Aug	Sep	Oct	Nov	Dec
COAST											•	
INTERIOR												• •

Distribution

The American Avocet breeds in alkaline marshes of western North America and winters on the coasts of southern United States and western Mexico.

Northwest Status

Much more wide-ranging than the Black-necked Stilt, this species breeds north in our region to central Washington, southern Idaho, and northwestern Montana, with former breeding records in northern Idaho. It has bred at least thrice in southern interior British Columbia, near Creston, Kelowna, and Kamloops (the latter two in 1987), and one to two pairs have bred since 1988 at the Serpentine Fen, the only such record in the coast subregion. The largest concentrations occur at the Oregon lakes, especially in the Klamath region, where thousands of birds breed. Tremen-

dous flocks have been reported in postbreeding concentrations in southern Oregon, but away from that area, a flock of over 50 would be unusual.

Spring migrants arrive by mid April in most parts of the region, occasionally considerably earlier. Most have departed from the interior by mid September, but later individuals are regularly seen. The latest record for Idaho is at Lewiston, November 10, 1953, yet thousands of birds were at Summer Lake on the same date in 1984, a very large number for so late in the season.

North of the breeding range, avocets are occasional spring (April and May) visitors to the interior of British Columbia. Elsewhere in the Northwest the species is a rare migrant, with stragglers west of the Cascades in southern British Columbia (north to Port Hardy and Vancouver), Washington, and Oregon. They are usually in very small numbers, but, unlike Black-necked Stilts, are about as likely to be seen in fall as spring. There have been no invasion years.

COAST SPRING. *Early date* Seattle, WA, March 31, 1988. *High counts* 10 at Auburn, WA, May 3, 1986; 39 at Sequim Bay, WA, April 28, 1987. *Late date* Lopez Island, WA, June 18, 1980.

COAST FALL. *Early date* Ridgefield Refuge, WA, July 2, 1976. *High count* 6 at Sauvie Island, OR, August 28-30, 1984. *Late dates* Fern Ridge Reservoir, OR, October 26, 1969; Coos Bay, OR, December 12, 1980 (2, not seen thereafter).

INTERIOR SPRING. *Early dates* Deer Flat Refuge, ID, February 15; Malheur Refuge, OR, March 11, 1980; Colville, WA, March 11, 1860. *High count* 4,277 at Summer Lake, OR, April 21-22, 1987.

INTERIOR FALL. *High counts* 2,600 at Ameri-

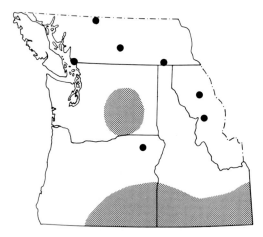

American Avocet breeding distribution

can Falls Reservoir, ID, August 18, 1987; 5,600 at Malheur Refuge, OR, September 1977; 10,000 at Harney Lake, OR, August 6, 1980; 50,000 at Summer Lake, OR, July 24, 1981; 2,350 at Summer Lake, OR, November 10, 1984; 16,000 at Summer Lake, OR, July 6, 1987. *Late date* Klamath Falls, OR, December 7-16, 1978 (2 in a frozen marsh).

Habitat and Behavior

Any shallow freshwater habitat in open country seems acceptable as avocet breeding habitat, whether the shore of a large lake or a small pond. Highly productive alkaline ponds and lakes are optimal, where there is a virtual soup of tiny crustaceans and insect larvae. Coastal migrants have occurred at both freshwater ponds and on mud flats.

Avocets are semicolonial breeders, often occurring in large numbers, and at their colonies are noisy and aggressive, divebombing visitors at high speed and producing considerably more apprehension than the equally noisy but nonaggressive stilts and Killdeers. Their *wheet* calls resound around breeding sites, but they are basically silent in migration, unlike many other shorebirds. Probably their vivid pattern and large flock size eliminate the necessity of a "flock with me" call.

Other than the much smaller phalaropes, avocets are the only shorebirds that swim regularly, feeding in deep water in this way or even resting on the water like a flock of ducks. Other large shorebirds swim only briefly to pass quickly through deeper water. Their submerged prey is probably detected both visually and tactilely, and, unlike stilts, avocets often feed in flocks. Such groups may drive prey in the manner of the pelicans that share their habitat. Avocet feeding style is distinctive, with lengthy scythelike sweeps of the upcurved bill as they move rapidly through the water, this feeding method duplicated at times only by yellowlegs and Wilson's Phalaropes and by the much larger spoonbills, which do not occur in our area. Unlike these other species, avocets do the same thing in soupy mud, and they also pick, probe, and chase larger organisms such as fish. They usually roost at or near their feeding areas, often in tight flocks.

Nests are often placed on small islands right out in the open, and incubating birds are conspicuous, as in stilts. The nest is a shallow depression with minimal lining. Four eggs is the typical clutch size, but some nests contain seven or eight, almost surely laid by two females. Both sexes incubate and care for the young and defend them very vigorously. It is interesting to compare the conspicuous distraction displays of the two recurvirostrids with each other and with the more familiar Killdeer.

Structure

Size large. Bill long and smoothly upcurved, somewhat flattened and abruptly broader at base. Sexes differ in bill shape more than any other bird of region. Male bills consistently longer and straighter than those of females (Figure 27), and this difference usually evident, especially when both sexes present for comparison. Legs long, toes fairly short with moderately developed basal webs. Wings extend beyond tail tip, with moderate to long primary projection. Tail square.

Plumage

Breeding. Bill black, legs bright blue-gray. Head and neck buffy orange except for white around bill base and eyes. Looks black and white at a distance, although colors include light gray and dark brown. Alternating bands of dark and light produced by whitish mantle; blackish brown upper scapulars; white lower scapulars; and blackish brown coverts, gray-brown tertials, and black primaries. Underparts white.

Lower back white, uppertail coverts and tail light gray. Wing blackish above except for white humerals and extensive white on rear edge of inner wing (innermost secondaries and tips of greater coverts white, becoming increasingly marked with black toward wing tip). From below axillars, most of underwing coverts, and secondaries (thus wing base) white. Primaries and outer primary coverts (thus wing tip) dark.

Dark colors of upperparts fade gradually from spring to autumn, looking more brownish and patchy, and tertials become paler and very worn. Wing molt on breeding grounds from July into September, possibly also on staging grounds; not known if any migrate before molt.

Nonbreeding. Head and neck gray to whitish. Molting takes place before spring migration, so

Fig. 27. Avocet sexual dimorphism. Female with slightly shorter and more curved bill than male.

birds arrive with orange heads. Body molt begins in August, migrants in September in full non-breeding plumage.

Juvenal. Head and neck partly to almost entirely pale pinkish orange, typically paler than in breeding adults but may look almost identical. Thought to represent mimicry of adults, perhaps so predators cannot home in on obviously young birds. Brown of upper scapulars, coverts, and tertials more uniform than that of very worn and patchy-looking adults, but adult and juvenile similar enough to warrant close checking. Young birds most easily recognized by fresh tertials and, at least at first, shorter bills. Orange replaced by gray during August, remaining longest on hindneck.

Immature. First-winter birds distinguished from adults by dark, worn tertials, retained from juvenal plumage; tertials of fresh-plumaged adults paler and grayer. Many one-year-old birds attain breeding plumage, might be distinguished on breeding grounds by primaries more worn than those of older birds. However, breeding specimens from Washington with very worn primaries do not show white-tipped inner primaries supposedly characteristic of one-year-old birds. Probably does not breed until second summer.

Identification

All Plumages. Whether in breeding or non-breeding plumage, the long neck, white underparts, striking *black-and-white pattern of upperparts, recurved bill, and long, bright blue-gray legs* stand out as a unique set of attributes. Occurring with STILTS, AVOCETS are readily identifiable at a distance because of their relatively shorter legs and heavier bodies and the *lack of black on head and neck.* Although vividly colored, they do not appear as contrasty as the more simply patterned STILTS.

Breeding and Juvenal Plumages. The *orange head and neck* in these plumages merely adds distinction to distinctiveness.

In Flight (Figures 8, 27)

From above, AVOCETS show a *largely white body with black scapular stripes* and *black wings with white basal patches fore and aft.* From below, the wings are *white-based and black-tipped.* The *long, upcurved bill* is evident at closer range. WILLETS are brownish to grayish and have a long, conspicuous white wing stripe bordered by black. No other big shorebirds in the Northwest have vivid wing patterns, except the very rare HUDSONIAN GODWIT and extremely unlikely

AMERICAN OYSTERCATCHER and BLACK-TAILED GODWIT. AVOCETS are much more powerful fliers than STILTS, almost ducklike.

Voice

AVOCETS are noisy "sentinel" shorebirds, with constant piercing *wheet* whistles as they fly about an intruder. Their calls are much higher-pitched than the comparable calls coming from STILTS in the area. Birds in migration are just about silent; the characteristic vocalizations are predator-alarm calls rather than flocking calls.

Further Questions

It will be worthwhile to look again and again at apparently mated pairs to test sex recognition by bill shape. Anyone becoming proficient at it should watch birds closely to see if males and females, with this unusual dimorphism, feed differently. Males take prey from lower in the water column than females, more often plunging their entire head and neck underwater, and their longer bills may facilitate this feeding mode; the differences in curvature remain unexplained.

Observers could watch juveniles closely in August and September to determine whether their heads become gray by fading (reddish color becoming paler) or molt (gray feathers interspersed with reddish), or perhaps both.

Avocets (and stilts) have the peculiar propensity in our area of two females laying in the same nest (a "dump nest"). Clutch size is typically four, but seven- and eight-egg clutches are not rare (five- and six-egg clutches are very unlikely). It would be of great interest to determine

how many birds and of what sex are associated with such a nest. In some gulls there is a surplus of females, and they pair with one another after being fertilized by males. If this were happening in avocets or stilts, perhaps (1) no male could be found, or (2) two females would be associated with a single nest. Unfortunately such nests rarely succeed, as the eggs cannot be uniformly incubated.

Notes

The flying avocet in the National Geographic Society (1987: 103) field guide shows too much white in the wing.

Photos

A good comparison between male and female avocets is in Johnsgard (1981: Pl. 3-4). Males are shown in Bull and Farrand (1977: Pl. 244), Udvardy (1977: Pl. 220), and Farrand (1983: 339 [breeding adult in water], 1988a: 155; 1988b: 140). Females are shown in Farrand (1983: 339 [in flight and winter adult]) and Hosking and Hale (1983: 156-57).

References

Baldassarre and Fischer 1984 (diet and foraging behavior in migration), Bucher 1978 (sex distinctions), Gibson 1971 (breeding biology), Gibson 1978 (breeding time budgets), Hamilton 1975 (behavior), Mahoney and Jehl 1985 (adaptations to salinity), Sordahl 1982 (chick behavior), Sordahl 1984 (breeding site fidelity), Sordahl 1990 (antipredator behavior), Wetmore 1925 (food habits).

SANDPIPERS Scolopacidae

This is a large family of shorebirds, comprising 88 species, that dominates northern shorebird faunas throughout the world. It is the largest family of shorebirds in the Northwest, with 34 species visiting the region on a regular basis and an additional 14 species recorded one or more times but not of annual occurrence.

Sandpipers span the entire Northwest shorebird size range, from Least Sandpiper to Far Eastern Curlew, 38 times as heavy. Most members of

the family move continually when feeding, searching for food visually or finding it by touch with their probing bills. The head is usually held up in longer-necked, visual foragers, which include all upland species, but it remains down in mud- and sand-probers except when they scan briefly for predators. Some of the visually foraging species are at times almost ploverlike in habits, standing for brief periods and then moving to capture prey or to change position. This is especially characteristic of upland species such as Upland and Buff-breasted sandpipers. As so many of them probe for food, sandpipers generally have longer, and usually much longer, bills than plovers. For the most part they are more complexly patterned, with marked seasonal changes in plumage. The juvenal plumage is typically distinct from that of either breeding or nonbreeding adults, in many species looking intermediate between the two adult plumages.

The typical sandpiper breeds in the Arctic or Subarctic and migrates at least to lower temperate-zone latitudes (for example the contiguous United States) for the winter. Many proceed farther, to the tropics or even well into the southern hemisphere. They are among the world's champion long-distance migrants, and they have been correspondingly successful at colonizing the highest northern latitudes.

Only four sandpipers are widespread breeders in the Northwest—the ubiquitous Spotted Sandpiper and Common Snipe and the drier-country Long-billed Curlew and Wilson's Phalarope. Four other species breed only locally in the region—Greater Yellowlegs, Solitary Sandpiper, Upland Sandpiper, and Willet.

A minority of species regularly winter in the region in large numbers. Dunlins and Sanderlings are the most abundant, Common Snipes are less common but widespread, and numbers of Greater Yellowlegs, Black Turnstones, Surfbirds, Long-billed Dowitchers, and Western, Least, and Rock sandpipers occur at least locally. A whole array of additional species have been observed less commonly in winter, especially toward the south on the Oregon coast but occasionally even north to British Columbia.

The majority of species of sandpipers are seen primarily during migration in the Northwest. Ruddy Turnstones and Red Knots are much more common in spring than fall, and other species—for example, Western Sandpipers and Short-billed Dowitchers—occur in higher concentrations during their relatively brief passage in spring than during their lengthy passage in fall. Many others are more likely to be seen in fall than spring, primarily because it is mostly the juveniles of these species that visit our region. This list includes Lesser Yellowlegs and Solitary, Semipalmated, Baird's, Pectoral, Sharp-tailed, Stilt, and Buff-breasted sandpipers, Ruff, and most of the very rare species. Thus fall migration is characterized by a considerably greater diversity of sandpipers than is spring, with August and September the best months for sandpiper searching and sorting.

Sandpiper Tribes

This family is large enough to warrant breaking it down into smaller, more manageable taxonomic groupings. It contains a series of tribes, each of which comprises a set of species with more-or-less similar anatomical modifications for particular ways of life.

TRINGINES Tringini

There are 18 species of tringines, of which 14 have been recorded from North America. All of the North American species are discussed here (Table 21), although only 8 have been observed in the Northwest.

The tringines as a group have always lacked a good common name, although they have been called "shanks." "Marshpiper" would be an appropriate name for the group, as most of the species in the group forage in water, and they certainly pipe, being among the noisiest of our shorebirds. However, the technical term "tringines" is common in the literature, and it will suffice.

Tringines are small to medium sandpipers (with the Willet just making it into the "large" category). The bill is relatively straight in most species (upcurved in Terek Sandpiper), varying from thin to thick and short to long. Most tringines forage by picking their prey from aquatic or terrestrial substrates rather than probing, although the Solitary Sandpiper and Willet probe as well as pick. The bills and especially the legs of the water-foraging species (all *Tringa*) are relatively long, while those that forage on land (*Actitis* sandpipers and tattlers) have shorter bills and legs as would be expected. Appendix 3 shows relative bill and leg length of tringines. The tribe divides well into a series of longer-legged (tarsus/wing 0.30-0.37) and longer-billed (bill/wing 0.28-0.35) species, and another of shorter-legged (tarsus/wing 0.19-0.24) and shorter-billed (bill/wing 0.22-0.24) species, with a distinct gap between them. The two conspicuous exceptions in the Northwest—the Lesser Yellowlegs, short-billed and long-legged, and the Terek Sandpiper, long-billed and short-legged—forage very differently from one another.

Tringines undergo moderate seasonal plumage changes, in some species subtle and only visible at closer ranges, with the Spotted Redshank a striking exception. Species of the group breed in subarctic and temperate latitudes for the most part, although a few are arctic breeders. Migratory movements range from relatively short (northerly wintering Spotted Sandpipers and Willet) to lengthy (tattlers and some *Tringa*). Most species change habitat between breeding and nonbreeding ranges. On the average these visual foragers occur in smaller flocks than most other sandpipers, spaced out over feeding grounds and quite aware of each other and of predators. Most members of the group are "sentinel" species, flying up with loud calls when potential predators appear. Some are highly territorial, with a typical flock size of one. The Willet is an exception in many ways, occurring in large, fairly dense flocks that may probe like dowitchers or godwits and not so vocal when flushed.

Table 21. Tringine Distribution and Habitat Preference

Species	Hemisphere	Breeding Range	Breeding Habitat	Winter Range	Winter Habitat
Common Greenshank	E	subarctic	tundra/taiga	equatorial/S	bay/marsh
Greater Yellowlegs	W	subarctic	taiga	N/equatorial/S	bay/marsh
Marsh Sandpiper	E	temperate	taiga/steppe	equatorial/S	marsh
Lesser Yellowlegs	W	subarctic	tundra/taiga	equatorial/S	bay/marsh
Spotted Redshank	E	arctic	tundra	equatorial	bay/marsh
Wood Sandpiper	E	subarctic	taiga	equatorial/S	marsh
Green Sandpiper	E	subarctic	taiga	equatorial	marsh
Solitary Sandpiper	W	subarctic	taiga	equatorial	marsh
Willet	W	temp./tropical	beach/marsh	N/equatorial	beach/bay
Wandering Tattler	W	subarctic	river	equatorial	rocks
Gray-tailed Tattler	E	subarctic	river	equatorial/S	beach/bay/rocks
Common Sandpiper	E	sub./temperate	lake/river	equatorial/S	bay/marsh/river
Spotted Sandpiper	W	sub./temperate	lake/river	N/equatorial	bay/marsh/river
Terek Sandpiper	E	subarctic	river/lake	equatorial/S	bay

COMMON GREENSHANK *Tringa nebularia*

The Common Greenshank has not been reported from our region, but the number of records from western Alaska—many more in spring than fall—indicate it at least as a possible vagrant from eastern Asia. It is a medium-sized species, virtually identical in size, shape, and habits to the GREATER YELLOWLEGS but with light green legs. Leg coloration is somewhat variable and can be altered by a coating of mud, so this field mark is most effective when calling attention to a bird that might otherwise be passed off as a YELLOW-LEGS.

Breeding-plumaged GREENSHANKS differ from YELLOWLEGS in being paler on the average, with finer streaking on head and neck and virtual lack of barring on the sides, but some dark GREENSHANKS are more like YELLOWLEGS, with dark, blotchy backs and darker, coarser markings all over; only paler individuals stand out as distinctly different. Nonbreeding adults are considerably paler than GREATER YELLOWLEGS, with an entirely white, unstreaked breast, a difference apparent at great distances. Juveniles also look strikingly different from juvenile YELLOWLEGS (Figure 29). They are slightly less heavily marked on the breast than are GREATER YELLOW-LEGS, and they are differently marked above, appearing striped rather than spotted because of pale edges on scapulars, coverts, and tertials (the latter are also dark-dotted). GREENSHANKS in all plumages have the basal part of the bill gray, the pale part often extending to midlength on the bill or farther, whereas in GREATER YELLOWLEGS the gray is usually confined to the basal third (occasionally more).

In flight a COMMON GREENSHANK stands out dramatically from a YELLOWLEGS, its flight pattern white-backed as in a flying DOWITCHER. Its loud calls are similar to those of a GREATER YEL-LOWLEGS but somewhat sharper-voiced and with two-noted as well as three-noted calls; they are all given on one pitch, while the call of a GREATER YELLOWLEGS descends slightly.

NORDMANN'S GREENSHANK should be briefly mentioned here, as it is migratory and could conceivably reach North America; its rarity as much as anything makes its occurrence very unlikely. In nonbreeding and juvenal plumages it looks much like a COMMON GREENSHANK but is obviously shorter-legged, with distinctly less of the tibiotarsus showing above the ankle. In breeding plumage the conspicuous black spots on the breast and sides are diagnostic.

References

Kieser and Kieser 1982 (identification), Nethersole-Thompson and Nethersole-Thompson 1979 (breeding biology), Tree 1979 (winter ecology).

GREATER YELLOWLEGS *Tringa melanoleuca*

This medium-sized shorebird with long, yellow legs is an attention-getter in all parts of the Northwest. It dashes through the shallows of ponds and flooded mud flats, stands at attention watching you approach, and flies away trailing white tail, bright legs, and piercing calls.

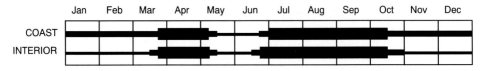

	Jan	Feb	Mar	Apr	May	Jun	Jul	Aug	Sep	Oct	Nov	Dec
COAST												
INTERIOR												

Distribution

The Greater Yellowlegs breeds in boreal-forest muskeg all across North America. It occurs continentwide in migration, and it winters in both fresh- and saltwater marshes from southern United States coasts south to the tip of South America.

Northwest Status

Nesting occurred during 1983-86 in the Downy

Lake area in northeast Oregon, far south of the nearest breeding localities in the Cariboo Parklands, just north of the region.

This species is a common migrant throughout the Northwest, most occurring from mid March to early May and again from late June well into October. Spring arrival occurs in coastal Oregon in mid March and in southern British Columbia in mid April. Rarely are more than a few dozen birds seen in a day at one location. Scattered birds remain through the summer on the coast, but most have departed by the second week of May. Overall numbers are similar in fall.

In the western lowlands, wintering is local although regular. Moderate numbers are reported on Christmas Bird Counts at Ladner, Victoria, Olympia, Leadbetter Point, and Coos Bay, but few or none are seen on many other counts in the intervening areas. Estuaries seem to be optimal wintering habitat for this species, yet it is absent from some of them. Five winter specimens from southwestern British Columbia and one from western Oregon are all first-year birds, but adults also winter in the region.

Migration dates are similar in the interior, early in both spring and fall. Numbers typically peak in the low dozens. Most have departed by early May, the peak period for many other shorebirds in the interior. Migrants are common from late June through October, most of them juveniles after July. A few birds have been seen throughout the winter in the interior, and even groups of up to a half-dozen appear willing and able to spend the winter in snowbound regions if open water, for example a stream mouth, is available. Two winter specimens from southeastern Washington are adults.

COAST SPRING. *High counts* 500 at New River, OR, April 27, 1984; 100+at Finley Refuge, OR, April 6, 1985; 100+near Newport, OR, April 24, 1985; 100 at Willapa Bay, WA, April 14, 1983; 100 at Iona Island, BC, April 12-15, 1983; 100 at Reifel Island, BC, April 16, 1986.

COAST FALL. *Juvenile early dates* Tacoma watershed, WA, July 22, 1961#; south jetty of Columbia River, OR, July 23, 1988. *High counts* 150 at Blackie Spit, BC, September 20, 1977; 175 at Reifel Island, BC, September 10, 1989; 190 at Ocean Shores, WA, September 13, 1980.

INTERIOR SPRING. *High counts* 175 at Lower Klamath Refuge, OR, March 27, 1982; 100 at Four Lakes, WA, April 14, 1957; 50+at Salmon Arm, BC, April 19, 1971. *Late dates* Malheur Refuge, OR, June 2, 1970; Osprey Lake, BC, May 21, 1967.

INTERIOR FALL. *Early dates* White Lake, BC, June 11, 1970. *Juvenile early dates* Okanagan Landing, BC, July 11, 1911; Willow Lake, WA, July 20, 1950#. *High counts* 25 at Creston, BC, September 22, 1948; 25 at Reardan, WA, July 16, 1966; 24 at Ninepipe Refuge, MT, September 6-8, 1977*.

| | Five-Year Averages | | |
CHRISTMAS BIRD COUNTS	74-78	79-83	84-88
Deep Bay, BC	10	10	5
Victoria, BC	49	38	19
Grays Harbor, WA	18	8	23
Leadbetter Point, WA	9	19	23
Coos Bay, OR	21	38	86
Eugene, OR	4	8	16
Yaquina Bay, OR	5	5	4

Habitat and Behavior

Greater Yellowlegs are as ubiquitous as any shorebird of the region, occurring in small numbers at almost any freshwater habitat, including rivers with moderate current. They are common at small ponds but appear to prefer extensive lake margins and mud flats. They spread out to feed, and single birds or groups of a few are usually seen, but high-tide roost sites, often at freshwater bodies rather than coastal islands, typically attract dozens to one place.

This species usually feeds in the water, commonly to several inches deep, by fairly rapid forward progression, picking at animals on or below the surface. Individuals often dash through the shallows chasing fish or other active aquatic animals, bringing to mind miniature Reddish Egrets. This characteristic behavior is occasionally seen in Lesser Yellowlegs but not in any other local shorebird, and the Greater shares the Lesser's avocetlike scything behavior as well.

Breeding occurs in coniferous-forest wetlands in the boreal-forest zone. The rising and falling display flight is spectacular, coupled with loud *whee-oodle, whee-oodle* vocalizations. The nest, with a clutch of four eggs, is well hidden in open

breeding

nonbreeding

Fig. 28. Breeding and nonbreeding Greater Yellowlegs. Breeding-plumaged adults more heavily patterned both above and below than same birds in winter.

sedges and grasses or among low shrubs in meadows or open woodlands, often near water and often partially screened by a log or vegetation. Probably both sexes incubate and care for the young, although this is poorly known.

Structure (Figure 30)

Size medium. Bill long and usually slightly upturned, legs long. Wings extend to or a little beyond tail tip, primary projection long in adults and juveniles. Tail squarish, with two central rectrices projecting slightly (more in some individuals than others).

Plumage

Breeding (Figure 28). Bill black, often with gray at base; legs bright yellow, varying to yellow-orange in a few individuals. Upperparts brownish gray, variably marked depending on how many feathers have been replaced. In fully developed breeding plumage, most head and neck feathers blackish with white edges, looking heavily streaked. Fairly conspicuous white eyering. Mantle feathers black with white fringes; scapulars, tertials, and coverts vary from black to gray-brown, tipped and notched with white. Most heavily marked birds quite dark above, heavily spotted with white. Individuals commonly retain a few worn and faded tertials and most coverts from the autumn before. Underparts white, heavily marked with black chevrons on breast and black bars on sides. Only throat and lower belly unmarked in most heavily patterned individuals.

Lower back blackish, feathers fringed with pale gray. Uppertail coverts white, tips narrowly barred with dark brown. Rectrices narrowly brown-and-white barred, light bars often tinged with light gray-brown. Wing plain brown above, fine markings not visible at a distance. From below secondaries barred and inner primaries mottled with light gray. Underwing coverts grayish white, densely and finely barred with gray-brown to brown. Axillars white, sparsely barred with brown.

Breeding plumage attained by early April. In

GREATER YELLOWLEGS

COMMON GREENSHANK

SPOTTED REDSHANK

LESSER YELLOWLEGS

WOOD SANDPIPER

SOLITARY SANDPIPER

Fig. 29. Juvenile *Tringa*. Juvenile *Tringa* fall into large and small size classes, with Lesser Yellowlegs somewhat larger than other two small species. Spotted Redshank (molting into first winter) distinctive with slender, slightly drooped bill and pale gray mantle as well as red legs. Greenshank distinguished from Greater Yellowlegs by striped rather than dotted upperparts and generally paler appearance. Lesser Yellowlegs from Greater by smaller size and small, straight bill. Wood Sandpiper from Lesser Yellowlegs by smaller size, shorter wings and legs, and more conspicuous supercilium. Solitary Sandpiper from Lesser Yellowlegs and Wood Sandpiper by darker color, short dark legs, and eye-ring. All species drawn with fully extended necks; Wood and Solitary usually look distinctly shorter-necked.

Fig. 30. Willet and Greater Yellowlegs shapes. Same length bill and legs on one-third larger bird makes Willet look distinctly heavier-bodied than Greater Yellowlegs.

returning fall migrants, white feather edges worn off; upperparts may look entirely blackish, with scattered small white spots, or blotched black and gray-brown. Some birds in this plumage at least through mid August.

Nonbreeding (Figure 28). Bill base with more gray than in breeding plumage. Compared with breeding plumage, breast finely and more obviously streaked and ventral markings much less extensive, with sides scarcely barred. White eye-ring and dark loral stripe well defined. Upperparts brownish gray, scapulars and tertials with black dots and narrow white fringes. Most adults that winter locally have attained full nonbreeding plumage by end of August. Wing molt in August and September, including some birds in areas where they do not winter (thus molting during migration).

Juvenal (Figure 29). Bill base most extensively gray of any plumage. Superficially like nonbreeding adult but more brightly marked, white- instead of black-dotted above. Upperparts gray-brown to brown, profusely spotted with small white to pale-buff dots on edges of mantle feathers, scapulars, coverts, and tertials. Breast crisply streaked with brown. Molt into first-winter plumage begins by early September.

Immature. First-winter birds distinguished from adults by jagged appearance of feathers, especially tertials, where juvenile dots wearing off; distinctive even into spring. First-summer birds vary considerably in extent of plumage change, some retaining most of nonbreeding plumage.

Identification

All Plumages. The *fairly long bill, long neck, and long, bright yellow legs* allow quick identification as a yellowlegs, and the only persistent problem is distinguishing the two species of yellowlegs, which is discussed under the LESSER YELLOWLEGS. Other shorebirds that could be confused with this one include the very rare SPOTTED REDSHANK and another Asiatic species, the COMMON GREENSHANK, which has not been recorded from our region but might occur.

No other shorebird of about this size has long, bright yellow legs, and other sandpipers that are gray to grayish brown above and white below with greenish yellow legs in nonbreeding or juvenal plumage—STILT SANDPIPER, WILSON'S PHALAROPE, DOWITCHERS, KNOTS, and TAT-

TLERS—are considerably smaller than GREATER YELLOWLEGS, with relatively shorter neck and legs.

In Flight (Figure 10)

Yellowlegs are *plain-winged and white-tailed* in flight, the white rump and tail contrasting strongly with the back. The *long, yellow legs* are usually plainly visible. As in most "white-tailed" shorebirds, the tail is actually strongly barred and thus darker than the rump, but this difference is obscure at a distance. The two species of yellowlegs appear identical in flight, but see under LESSER YELLOWLEGS for differences. See COMMON GREENSHANK and SPOTTED REDSHANK, possibly occurring species that show a white back in flight.

Voice

The flight call of this species is a loud series of descending notes, *whew whew whew*, or a higher-pitched *klee klee klee*, given usually in threes or fours. This striking call is often given from high in the air, perhaps to stimulate a response from other yellowlegs. Both LESSER YELLOWLEGS and SHORT-BILLED DOWITCHER have similarly repeated calls, but in these species they are shorter, faster, not descending, lower-pitched, and less musical—usually of two syllables in the LESSER YELLOWLEGS and two or three in the DOWITCHER. WHIMBRELS give repetitive whistled notes, but each is shorter and lower-pitched, and there are usually five or more of them in the series.

Both YELLOWLEGS, but especially GREATER,

can be heard giving their breeding-ground song in migration in the Northwest.

Further Questions

We are still not sure in what proportions individuals of both age classes winter in this region. Careful observation of plumage features on wintering populations after September would answer these questions; presence or absence of primary molt would be the easiest criterion. Also, additional breeding populations should be sought between the Cariboo Parklands and the Oregon breeding locality. An indefatigable spirit would be needed to search for nests—very difficult to find—in the insect-ridden muskegs, but persistently alarm-calling or tree-perching birds in late May or June would presumably be breeding.

Notes

Several identification guides have stressed the two-toned bill of the Greater Yellowlegs. Bear in mind that breeding individuals have entirely black bills, just as in the Lesser. The yellowlegs in Peterson (1990: 135) may be misleading, as each bird is an amalgam of plumages.

Photos

The "winter plumage" individual in Farrand (1983: 345) is in first-winter plumage, as indicated by its white-dotted tertials.

References

Brooks 1967 (prey choice in migration), Buchanan 1988b (Northwest seasonality and movements), Wilds 1982 (identification).

LESSER YELLOWLEGS *Tringa flavipes*

Reduced in both size and voice, this species is a pint to the Greater Yellowlegs' quart. It shows all of the larger species' admirable qualities: sleek and elegant posture, active foraging behavior, and unending alertness. When by itself its smaller size may be difficult to ascertain, but its relatively smaller and more slender bill and terse two-noted call are distinctive. Being a typical noisy tringine, it calls frequently enough to alleviate identification anxiety, and, having learned what it is, you can enjoy watching it.

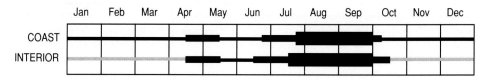

Distribution

The Lesser Yellowlegs breeds farther north in Canada's and Alaska's boreal forests than does the Greater, even above the stunted forests on the more southerly tundra. It then migrates farther south in winter, spending that season from the southernmost United States coasts south throughout the rest of the hemisphere.

Northwest Status

Less in evidence than its larger relative, this yellowlegs is an uncommon spring and common fall migrant in the region. Only small numbers are seen anywhere in the region in spring, primarily from mid April to mid May; the 200+at Iona Island in April were quite unusual. Individual birds occasionally remain through June. Small numbers of adults again pass through the region during July, followed by juveniles from late July to early October.

Lesser Yellowlegs are rarely seen after early October, and in fact very few winter even as far north as California. Nevertheless, there are occasional midwinter reports, usually of single birds and often with Greaters, from coastal British Columbia, Washington, and Oregon, with apparently regular occurrence at Coos Bay. Surprisingly, a few spent at least parts of three winters at Klamath Falls, in the interior, between 1978 and 1987.

COAST SPRING. *High counts* 11 at Kent, WA, April 11, 1976; 200+at Iona Island, BC, April 6, 1976; 25 at Iona Island, BC, April 24, 1981.

COAST FALL. *Juvenile early date* Comox, BC, July 18, 1931#. *High counts* 383 at Iona Island, BC, August 18, 1976; 543 at Iona Island, BC, August 20, 1977; 65 at Ocean Shores, WA, September 11, 1975; 96 at Nehalem, OR, August 23, 1980; 85 at Tillamook, OR, August 18, 1985.

INTERIOR SPRING. *Early dates* Fortine, MT, April 12; Malheur Refuge, OR, March 29, 1972; Potholes, WA, April 10, 1954. *High counts* 7 at Lee Metcalf Refuge, MT, April 29, 1981*; 30 near Acequia, ID, April 30, 1963. *Late date* Lewiston, ID, June 4, 1954.

INTERIOR FALL. *Early dates* Potholes, WA, June 11 (thought to represent fall arrival); Malheur Refuge, OR, June 22, 1940; Camas Refuge, ID, July 2, 1961. *Juvenile early date* Okanagan Landing, BC, July 28, 1915#. *High* *counts* 100+at Okanagan Landing, BC, September 2, 1933; 150+at Okanagan Landing, BC, August 29, 1935; 200+at Salmon Arm, BC, August 25, 1977; 150 at Reardan, WA, August 17, 1985; 38 at Summer Lake, OR, August 15, 1986; 60 at Pablo Refuge, MT, August 21, 1979*. *Late dates* southeastern WA, November 4; Malheur Refuge, OR, November 7, 1987.

Habitat and Behavior

The Lesser Yellowlegs is found in the same habitats as the Greater, both freshwater ponds and lake shores and coastal mud flats, and the two species occur together regularly. However, there are some average differences in habitat preference, with separate concentrations of each species in a given area. The Lesser typically occurs in more protected situations, a bit more common at small ponds and less so on extensive mud flats. Frequently it exceeds the Greater in numbers, at least during the juvenile migration. The Lesser occurs in tighter, often larger, flocks than the Greater, both while feeding and in flight.

Lesser Yellowlegs typically feed in an inch or two of water, picking in tringine fashion just as their larger relative. They also move forward rapidly while scything the bill back and forth like a little avocet. Greater Yellowlegs at times feed in the same fashion, but I have seen Lessers do so more often. Wilson's Phalarope completes the list of the three sandpipers that feed in this manner in our region.

Structure

Size medium-small. Bill medium (to long) and straight, legs long. Wings extend a little beyond tail tip, primary projection long in adults and juveniles. Tail squarish, two central rectrices projecting slightly.

Plumage

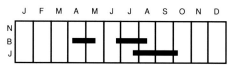

Breeding. Bill black; legs usually bright yellow, occasionally yellow-orange. Adults in spring crisply marked with black, brownish gray, and white. Dark markings less intense and less extensive than in Greater but still distinctive of this plumage. Upperparts basically brownish gray; scapulars, coverts, and tertials variably barred with black and sometimes with whitish between black bars (white averages less prominent than in Greater). Head finely streaked black; anterior supercilium, narrow eye-ring, and throat white. Underparts white, breast suffused with gray and heavily streaked with black. Streaks on sides of breast intermixed with bars, sides sparsely barred.

Lower back blackish, feathers fringed with pale gray. Uppertail coverts white, tips narrowly barred with dark brown. Rectrices narrowly barred with brown and white, light bars often tinged light gray-brown. Wing plain brown above, fine markings not visible at a distance. Underwing coverts grayish white, finely and densely barred with gray-brown to brown. Axillars white, sparsely barred with brown.

In fall upperparts dull and dark, as pale tips have worn off, but as season progresses, new, plain nonbreeding feathers begin to prevail. Most adults pass through Northwest while still largely in breeding plumage.

Nonbreeding. A plain plumage, with upperparts entirely brownish gray, underparts white, and breast finely streaked with brown. Scapulars and tertials obscurely dotted with dusky brown and faintly fringed with white, never so distinctly marked as in breeding adults and juveniles. Distinct white eye-ring and anterior supercilium, dark loral stripe. Birds in this plumage unlikely to be seen in Northwest, other than exceptional wintering individuals.

Juvenal (Figure 29). Heavily spotted above with whitish dots on edges of mantle feathers, scapulars, coverts, and tertials; much more regularly patterned than mottled fall adults. Breast grayish brown, finely and obscurely streaked with brown, markings more obscure than in adults; posterior underparts white. Distinct white eye-ring and anterior supercilium, dark loral stripe. By late September many juveniles well along in molt from juvenal to first-winter plumage.

Immature. First-winter birds distinguished from nonbreeding adults by clearly white-dotted tertials.

Identification

All Plumages. With *slender bill, long neck, and long, bright yellow legs*, this species is immediately distinguishable as a yellowlegs. Beginners can usually quickly learn to recognize birds as yellowlegs, but distinguishing between the two species when they are seen in isolation is a persistent challenge, at times even to experienced observers. Although LESSER is slightly different in each plumage than GREATER, the similarities far outweigh the differences (Table 22).

• This species is *considerably smaller* than the GREATER YELLOWLEGS, about half the weight of that species (therefore much smaller in bulk) and standing about two inches shorter. With experience the size difference will become more obvious, even in lone birds, but determining it is much easier when there are other shorebirds around for comparison. If a KILLDEER is in the area, it is a fixed size scale for yellowlegs identification. LESSERS are very slightly smaller than KILLDEERS in body bulk, although they look larger because of their long legs and neck; GREATERS are much larger than KILLDEERS. DOWITCHERS are also good scales: GREATERS are considerably bigger, LESSERS slightly smaller in weight.

• Proportioned virtually the same, the two species differ mostly in relative bill size. In LESSER the *bill is about the same length as the head*, often very slightly longer, while GREATER'S bill is distinctly (15 percent or more) longer than the head length. Another comparison can be made with the tarsus: in LESSER the bill looks about two-thirds to three-fourths the length of the tarsus, in GREATER only slightly shorter (see how much of their time yellowlegs spend in the water!).

• LESSER'S *bill is slender and straight*, while that of GREATER is distinctly thicker at the base and, in many individuals, appears slightly upcurved.

• LESSER'S legs are slightly longer relative to its body size than are GREATER'S, but this is not a striking field mark. Perhaps the shorter bill of the LESSER makes its legs look even longer.

• A slight structural difference is the "knobby-kneed" look of GREATER, its ankle joint more prominent than that of LESSER.

• Finally, LESSER has relatively longer wings than GREATER, the primaries typically extending

Juvenile Greater and Lesser Yellowlegs. Yellowlegs are wading shorebirds with long, yellow legs and medium-length bills; pale dots on upperparts diagnostic of plumage; Greater is larger, with relatively larger gray-based bill and more contrasty breast streaks. Juneau, AK, August 1977, Robert Armstrong

one to two centimeters beyond the tail tip; in GREATER the primaries typically extend just beyond the tail tip. This is a good clue, although not definitive, as there is overlap and the proportion varies with how the wings are held.

The WOOD SANDPIPER is very similar to this species and should be considered if a LESSER YELLOWLEGS looks small or short-legged or sounds wrong when flushed.

Breeding Plumage. Breeding adult LESSER are *less heavily marked beneath* than GREATER, with little or no barring on the sides, and tend to be less heavily marked above.

Nonbreeding Plumage. Nonbreeding adults are slightly *plainer above* than are GREATERS, especially on the crown and mantle. In LESSER the *bill is entirely black*, in GREATER broadly gray at the base.

Juvenal Plumage. In juvenile LESSER the *breast is more obscurely marked*, the stripes relatively broad and indistinct gray. In GREATER the stripes are narrow, sharply defined, and darker than in LESSER, and this difference extends to the neck as well. This characteristic is useful, but there is enough variation in both species so that caution is recommended. The bill color in juveniles of the two species differs in the same way as nonbreeding adults.

Two birds that are not closely related to the LESSER YELLOWLEGS nor to each other but are nevertheless potentially mistaken for it in the field are STILT SANDPIPER and WILSON'S PHALAROPE (juveniles and nonbreeding adults of both species). The STILT SANDPIPER is most similar in shape, with long legs for a member of the *Calidris* group, and from a distance it looks about

Table 22. Yellowlegs Identification

	Greater	Lesser
All Plumages		
Size	170 g (much larger than Killdeer, dowitcher)	76 g (slightly smaller than Killdeer, dowitcher)
Bill length*	>> head length	± head length
Bill shape	slightly upturned, heavy base	straight, slender
Wing projection	shorter	longer
Flight call*	3-4 noted, rich	2-noted, sharp
Season	common spring, common fall local winter	uncommon spring, common fall
Breeding		
Sides	heavily marked with black	lightly marked with black
Juvenal		
Bill color	gray-based	all black
Breast	fine black streaks	blurred gray streaks
Nonbreeding		
Bill color	gray-based	all black
Upperparts	more patterned	plainer

*most diagnostic characters

like a yellowlegs, with gray-brown back and breast and white belly. Its legs are greenish yellow rather than bright yellow, its bill is distinctly drooped at the tip, and its upperparts are either plain gray-brown (adult) or brown fringed with buff and white (juvenile), quite different from the dotted effect of a breeding or juvenile YELLOWLEGS. In addition, the STILT SANDPIPER normally feeds like a DUNLIN or DOWITCHER, with probing motions entirely unlike the surface-picking of a YELLOWLEGS.

The WILSON'S PHALAROPE has considerably shorter legs than the YELLOWLEGS and is virtually unmarked above and below (juvenile obscurely scalloped above), its legs vary from yellow to olive, and its bill is even more slender and needlelike. It forages by swimming or by running erratically along the shoreline, with typical jerky phalarope movements, although both species feed by scything at times.

In Flight (Figure 10)

As in its larger relative, this is a *plain-winged and white-tailed* species. At close range, the shorter, more slender bill may be seen in flight, allowing distinction from the GREATER. Very slightly more of the LESSER'S foot projects beyond the tail in flight, including the end of the tarsus, while in the GREATER the tarsus does not show. Thus in LESSER the *bill and projecting feet are about the same length*, while in GREATER the bill is considerably longer than the foot projection. The underwing pattern may be a useful field mark: in LESSER the coverts are paler than the flight feathers, while this is not the case in GREATER.

The STILT SANDPIPER and WILSON'S PHALAROPE, of similar size, are again similar in flight to the LESSER YELLOWLEGS, with plain wings and white tail. In breeding plumage the STILT is dark below and the PHALAROPE has a brightly patterned head and neck, but in nonbreeding and juvenal plumages they look much more like YELLOWLEGS. The STILT SANDPIPER, much rarer in the region, is very similar in flight, with foot projection as in the LESSER. Its droopy-tipped bill and usually duller yellow legs will have to be observed well to distinguish the two species in fall and winter. The STILT also shows an indistinct wing stripe at close range. The PHALAROPE has much shorter legs, only the tips of the toes projecting beyond the tail tip, and in fall its over-all coloration is paler, gray above and entirely white below.

Voice

The best flight characteristic is the usually two-noted (but regularly three-noted) flight call, sharper, more rapid, and less musical (*tu tu*) than the three- or four-noted GREATER YELLOWLEGS call. Once learned, the calls distinguish these two species readily, and the listener may be more puzzled by unseen SHORT-BILLED DOWITCHERS, the calls of which sound more like those of LESSER YELLOWLEGS. The DOWITCHER call is even quicker and sharper than that of the YEL-LOWLEGS but slightly more musical. Both WILSON'S PHALAROPE and STILT SANDPIPER are quiet in flight compared with a YELLOWLEGS and have muted, single-noted calls.

Both YELLOWLEGS typically give single, sharp, whistled calls at intervals as they feed, probably low-level alarm calls brought about by a human intruder. That of the LESSER is a bit flatter and less musical, just as in the flight calls. These calls may accelerate on closer approach, underscoring the point that the closer we get to an animal to appreciate it the more we influence its behavior!

Further Questions

As yet we have no information about what age stage winters in the region. Given the general rarity of adults on the coast, first-year birds seem likely, but this should be confirmed by close study of the tertials of wintering individuals.

Some enterprising field observer might take on an interesting project with the two yellowlegs species—to record, every time a yellowlegs calls, how many notes are in the series. This way we could actually quantify the differences between the two species, instead of statements like "sometimes gives three or four notes" or "occa-sionally calls like the other species." This may be one of the directions taken, for better or worse, by computer-age bird identification.

Notes

I consider inadequately documented the follow-ing published winter records from the interior: Lowden, January 18 (12 birds) and February 6, 1975; Asotin, January 22, 1978 (4 birds); and Irrigon, January 12, 1978.

The Lesser Yellowlegs' legs are relatively longer than those of the Greater, contrary to Wilds (1982: 172). Wilds surely made a slip of the pen, as she correctly stated the difference in Farrand (1983: 346).

Photos

Keith and Gooders (1980: Pls. 137-39) furnish a fine comparison, with a juvenile Lesser Yellow-legs between juveniles of Greater Yellowlegs and Wood Sandpiper. The juvenile in Farrand (1983: 347) is not typical, perhaps because of wear. The bird in "adult winter plumage" in Pringle (1987: 227) is a juvenile. The "winter plumage" indi-vidual in Farrand (1988a: 163) is in first-winter plumage, as indicated by its white-dotted tertials. The "Lesser Yellowlegs" in breeding plumage in Farrand (1988a: 163) is a Greater from its long bill and heavily marked sides; compare it with the more lightly marked Greater on the opposite page. The yellowlegs shown with Black-necked Stilts in Hosking and Hale (1983: 44) are Lesser, not Greater as indicated.

References

Baker 1977 (summer diet), Baker and Baker 1973 (summer and winter foraging behavior), Baldassarre and Fischer 1984 (diet and foraging behavior in migration), Brooks 1967 (diet in mi-gration), Jackson 1918 (molt), Wilds 1982 (identification).

MARSH SANDPIPER *Tringa stagnatilis*

Breeding primarily in the interior of central Asia and wintering mainly in Africa and India, the Marsh Sandpiper is also fairly common as far east as northern Australia. It has been recorded only once on the American side of the Pacific, a juve-nile in the western Aleutian Islands in early Sep-tember, and is of unlikely occurrence in the Northwest. Where it is common it usually occurs in freshwater marshes, feeding in the water like a miniature Black-necked Stilt. It runs about as actively as any yellowlegs.

The MARSH SANDPIPER is a distinctive bird, somewhat like a small COMMON GREENSHANK but with a character all its own. Medium-small and about the size of a LESSER YELLOWLEGS, it has slightly longer, greenish legs and a distinctly longer, straight, and very thin bill rather like that of a WILSON'S PHALAROPE. The bill is an adaptation to feeding on the water's surface, just as in phalaropes and stilts.

Birds in breeding plumage have a sandy brown back finely marked with black and a finely streaked breast. Juveniles are gray-brown above with all feathers fringed with pale buff to white and entirely white below, and nonbreeding birds are similar but unmarked above. In all plumages it looks paler than a YELLOWLEGS. In flight it looks like a miniature GREENSHANK, with long, greenish legs and white lower back, rump, and tail. Only DOWITCHERS of our common shorebirds have a similar flight pattern. The entire length of the toes projects beyond the tail tip, only about two-thirds of their length in a GREENSHANK. The flight call is a sharp *tew*, like a single-noted LESSER YELLOWLEGS, although soft, trilled calls have been noted as well.

References

Kieser and Kieser 1982 (identification).

SPOTTED REDSHANK *Tringa erythropus*

A spring individual in virtually black plumage might register as a hallucination to an observer unaware of the existence of such a bird, whether the red legs were seen or not. A more likely individual of this casual visitor would look like a slender-billed, gray yellowlegs with those same startling red legs.

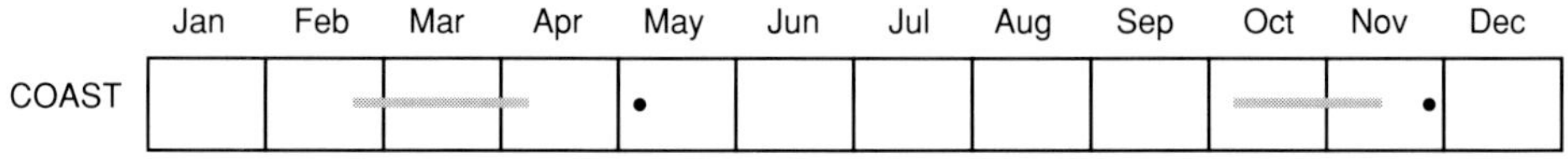

Jan
Feb
Mar
Apr
May
Jun
Jul
Aug
Sep
Oct
Nov
Dec
COAST

Distribution

Spotted Redshanks breed on tundra all across the Eurasian Arctic and winter on fresh and salt water in southern Europe, equatorial Africa, and the mainland of southern Asia.

Northwest Status

There are surprisingly many records of this Old World species from the region, considering it is only an occasional migrant, particularly in fall, across the Bering Sea. Perhaps only a few birds produced records in three autumns (October 9 to November 29), one spring (early May), and perhaps a winter (February 21 to April 1) in southwestern British Columbia and Oregon. Two fall (October 25 to November 20) and two spring (April 30 to May 15) records from California furnish further evidence of continued presence on the Pacific coast and probable wintering in this hemisphere.

COAST SPRING. South jetty of Columbia River, OR, February 21 to March 1, 1981; Reifel Island, BC, early May 1971, March 1 to April 1, 1981.

COAST FALL. Reifel Island, BC, October 17 to November 11, 1970, November 29, 1980; Surrey, BC, October 9-17, 1982.

Habitat and Behavior

These birds have been reported from freshwater ponds and less often from coastal estuaries in our region. The 1970 bird persistently roosted with Long-billed Dowitchers. Spotted Redshanks occur typically by themselves or in small groups, usually in marshes or protected estuaries. They feed in water as deep as that frequented by any *Tringa*, readily tipping up in the manner of a dabbling duck. Like yellowlegs, they are active feeders, dashing about after relatively large prey. They also probe more than other *Tringa*, their bill shape reminiscent of probing *Calidris*.

Structure

Size medium. Bill long, slender, and slightly decurved at tip, legs long. Wings extend about to tail tip, primary projection short to moderate in adults and juveniles. Tail squarish, central pair of rectrices projecting beyond others.

Plumage

Breeding. Bill black with base of lower mandible orange-red; legs dark red, much suffused with black or largely black. Looks almost entirely black (actually dark brown above) with white-dotted upperparts. Undertail coverts and lower belly in some birds barred dark gray and white. Females average paler than males on head and breast, with more white markings.

Lower back white, uppertail coverts narrowly barred black and white; tail gray-brown, obscurely barred with dark brown. Wing brown above, secondaries and innermost primaries conspicuously barred with white. Axillars, underwing coverts, and inner webs of primaries white; secondaries barred, contrasting vividly with dark underparts. The spring British Columbia bird had begun to molt into this plumage before it disappeared.

Nonbreeding. Legs orange-red to dull red, much paler and brighter than in breeding adults. Gray above with pale gray wash on breast, remainder of underparts white. White supercilium from bill to eye and dark loral stripe. Molt into this plumage early, beginning in late June and mostly completed by end of August.

Juvenal (Figure 29). Legs as in nonbreeding adult. Plumage much darker than nonbreeding adult, brown above with white-dotted scapulars, tertials, and coverts and brownish-gray washed underparts, lightly to heavily and finely barred with darker gray. Darkest individuals reminiscent of breeding-plumaged adults, but supercilium distinctive. Tail bars slightly more conspicuous than those of adults. Juveniles molt fairly rapidly into first-winter plumage during migration, and a number of Northwest birds have been in this transitional state.

Immature. First-winter birds distinguished by their combination of gray mantle and scapulars, strongly dotted tertials, and entirely white underparts.

Identification

All Plumages. When they can be seen, the *entirely red legs* (darker in breeding adults) allow quick identification. Occasional YELLOWLEGS with red legs are reported, the color either some staining effect from the substrate or perhaps an actual modification of the carotenoid pigments of the legs. I have seen a pond full of LESSER YELLOWLEGS with red-orange legs provided by the organic ooze into which they sank at every step.

The SPOTTED REDSHANK is slightly smaller than a GREATER YELLOWLEGS, but this difference would be evident only in comparison. Being slightly smaller, with a bill of the same length, the *bill is relatively longer*, this effect enhanced because it is also more slender than that of the YELLOWLEGS, especially at the base; it is also slightly drooped at the tip. The *base of the lower mandible is red* in this species, unlike any YELLOWLEGS.

Breeding Plumage. In breeding plumage this species is one of two Northwest shorebirds that look essentially *all black*—the other being the much larger BLACK OYSTERCATCHER. Mostly black male RUFFS with bright reddish legs have been mistaken for this species but are very differently shaped.

Nonbreeding Plumage. The SPOTTED REDSHANK'S nonbreeding plumage is very *plain gray*, less contrasty than any YELLOWLEGS plumage, with *unstreaked crown* (distinctly streaked in YELLOWLEGS) and *obscurely marked gray breast* (prominently streaked in YELLOWLEGS). In addition, the white supercilium is more conspicuous on the average in the REDSHANK than in the YELLOWLEGS, extending from bill to eye and bordered below by a dark loral stripe; some YELLOWLEGS are as sharply marked, however. Bear in mind that first-winter birds retain the dotted upperparts of the juvenal plumage late in fall.

A bird in full nonbreeding plumage looks somewhat like a DOWITCHER, with plain back and breast, unstreaked crown, and conspicuous supercilium, but the REDSHANK is much more elongate (longer-necked and -legged) with red legs and bill base. In a mixed roosting flock (as has been seen in our area) it also appears paler. The REDSHANK may feed at times in DOWITCHER fashion, unlike YELLOWLEGS.

Juvenal Plumage. The juvenal plumage is darker and more like a YELLOWLEGS than that of the adult, with heavily dotted coverts and tertials,

Juvenile Spotted Redshank. Tringine with long, orange to red legs and long, slender, slightly drooped bill with red-based lower mandible; tiger-striped flanks and overall heavy markings below diagnostic of plumage. Japan, September 1984, Urban Olsson

but the plain head, unstreaked breast, *obscurely to heavily barred underparts*, and of course bill and legs are still sufficient field marks. Fresh-plumaged juveniles, so far not seen in this region, are very dark, impossible to confuse with any other sandpiper.

In Flight (Figure 10)

Plain-winged like a YELLOWLEGS, it is more typical of Old World *Tringa* in being *white-backed* like a DOWITCHER. The tail is slightly darker than that of a YELLOWLEGS, so the rump contrasts more with the tail than it does in the latter. The barred secondaries and inner primaries appear paler than the rest of the wing in flight. From a DOWITCHER it can be distinguished by *long, red legs* projecting well behind the tail tip, longer neck, and shorter bill. The COMMON GREENSHANK has a similar flight pattern but greenish legs. The underwings are mostly white and would look paler than the underparts of a bird in breeding or juvenal plumage, while the under-

wings of a YELLOWLEGS are always darker than its white underparts.

Voice

The flight call is a loud double whistle—*tew-wit*—rather like a SEMIPALMATED PLOVER and nothing like the repeated calls of a YELLOWLEGS.

Photos

Hosking and Hale (1983: 26) show a nonbreeding adult, Keith and Gooders (1980: Pl. 111) a pale juvenile and Chandler (1989: 160) a dark one. Hammond and Everett (1980: 119) illustrate the similarity between a breeding adult and a juvenile, apparently assuming the latter was a molting adult.

References

Clark 1977 (size variation and molt), Hildén 1979a (migration schedule in Finland), Taverner 1982 (feeding behavior in migration).

WOOD SANDPIPER *Tringa glareola*

The Wood Sandpiper is an Old World species so far unreported from the Northwest but a likely candidate for occurrence, as it is one of the most common Siberian shorebirds in Alaska (large numbers in spring in the western Aleutians, much less common in fall; has bred). There are two sight records south of there, one in northern California on August 20, 1985 (seen briefly and not accepted by the California Bird Records Committee), and one at Tokeland, Washington, on October 9, 1988 (photographs taken are not definitive).

This species is more similar to the LESSER YELLOWLEGS than other birds of our region but is a bit smaller, closer to the size and shape of a SOLITARY SANDPIPER. It typically occurs in freshwater habitats, feeding in water like these close relatives, but it is also common in salt-water estuaries. Bear in mind that both WOOD and GREEN SANDPIPERS are more likely to be seen on salt water than is the SOLITARY.

The WOOD SANDPIPER has longer legs than the SOLITARY and is colored like a YELLOWLEGS, brown and profusely marked with light dots above and with a whitish rump and tail and pale wing linings. Its legs and neck are shorter than those of YELLOWLEGS, and the legs are often greenish yellow or dull yellowish brown, but they may be as bright as those of a YELLOWLEGS.

Breeding adults are brown with distinct and fairly large white dots and scattered dark bars above, presenting a more brightly spotted appearance than do YELLOWLEGS at that season. The sides are sparsely barred and the breast is more finely streaked than that of a YELLOWLEGS. Nonbreeding adults have the pale-dotted and dark-barred scapulars and tertials of breeding adults but a virtually plain mantle. Juveniles are dotted with buff above, both the pale spots and the ground color slightly darker than in YELLOWLEGS. The white supercilium is more prominent than in a YELLOWLEGS, especially behind the eye, but occasional YELLOWLEGS are similar.

Several minor differences may be used to separate the two species, even if they are not together for size comparison (Figure 29).

• In the WOOD SANDPIPER the legs are of medium length, about the same as the midbody width. In the YELLOWLEGS the legs are distinctly longer than the body width.

• The bill of the WOOD SANDPIPER is very slightly shorter and thicker than that of the YELLOWLEGS and is gray at the base in juveniles (and some adults), the color difference more noticeable than the size difference.

• The wings are shorter in the WOOD SANDPIPER, just reaching the tail tip in that species but surpassing it in the YELLOWLEGS. Perhaps an even more obvious difference is the primary projection, short in the WOOD SANDPIPER and long in the LESSER YELLOWLEGS. The latter is the most definitive close-range field mark.

In flight the WOOD SANDPIPER looks very much like a LESSER YELLOWLEGS but with shorter legs, only about half the length of the toes extending beyond the tail tip (the entire toes and a bit of the tarsi in the YELLOWLEGS). The flight call is an emphatic series of high-pitched notes, usually three or more (*pee pee pee*), sounding something like the flight call of a LONG-BILLED DOWITCHER. The WOOD SANDPIPER regularly flies high when flushed, like the characteristic "towering" of a SOLITARY SANDPIPER or SNIPE, and its flight is typically fluttery, with a faster wingbeat than a YELLOWLEGS. It may fly with wings fluttered below the body axis like a SPOTTED SANDPIPER, something I have not observed in YELLOWLEGS. Unlike American *Tringa* (but like the GREEN SANDPIPER), this species at times teeters the rear end of its body in the manner of a SPOTTED SANDPIPER.

References

Tree and Kieser 1982 (identification).

GREEN SANDPIPER *Tringa ochropus*

Very rare in Alaska, with only a few spring records and a single fall record from the far western Aleutian Islands, this Old World species is an unlikely visitor to our area. It is much like the SOLITARY SANDPIPER in general appearance, dark coloration, dark wing linings, olive legs, freshwater habitat, and behavior and is best distinguished in flight. The rump and tail look entirely white as in a YELLOWLEGS, quite different from the SOLITARY's dark tail with white, heavily barred outer feathers. The upperparts are heavily dotted with white in breeding plumage, more sparsely dotted in nonbreeding and juvenal plumages. The breast is more conspicuously striped in breeding than in nonbreeding and juvenal plumages.

Even when perched the GREEN SANDPIPER is distinguishable from the SOLITARY with some attention to detail. It is about 50 percent larger than the SOLITARY, and distinctly larger than the SPOTTED, while SOLITARIES look about the size of SPOTTEDS. A behavioral characteristic that may allow distinction of GREEN SANDPIPERS is their common habit of teetering like a SPOTTED SANDPIPER, an action not typical of the SOLITARY. The flight call, a series of high whistles, is much like that of the SOLITARY, but commonly low or liquid notes are mixed with the high ones—for example, *weet-oo-weet* or *weet-ooeet-weet*.

Note

The "Green Sandpiper" in Hammond and Everett (1980: 117) is a Common Sandpiper.

SOLITARY SANDPIPER *Tringa solitaria*

Resembling a small yellowlegs, but dark with greenish legs, this puddle-jumper is as solitary as its name implies. Up like a rocket when flushed from a freshwater pond, it circles above, shows its dark underwings, and drops a series of high whistles as it departs.

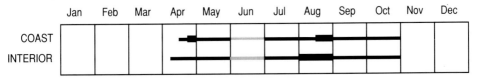

Distribution

Solitary Sandpipers breed throughout the North American Subarctic at boreal-forest ponds north to tree line. They migrate through freshwater habitats to winter in the same sorts of places in the humid tropics from Mexico and the West Indies south through much of South America.

Northwest Status

From the number of pairs seen during the summer in southern British Columbia in recent years, the Solitary may breed fairly commonly in the Cariboo Parklands just north of the region and sparingly in the Okanagan valley (near Dee Lake) within the region. Suspected breeders have also been seen in Manning Park in that province. Much farther south, a pair has bred most summers from 1981 to 1987 at Gold Lake Bog in the central Oregon Cascades. Finally, possible breeding individuals have been seen at Downy Lake, in northeast Oregon, the only regional nesting locality for the Greater Yellowlegs.

Otherwise, Solitaries occur throughout the region as migrants both spring and fall. In the coastal subregion a few are seen, usually singly, from mid April to mid May (primarily in May in southern British Columbia) during the northward passage. Adults appear again in July and are uncommon. Juveniles are more common, occurring from late July through August and Septem-

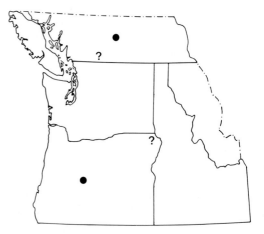

Solitary Sandpiper breeding distribution

ber, rarely into early October. Migration in the interior is similar to that in the coastal subregion, with apparently fewer adults in spring (highest counts no more than two per day) and larger numbers of juveniles in fall.

COAST SPRING. *Early date* Finley Refuge, OR, April 8, 1983. *High counts* 7 at Gaston, OR, April 26-27, 1981; 4 at Banks, OR, April 28, 1986; 3 at Federal Way, WA, April 26, 1987. *Late date* Iona Island, BC, June 10, 1967.

COAST FALL. *Early date* Saanich, BC, June 27, 1978. *High counts* 7 at Iona Island, BC, August 15, 1974; 5 at Ridgefield Refuge, WA, August 25, 1975; 4 near Shady Cove, OR, August 29, 1961. *Late date* Duncan, BC, October 26, 1978.

INTERIOR SPRING. *Early date* Rupert, ID, April 9, 1920. *Late date* Grays Lake, ID, May 30, 1950; Malheur Refuge, OR, May 30, 1987.

INTERIOR FALL. *Early dates* Wye Lake, BC, June 25, 1929; Vaseux Lake, BC, June 25, 1976. *Juvenile early date* Okanagan Landing, BC, July 21, 1915#. *High counts* 12 at Vernon, BC, August 8, 1911; 7-8 at Reardan, WA, August 3, 1965. *Late date* Potholes, WA, October 26.

Habitat and Behavior

This species is very much a bird of fresh water, with saltwater records scarce. Even at the coast, where it is rare, it is usually seen at freshwater ponds or sewage lagoons. It especially favors muddy ponds that dry up in late summer and can be found literally in pig wallows, a stark contrast in aesthetics considering the beauty and grace of the Solitary. It is one of the few shorebirds, along with the Spotted Sandpiper, that are at home on the margins of heavily forested ponds and lakes, and, like the Spotted, occurs well up in the mountains in migration. True to its name it is not a flocking species, although one might see several around the shore of a pond or lake during migration peaks.

Like other tringines, Solitary Sandpipers bob the front part of their body up and down at intervals. Yellowlegs have exactly the same habit, not shared by any other common Northwest sandpiper (tattlers occasionally do it) and distinct from the Spotted Sandpiper's habit of moving the *rear* part of its body up and down. Solitaries feed typically in shallow water, moving about fairly

actively and picking invertebrates from the surface. They also probe vigorously in water or even in mud, foraging by touch like a miniature Willet.

Breeding localities are in boggy areas, typically near open water, in semiopen coniferous forest. The song is a repetitive series of whistled notes given from a perch or during a display flight less spectacular than that of the yellowlegs. The four eggs are laid in old nests of passerine birds, rather than on the ground like most shorebirds. American Robin and Rusty Blackbird nests are commonly used, probably because they persist well from the year before.

Structure

Size small. Bill medium, legs short. Wings project to or just beyond tail tip, primary projection short in breeding adults and moderate in juveniles. Tail rounded.

Plumage

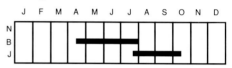

Breeding. Bill black at tip, grading into gray at base; legs greenish. Upperparts dark brown; edges of mantle feathers, scapulars, coverts, and tertials heavily dotted with white. Faintly indicated white anterior supercilium, conspicuous white eye-ring, and dark loral stripe. Head and hindneck coarsely striped and throat, foreneck, and breast finely striped with dark brown.

Central pair of rectrices colored like back, with white notches in fresh plumage; remainder of feathers broadly and vividly barred black and white. Wing dark brown above and below, relieved above by fine pale dots on secondary coverts and below by fine gray bars on lesser underwing coverts, none of them visible at any distance. Axillars evenly barred brown and white.

Southward-bound adults pass through area before molting, spots on upperparts reduced in size and head streaks obscured compared with spring birds.

Nonbreeding. In this plumage, probably not seen in Northwest, upperparts with scattered or no white dots, head and breast striping obscured.

Juvenile Solitary Sandpiper. Dark freshwater tringine with short, straight bill, prominent eye-ring, and short, greenish legs; brownish-washed breast and small, regular pale dots diagnostic of plumage; buff dots typical of *T. s. cinnamomea*. Victoria, BC, August 1985, Tim Zurowski

Juvenal (Figure 29). Head and neck virtually unstriped, plainer than in nonbreeding adult; plain crown and brownish wash on sides of breast distinctive. Dotted on mantle, scapulars, tertials, and coverts like fresh breeding adult, dots varying from whitish to buff. More conspicuously and evenly dotted than worn adults with which they might occur.

Immature. Dots on upperparts wear off during first winter. First-summer birds similar to adults but with very worn primaries. Age at maturity not known but may breed in first year.

Subspecies

T. s. cinnamomea breeds in the northern part of the range of the species, from Alaska and northern British Columbia to Hudson Bay, while *T. s. solitaria* breeds south of that, from southern British Columbia to Labrador. The winter distribution of the subspecies is not well understood. Typically *cinnamomea* averages slightly larger than *solitaria*, with slightly paler, more olive-brown, upperparts in breeding plumage and buff- instead of whitish-dotted upperparts in juvenal plumage. Both subspecies migrate through the Northwest, and their migration schedules appear to be similar from the specimen record. Fall specimens from interior Washington that I examined included one adult and seven juvenile *cinnamomea* and three juvenile *solitaria*, but Burleigh (1972) considered the two of equal abundance in Idaho, based on the examination of a much larger series of specimens. Juveniles at close range should be identifiable to subspecies by the color of their dots, but adults are indistinguishable in the field.

Identification

All Plumages. SOLITARY SANDPIPERS are about the size of the SPOTTED SANDPIPERS with which they often occur but look more like miniature LESSER YELLOWLEGS, with their *light-spotted back, dusky, streaked breast, and slender, straight bill*. They differ from that species by being *smaller, with relatively shorter neck*, a bill slightly shorter than the head (about the same length as head in LESSER YELLOWLEGS), and *shorter, greenish legs*. On a standing or walking SOLITARY the leg length is about the same as or slightly less than the body diameter, while in a LESSER YELLOWLEGS the leg length is obviously greater than the body diameter.

SOLITARY SANDPIPERS can be distinguished from similar-sized SPOTTED SANDPIPERS even when the latter lack their diagnostic spots. Although characteristic of tringines in general, the eye-ring is complete in SOLITARY, incomplete in SPOTTED. SOLITARIES in our region vary from lightly to heavily dotted on the back, coverts, and tertials; SPOTTEDS are plain above or barred on the coverts. The characteristic body movements of the two species are different (see above under habitat and behavior), and their different appearance in flight cinches any tentative identification.

See the very similar but unlikely GREEN SANDPIPER. See also WOOD SANDPIPER, a similar-sized *Tringa* that might be mistaken for this species. It is paler, more heavily marked, and longer-legged than SOLITARY and has a different flight pattern, in all attributes more like a YELLOWLEGS.

Breeding Plumage. Breeding adults differ from LESSER YELLOWLEGS in the same plumage in being *plain dark brown with pale dots above*, while YELLOWLEGS are heavily marked with blackish on gray.

Juvenal Plumage. Juveniles are somewhat like juvenile LESSER YELLOWLEGS, but the ground color of the SOLITARY is darker than that of the YELLOWLEGS, and a *complete eye-ring* is conspicuous against this darker color and is often one of the immediately obvious marks of this species.

In Flight (Figure 10)

Often SOLITARIES will flutter up into the air in a series of short ascents when flushed, only to drop down again at the pond edge, but as commonly they *fly high into the air*. This behavior, called "towering," is characteristic of only a few shorebirds and may be an adaptation to living on wooded bodies of water, where upward flight is the only option, especially to escape predators. Even the normal overhead flight is distinctive, almost swallowlike. The appearance in flight is also distinctive—*all dark with a tail that is dark in the center and white, barred with black, on the sides*. From below the *dark wing linings* contrast vividly with the white belly, a pattern unique among regularly occurring American tringines and only approximated by the COMMON SNIPE and the very different BLACK-NECKED STILT of our shorebirds. SOLITARIES typically hold the

wings aloft for a second or two after landing, showing again their characteristic dark undersides.

Voice

SOLITARIES are noisy, often giving high-pitched calls when they flush or in cross-country flight. The call, typically with three notes (*wheet wheet wheet*), is reminiscent of the call of a SPOTTED SANDPIPER but on a more even pitch, the SPOTTED'S notes slightly wavering and often dropping in pitch toward the end of the series.

Further Questions

Many small, boggy lakes in upland parts of our region look like ideal Solitary breeding habitat, and a person with considerable mosquito tolerance might be able to confirm this. Nesting in trees in cup nests of passerines such as robins, Solitaries are more amenable to proof of nesting than those master nest-hiders, the yellowlegs. A pair on territory during the summer would be good evidence; breeding birds commonly perch in trees, while migrants do not. All specimens from the region, especially adults and birds from west of the Cascades, should be determined to subspecies if possible.

Notes

A published record from Mount Vernon, December 1, 1963, is considered dubious.

The flight illustration in Farrand (1983: 350) shows the legs much too long; only the toe tips extend beyond the tail at most.

Photos

The "winter plumage" bird in Farrand (1983: 351) is a juvenile, as is the bird in Bull and Farrand (1977: Pl. 216). Hosking and Hale (1983: 119) show a bird that appears to be a very plain juvenile, probably in early winter with its pale markings worn off.

References

Conover 1944b (subspecies distinctions), Oring 1968 (vocalizations), Oring 1973 (breeding behavior), Swarth 1935 (critique of subspecies distinctions).

WILLET *Catoptrophorus semipalmatus*

A big and heavy tringine, the Willet seems scarcely related to its smaller, more elegant relatives. Plain and gray with gray legs, it is not abundantly endowed with field marks, but when it springs into the air on flashy black-and-white wings, it leaves no doubt about its identity. Look for it on coastal mud flats or, more striking, on fence posts in its southern Oregon and Idaho breeding range.

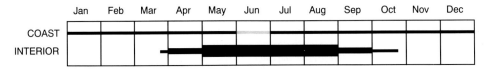

Distribution

The Willet is a widespread breeder in the Great Plains and Great Basin, and a resident on the Atlantic and Gulf coasts, locally on the Pacific coast of the United States, and on West Indian and Mexican coasts. Birds from the interior population migrate into the range of resident populations during the nonbreeding season.

Northwest Status

This species breeds fairly commonly in the Klamath and Malheur lakes region of south-central Oregon and the Grays Lake and Camas refuges of southern Idaho and in smaller numbers at other lakes and marshes of those areas. Breeding habitats include shallow marshes and lake shores, also moist upland meadows some distance from water. It is an uncommon to rare migrant throughout the rest of the region, with fewest records from northern Idaho (one) and interior British Columbia (five), and more from interior Washington, western Montana, and throughout the breeding range. The big alkaline lakes at Malheur Refuge are used as staging areas in early fall. Interior records away from breeding localities come mostly from mid April to late May, with a few in June, and again from mid August to late September (probably mostly juveniles).

Most Willets west of the Cascades occur on the coast, but a few birds have been seen in the southern Willamette valley in spring. In British Columbia, well to the northwest of its breeding range, records are few, and it is only rarely recorded even in well-scrutinized areas. Coastal migrants are seen from mid April to late May, with a few into June, and again from early July to mid October in areas in which they do not winter. Enough June birds have been recorded to confuse

the issue of earliest or latest migration dates, although fall migrants are more likely in that month than spring migrants. Single birds or groups of up to four have been recorded, except for the extraordinary spring record at Oak Bay, where three flocks flew past the observer.

In winter most Willets have moved south to California and beyond, but flocks numbering one to two dozen birds overwinter regularly in protected estuaries on our coast, including Willapa Bay, Yaquina Bay, Coos Bay, and Bandon. Occasional individuals are seen elsewhere during that season, as in migration, and there are four December-January records from British Columbia. The only winter specimen examined, from Comox, January 26, 1950, is a first-year bird.

COAST SPRING. *High count* 50 at Oak Bay, BC, April 29, 1945.

INTERIOR SPRING. *Early date* Malheur Refuge, OR, March 21, 1958. *High count* 139 at Summer Lake, OR, May 1, 1987.

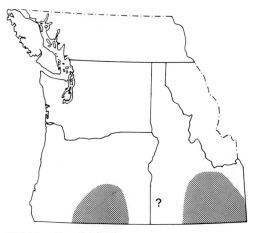

Willet breeding distribution

INTERIOR FALL. *High count* 3,000+at Stinking Lake, OR, mid July. *Late dates* Wenas Lake, WA, September 29, 1976; Malheur Refuge, OR, September 5, 1964; Rupert, ID, October 20.

Habitat and Behavior

Willets breed in freshwater marshes in our region and are seen in migration on the shores of reservoirs and on coastal mud flats. They feed in flocks, some of them of large size, farther south, but in our area they are scarce generally and often seen by themselves. They are not as gregarious as godwits but more so than yellowlegs, and they feed by probing more than do most tringine sandpipers.

Both their loud vocalizations and their vivid wings, used in both aerial and ground displays, call attention to Willets on their breeding territories. Where common, as around Burns, Oregon, displaying birds at times seem to fill the air. The nest is a shallow depression in an open area, near water or up to several hundred yards away from it. It is usually well concealed in grasses or associated with larger objects if entirely in the open. The clutch size is four, and both parents incubate the eggs and care for the young.

Breeding Willet. Bulky tringine with heavy, straight bill and gray legs; heavy dark barring on upperparts and underparts diagnostic of plumage; sparseness of markings on underparts and tail indicate *C. s. inornatus.* Burns, OR, 16 June 1983, Dennis Paulson

Structure (Figure 30)

Size at low end of large. Bill long and heavy, legs medium to long. Wings reach about to tail tip, primary projection short in adults. Web between outer and middle toes better developed in this species than in any other tringine. Tail squarish, central pair of rectrices projecting.

Plumage

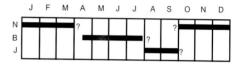

Breeding. Bill black with brownish to gray base; legs blue-gray, in some individuals brownish gray. Upperparts basically gray-brown, breast and belly white. Head and neck heavily but finely striped with darker brown, with faintly indicated darker loral stripe. Breast and sides barred with dark brown, often with warmer brown suffusion in same area. Scapulars, coverts, and tertials variably streaked and barred with blackish. Intensity and distribution of markings vary more than in most shorebirds.

Uppertail coverts white, longest ones entirely gray or finely barred with gray at tips. Rectrices light gray-brown, either unmarked or obscurely to fairly distinctly barred with darker gray-brown, especially inner pairs. Wing gray-brown above with mostly white inner secondaries, white-tipped greater secondary coverts, and white primary bases forming wide, conspicuous wing stripe. Rest of primaries and primary coverts dark brown, contrasting with generally paler inner wing. From below similar white wing stripe contrasts with dark brown primary tips and coverts; marginal coverts on inner wing white. Axillars dark brown.

Nonbreeding. Lacks complex markings of breeding adult. Plain gray-brown on upperparts and breast and white on belly. Breast and coverts in some individuals with fine dark shaft streaks, coverts also with narrow white fringes. White supercilium extends back to eye, bordered by dark loral stripe. Breast often with fine black dots.

Juvenal. Like nonbreeding adult but distinctly browner, with buff to whitish fringes on all feathers of upperparts. Scapulars and tertials further complexly marked with buff notches, dark shaft

Juvenile Willet. Dark subterminal fringes on mantle, scapulars, and coverts diagnostic of plumage; note freshly regurgitated pellet. Lubec, ME, 9 August 1985, Wayne Petersen

streaks, and subterminal brown fringes that may be interrupted by the notches. White wing stripe appears to average broader in juveniles than in adults.

Immature. First-winter plumage looks like that of nonbreeding adults, but some coverts retain distinctive juvenile markings; subterminal bars especially visible. First-summer plumage worn on wintering grounds by one-year-old birds; like nonbreeding plumage but primaries in molt in midsummer.

Subspecies

The western (Pacific coast and interior) subspecies *C. s. inornatus* occurs in the Northwest, and the eastern (Atlantic and Gulf coast) subspecies *C. s. semipalmatus* is quite unlikely in the region as it is basically a coastal bird. Eastern birds are smaller and in breeding plumage average more heavily barred all over, including strongly barred central rectrices, while western birds have relatively less barring and usually weakly barred or unbarred central rectrices. Unfortunately these are not clear distinctions, and breeding Willets from the West may be as heavily marked as any Atlantic coast birds; the illustration in Hayman

(1986: 155, Fig. 146b) is misleading in this regard. Thus it will be impossible to determine individual birds in the field to subspecies, although sexed specimens might be determined in the hand by measurements.

Identification

All Plumages. WILLETS are *large* shorebirds, about the body bulk of WHIMBRELS and distinctly larger than GREATER YELLOWLEGS. They regularly associate with both of these species and are much paler than WHIMBRELS and straight-billed. Their relatively shorter gray legs and greater bulk (Figure 30) allow easy distinction from GREATER YELLOWLEGS. A nonbreeding adult HUDSONIAN or BAR-TAILED GODWIT, unlikely in our region, would be more or less similar in appearance but would have a slightly upcurved, usually longer bill with at least some pink color on its base and dark legs, while the WILLET has a *fairly heavy, straight black bill and gray legs*.

In Flight (Figure 8)

As soon as a WILLET springs into flight its best identification features are revealed—the *largely black wings with very wide, white wing stripes.*

Most of the wing looks black (actually very dark brown) and white, only that part that is visible when it is folded (secondary coverts and tertials) gray or mottled.

The white is much more extensive than in the very rare HUDSONIAN GODWIT, the only other large shorebird with white wing stripes. In the GODWIT the upperwing is gray to brown with the white stripes much less conspicuous than in the WILLET, and the underwing looks largely dark, with whitish areas in the same places as in the WILLET but less flashy; a nonbreeding or juvenile bird flying directly overhead might be confusing. In the WILLET the white under the primaries occupies more than half of each feather, in the GODWIT less than half. In addition, the WILLET's tail is whitish, the GODWIT's black with a white base. The tail character is obvious in a bird flying away. An AVOCET in the distance has a similarly flashy pattern, but the primaries are entirely black, only the base of the wing showing extensive white.

Voice

WILLETS in migration are not very vocal, but they are quite noisy on their breeding grounds. The song, usually given in flight, is a ringing *pill will willet*, and two different alarm calls are frequently heard, a fairly musical *kay-whuh* and, at higher intensity, a series of loud *kip kip kip* calls, in some cases each note breaking in the middle. With the loud and distinctive voices of many shorebirds, it is surprising that this species and the Killdeer are the only shorebirds with onomatopoetic species names, although the group name "curlew" certainly derives from the flight calls of the Eurasian Curlew. Although not so considered in the etymological literature, it seems possible that "godwit" and "snipe" could be similarly derived.

Further Questions

So far, few Willet observations in the Northwest have been accompanied by age designations, so we have minimal information about age-class representation among either migrant or wintering birds. The Willapa Bay flock should be scrutinized when it arrives in fall. More studies of the breeding biology of western birds would be of value to contrast with the excellent studies of Atlantic Willets.

Notes

Contrary to the statement in National Geographic Society (1987: 114), the Willet's bill is typically as long as or longer than that of the Greater Yellowlegs. Western Willets have distinctly longer bills than yellowlegs.

References

Burger and Shisler 1978 (nest site selection), Higgins et al. 1979 (breeding ecology), Howe 1974 (wing displays), Howe 1982 (social behavior and nesting), Kelly and Cogswell 1979 (winter movements and habitat use), Ryan and Renken 1987 (breeding habitat use), Sordahl 1979 (breeding behavior and vocalizations), Stenzel et al. 1976 (winter feeding behavior and diet), Tomkins 1965 (breeding behavior and ecology), Vogt 1938 (behavior and ecology), Wilcox 1980 (sex differences, population biology, and breeding biology).

WANDERING TATTLER *Heteroscelus incanus*

Any creature wandering annually between Mount Denali and Polynesia must live a special kind of life, and this is a special shorebird, with high, ringing calls, wild flights into the sky, and a penchant for solitude. Nervous and alert, it is quick to announce its presence by noisily announcing *your* presence when its rocky coastal habitat is invaded. No other shorebird of the region is plain gray above, nor heavily barred below in spring and summer.

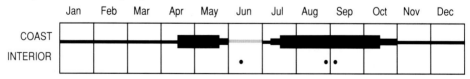

Distribution

The Wandering Tattler breeds on mountain streams from southern Alaska to northern British Columbia and in far northeastern Siberia. Its winter haunts are far-flung, from central California down the Pacific coast of South America and all across the Pacific Ocean in small numbers to eastern New Guinea and Australia. It probably visits every little island of the scattered archipelagos.

Northwest Status

This species is a common migrant on the wave-washed outer coast. Considerably less numerous on protected marine shorelines, it is regular on both coasts of Vancouver Island but quite uncommon to the south in Puget Sound. Spring migrants are seen from late (rarely mid) April through May, perhaps into June. There are enough June records to indicate occasional summering in both British Columbia and Washington.

The first adults return in mid (rarely early) July and are common into August (a few to the end of that month), when they are quickly replaced by juveniles. The latter first arrive about the second week of August and remain common through September and less common through October. The species winters primarily south of the United States and only rarely north of the southern California coast. Nevertheless, four birds have been seen in winter on British Columbia and Washington coasts, a few more have occurred in Oregon during December and January, and a surprising dozen birds wintered from Tillamook to Gold Beach in 1977-78.

Four tattlers have been reported from the interior: one in June and three in fall. The Osoyoos Lake bird was an adult in breeding plumage, and the Wenas Lake and Crater Lake birds were juveniles, as probably was the third fall bird. A bird about 12 miles up the Hamma Hamma River, August 23, 1986, was much closer to the coast but still an unusual sighting.

COAST SPRING. *High counts* 60 at south jetty of the Columbia River, OR, May 8, 1977; 30 at Destruction Island, WA, May 11, 1973; 50 at Cleland Island, BC, May 12, 1976.

COAST SUMMER. Flattery Rocks, WA, June 2-18, 1914; Cleland Island, BC, first 3 weeks of June, 1970.

COAST FALL. *High counts* 40 at Cleland Island, BC, July 24, 1967 (presumably all adults); 50 at south jetty of the Columbia River, OR, August 13, 1988. *Juvenile early date* Destruction Island, WA, July 24, 1963#.

COAST WINTER. Victoria, BC, February 25, 1943; Vancouver, BC, December 26, 1963; Great Chain Island, BC, February 13 to March 2, 1974; Ocean Shores, WA, February 26, 1977.

INTERIOR SPRING. Osoyoos Lake, BC, June 8, 1985.

INTERIOR FALL. Wenas Lake, WA, September 3-5, 1982; Crater Lake, OR, August 25, 1882; Upper Klamath Lake, OR, September 3, 1979.

Habitat and Behavior

The Wandering Tattler is a bird of rocky shores and jetties, very unlikely to be seen in any other habitat. Rare individuals, both adults and juveniles, have been seen on a mud flat or sand beach or at a sewage pond, and the few interior birds were at lake shores. Tattlers and other rock shorebirds regularly feed on mud flats away from rocks at Bandon, apparently lured by rich feeding opportunities.

Tattlers bob like *Tringa* and teeter like *Actitis* but do neither as consistently as their relatives. They move fairly rapidly over rocky substrates, probing into mats of mussels, barnacles, and algae and picking active prey from the surface. Typical tringines, they cover more ground while foraging than the calidridine rock shorebirds. Their choice of surface-active prey may limit the northern extent of their winter range. Although several are often seen in the same area, even roosting together or feeding near one another, the Wandering Tattler is basically a solitary species, each bird flushing independently and flying off in its own direction.

Structure (Figure 31)

Size medium-small. Bill medium and straight, legs short and relatively thick. Wings extend just beyond tail tip, with long primary projection in breeding adults and juveniles. Tail approximately square.

WANDERING TATTLER   GRAY-TAILED TATTLER

Fig. 31. Tattler bills. Wandering with longer nasal groove, extending well beyond midlength of bill, while that of Gray-tailed just reaches midlength.

Plumage

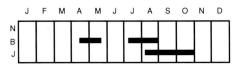

Breeding (Figure 32). Bill largely black or with greenish to gray base of lower mandible, more extensively gray at base in some individuals. Legs yellow, varying to greenish yellow. Upperparts, including rump and tail, medium gray. Underparts white, heavily barred with gray; bars extend to belly and undertail coverts, although some individuals with much of lower belly unbarred. White supercilium moderately distinct, mostly before eye. Loral stripe dark gray, cheeks and throat streaked with gray.

Wing gray above, slightly paler below, unmarked except for narrow white tips on some greater coverts forming very indistinct wing stripe. Axillars gray. Almost all tattlers appear to pass through our area in fall without molting, but at least a few undergo partial molt into nonbreeding plumage (specimen from Ocean Shores, August 20, 1987). Molt typically takes place on winter range, body feathers mostly during October and primaries during November and December.

Nonbreeding (Figure 36). Differs from breeding adult in being entirely unbarred beneath; breast and sides same color as upperparts, belly white. Loral stripe more distinct than in breeding adults. Unlikely to be seen in Northwest.

Juvenal. Differs from nonbreeding adult in having indistinct pale fringes on scapulars and tertials, often accompanied by alternating paler and darker dots, and a network of fine, pale bars on breast, visible only at very close range. Uppertail coverts plain or obscurely barred or tipped with light gray. Gray on bill base often more extensive than in adults.

Immature. First-winter individuals much like nonbreeding adults after autumn molt. One-year old birds remain on wintering grounds in worn nonbreeding plumage.

Identification

All Plumages. Among not only the rock dwellers, but all of our regularly occurring shorebirds, this species is unique in its *entirely plain gray upperparts* with no hint of brown or markings. See under GRAY-TAILED TATTLER for distinction from that species.

Breeding Plumage. Birds in this plumage are again unique in the region by virtue of their *heavily barred underparts*.

Nonbreeding and Juvenal Plumages. Juveniles, which lack ventral barring, may be confused with the three other rock-inhabiting shorebirds, all of which have grayish backs and breasts. The TATTLER is the only one with *gray sides*, however. Of these four species the TATTLER and ROCK SANDPIPER have *bills about as long as the head*. TATTLERS are larger than ROCK SANDPIPERS, and their bill is slightly longer, straight and only slightly tapered, compared with the strongly tapered and slightly drooped bill of the sandpiper. Both species have yellowish legs as does the SURFBIRD. The SURFBIRD is larger and about the same shade of gray, the BLACK TURNSTONE smaller and considerably darker, looking black at a distance; both of these species have much shorter bills, only half the head length.

Of other short-legged "gray" shorebirds in their nonbreeding plumages, usually of other habitats but at times occurring on rocks, RED KNOTS are paler with spotted or streaked sides and greenish legs, and SANDERLINGS are much paler and smaller with black legs. The WILLET is considerably larger, with gray legs. None of these species is the gunmetal gray color of a WANDERING TATTLER.

GRAY-TAILED TATTLER

Extreme Variant

WANDERING TATTLER

Typical

Fig. 32. Breeding tattlers. Gray-tailed paler above and less heavily marked below than Wandering, with more distinct supercilium. Wandering can have unbarred central underparts.

In Flight (Figures 9, 37)

This species is even more distinctive in the air than at rest, as it is the only shorebird that is *entirely plain gray above* in flight. The sides are also medium gray, contrasting with the white belly. All the species that share its habitat have conspicuous wing stripes, the SURFBIRD and TURNSTONES with white-patterned tails as well. This species seems to be more likely than any other of our region to perform "crazy flights," shooting up into the air and zigzagging in a way that no aerial predator could follow. With no benefit from flocking, solitary species such as this one are more likely to benefit from this behavior than are social ones. Not only is a predator unlikely to be able to capture such a bird, but it is also informed of the bird's agility in flight.

Voice

The flight call of this species, often given, is a fairly high-pitched, ringing series of notes easily heard over crashing waves. Only the WHIMBREL among our shorebirds has a similarly loud, extended call, but the notes of its call are lower and farther apart and all on one pitch, while the TATTLER'S series gives the impression of alternately slightly lower and higher notes. I have heard this species give its series of calls in pairs— *too-ee, too-ee, too-ee*—which should be kept in mind when considering the possibility of a GRAY-TAILED TATTLER.

Further Questions

Because autumn adults and juveniles are easily distinguished, and, because they feed essentially independently of one another, this would be a good species to study to answer questions about differences in behavior in adults and juveniles. Information accessible to a patient observer could include places and methods of feeding, amount of time spent feeding versus resting, foraging rate and success, response to predators (as exemplified by the approaching observer), type and frequency of vocalizations, and results of any

Juvenile Wandering Tattler. Rock-frequenting tringine with plain gray upperparts and sides, short bill, and short, yellow legs; nasal groove extends more than one-half length of bill; unbarred underparts and pale fringes on scapulars, coverts, and tertials diagnostic of plumage. Ocean Shores, WA, late August 1984, Urban Olsson

interactions (particularly between adults and juveniles).

As flight calls in shorebirds often serve to bring the members of a flock together, their function is puzzling in species such as this one and the Solitary Sandpiper, in both of which they are especially loud. It would be of interest to determine whether tattlers in different situations (in groups or singly, for example) are more or less likely to call when they fly.

Notes

The drawing of tattler heads in Johnsgard (1981: 334) has the species reversed.

Photos

The "winter adult" in Farrand (1983: 353) is a juvenile.

References

Dixon 1933 (breeding biology), Johnson 1973 (reproductive condition of summering birds), Johnson 1977 (plumage and molt of summering birds), Johnson 1979 (summary of 1973/1977 papers), Murie 1924 (breeding biology), Paulson 1986 (identification), Prater and Marchant 1975 (primary molt), Weeden 1959 (breeding biology), Weeden 1965 (breeding biology).

GRAY-TAILED TATTLER *Heteroscelus brevipes*

Tattlers away from rocks are worth a second look. A pale-looking one with fine, relatively sparse barring in breeding plumage or with white sides and heavily marked back in juvenal plumage might be this species. If its flight call is a two- or three-noted whistle, Gray-tailed is confirmed.

	Jan	Feb	Mar	Apr	May	Jun	Jul	Aug	Sep	Oct	Nov	Dec
COAST										•		

Juvenile Gray-tailed Tattler. As juvenile Wandering, but with white sides and more extensively fringed and dotted upperparts; nasal groove extends less than one-half length of bill. Kyushu, Japan, mid September 1984, Urban Olsson

Distribution

The Gray-tailed Tattler is the Siberian counterpart of the Wandering, with breeding range on the other side of the Bering Strait, on upland rivers in northern Siberia. Its overlap there with the Wandering Tattler is poorly understood. In migration and winter it frequents the shores of the Asian and Australian continents and also occurs in the western Pacific islands, where it overlaps extensively with the Wandering.

Northwest Status

There is one record of the species in the Northwest, a juvenile photographed on the Washington coast. An adult at Lancaster, California, on July 23, 1981, furnished the only other American record south of Alaska. The Gray-tailed Tattler is a regular spring and fall migrant through western Bering Sea islands and is a likely candidate for recurrence in the region.

COAST FALL. Leadbetter Point, WA, October 13, 1975.

Habitat and Behavior

This species frequents a wider range of environments than the Wandering and is characteristic of mud flats, especially the channels that dissect them, but also occurs on sand beaches and rocks. A vagrant individual probably could turn up any-where, but the Washington bird was on a mud flat. Its behavior is like that of the other tattler, each bird foraging by itself but gathering into flocks (typically larger flocks than Wandering) while roosting. Probably because of their more easily traversed habitats, birds forage more rapidly than Wandering Tattlers, often running. At times they elevate their rear end like a Terek Sandpiper, with some teetering as in that species and the related *Actitis* sandpipers. Relatively large, mobile prey is detected visually, and most feeding is accomplished by picking, although they sometimes probe briefly in shallow water. Crabs are a major part of the diet in some areas, captured from mud and taken to the water's edge to be washed.

Structure (Figure 31)

Size medium-small. Bill medium and straight. See below under Identification for difference in bill structure between tattlers. Legs short and relatively thick. Wings extend just to tail tip, primary projection moderate to long in breeding adults and long in juveniles. Tail approximately square.

Plumage

Breeding (Figure 32) Bill black grading into gray, grayish brown, or grayish pink at base; legs yellow. Brownish gray above, tail slightly paler. White supercilium and dark loral stripe distinct, cheeks and neck streaked with gray, underparts white with fairly fine gray barring on breast and sides. Uppertail coverts tipped or obscurely barred with light gray. Wing grayish brown above, slightly paler below, unmarked except for whitish tips of greater coverts forming indistinct wing stripe. Axillars gray.

Plumage changes in this species much as in Wandering, adults molting largely after completing southward migration. Breeding plumage acquired by mid April and lost by mid October.

Nonbreeding. Like breeding adult but without bars underneath; breast gray, posterior underparts white. Eye stripe a bit paler in this plumage, contrasting less with supercilium. Less likely to be seen in our region than other two plumage types.

Juvenal. Similar to nonbreeding adult but with fairly conspicuous pale fringes on coverts, scapulars, tertials, uppertail coverts, and rectrices; typically with pale dots edging tertials and, in some individuals, rectrices. Breast slightly darker than that of adult, with indistinct fine light bars visible at very close range.

Immature. As Wandering Tattler, although may often mature in one year (a smaller proportion of population seems to summer in winter range).

Identification

All Plumages. This species and the TEREK SANDPIPER are about the same size and similar in overall coloration and yellow to orange legs, and both run about actively on mud flats. Their quite different bills and the TATTLER's duller yellow legs and darker coloration should distinguish it. The only real identification problem will be between it and the WANDERING TATTLER.

• GRAY-TAILED is *slightly paler* on back and breast than WANDERING, a distinction that would be evident if they were seen together. In addition, it is *brownish gray* instead of neutral gray, apparent in museum specimens and obvious the only time I saw both species on the same day.

• The rump and tail are slightly paler and grayer than the back, the two areas colored the same in WANDERING. This is due to paler feathers rather than fine pale tips, contrary to statements in the literature.

• Tattlers have a well-defined nasal groove, the linear depression along the upper surface of the bill containing the nostril, and in the hand it can be seen that this groove runs from the base of the bill to about half length in GRAY-TAILED and to almost two-thirds length in WANDERING. The groove tapers and may appear to end before it actually does. Under very favorable conditions in the field this can be used as an identification mark. Close scrutiny of the nasal grooves of WANDERING TATTLERS is a good way to prepare in advance for that unlikely but not impossible GRAY-TAILED.

• GRAY-TAILED forages on mud and sand as well as rocks, and tattlers seen *away from rocks* should be especially closely inspected. Bear in mind that WANDERING TATTLERS wander away from their preferred habitat at times.

Breeding Plumage. In breeding plumage the barring on the GRAY-TAILED is not only *less extensive* but each bar is paler and narrower, giving the overall underparts a paler appearance. However, occasional WANDERING TATTLERS are almost this sparingly barred (Figure 32).

Nonbreeding Plumage. In this plumage the two species are quite similar, but there is at least one absolute distinction, the side color.

• The *sides are white* in GRAY-TAILED, gray in WANDERING. If it were not for this character, field identification would be difficult enough so that it should be based on call.

• The *breast is paler* in GRAY-TAILED, and this lessened contrast between breast and belly may be more distinctive than the pale upperparts.

• The white supercilium averages broader in GRAY-TAILED, and the supercilia from either side often meet over the bill. This is rarely the case in WANDERING, which usually has a narrower supercilium and entirely dark forehead.

Juvenal Plumage. The differences between the species in nonbreeding plumage are all appropriate to distinguish juveniles as well. In addition, GRAY-TAILED has on the average *more fine pale markings above* than WANDERING, including fringes and dots on the scapulars and tertials and even the rectrices, but this is highly variable. A very heavily marked juvenile warrants a closer examination. In adult plumages GRAY-TAILED is also slightly more heavily marked above, but in juveniles these fine markings are especially conspicuous.

In Flight

In good light the GRAY-TAILED'S *slightly paler tail* may show up as such (but it does not always), and its *brownish cast* should be apparent. In addition, it shows, at least in autumn adults, a slightly more obvious *wing stripe* than in WANDERING. The *entirely white underparts* distinguish nonbreeding adults and juveniles in flight from the conspicuously gray-sided WANDERING, but breeding-plumaged birds are more similar. Flushing a bird usually serves to stimulate it to call, the most useful "field mark" in this species.

Voice

As the other tattler, the GRAY-TAILED is quite vocal. Its call is a fluid double whistle—*tuweet*, quite different from the WANDERING'S series of notes and reminiscent of a SEMIPALMATED PLO-

VER. It may be extended to a three-noted, rolled *tuduweet* or a doubled *tuwee tuwee*.

Notes

Relative wing length has been stated to be a distinguishing mark between the species of tattlers (Hayman et al. 1986), the wings extending to the tail tip in Gray-tailed and well beyond it in Wandering, but the two species overlap widely in this characteristic.

Photos

Pringle (1987) contains many excellent photographs of this species, although one of the standing birds and all those in flight in the photo on the bottom of page 230 are Great Knots.

References

Hindwood and Hoskin 1954 (winter ecology), Keast 1949 (molt and winter feeding behavior), Neufeldt et al. 1961 (breeding biology), Paulson 1986 (identification), Prater and Marchant 1975 (primary molt).

COMMON SANDPIPER *Actitis hypoleucos*

This species, *the* sandpiper of Europe, has not been recorded from the region, but it is of regular occurrence in both spring and fall on Bering Sea islands, especially the western Aleutians (with even a breeding record), and could stray in our direction. It is much like SPOTTED in general aspect, only very slightly larger in all dimensions and virtually identical at a distance. Both species may be colored identically above, although SPOTTED does average slightly grayer, COMMON slightly browner. Breeding adults have no indication of spots below but instead have a dusky breast with fine but clearly visible streaks. The bill is dark, the base often pinkish, and the legs are grayish olive or dull yellow.

Juvenile COMMON SANDPIPERS differ from SPOTTED in a number of plumage characteristics, most of them discernible only at close range and under favorable lighting conditions (Figure 33):

• The tertials of COMMON possess black dots all along their edges, usually running the length of the exposed parts of the feathers, while those

of SPOTTED have no dots or, in extreme cases, possess as many as six dots along the longest feathers (there were three or more dots—comparable to the COMMON—on the longest tertial in several birds of a series of specimens examined). The greater coverts differ in the same way, usually dotted in COMMON but not in SPOTTED ("*not* dotted in SPOTTED" would have to be the mnemonic device here). The bird must be a juvenile to use this characteristic, as nonbreeding adults and first-winter birds of both species have plain tertials.

• The middle and lesser coverts of juvenile COMMON are barred with relatively narrow black bars and finely tipped with lighter buff, the pattern showing as brown/black/white and duplicated to some degree on the back and scapulars, while in SPOTTED the coverts are evenly barred with slightly broader black and buff bars, the pattern showing as alternating black/buff and more strongly marked than the scapulars (which are marked about as in COMMON). The back ver-

Fig. 33. Juvenile Spotted and Common sandpipers. Spotted with shorter tail; more contrast between barred coverts and plainer back; rather plain tertials; and paler, brighter bill and legs. Common with longer tail; more-or-less evenly marked back and coverts; conspicuously dotted tertials; and duller bill and legs.. In flight, shorter white wing stripe of Spotted does not reach wing base, Common's wing stripe does so conspicuously; Common's longer tail less obvious field mark.

sus coverts contrast is also variable in both species but serves as a good average difference.

• The breast in COMMON has well-defined dusky patches at the sides that are often faintly barred. In addition, there are very fine streaks all across the breast. In SPOTTED the patches are paler, without barring, and there are no streaks on the breast. Unfortunately, nonbreeding adult SPOTTEDS occasionally show both faint bars and fine streaks on the breast, so the bird in question must be determined to be a juvenile before these characteristics can be used.

• Although the underwings are not often seen, both birds display them briefly during interactions with others, and the difference in the amount of white is apparent (a photo of this would assure definitive identification). In SPOTTED the rear edge of the wing is brown (except for the narrow trailing edge), the brown area of about equal width all the way to the wing base, while in COMMON it narrows toward the wing base, where there is much more white.

• The bill is dark in COMMON, paler in SPOTTED (often pinkish with dark tip), but there is overlap. A bird with a bright bill is probably a SPOTTED, but one with a dull bill could be either species.

• The legs are greenish to pale straw color in COMMON and often but not always brighter in SPOTTED, usually yellow. As with the bill, a bird with brightly colored legs is probably a SPOTTED, but one with dull legs could be either species.

• Finally, perhaps the best character is tail length, which makes the two species look a bit different in any plumage. COMMON SANDPIPERS have a longer tail than SPOTTEDS, averaging 6 mm longer in males and 5 mm longer in females. This difference, although not much, can be assessed in the field by determining how far beyond the wing tips the tail projects. In COMMON the tail usually projects beyond the wing tips a distance greater than half the bill length, in SPOTTED no more than half. This is more of an indication than an absolute criterion, as a small percentage of individuals of each species would be misidentified by using only this character. Watch the bird for a while, as this proportion varies depending on how the wings are held, and attempt to confirm by the other characteristics listed.

In flight two additional differences may be noted:

• The extent of white at the wing base is diagnostic and easily seen in flight. In COMMON the white wing stripe extends all the way to the base of the wing, where it just touches the trailing edge. In SPOTTED the stripe ends well short of the wing base.

• COMMON averages more white in the outer tail feathers, with the outer webs of the two to three outer feathers white versus the same on one to two outer feathers in SPOTTED.

The usual flight calls of the two species are distinctive, the COMMON sounding more like a WOOD SANDPIPER than a SPOTTED. The multiple notes of COMMON are loud and shrill, less musical and higher-pitched than those of SPOTTED, with the terminal consonant of each note less evident. They are usually given in threes or fours, those of SPOTTED in pairs. Each has additional calls that are not typical of the other. Familiarity with the range of calls possible in SPOTTED SANDPIPERS prepares one to single out the different calls of COMMON.

The habitat preferences and behavior of these two hemispheric replacements are very similar, although I have seen COMMON foraging more often in dry upland areas in its winter range than SPOTTED.

Note

The photos in Chandler (1989: 179, 181) are excellent for comparing these two species.

References

Holland et al. 1982 (breeding biology), Kieser 1983 (identification), Madge 1977 (identification), Oddie 1980 (identification), Wallace 1970 (identification).

SPOTTED SANDPIPER *Actitis macularia*

This is a familiar bird to recreation-minded Northwesterners, nesting on almost every lake and river shore in the region. Teetering as it moves along the shore, fluttering as it flies away with high whistles, its behavior is often adequate for identification. Spots in summer or white shoulder marks in winter will provide certainty.

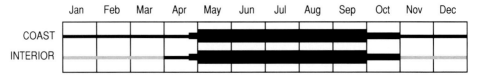

	Jan	Feb	Mar	Apr	May	Jun	Jul	Aug	Sep	Oct	Nov	Dec
COAST												
INTERIOR												

Distribution

The Spotted Sandpiper breeds throughout the northern part of the North American continent from tree line in the Subarctic south to the Southwest and almost the Deep South. It winters from the southern edge of its breeding range down through the Tropics to southern Brazil.

Northwest Status

Another unusually ubiquitous shorebird, this species breeds throughout the Northwest, arriving everywhere from the last week of April (Oregon) to the first or second week of May (Washington and British Columbia). The average arrival date over 34 years at Fortine was May 16, the range April 27 to May 29. Most adults and juveniles vacate their breeding grounds during August, the adult females perhaps in late July. Small young have been found well up in the mountains as late as mid August, however, presumably not to fledge until the end of that month. Some birds, probably all young of the year, stay on the breeding grounds into September or rarely as late as October. Birds that are clearly migrants appear on the coast uncommonly from late April to mid May and in slightly larger numbers from mid July to September. All I have seen at Grays Harbor in September have been juveniles.

Wintering birds occur mostly along the coast, often in coastal rivers or up to a few dozen miles inland. In most winters they are confined to Puget Sound, the Willamette Valley, and protected bays of Oregon, with very few as far north as southern British Columbia. The only winter specimens examined are two adults from British Columbia: Comox, December 25, 1925, and Jordan River, November 23, 1945. There are also occasional midwinter records from interior localities in Brit-

ish Columbia, Washington, and Montana and, with slightly greater frequency, from Oregon and southern Idaho.

COAST SPRING. *High count* 20 at Victoria, BC, May 20, 1974.

COAST FALL. *High counts* 15 at Saanich, BC, August 3, 1975; 19 at Long Beach, BC, September 8, 1983.

INTERIOR SPRING. *High count* 17 at Osoyoos Lake, BC, May 23, 1974.

INTERIOR FALL. *High counts* 35 at Summit Creek, BC, August 13, 1947; 60 at Salmon Arm, BC, August 24, 1973.

INTERIOR WINTER. Edgewood, BC, March 7, 1924; Summerland, BC, December 20-27, 1969; Kelowna, BC, December 19, 1970; Walla Walla, WA, February 28, 1981 (2-3); Stevensville, MT, December 18, 1982.

Habitat and Behavior

This species inhabits shores of lakes, ponds, marshes, rivers, and streams from sea level up to 6,000 feet or more in the mountains. No other shorebird is so ubiquitous at that time; Killdeers have even broader habitat preference but remain at lower elevations. Forested and open-country freshwater bodies are equally occupied as long as there is shore for foraging and herbaceous vegetation for nest concealment. Migrants prefer freshwater ponds with open shores but are regularly found on coastal mud flats and, more often than most other non-rock species, on rocky shores and jetties.

Spotted Sandpipers are very distinctive behaviorally, as they often "teeter" the rear end of their body up and down, presumably a social signal highly visible along rivers, where the prevailing motion is horizontal. Teetering is not mandatory, however, and they may perform only bobbing of the foreparts like a *Tringa* or combine the two motions. They forage visually, picking invertebrates from dry or moist substrates at the junction of water and land, often on complex topography. They rest and forage on floating logs or stream-washed rocks, and on the breeding grounds they are just as likely to move away from the water, picking their way through semiopen vegetation like a pipit.

The Spotted is a highly territorial species at all times, and single birds are the rule, except for pairs (or trios) in summer or family parties in early fall. Family groups are encountered, sometimes in surprising numbers, on rivers and at their saltwater mouths in early fall, with local concentrations of a half-dozen or more birds. This species is known to be polyandrous (each female mating with more than one male) where it has been studied in the East, and it is the slightly larger females that perform courtship displays and chase each other around aggressively. Both sexes sing loudly as they interact, and such vocalizations should call attention to the females' impressive displays, both aerial and terrestrial.

Nests are usually near water, partly concealed by projecting objects on sparsely vegetated gravel substrates or well hidden among grasses and sedges. Four is the usual clutch size, and either male, female, or both parents may incubate and care for the young from a given nest.

Structure

Size small. Females slightly larger than males; sexes of birds in pairs should be distinguishable. Bill short and straight or slightly decurved, fairly thick at base; legs short. Wings fail to reach tip of fairly long tail, primary projection short or lacking in adults and juveniles. Tail rounded.

Plumage

Breeding. Bill pinkish to yellow-orange to yellow with black tip, legs yellowish to orange. Entirely medium brown above from crown to tail, variably and in some birds profusely marked with

Nonbreeding Spotted Sandpiper. Tringine with brown upperparts, prominent supercilium, short bill, and short, yellowish legs; complete lack of ventral spots and extensively barred coverts indicate first-winter plumage. Miami, FL, December 1980, Jim Erckmann

dark brown streaks and bars. White, heavily spotted with black, below. Spots in females average slightly larger and more profuse, but much overlap between sexes. White supercilium and eye-ring and dark eye stripe fairly prominent, cheeks finely streaked with black.

Rectrices tipped with white, outer one or two pairs barred brown and white. Wing brown above, marked by incomplete white wing stripe formed from white bases of secondaries just showing behind coverts and white marks on inner webs of seven inner primaries; secondaries white-tipped. From below complex wing pattern formed by these white markings: marginal coverts, greater and median secondary coverts, basal half and extreme tips of secondaries, narrow tips of primary coverts, and extensive markings on innermost eight primaries, moving from basal half to center of feathers with progression toward wing tip. Axillars white.

Adults leave region while still in breeding plumage, which they retain at least until mid August.

Nonbreeding. Bill yellowish to pinkish to horn color, rarely black. Legs may be as bright as in breeding adults or may vary to greenish or even gray. Upperparts plain brown, many feathers with dark streaks, coverts finely barred with black. Supercilium slightly less distinct than in breeding adult, bringing eye-ring into greater prominence. Underparts white with pale gray patches on either side of breast. Typically a few black spots retained on undertail coverts and more rarely elsewhere.

Juvenal (Figure 33). Bill and leg colors as in nonbreeding adults. Streaks of upperparts lacking, wing coverts more strongly barred buff and black than in adult. Some scapulars and tertials tipped or fringed with pale buff, often with fine dark subterminal bar. Tertials on some individuals sparsely dotted with dark brown.

Immature. First-winter birds should be distinguishable from adults by lack of undertail spots and perhaps more heavily barred coverts. Most birds mature and attempt to breed in first summer.

Identification

All Plumages. The SPOTTED is about the same size and overall coloration as a nonbreeding-plumaged DUNLIN, but its *teetering* behavior identifies it from a great distance. Only WANDERING TATTLER among our regularly occurring shorebirds ever acts similarly, and it is a much larger, gray species.

Breeding Plumage. The SPOTTED stands out in breeding plumage because of its *brightly colored bill and legs and heavily black-spotted underparts*. No other shorebird possesses anything remotely resembling these characteristics. Don't be confused by the heavy spotting on the sides of the very dissimilar SURFBIRD.

Nonbreeding and Juvenal Plumages. Because most of us are familiar with SPOTTEDs in the context of their breeding grounds, where they are noisy and active, the unobtrusive juveniles and wintering birds may be puzzling when they occur on a mud flat or jetty with other, more social species. They can be recognized by being *plain above with white eye-ring and supercilium and barred coverts*. They are white below with a brown-smudged breast and an *extension of white in front of the bend of the wing*. The legs and often the bill base are pale.

The somewhat similar SOLITARY SANDPIPER, often occurring with the SPOTTED, is dotted above with buff or white and lacks the white supercilium and shoulder mark. Note that BAIRD'S and SPOTTED may also be seen together in late summer, as the former occurs in high mountain ponds as well as sharing other interior habitats with SPOTTED.

Winter DUNLINS are dumpier and very plain, without the SPOTTED's barred coverts, head markings, or shoulder mark and with a longer, droopy bill and black legs. A winter ROCK SANDPIPER is again like a DUNLIN but with pale bill base and legs; it is darker gray than the SPOTTED, with heavily spotted sides. Both DUNLIN and ROCK SANDPIPER have sharply defined pale edges to coverts and tertials, a very different patterning from that of the SPOTTED with its plain back and finely barred coverts.

In Flight (Figures 11, 33)

In flight SPOTTEDs move off in a beeline with *rapidly fluttering wings held bowed below the horizontal*, each wing tip covering a very short arc. No other Northwest shorebird flies like this, other than during display flights on their breeding grounds (away from our region). SPOTTEDs on long-distance flights fly like any other species, with deep wingbeats. They could scarcely migrate thousands of miles with their everyday flight, but birds flying like other shorebirds are likely to be puzzling. They are plain brown above, with an *abbreviated wing stripe*, the stripe well separated from both the tip and base of the wing. All other regional shorebirds with conspicuous wing stripes have them extending to the wing base. The *tail is white-edged* rather than white-striped as in the small *Calidris* with which this species might be confused.

Voice

The flight calls, given commonly in summer and much less so in migration, are loud, repeated whistles, high-pitched and often disyllabic, *peet weet weet weet* or *tuweet tuweet weet weet* being only two variations among several. Often the series drops in pitch and volume toward the end. Only the SOLITARY has a similar call among regularly occurring species. Adults on the breeding grounds sing *pit-a-weet, pit-a-weet* etc.

Further Questions

It would be of interest to determine whether our populations of this species are polyandrous, but birds would have to be color-banded and closely studied to do so. Also, close observation might determine whether adults or first-year birds were predominant among the small number that winter in the region, at the very least checking birds for undertail spots.

Notes

The drawing of Spotted and Common sandpiper heads in Johnsgard (1981: 339) has the species reversed.

Photos

The "winter plumage" bird in Udvardy (1977: Pl. 232) is a juvenile.

References

Dwight 1900 (molt and plumages), Gochfeld 1971 (winter roosting), Hays 1972 (breeding biology and mating system), Miller and Miller 1948 (breeding biology), Oring and Knudson 1972 (breeding biology and mating system), Oring and Lank 1982 (population biology), Oring and Maxson 1978 (polyandry), Oring et al. 1983 (population biology), Pickett et al. 1988 (interspecific interactions).

TEREK SANDPIPER *Xenus cinereus*

This accidentally occurring sandpiper is undistinguished until it exposes its long, upturned bill, when all doubt is removed. On the move it passes other shorebirds as if they were standing still. About the size of a Lesser Yellowlegs with a bill almost as long as that of a Greater and with short legs, this bird could best be described as a "squashed yellowlegs." Unlike any yellowlegs, it teeters frenetically, at its extreme putting a Spotted Sandpiper to shame.

	Jan	Feb	Mar	Apr	May	Jun	Jul	Aug	Sep	Oct	Nov	Dec
COAST												

Distribution

The Terek Sandpiper breeds on the banks of rivers and lakes from northern Europe across most of Siberia and winters on the coasts of Africa, southern Asia, and Australia.

Northwest Status

This Siberian species is a rare migrant in the Bering Sea region of Alaska, much more often reported in spring than fall, with three records as far east as Anchorage. South of Alaska, a single bird—an autumn adult on the British Columbia coast—is photographically documented from the Northwest, and another adult was found on the coast of central California from August 28 to September 18, 1988.

COAST FALL. Sooke, BC, July 21 to August 6, 1987.

Habitat and Behavior

The Terek Sandpiper characteristically occurs on mud flats during migration and winter, less often on sandy beaches and only rarely at fresh water. It usually forages singly where it is common, but at times several birds may forage near one another, and it joins mixed roosting flocks. It is

certainly likely to be by itself in the Northwest. It forages actively just above the water line on beaches or anywhere on wet mud flats, running rapidly to capture larger prey items such as crabs or stopping to probe vigorously and briefly. The dashing gait is unmistakable, the bird often with head down and extended forward, as if off balance; inserting the bill in the substrate stops it abruptly. The long bill seems adapted both for reaching out on the run and for probing.

Related to both Wandering Tattler and Spotted Sandpiper and between them in size, it similarly forages on land and has some behaviors reminiscent of both of these species. It teeters much like Spotted and Common sandpipers and, active bird that it is, even appears to outdo those species in the fervor of its movements. Teetering seems to be most intense when the bird is "nervous," i.e., when approached by the sandpiper enthusiast.

Structure (Figure 34)

Size medium-small. Bill long and evenly and slightly upcurved, looking almost straight from a distance in some birds. Male bills average slightly shorter than those of females, and juvenile bills average slightly shorter than those of

Fig. 34. Terek Sandpiper shape. Typical foraging by running with head down, tail up, and long bill projecting way out in front.

adults until midwinter. Legs short. Wings just reach or fall short of tail tip, primary projection short. Tail slightly rounded, almost square.

Plumage

Breeding. Bill black, in some individuals with light brownish base; legs orange-yellow. Upperparts pale gray-brown, with blackish streaks on many feathers. Upper scapulars vary from black-spotted to entirely black, forming pair of wide black lines down either side. Fairly prominent whitish supercilium may or may not extend behind eye, brownish eye stripe typically darker on lores. Cheeks and side of neck whitish, finely streaked with brown. Underparts white, breast washed with gray-brown on sides and finely streaked with dark brown.

Uppertail coverts dotted with brown and fringed with gray, rectrices may be similarly although less conspicuously marked. Wing gray-brown above, broad white secondary tips form conspicuous stripe on rear edge of inner wing. Primaries and primary coverts darker than rest of wing. Underwing pale, axillars and all coverts white except brown marginals, forming dark fore edge of wing.

Nonbreeding. Differs from breeding plumage above by suppression of black streaks and lack of black markings on scapulars. Some individual birds apparently retain a few black-marked feathers. Uppertail coverts less heavily marked than in breeding plumage, breast less conspicuously streaked and usually paler, and head markings may be less distinct, but scapular markings represent most obvious difference.

Juvenal. Averages slightly darker than adults, and many feathers of upperparts, in particular

coverts, narrowly fringed with gray-buff. Black scapular markings often present but less well developed than those of breeding adults. Birds in this plumage may be distinguished from autumn adults as much by their unworn tertials as their minor plumage differences.

Immature. Many birds remain on wintering grounds in nonbreeding plumage with worn primaries; others may molt into breeding plumage in first summer.

Identification

All Plumages. The *long, upturned bill* looks malproportioned on such a small sandpiper, and it and the *short, orange-yellow legs* present a unique combination. From a great distance TEREK SANDPIPERS may be recognized by their *active foraging behavior*, dashing through flocks of slower-feeding species. The *plain, pale-brown upperparts*, slightly paler than a SPOTTED SANDPIPER, and black scapular stripes are distinctive in breeding adults and juveniles.

In Flight

The *white stripe on the rear edge of each wing* is distinctive, coupled with the *patterned look* caused by the contrast between blackish primaries and gray lesser and median wing coverts. Some birds look plain-winged in flight, however, so the absence of this pattern does not eliminate the species from consideration. The flight varies from rapid and erratic with deep wingbeats to

Breeding Terek Sandpiper. Tringine with long, upturned bill and short, bright yellow legs; dark scapular stripe together with worn tertials diagnostic of plumage. Sooke, BC, 21 July 1987, Tim Zurowski

straighter with shallower wingbeats, the latter recalling SPOTTED SANDPIPER. At times birds fly in a distinctive way, with neck extended and wings bowed downward and fluttered irregularly.

Voice

The flight call is a rolling series of short whistles, like a subdued WHIMBREL but with fewer notes and quite different from the multiple call of a WANDERING TATTLER. As other tringines, the species is quite vocal.

Notes

An unpublished record from Dungeness, October 27, 1972, is not documented sufficiently to be acceptable.

References

Ferguson-Lees 1959 (photographic studies), Wilson and Harriman 1989 (identification, North American occurrence).

CURLEWS Numeniini

There are nine species worldwide, all of which have been recorded from North America. All but the Slender-billed and Eurasian curlews are discussed here (Table 23).

Curlews are medium to very large shorebirds with moderate to long, conspicuously decurved bills, with the exception of the Upland Sandpiper, a smaller species with straight bill. The gap between that species and the Little Curlew, however, is very slight, and the group may have evolved from an ancestor similar to the sandpiper. Bill length in the group varies remarkably (Appendix 3, Table 24), with four distinct subgroups differing in size and relative bill length; leg length is relatively invariable in comparison (Appendix 3). The tribe is distinguished by the overall brown coloration and lack of seasonal plumage change typical of grassland shorebirds,

although not all of them inhabit grasslands. Breeding and nonbreeding adults look essentially the same, although there may be a partial molt in spring. Juveniles can be distinguished from adults at close to moderate ranges by the broader buff edges of the feathers of their upperparts.

All curlew species breed in the northern hemisphere, from the Arctic to central North America and Eurasia, and winter from lower northern latitudes (north to the Pacific Northwest and Britain) all the way to the ends of the southern continents. The constant plumage points to similar breeding and nonbreeding habitats, and the three smallest species tend to be confined to grassland. However, the larger species winter also on sand beaches and mud flats and appear to be equally at home in upland or intertidal situations.

Table 23. Curlew Distribution and Habitat Preference

Species	Hemisphere	Breeding Range	Breeding Habitat	Winter Range	Winter Habitat
Upland Sandpiper	W	subarctic/temp.	grassland	S	grassland
Little Curlew	E	arctic	tundra	equatorial/S	grassland
Eskimo Curlew	W	arctic	tundra	S	grassland
Whimbrel	both	arctic	tundra	N/equatorial/S	beach/bay
Bristle-thighed Curlew	W	arctic	mountain tundra	equatorial	beach
Far Eastern Curlew	E	subarctic	taiga	equatorial/S	bay/beach
Long-billed Curlew	W	temperate	grassland	N	bay/grassland

UPLAND SANDPIPER *Bartramia longicauda*

With short bill, long neck, and long tail, this graceful sandpiper looks like a curlew with a plover's bill. Striding deliberately through breeze-ruffled grass, mottled brown throughout the year, and shunning the waterside entirely, it represents the epitome of an upland shorebird. Sadly, it is a vanishing species in the Northwest, restricted to a few interior localities as a breeder.

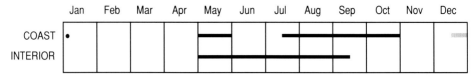

Distribution

Upland Sandpipers breed in open grassland and sizable meadows across North America from central Alaska to the Atlantic coast and well south on the Great Plains but have decreased in many parts of their range in recent years. They winter in similar habitats in the pampas of southern South America, adults migrating primarily up and down the Plains and juveniles spreading also to the Atlantic coast.

Northwest Status

Once more widespread in grassland regions, this species has disappeared from or become rare in many parts of its range in North America, and the Northwest is no exception. Never abundant in the region but probably breeding widely east of the Cascades, it is now restricted to tiny relict populations. During the last few years at least a few birds have bred east of Spokane; just across the state line near Hauser Lake, Idaho (probably extirpated); in Long Valley, Idaho; near Ovando, Montana; and scattered from northeastern to southcentral Oregon.

The Oregon populations are the largest, with a few dozen pairs in Bear and Logan valleys and additional populations with one to a few pairs each near Enterprise, La Grande, Ukiah, Prineville, and Sycan Marsh. Oregon breeding habitats lie between 3,500 and 5,000 feet elevation. Typical of shorebirds nesting in arid environments, they are present for only a short time on their breeding grounds, from May through July, with latest dates near breeding localities at Turnbull Refuge, August 2, 1929, and Logan Valley, August 11, 1980.

Upland Sandpipers have been observed about once a year as migrants in the region in the past three decades, although only rarely in spring. In that period there were seven spring records (three coastal and four interior), mostly during May, and 36 fall records (20 coastal and 16 interior), mostly during August in the interior and late August and September on the coast. Coastal records come primarily from the heavily scrutinized Vancouver region, with fewer from Washington and only one from Oregon. From the dates, most of the fall records are of juveniles. Most of our migrants may be from the somewhat isolated Alaska/Yukon breeding population, which is more secure than our local breeding birds.

A single winter record from Victoria is highly anomalous in a species that normally winters far to the south. Although almost three months later than the next-latest fall migrant, it apparently was still wandering.

COAST SPRING. *Early date* Saanich, BC, May

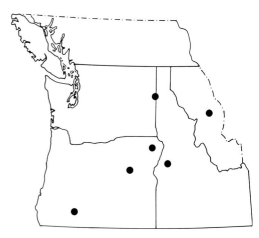

Upland Sandpiper breeding distribution

5, 1979. *Late date* Cleland Island, BC, June 3, 1976.

COAST FALL. *Early dates* Iona Island, BC, July 16, 1983; Newport, OR, July 23 to August 3, 1987. *High counts* 4 at Richmond, BC, August 30, 1979; 4 at Richmond, BC, September 7, 1985. *Late date* Grays Harbor, WA, October 6, 1963.

COAST WINTER. Victoria, BC, December 20, 1969, and January 4, 1970 (2 localities).

INTERIOR SPRING. *Early dates* Robinson Lake, ID, April 30, 1977; Spokane Valley, WA, May 1, 1972; Yoho National Park, BC, April 24, 1975. *Late date* Manning Park, BC, May 31, 1981.

INTERIOR FALL. *Early date* Ellensburg, WA, July 30, 1980. *Late date* Oliver, BC, September 13, 1980.

Habitat and Behavior

Grassland very little altered by grazing may be a requirement for successful nesting; the birds feed in short grass but apparently need taller grass and broad-leaved forbs for nest concealment. This species is absolutely committed to a life away from water, feeding in pastures, native grassland, and plowed fields during migration. It shares these habitats with an array of species that are somewhat similar, especially golden-plovers and Buff-breasted Sandpiper. On its breeding grounds it commonly perches on fence posts and even utility poles, from which males give their long whistles. It feeds by picking from the surface, walking through grass or over open ground fairly rapidly with jerky head movements accentuated by the long neck.

Display flights are lengthy and high in the air, accompanied by long, rolling whistles. The nest is placed among dense herbaceous vegetation, usually well hidden from above. Four eggs are laid, and both sexes incubate and care for the young.

Structure

Size medium. Bill short, legs medium. Wings fairly long but extend well short of tail tip; primary projection short in adults, short to moderate in juveniles. Tail long, graduated.

Plumage

Adult. Bill mostly yellowish, with tip and upper edge black. Legs greenish yellow to fairly bright yellow. Typical upland shorebird, mottled with shades of brown above and, as in closely related curlews, scarcely changing plumage. Crown dark brown; feathers narrowly edged with light brown, with narrow but well-defined pale buff median crown stripe. Hindneck and mantle dark brown, feathers narrowly edged with light brown. Scapulars and tertials brown with darker brown bars and narrow buff fringes, coverts light brown with darker brown bars and streaks. Face and neck light brown, streaked with black and contrasting with pale sandy buff throat; prominent paler eye-ring. Breast and sides light brown, heavily marked with narrow black chevrons; rest of underparts pale sandy buff.

Lower back and central uppertail coverts blackish, outer uppertail coverts black-and-white barred. Central one or two pairs of rectrices medium brown, strongly barred with black. Outer rectrices light reddish brown and less heavily barred with black, with wider subterminal bar and conspicuous white tips; outermost pair with white edges. Wing brown above, secondaries and inner primaries obscurely barred with whitish and darker brown; secondaries narrowly tipped white. Dark primary coverts contrast with paler, patterned secondary coverts. All underwing coverts, secondaries, and much of primaries barred with light gray-brown, overall effect fairly dark underwing. Axillars evenly barred brown and white.

Juvenal. Scapulars and tertials darker than those of adults, not showing adults' barred pattern, and conspicuously fringed with pale buff or white; overall effect a bird more obviously striped than barred above. Tertials dotted with black just inside their pale fringes, while in adults crossbarring extends to edges of each feather. Otherwise age classes look much alike, although some juveniles washed with rich buff all over.

Immature. First-winter birds look like adults by mid winter. First-summer birds distinguished from adults only by their much more worn flight feathers. Age at maturity not known.

Identification

All Plumages. The *bill is very short* for a sandpiper, about the length of the head, but the *long neck and small head* eliminates from consideration any plover, for example GOLDEN-PLOVERS, which are colored somewhat similarly in fall. And of course all plovers have still shorter, thicker bills.

The UPLAND SANDPIPER has the *longest tail* of any sandpiper, the tail projecting about an inch beyond the wing tips, while the two are about even or the wings a bit longer in most species. This, together with the long neck, contribute to the distinctive elongate shape.

Probably the BUFF-BREASTED SANDPIPER is the most superficially similar to the UPLAND, with similarly yellowish legs, but it is much smaller, shorter-necked, and shorter-tailed, and has its entire underparts buffy to whitish, with very little striping on the breast or barring on the sides. PECTORAL SANDPIPERS are also brown with yellowish legs but are even less similar, also smaller and with brown, striped breast sharply demarcated from white belly.

In Flight (Figure 9)

Entirely brown from above, the *long tail* gives it a distinctive shape in flight, and the *short bill* distinguishes it from the larger CURLEWS and GODWITS. The darker primaries and primary coverts of the wing tip contrast with the brown wing base, more in juveniles (fresh primaries) than in summer adults (worn, paler primaries). The tail is bright reddish-buff color on the outer feathers, with a dark subterminal band, white tip, and white edges, reminiscent of but not as colorful as a KILLDEER'S tail and even more like the tail of a COMMON SNIPE. From beneath, the strongly barred underwings may be seen, and they are distinctive when the bird lands and holds its wings up as it often does. Birds on the breeding grounds often fly with stiff, bowed wings reminiscent of a SPOTTED SANDPIPER.

Voice

One of the flight calls is WHIMBREL-like but not so loud and with the notes more run together. Where the bird is common in migration, its calls are often heard in the night sky, but it would be good fortune indeed to hear one in the Northwest.

Adult Upland Sandpiper. Grassland sandpiper with relatively short bill, long neck, long tail, and muted brown plumage; prominent barring on tertials and coverts diagnostic of adult. Attwater Prairie Chicken NWR, TX, spring 1984, Linda M. Feltner

Another flight call is a liquid *pulip*, sort of a gurgled whistle, and I have not been able to sort these two call types into different contexts. The breeding-ground song is one of nature's many magical sounds. It is a long whistle, trilled at the beginning, ascending and then descending, *whrrreeeee-wheeeeyuuuuuuu*, like a "wolf whistle of the gods."

Further Questions

This bird is endangered throughout the region, and all existing populations badly need protection. New populations should be sought everywhere in our wide-open spaces and any discovered brought immediately to the attention of conservation agencies. This sandpiper is an indicator species for undisturbed grasslands, which are extremely rare in the Northwest and must be given protection.

Notes

Much has been made of the "small head" of this species in field guides (and it unquestionably looks small-headed), yet its skull is actually about the same size as that of the similar-sized Greater Yellowlegs, which has not been called microcephalic. Perhaps the short bill and long neck create an illusion. Also, the short-billed plovers have relatively larger heads than sandpipers, and the ploverlike bill of this species may impart the feeling that it *should* have a larger head. Field guides also emphasize the relatively large eyes of the Upland Sandpiper, and indeed the eyes do look large compared with those of the yellowlegs. An entirely visual feeder like the plovers, this bird may have large eyes for the same reason.

References

Buss 1951 (breeding biology), Buss and Hawkins 1939 (breeding biology), Dorio and Grewe 1979 (nesting habitat), Higgins and Kirsch 1975 (breeding biology), Kirsch and Higgins 1976 (breeding biology and management), Stern and Rosenberg 1985 (breeding in Oregon), White 1988 (wintering grounds and migration).

LITTLE CURLEW *Numenius minutus*

This species would not have been considered for this book but for the occurrence of a juvenile that spent September 16 to October 14, 1984, in flooded fields in southern California, and another bird in the same area September 23-24, 1988. These are the only North American records of a species that breeds locally in interior Siberia and winters largely in northern Australia. It is primarily an interior migrant, is uncommon along the Asian Pacific coast, and is unlikely to occur in our region.

The LITTLE CURLEW is small indeed for a curlew, a medium-sized sandpiper with a body bulk not much greater than an UPLAND SANDPIPER, for which it could easily be mistaken at first glance. Both species occur in grasslands and are basically similar in coloration, although the SANDPIPER's legs are yellow, those of the CURLEW bluish gray, greenish, straw-colored (almost yellow), or even orange- or pink-tinged. The legs are about the same length, the bill is half again longer, and the tail is about one-eighth shorter in the CURLEW; thus the birds would look similar at a distance. The UPLAND SANDPIPER's tail projects substantially beyond its wing tips, that of the CURLEW barely. The CURLEW's bill is so slightly curved that it looks almost straight, but a closer look will reveal the distinct curvature as well as the pink instead of yellow basal coloration. Furthermore, the CURLEW has a well-developed eye stripe, lacking in the SANDPIPER.

Because of its small size and slightly curved bill, it should not be mistaken for the much larger WHIMBREL. Very short-billed juvenile WHIMBRELS still have bills half again as long as those of LITTLE CURLEWS, but because the WHIMBREL is much larger, the bill of a juvenile can look surprisingly short (see discussion of bill length under that species). Fine points to be looked for include the loral stripe, which in the LITTLE CURLEW does not reach the bill and is expanded vertically into a blotch; it is a standard stripe in the WHIMBREL.

In flight the LITTLE CURLEW is only about half the bulk of a WHIMBREL. It might more likely be mistaken for an UPLAND SANDPIPER or even a GOLDEN-PLOVER, and its bill, considerably longer than either of those species and with slight curvature, would have to be seen clearly. The flight call is somewhat like that of a WHIMBREL but shorter, often with three notes, and distinctly harsh or hoarse. One of the UPLAND SANDPIPER flight calls is also a series of whistles, but rather musical and more of them.

See ESKIMO CURLEW for an easily confused but even less likely species to occur in the Northwest.

References

Boswall and Veprintsev 1985 (biology and identification), Crawford 1978 (behavior in winter), Farrand 1977 (identification), Labutin et al. 1982 (breeding biology), Lehman and Dunn 1985 (California record and identification), McGill 1960 (behavior in winter).

ESKIMO CURLEW *Numenius borealis*

This species was not recorded on the Pacific coast even when it was abundant in North America but, as an interior and eastern migrant, certainly would have occurred in this region as a rare migrant or vagrant. It was common as an arctic breeding species from the Northwest Territories to western Alaska, with a few records of spring migrants as far west as Bering Sea islands. South of Alaska there are no records of migrants anywhere west of Colorado. Its migratory pathways and wintering grounds were much like those of the American Golden-Plover, but it has been very rare for decades and had been considered possibly extinct. Repeated sightings of migrants in recent years keep alive the hope of the optimists among us that it might increase again, but it has had many years to recover from the difficult years of the pot-hunter and has failed to do so, perhaps because factors other than hunting were also involved in its precipitous decline.

The ESKIMO CURLEW looks much like the LITTLE CURLEW but is slightly larger, has a slightly longer bill and slightly shorter legs (tarsus shorter than bill), and its wing tips project well beyond the tail (typically just to the tail tip in LITTLE). The toe tips project slightly beyond the tail tip in LITTLE in flight, not at all in ESKIMO. The wing linings are cinnamon-buff rather than the paler buff of the LITTLE CURLEW. The ESKIMO is more heavily marked below, with strong chevrons on its lower breast and flanks, while the LITTLE has a finely streaked breast and scattered narrow bars on its flanks. Like the WHIMBREL, the ESKIMO has a loral stripe rather than the loral blotch of the LITTLE. The ESKIMO is said to have less pink color on the bill than the LITTLE (less than half the bill length versus half or more), but the great variation in this characteristic in WHIMBREL and BRISTLE-THIGHED CURLEW and the scarcity of ESKIMO CURLEWS to check for similar variation causes me to be skeptical of its value. Some LITTLE CURLEWS appear to have very little pink on the bill, in fact.

This species is a bit closer than the LITTLE CURLEW to the size of a WHIMBREL, and its bill measurements just about reach those of a young juvenile of that species. With larger body, the WHIMBREL would actually look relatively smaller-billed than the ESKIMO, so plumage characters will always be necessary to distinguish them. ESKIMO CURLEW flight calls have been described as soft, repeated, melodious whistles, clearly not as loud and striking as those of the WHIMBREL. From published accounts it is not clear how the ESKIMO and LITTLE CURLEWS could be distinguished vocally, although the word "harsh" has not been applied to the call of the ESKIMO.

References

Banks 1977 (discussion of extinction), Farrand 1977 (identification), Gollop et al. 1986 (general biology and historical status), Lehman and Dunn 1985 (identification).

WHIMBREL *Numenius phaeopus*

Every spring and fall, big shorebirds in plain brown wrappers are seen roosting in tight flocks or well spaced across coastal mud flats. As they are approached, they chase one another or fly up with loud, repeated notes. At closer range the decurved bill and dark head stripes confirm they are indeed Whimbrels.

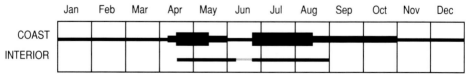

	Jan	Feb	Mar	Apr	May	Jun	Jul	Aug	Sep	Oct	Nov	Dec
COAST												
INTERIOR												

Distribution

The Whimbrel breeds around the world on arctic tundra, winters on all tropical coasts, and migrates coastally in-between. Surprisingly rare in the interior, it occurs there most often in spring.

Northwest Status

Whimbrels have been recorded in every month of the year in our region but are common only during migration. That period extends from mid (rarely early) April through May and from late June well into October, peaking in late April and early May and from the end of June to mid August on the coast. A specimen from Comox, June 17, 1940, is an adult, probably a migrant, but it could have been in either north- or southbound migration. Adults attain much higher concentrations in fall than juveniles, which occur in numbers of no more than 50 to 100 birds per day, perhaps because of their lengthier migration period. The two age classes overlap somewhat in August.

A few birds winter on the coast, occasionally as far north as southern Vancouver Island. Maximum numbers of 20 have wintered near Vancouver, six at Victoria, five at Port Angeles, and 14 at Yaquina Bay. Whimbrels also summer on the coast, with widespread occurrence in very small numbers but with flocks of up to several dozen birds at Grays Harbor, Leadbetter Point, and Yaquina Bay. Oddly, the birds are not first-year birds, as is the case with many other summering shorebirds.

In the interior the Whimbrel is a rare migrant, with very few records for interior British Columbia, western Montana, and northern Idaho, and a few more for interior Washington and Oregon; there are about half as many fall as spring records. More than two birds have been seen together only in southern Oregon. The few fall records are concentrated early in that season (over half in June and July), indicating that adults rather than juveniles wander into the interior, off the main migration route. A few birds seen in late August may have been juveniles.

Furthermore, it appears that Whimbrels are more common in spring than in fall in protected waters (Bellingham to Vancouver and the San Juan Islands), but that the reverse is true in coastal estuaries (Grays Harbor and Willapa Bay). This also leads to the conclusion that adults disperse inland from the coast more than do juveniles, a most unusual pattern for a coastal migrant shorebird, but it is further supported by evidence from the interior of the continent and elsewhere on the Pacific coast.

COAST SPRING. *High counts* 350 at Tillamook, OR, May 5, 1989; 750 at Skagit Flats, WA, May 5, 1990; 450 near Elma, WA, May 17-18, 1990; 475 at Tofino, BC, May 13, 1976; 700 at Tofino, BC, April 29, 1978.

COAST FALL. *Adult high counts* 750 at Cleland Island, BC, July 23, 1982; 400 at Leadbetter Point, WA, August 13-14, 1978; 480 at Ocean Shores, WA, July 26, 1987; 600 at Ocean Shores, WA, July 13, 1988; 155 at Yaquina Bay, OR, July 11, 1982. *Juvenile early date* Ocean Shores, WA, August 26, 1982#. *Juvenile high count* 100 at Ocean Shores, WA, September 19, 1976.

INTERIOR SPRING. *Early date* Summer Lake, OR, April 9, 1974. *High counts* 13 at Lower Klamath Lake, OR, May 5-6, 1979; 13 at Lower Klamath Refuge, OR, April 29, 1982; 24 at Malheur Refuge, OR, May 20-21, 1989. *Late date* Potholes Reservoir, WA, June 11, 1951.

INTERIOR FALL. *Early date* Spokane Valley, WA, June 21, 1982. *High count and late date* 3 at Walla Walla River delta, WA, August 30, 1987.

Habitat and Behavior

Whimbrels are birds of many habitats, from sand dunes and outermost ocean beaches to salt marshes and mud flats. They are the only North American curlews that regularly feed on rocky shores, and birds have wintered in the San Juan Islands on small, isolated rocks. They are most common on mud flats, where they scatter and maintain feeding territories through the low-tide period, then aggregate at high-tide roosts in dense flocks that may number in the low hundreds. During wet springs, they concentrate in flooded fields near the coast in large numbers. They fly long distances between roosting and feeding areas, often high in the air. Other large shorebirds often associate with them, and conversely, individual Whimbrels have been found in the interior with groups of Long-billed Curlews on several occasions. Adults are highly territorial when they arrive here from the north in midsummer, chasing one another frequently and giving calls like those of the breeding grounds.

Whimbrels often feed on drying expanses of mud that are shunned by other shorebirds. Although long-billed, they feed visually and pick prey from the surface or just beneath it as they move fairly rapidly over a variety of substrates. In some regions they specialize on small crabs, which they extract from their burrows with a twist of the bill.

Structure

Size large. Bill long and decurved. Males slightly smaller than females; bills average 8 mm shorter, but field differentiation would be virtually impossible. Juveniles with shorter bills than adults, difference averaging about 10 mm. Bill length varies from at least 60 to 99 mm in American Whimbrels (Table 24), variation apparent in field. Legs short to medium, outer and middle toes connected by evident web at base. Wings reach about to tail tip, primary projection moderate to long in adults and juveniles. Tail rounded.

Plumage

Adult. Very plain bird, head stripes only markings conspicuous at a distance. Bill color varies from entirely black to reddish at base, at maximum extent reddish involving basal half of lower mandible. Variation not associated with age, sex, or seasonal change; extremes observed in group of ten adults in July. Legs dull blue-gray. Head and neck finely streaked with brown, conspicuous eye stripe and broader lateral crown stripes dark brown. Mantle feathers brown with gray-brown fringes; scapulars, tertials, and coverts gray-brown with darker brown bars and light brown notches. Underparts whitish; neck and breast striped with dark brown, sides and undertail barred with same color.

Lower back, uppertail coverts, and tail gray-brown with narrow dark brown bars. Wing looks brown at a distance, at close range secondaries and inner primaries conspicuously barred and notched with dark brown and light buffy brown. Dark outer primaries and especially primary coverts contrast strongly with rest of wing. Underwing and axillars slightly paler, entirely barred gray-brown and buff.

Fresh-plumaged adults, not likely to be seen in Northwest, slightly more brightly marked than lightly worn birds in spring and more heavily worn birds in early autumn. Very little body molt takes place in adults moving through region.

Juvenal. Somewhat more brightly patterned than adult, with dark brown mantle feathers,

Table 24. Curlew Bill Lengths

	Adult		Juvenile	
Species	Male	Female	Male	Female
Little[a]	38-46	40-48		
Eskimo*	48-53	47-60	44-53	42-57
Whimbrel	74-89	88-99	60-99[b]	
Bristle-thighed[a]*	69-88	83-96		
Far Eastern*	128-170	154-201	107-132	116-159
Long-billed[a]	105-173	144-219		

[a]juveniles included in adult measurements
[b]sexes combined
*small sample
Sources: Prater et al. 1977, Cramp and Simmons 1983, museum specimens.

scapulars, and tertials dotted and notched with pale buff to whitish. Coverts medium brown with broad buff notches and fringes. At a distance more contrast between "back" (scapulars) and "sides" (coverts) in juveniles than in adults. Lower back and rump barred dark brown and buff; tail evenly barred gray-brown and dark brown, thus rump slightly more distinctly marked than tail.

Immature. First-year birds should be distinguishable from adults in spring by very worn primaries and in summer by molting primaries.

Subspecies

The American subspecies (*N. p. hudsonicus*) is the regularly occurring form. The Siberian subspecies (*N. p. variegatus*) is a regular spring and fall migrant in far western Alaska. There are two sight records of this subspecies from the Northwest: an adult at Ocean Shores on May 16, 1987, and a juvenile at Clatsop Beach on September 25, 1985 (photographed). Another was recorded on the north coast of California, October 29 to November 1, 1981. This subspecies can be distinguished easily from the American Whimbrel by its white lower back, which shows up as a contrasty pale spot in flight even at great distances. It also averages slightly smaller, which might be evident in direct comparison.

European Whimbrels (*N. p. phaeopus*), which have been recorded from the Atlantic coast of North America, are typically pure white on the

Adult Whimbrel. Dull brown curlew with conspicuous dark head stripes and barred sides; uniform upperparts characteristic of adult. Churchill, MB, 9 June 1982, Dennis Paulson

Juvenile Whimbrel. As adult, but mantle, scapulars, and tertials dark, with pale notches. Ocean Shores, WA, September 1988, Robert Ashbaugh

Adult Bristle-thighed Curlew. Much like Whimbrel, but washed with buff, contrastily patterned above, and lacking bars on lower sides; coarseness of markings diagnostic of adult. St. Paul Island, AK, 31 May 1989, Robert Sundstrom

lower back, while Siberian Whimbrels are spotted with brown in the same area. Nevertheless, in my experience *variegatus* look white-backed in the field, contrary to some published accounts (for example, Hayman et al. 1986: 138 ["often little contrast between back and rump"], 139, Fig. 129h). All three birds sighted on the American Pacific coast were described as conspicuously "white-rumped." Asiatic birds are more heavily marked below than American ones, with less of the belly unmarked; the underwings are more heavily barred, and the pale underparts and markings on the upperparts of juveniles are whitish rather than buff. Asiatic birds behave and sound like their American counterparts, although on Australian barrier-reef islands, some of them forage within low, dense forest and perch in canopy trees!

Identification

All Plumages. WHIMBRELS are *large, brown shorebirds with downcurved bills*. At a distance this species looks medium brown, the monotony of its plumage relieved only by distinctly *paler, almost white, belly*. At closer range it is more complexly patterned with stripes and bars but still plain compared with many shorebirds. The contrasty *dark and light head stripes* place it as a WHIMBREL (but see the similar but very rare BRISTLE-THIGHED CURLEW). The LONG-BILLED CURLEW is the only common look-alike, and its longer bill, plainer head, and reddish tones allow easy distinction.

In a roost, where bills are frustratingly hidden in scapulars, WHIMBRELS stand out by their plain brownness against other similar-sized birds, all of which are less common on our coast—the reddish LONG-BILLED CURLEW and MARBLED GODWIT and the grayish WILLET. All of these birds are more uniformly colored as well, different from the WHIMBREL with its pale belly contrasting with brown back, breast, and sides.

In Flight (Figure 8)

Again, the very *plain brownness* of this bird is distinctive, even when the *long, decurved bill* is not apparent. The wings are plain and without stripes (strongly barred underwings and barred flight feathers visible at close range), and the back and tail look similarly colored at any distance. MARBLED GODWITS and LONG-BILLED CURLEWS are more reddish overall, especially their richly colored wings, and again they lack the contrast of the WHIMBREL'S *paler belly*. These larger shorebirds often fly in lines or vee formations, while smaller species fly in more compact flocks.

Voice

WHIMBRELS frequently vocalize in flight, with loud repeated whistles *whi whi whi whi whi whi whi* (typically five to seven notes) unlike any other regional species. WANDERING TATTLER voices are much higher, the individual notes not exactly alike as in WHIMBREL. Only the rare UPLAND SANDPIPER has a similar voice, and its individual notes are sharper and more run together.

Further Questions

It would be interesting to monitor territorial interactions to see if they decrease as the season progresses. Also, are juveniles as aggressive as adults are toward one another? Further study might show that the variation in bill color in this species conforms to some pattern of age, sex, or season. We do not know whether adults, immatures, or both winter in the Northwest, nor why our summering individuals are not first-year birds, as in other species.

Photos

Photos of adults are in Bull and Farrand (1977: Pl. 245), Udvardy (1977: Pl. 216), Armstrong (1983: 119), and Farrand (1983: 361; 1988a: 151; 1988b: 142), of a juvenile in Terres (1980: 791). A juvenile *variegatus* is shown in Pringle (1987: 200), and a good comparison of *hudsonicus* and *variegatus* in flight is in Pringle (1987: 204).

References

Bayer 1984 (summering on Oregon coast), Handel and Dau 1988 (fall migration in Alaska), Johnson 1973 (reproductive condition of summering birds), Johnson 1977 (plumage and molt of summering birds), Johnson 1979 (summary of 1973/1977 papers), Mallory 1982 (winter territoriality), Skeel 1978 (vocalizations), Skeel 1982 (sex determination in the hand), Skeel 1983 (breeding biology), Williamson 1946 (breeding biology).

BRISTLE-THIGHED CURLEW *Numenius tahitiensis*

This rare Pacific wanderer seldom alights on continental shores during migration but should be kept in mind nevertheless. A virtual facsimile of the Whimbrel, its warm buffy rump and tail should be seen and its distinctive call heard for definitive identification.

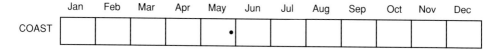

	Jan	Feb	Mar	Apr	May	Jun	Jul	Aug	Sep	Oct	Nov	Dec
COAST					•							

Distribution

The Bristle-thighed Curlew has one of the most restricted breeding ranges of any sandpiper, known so far only from hill-country tundra on Alaska's Seward Peninsula and Yukon River delta. During winter it ranges across the Micronesian and Polynesian islands of the South Pacific.

Northwest Status

Moving back and forth between far-flung south Pacific islands and a very restricted Alaskan breeding range, the populations of this species are so small that records out of range should be few and far between. It is rarely recorded even in Alaska south of its breeding range and staging grounds on the nearby coast; most birds are seen in autumn migration on the Yukon-Kuskokwim delta. Nevertheless, some birds winter as far east as the longitude of Vancouver Island, and, migrating across a region of westerlies, should reach our coast occasionally. There is only one unequivocal record from our region, a bird found at the northern tip of Vancouver Island on May 30, 1969, and collected the next day.

COAST SPRING. Grant Bay, BC, May 30-31, 1969#.

Habitat and Behavior

The Grant Bay bird was by itself on an ocean-front beach, and the species may be most likely to occur in such a situation in this region, as it inhabits beaches throughout its winter range. It also forages on mud flats, coral reefs, and grasslands, perhaps a broad habitat spectrum because of the restricted size of the islands on which it winters. It occurs singly and in small flocks and forages for crabs and other invertebrates much as do other curlews.

Structure

Size large. Bill long and curved; legs short to medium, outer and middle toes connected by evident web at base. Wings reach tail tip, primary projection moderate. Tail rounded. Peculiar among shorebirds in having long, stiff "bristles" (unbranched contour feathers) at base of legs, in juveniles as well as adults.

Plumage

Adult. Bill all black or with reddish base that may occupy as much as half or more of lower mandible and extend onto upper. Females may have more pale coloration on bill than males. Legs dull blue-gray. Coarsely marked with dark brown and buff above. Head and neck generally pale, finely streaked with brown. Conspicuous eye stripe and slightly broader lateral crown stripes dark brown. Mantle feathers, scapulars, tertials, and coverts dark brown with bright to pale buff fringes and notches. Throat whitish, foreneck and breast streaked with brown, sides of breast barred with brown. Underpart markings variable in intensity, from somewhat obscure to quite dark and conspicuous. Remainder of underparts at darkest buff, paler on center of belly; at lightest mostly whitish.

Lower back brown, uppertail coverts unmarked buff. Tail pale rufous to buff (in some individuals clouded with brownish) with narrow black bars. Wing looks brown at a distance, but secondaries and inner primaries barred and notched brown and buff; dark primary coverts contrast with rest of wing. From below wing entirely barred brown and buff, except outermost four or five primaries dark with light mottling. Axillars buff with narrow brown bars.

As in other curlews, scarcely any change between nonbreeding and breeding plumage in adults, but of course very worn birds in autumn duller than same individuals in spring.

Juvenal. Similar to breeding adult and only slightly more brightly patterned, in particular on tertials, than adults with which it would occur in autumn. Some indication that juveniles have more pale coloration on bill than adults. By late October on wintering grounds much duller and darker above, with substantial percentage of light markings of upperparts worn off. Little indication of buff except on rump and tail.

Immature. Immatures spend first one or two years on island wintering grounds, distinguishable from adults in spring by extremely worn primaries (and generally scruffy appearance) and in summer by molting flight feathers. Some South Pacific summering birds much more heavily marked than others and, in general, substantial amount of variation in comparison with other curlews.

Identification

All Plumages. This bird is amazingly WHIMBREL-like: the two species are identical in size and proportions and at any distance look about the same, the distinguishing marks subtle and becoming obvious only at closer range. This species differs from the LONG-BILLED CURLEW in about the same ways a WHIMBREL does: smaller size, shorter bill, conspicuous head stripes, and lack of cinnamon on the wings. It is paler below than the LONG-BILLED, the pale area even more extensive than in a WHIMBREL.

BRISTLE-THIGHED should be distinguishable from WHIMBREL in the following ways at close range:

• BRISTLE-THIGHED has a *bright buff rump and tail, the rump unmarked*, while WHIMBREL has a heavily barred rump and tail the same color as the back. This is perhaps the only diagnostic mark, short of seeing bristles, to distinguish the two species, but it is not easily seen on birds at rest.

• The brightest BRISTLE-THIGHED are *buffy all over*, brighter on rump and tail. Some BRISTLE-THIGHED, however, are not much brighter than typical juvenile WHIMBRELS.

• BRISTLE-THIGHED is *more coarsely and contrastingly marked above*, with feathers about half dark (almost black) and half buff, whereas adult WHIMBRELS are very plain brown, almost unmarked above, and juvenile WHIMBRELS have finer spotting on back and wings. Nevertheless, juvenile WHIMBRELS almost overlap with BRISTLE-THIGHED in the variegation of their upperparts.

• The *barring on the sides is less extensive* in BRISTLE-THIGHED, leaving the pale color of the belly more prominent than in WHIMBREL (often no bars at all are visible on BRISTLE-THIGHED). Some WHIMBRELS are lightly barred, however.

• At very close range the *unmarked undertail coverts* of BRISTLE-THIGHED is a diagnostic mark, contrasting with the substantially (but not always conspicuously) barred undertail coverts of WHIMBREL.

• At still closer range long, bristlelike feathers may be visible at the base of the legs of BRISTLE-THIGHED; WHIMBREL lacks them. Identification would be cinched by a good photograph of the bristles!

A few additional characteristics are helpful but not definitive; seeing more of them adds confidence in identification.

• The breast streaks are less extensive in BRISTLE-THIGHED, again contributing to the impression of more extensively pale underparts.

• The axillars, underwing coverts, and inner flight feathers are similarly barred in both species, but the BRISTLE-THIGHED has its outer primaries mottled rather than notched as in the WHIMBREL. This could be photographed on a bird that had lifted its wings but is unlikely to be a helpful field mark.

• The bill on the average has more pale color at the base in BRISTLE-THIGHED, but the overlap is too extensive for this to be as useful a field mark as has been claimed. Birds with about half of the bill reddish, the color extending onto the upper mandible, are likely to be this species.

In Flight

The *long, decurved bill* marks it as a curlew, and the overall buffy cast and especially the *bright buff rump and tail*, the rump somewhat paler, should allow distinction from WHIMBREL if well seen. Otherwise, from any distance the two species look essentially identical.

Voice

As is often the case with similar shorebirds, the flight calls are not only important in calling attention to the bird, but ultimately they may be the best "field marks." The BRISTLE-THIGHED CURLEW's flight call is a two- or three-noted whistle *too-ee*, *tee-oo-wit* or *wheet-o-weet*, similar to but less plaintive than the familiar call of the BLACK-BELLIED PLOVER. The call may be repeated several times by a flying bird but is very different from the WHIMBREL's series of identical notes.

Further Questions

An overall census of this rare species on either breeding or wintering grounds is a long-overdue but very difficult endeavor. From the numbers in the South Pacific in winter, the species seems more numerous than is evident from searches of potential breeding grounds in Alaska; does it breed in Siberia as well? Differences between males and females and between adults and juveniles in the amount of reddish coloration on the bill need to be confirmed. Also, variation in coloration in specimens is greater than in other curlews, and a further analysis and understanding of this would be of value. It is possible that an undiscovered breeding population is subspecifically different from that breeding on the Seward Peninsula. I leave it to the reader to attempt to determine the function of the structures that give this species its name.

Notes

Spring sight records of this species from the region include two from Leadbetter Point, May 18, 1980, and one there, May 1, 1982; and one from Blackie Spit, May 13-14, 1983. There are also three sight records in fall, each of them accompanied by photographs that are inconclusive: near Tofino, September 1, 1982; Sidney Spit, September 11, 1986; and two from Bandon, September 16, 1981. The photograph from Tofino is excellent but still not definitive, pointing out the great difficulty in distinguishing this species from the Whimbrel.

I consider the following characteristics, mentioned elsewhere as field marks for Bristle-thighed, not very useful: (1) Breast streaked instead of marked with chevrons as in Whimbrel (Armstrong 1983). In fact both species look streaked at any distance, and only in the hand does the preponderance of chevron-shaped marks on the Whimbrel's breast become apparent. (2) Wing linings cinnamon (Johnsgard 1981). Johnsgard (1981) shows in error a bird with unmarked underwings and axillars and far too many tail bars, and Hayman et al. (1986: 139) incorrectly show the underwing with notched outer primaries.

Photos

The photograph of a "Bristle-thighed Curlew" in Terres (1980: 791) is an adult Whimbrel. Armstrong (1983: 120) shows a worn adult on breeding grounds, Farrand (1983: 361) a worn juvenile on wintering grounds.

References

Allen and Kyllingstad 1949 (nesting), Conover 1926 (general biology), Gill et al. 1988 (status in Alaska), Handel and Dau 1988 (fall migration in Alaska), Johnson 1973 (reproductive condition of summering birds), Johnson 1977 (plumage and molt of summering birds), Johnson 1979 (summary of 1973/1977 papers), Marks et al. 1990 (longevity and flightlessness), Widrig 1983 (Northwest records).

FAR EASTERN CURLEW *Numenius madagascariensis*

Brown like a Whimbrel but with Long-billed Curlew head markings, size, and long bill, this very large Siberian curlew seems to combine characters of both species. Its heavily streaked underparts should facilitate recognition if it occurs again in the Pacific Northwest.

	Jan	Feb	Mar	Apr	May	Jun	Jul	Aug	Sep	Oct	Nov	Dec
COAST									•			

Distribution

The Far Eastern Curlew breeds in the subarctic muskegs of Siberia and migrates along the western shore of the Pacific and over the ocean to its wintering grounds on the coasts of Indonesia, New Guinea, and Australia.

Northwest Status

This east Asian species occurs on western Bering Sea islands in small numbers both spring and fall, and the prediction that occasional individuals might turn up in our region was confirmed when a juvenile was photographed near Vancouver.

COAST FALL. Boundary Bay, BC, September 24, 1984.

Habitat and Behavior

Far Eastern Curlews forage in a dispersed fashion on mud flats in Asia and Australia but roost and fly in flocks like Whimbrels. In our area individuals would probably be by themselves (as was the Boundary Bay bird) or with other large species such as Whimbrels. They move at a steady pace over the substrate, picking from the surface or probing into crustacean burrows. Crabs are common prey items, as in other curlews.

Structure

Size very large (huge relative to most other shorebirds). Bill very long and highly curved. Bill averages shorter in males than females and shorter in juveniles than adults (Table 24), probably about as in Long-billed Curlew, although good series of measurements distinguishing sex and age categories not available. Legs medium, outer and middle toes connected by evident web at base. Wings reach tail tip, primary projection short in adults and moderate in juveniles. Tail rounded.

Plumage

Adult. Bill black with varying amounts of pink to reddish on base of lower or both mandibles (at extreme of paleness basal two-thirds). Legs dull gray. Head and neck pale buff, finely striped with dark brown. Feathers of upperparts vary from gray-brown to warm reddish brown, with dark brown central stripes; scapulars and tertials also regularly barred with medium brown. Underparts

Juvenile Far Eastern Curlew. Long-billed curlew with prominently striped upperparts; relatively short bill for species and muted striping below diagnostic of juvenile. Kyushu, Japan, mid September 1984, Urban Olsson

sandy buff to whitish, heavily striped with brown; sides brown-barred.

Lower back and uppertail coverts buffy brown, striped and barred with brown. Tail evenly barred with gray-brown and dark brown, pale bars tinged with buff. Wing brown at a distance, from above inner primaries and secondaries barred and notched brown and whitish; dark primary coverts contrast with rest of wing. Underwing coverts and axillars conspicuously barred brown and white.

Breeding and nonbreeding plumages of adults similar but autumn birds duller, less likely to be reddish above than those in spring.

Juvenal. More distinctly marked above than adult. Scapulars dark brown with buff notches, tertials brown with buffy-brown notches. Averages less heavily striped below than adult, with finer breast stripes. Some overlap between age classes, probably best distinguished in fall by plumage wear.

Immature. First-year birds distinguishable

from adults by very worn primaries in spring and molting flight feathers in summer.

Identification

All Plumages. This bird matches the LONG-BILLED CURLEW in *large size, long bill, and fine head streaking*, with similarly great variation in bill length. Note that very short-billed birds have bills little longer than those of the longest-billed WHIMBRELS. FAR EASTERN is *duller brown* than LONG-BILLED, drab sandy or pale buff with dark brown markings rather than the rich cinnamon-buff of the American species. It is furthermore *heavily striped above*, more heavily marked than the two common curlews and approached only by BRISTLE-THIGHED, and *heavily striped below*, even the juveniles more heavily marked than any other Northwest curlew. The dark striping covers much of the underparts in many individuals, and the side feathers are both striped and barred.

In Flight

The *very long, decurved bill* places it as a curlew. The coloration is like that of other curlews, plain brown at a distance, but with *no hint of reddish* as in the LONG-BILLED. The underwings are barred brown and white, even more heavily than those of a WHIMBREL. Distinction from a WHIMBREL at a distance would be difficult except by larger size and longer bill.

Voice

The usual flight call is a plaintive *curee, curee, curee* given fairly often, quite distinct from calls of the common curlews of the region.

References

Hindwood and Hoskin 1954 (winter ecology), Piersma 1986 (feeding habits in migration), Robertson and Dennison 1979 (feeding ecology in winter).

LONG-BILLED CURLEW *Numenius americanus*

These biggest of North American shorebirds make the spring prairie air ring when they launch into their spectacular roller-coaster display flights. Unobtrusive by late summer, they disperse quickly to the coast and points south, and only a few are seen away from interior breeding grounds. The long bill is especially so in females, but they use it with as much finesse as any other shorebird.

Adult Long-billed Curlews. Long-billed curlew with mottled and barred upperparts and cinnamon underparts; crossbarred coverts diagnostic of adults; male (shorter bill) on left. Port Isabel, TX, 25 January 1991, Dennis Paulson

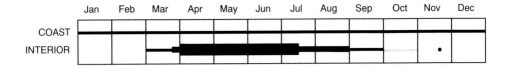

	Jan	Feb	Mar	Apr	May	Jun	Jul	Aug	Sep	Oct	Nov	Dec
COAST												
INTERIOR											•	

Distribution

The Long-billed Curlew breeds in the interior short-grass prairies of the western Great Plains and Great Basin. Its populations move to dry southwestern grasslands and the California and Mexican Pacific coasts in winter, with much smaller numbers to the Gulf Coast of the United States.

Northwest Status

This shorebird is never associated with the shore in its breeding habitat, the grass and sagebrush steppes from the Cariboo Parklands of southern British Columbia through the Columbia Basin of Washington, central Oregon, and southern Idaho. It is not known to nest in the moister grasslands of northeastern Washington or northern Idaho but does so locally in western Montana (recently around Eureka). Overall the species has decreased substantially as a nesting bird in the region with the constant attrition of grassland to agriculture. For example, the thousands that nested around Pocatello in the 1960s were reduced to a few in the 1970s. In areas not so altered it remains common, as for example 3,000 birds estimated in the Harney Basin in 1979

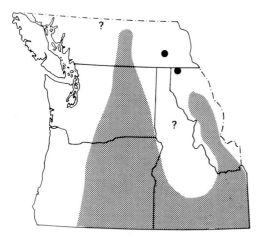

Long-billed Curlew breeding distribution

(1,500 estimated at Malheur Refuge in more recent years). Local populations total in the dozens in most other parts of the breeding range.

The Long-billed Curlew is one of the earliest shorebirds to appear on its breeding grounds, arriving in late March throughout the region, even as far north as the Okanagan Valley. After breeding the adults disappear quite early, usually by mid July from British Columbia and Washington. The earliness of its season is appropriate for the dry habitats it frequents, which become less and less productive as summer proceeds. Migrants gather into flocks in mid summer, and juveniles are found at the waterside with other migrating shorebirds until at least mid August. There are few records after August, but some of them are surprisingly late.

The vast majority of our Long-billed Curlews fly directly between breeding grounds in the Northwest interior and wintering grounds in California and western Mexico. Spring migration is little in evidence on the coast, but birds have appeared between late March and early June in areas in which they did not winter. The later spring records from the coast and Willamette Valley almost surely indicate a restricted northward movement of subadult birds well after breeding birds have begun nesting in the interior. June records are difficult to allocate; a bird at Blackie Spit, June 9 and 15, 1982, may have been summering, and birds have summered on the Oregon coast. Fall migration probably begins in mid June, and seven to eight were at Willapa Bay, June 16-24, 1978, and four at Ocean Shores, June 17-18, 1988. A few migrants have been encountered in meadows high in the Cascades in mid and late July. A few may be seen throughout the fall to early October on outer coast and protected shores. There are about two dozen records in the Vancouver area and about a half-dozen from southern Vancouver Island.

A few birds are recorded every winter on the Northwest coast, but only at Tokeland in Willapa Bay is there a wintering flock, usually with Marbled Godwits and consisting of up to about 80 birds. Most individuals of this flock have departed by early April.

COAST SPRING. *Early date* Westport, WA, March 27, 1982 (20 with Whimbrels). *Late date* Beach Grove, BC, June 7, 1981.

COAST FALL. *Juvenile early date* Sea Island, BC, August 15, 1966#. *High count* 110 at Long Beach Peninsula, WA, August 30, 1989.

COAST WINTER. *Northernmost record* Boundary Bay, BC, December 29, 1979, to February 16, 1980. *High counts* 80 at Tokeland, WA, February 6, 1983, and August 18, 1987.

INTERIOR SPRING. *Early dates* Grangeville, ID, March 14; Malheur Refuge, OR, March 20, 1980; Prescott, WA, March 23, 1905; Vernon, BC, March 3, 1949.

INTERIOR FALL. *High counts* 50 at Vernon, BC, July 14, 1959; 600 at Boardman, OR, July 8, 1980; 454 at Nampa, ID, July 17, 1976. *Late dates* Okanagan Landing, BC, October 29, 1902, October 11, 1919; Malheur Refuge, OR, October 15, 1964; Klamath Lake, OR, November 19.

Habitat and Behavior

Although basically natural-grassland breeders, Long-billed Curlews have adapted locally to human alteration of the landscape by nesting in grain fields and pastures. They are in fact abundant in short-grass pastures in some parts of the region, as much so as ever in native grassland. In migration they occur in moist upland environments, on shores of reservoirs, and on coastal mud flats, where they associate with Whimbrels and Marbled Godwits. They are surprisingly adapted to grasslands, wintering in larger numbers in the interior of California than on the coast. They feed by striding rapidly over the substrate, reaching ahead with long bill to probe into mud for crabs and other large invertebrates or capture insects among the prairie plants.

Nesting sites are usually on dry grassland, in some cases near moister areas where the young can find adequate food during the dry early summer. Nests are in open grass, often adjacent to an object such as a rock or dried cow dung. They are very difficult to find, as incubating birds flush either at a great distance or almost underfoot. Typically four eggs are laid, with both sexes incubating and caring for the young.

Structure

Size very large. Bill very long and gradually decurved. Females slightly larger, bills average at least 30 mm longer than those of males, so individuals in mated pairs easily sexed. Although sexes overlap, very long-billed individuals should always be females and very short-billed individuals males. Variation impressive, with longest-billed birds of astonishing proportions. Juveniles average shorter-billed than adults (Table 24). Difference in bill length probably produces subtle difference in bill shape that also distinguishes sexes—male bills look evenly curved, female bills look straighter at base and more curved toward tip. Legs medium, outer and middle toes connected by evident web at base. Wings reach tail tip, primary projection short to nonexistent. Tail rounded.

Plumage

Adult. Bill black with reddish base to lower mandible, legs blue-gray. Overall appearance sandy to warm reddish buff, heavily marked above and sparsely marked below with brown. Upperparts dark brown; mantle feathers and scapulars notched with buffy brown, tertials with gray-brown; coverts cinnamon-buff with dark brown streaks and bars. Crown dark brown, feathers finely edged with light buff. Faintly indicated darker eye stripe and conspicuous white eye-ring. Underparts rich cinnamon-buff; face, neck, and breast finely striped and sides sparsely barred with brown.

Lower back, uppertail coverts, and tail light reddish brown barred with dark brown. Wing rufous at a distance; from above inner primaries and secondaries light rufous, barred with brown. Outermost primaries dark; primary coverts even darker, contrasting with rest of wing. Underwing light rufous, flight feathers and greater and middle coverts narrowly and relatively sparsely barred with brown; lesser and marginal coverts darker and brighter, each with fine dark shaft streak. Axillars rufous, very sparsely barred with brown.

Adults look similar all year, duller and more worn just before autumn molt. Some tertials molted in spring, imparting more variegated look to these birds.

Juvenal. Tertials more brightly marked in fall

than those of worn adults, with darker, wider stripes and bars on cinnamon-buff as opposed to grayish buff ground color. Coverts typically with wide teardrop-shaped marks along shafts and without bars or notches, quite distinctive from adult markings on folded wings. Also on average finer breast stripes and more lightly striped breast than adult. Condition of tertials should be carefully noted to confirm age distinction in fall.

Immature. One-year-old birds, most of which remain on wintering grounds, distinguished in spring and early summer by very worn primaries compared with those of adults and by wing molt during May to July. Two-year-olds return to breeding latitudes but apparently do not breed; probably not recognizably different from adults.

Subspecies

Most populations of the Northwest are considered to be the northerly breeding *N. a. parvus*, but differences between it and the more southerly breeding *N. a. americanus* (listed only from southern Idaho) are very slight, involving smaller size in the former, and it is not clear where the break between the two subspecies occurs. Nevertheless, the bill length of northern birds averages about an inch shorter than that of southern birds.

Identification

All Plumages. Very large, with very long decurved bill, this largest of Northwest shorebirds could only be mistaken for another curlew or, if the bill were not visible, a MARBLED GODWIT. The WHIMBREL and the very rare BRISTLE-THIGHED CURLEW are much smaller with much shorter bills, no more than twice the head length, while the LONG-BILLED CURLEW bill is usually substantially more than twice the head length. Shortest-billed LONG-BILLEDS approach the other species in bill length, and their *even, fine head streaking* is then the best mark, as well as their *more reddish coloration* overall. Some LONG-BILLED CURLEWS have a dark loral stripe and a darker lateral crown stripe, surprisingly like a WHIMBREL, but the top of the head is finely and evenly streaked. These birds could be confusing enough, especially if short-billed juveniles, to warrant using other plumage features for identification. See the accidental FAR EASTERN

CURLEW, a similar-sized but differently colored species.

In a roost, with shorebird bills frustratingly hidden, LONG-BILLED CURLEWS can be distinguished from WHIMBRELS by their larger size, overall reddish-brown coloration, and uniformly shaded reddish belly; a WHIMBREL's belly is paler than its breast and sides. MARBLED GODWITS are similarly cinnamon and uniformly shaded but smaller, Whimbrel-sized, with largely crossbarred or unmarked breast, LONG-BILLED CURLEW with striped breast. These two species are difficult to distinguish at a distance without size comparison.

In Flight (Figure 8)

The *large size and cinnamon coloration*, especially of the underwings, furnish easy identification features, once the *very long curlew bill* is seen for distinction from MARBLED GODWIT. In the CURLEW more of the outer primaries are dark than in the GODWIT, and there is less contrast between the front and rear of the wing, as the flight feathers are more heavily patterned. See FAR EASTERN CURLEW for characteristics of that species in flight.

Voice

The flight call, a loud *curlee,* is less frequently given by this species than is the case in the noisy multi-noted WHIMBREL. During the display flight, loud, repeated *curlee* calls are followed by a descending series of sharp whistles.

Further Questions

The sex of wintering Long-billed Curlews should be determined from some assessment of bill length if possible. The age classes represented also have not been determined. It would be of interest to look closely at late-spring migrants and summering birds on the coast to see if they show the very worn or molting primaries characteristic of first-year birds. A major question, of course, is the destination of these birds, on the move well after adults have established breeding territories in the interior.

The sexual difference in bill length surely correlates with some difference in feeding methods, as has been shown for Eurasian Curlews, in which males were more likely than females to feed in upland situations (Townshend 1981); observers should watch for this where numbers of birds are present. Are the birds that winter on grasslands and mud flats—these very different environments—different sexes? And what is special about the north end of Willapa Bay that allows a large flock of Long-billed Curlews and Marbled Godwits to winter there, while they are absent from most of the region at that season? Examination of additional specimens or captured birds will be necessary to clear up questions about the validity and occurrence of the two subspecies of this species.

Notes

The sexes of this species have been distinguished by bill shape (Allen 1980, Bucher 1978), with female bills relatively straight at the base and male bills evenly curved throughout. In fact, this distinction is difficult to use in the field, and longer-billed males and shorter-billed females probably have bill shapes approaching those of the other sex.

Photos

Bull and Farrand (1977: Pl. 246), Udvardy (1977: Pl. 217), Farrand (1983: 363; 1988a: 150; 1988b: 143), Hosking and Hale (1983: 113), and Chandler (1989: 157) show adult females. Perhaps photos of males have been passed over as not definitive enough. The photos in Chandler (1989: 157) show the difference in adult and juvenile covert markings.

References

Allen 1980 (breeding biology), Bucher 1978 (sex distinctions), Forsythe 1970 (vocalizations), Forsythe 1973 (chick growth), Graul 1971 (nesting behavior), Grinnell 1921 (comments on subspecies), Redmond and Jenni 1982 (fidelity to breeding area), Redmond and Jenni 1986 (population ecology), Sadler and Maher 1976 (nesting and foraging), Stenzel et al. 1976 (winter feeding behavior and diet), Sugden 1933 (breeding biology).

GODWITS Limosini

There are four species of godwits, all of which have been recorded from North America; all are discussed here (Table 25).

Godwits are large shorebirds with only moderately long legs but with quite long, straight to slightly upcurved bills. Relative bill and leg lengths are fairly similar in all four species (Appendix 3, Table 26), although the Marbled is relatively long-billed and the Hudsonian relatively short-billed, the Black-tailed relatively long-legged and the Bar-tailed relatively short-legged. If these differences are reflected in differences in foraging behavior, it is not immediately evident, although the long legs of the Black-tailed may better adapt it for freshwater life.

Godwits typically breed on tundra and grassland from arctic to north-temperate latitudes, wintering on coastal beaches and mud flats from warmer temperate latitudes through the tropics to the ends of the southern continents. All species

Table 25. Godwit Distribution and Habitat Preference

Species	Hemisphere	Breeding Range	Breeding Habitat	Winter Range	Winter Habitat
Black-tailed	E	sub./temperate	marsh	N/equatorial/S	bay/marsh
Hudsonian	W	subarctic	taiga	S	bay/marsh
Bar-tailed	E	arctic	tundra	N/equatorial/S	bay
Marbled	W	temperate	grassland	N	bay

are adapted for probing in mud in the nonbreeding season, often while wading in shallow water. Three of the four undergo substantial seasonal plumage change: brown-speckled above and bright rufous below in breeding season, gray-brown above and white below in nonbreeding season. Males are conspicuously brighter than females in breeding plumage. The Marbled is an exception, curlewlike in plumage with substantially less seasonal change and sexual dimorphism.

Table 26. Godwit Bill Lengths

Species	Adult		Juvenile	
	Male	Female	Male	Female
Black-tailed[a]	67-88	73-93		
Hudsonian	69-84	74-96	64-80*	83-93*
Bar-tailed	75-94	102-119	72-88	96-116
Marbled	82-122	90-130	87-114	96-126

[a]juveniles included in adult measurements
*small sample
Sources: Prater et al. 1977, Cramp and Simmons 1983, museum specimens.

BLACK-TAILED GODWIT *Limosa limosa*

Individuals of the Siberian population (subspecies *melanuroides*) of the Black-tailed Godwit are rare spring migrants through western Alaska and might occur in our region, as likely in freshwater wetlands as coastal estuaries. This is a tall species, with longer neck and legs than other godwits. The bill is relatively short and straight, quite distinct from those of BAR-TAILED and MARBLED GODWITS. It is different in proportions from these two species, but it is rather similar to HUDSONIAN in all plumages, and a good look in flight would be necessary for certain identification.

Breeding adults are often paler and more heavily barred with black below than are HUDSONIANS, the undertail coverts are usually white (typically reddish in HUDSONIAN), and the back is often marked with rich rufous (brown to blackish in HUDSONIAN), the latter characteristic diagnostic of BLACK-TAILED. The bill is usually more extensively pink at its base than is the case in HUDSONIAN, the bright color extensive on both mandibles.

Nonbreeding BLACK-TAILED are quite plain gray-brown above, with a gray-brown breast and white belly. Eastern Asian birds are darker than those from Europe and look very much like HUD-

SONIAN, even to the variation in the extent of the supercilium (used as a field mark by some authors). Juveniles are more richly colored, bright buff on the neck and breast and more strongly patterned above, with black-centered scapulars and coverts and more brightly notched and fringed tertials. Juvenile BLACK-TAILED and the more brightly marked juvenile HUDSONIAN present an overall impression of being spotted, in contrast with the striped look of BAR-TAILED and MARBLED juveniles. The straight bill is a good mark for BLACK-TAILED in any plumage.

In flight this species differs conspicuously from the HUDSONIAN by its white rather than black wing linings, even more vivid white wing stripes (almost WILLET-like from above), and more white in the tail base. The legs project farther beyond the tail tip, with the tarsi showing in BLACK-TAILED but not HUDSONIAN. It is typically silent away from the breeding grounds, occasionally giving single, low-pitched contact notes.

References

Lind 1961 (breeding biology), Ward and Bullock 1988 (winter feeding ecology).

HUDSONIAN GODWIT *Limosa haemastica*

This bird is almost mystical to shorebirders, as its conservative migration strategies keep it out of many parts of North America. Like Willets, nonbreeding birds are gray on the ground and flashy in flight, easily recognized by a combination of long, pink-based bill, conspicuous wing stripes, black underwings, and black-and-white tail. It is a fortunate observer who comes across one in the Northwest, where drab juveniles are more likely to be seen than colorful breeding-plumaged adults.

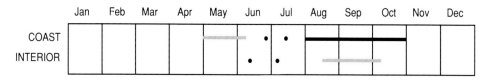

Distribution

The Hudsonian Godwit breeds in scattered populations on arctic tundra from Alaska to Hudson Bay, in small numbers in most areas. In fall most of the population heads off the Atlantic coast to South America, while in spring most birds return up the middle of the continent. Wintering grounds are fairly restricted to pampas marshes and coastal mud flats of the Atlantic side of southern South America.

Northwest Status

A rare migrant, this species has been reported at least 45 times in the coastal subregion through 1990, primarily in British Columbia (31 records) but including 5 times in Washington and 9 in Oregon. In the interior there are 13 records: 5 from Washington, 3 from Oregon, 2 from British Columbia, 2 from Idaho, and one from western Montana. The majority (78 percent) of records are thus coastal, although this frequency is strongly influenced by observer concentration. Of the coastal records, two-thirds are concentrated in protected waters, the other third on the outer coast.

Based on its migration patterns, spring records should predominate, but our records are concentrated in fall, the great majority juveniles. As the juveniles typically follow the adults in fall and are rare anywhere in the interior, the preponderance of fall records in the Northwest may indicate the existence of a small Pacific Ocean population, perhaps birds that breed locally in southern Alaska and move south across the ocean to a yet-undiscovered wintering ground.

Coastal spring occurrences are primarily in May: once each on British Columbia's, Washington's, and Oregon's outer coasts, once on southern Vancouver Island, and four times near Vancouver. The origin of a bird in breeding plumage on June 17-18 is equivocal, but I am considering two other early fall records from the southern British Columbia mainland as probably southbound adults. Other coastal records involve mostly if not entirely juveniles, primarily in August and September.

There are only three records of adults from the interior, two in spring and one in early July, and not many of juveniles, mostly in September. Almost invariably single or a few birds have been seen, except for a flock of 16 on the Oregon coast.

COAST SPRING. *Early date* Ucluelet, BC, April 30, 1983. *Late date* Vancouver, BC, June 8, 1985.

COAST SUMMER. Vancouver, BC, June 17-18, 1978.

COAST FALL. *Adults* Crescent Beach, BC, June 22-29, 1974; Vancouver, BC, July 15, 1983. *Juvenile early date* Iona Island, BC, August 4, 1970. *High counts* 3 at Boundary Bay, BC, September 13, 1987; 3 at Ocean Shores, WA, September 24, 1966; 16 at Tillamook Bay, OR, August 17, 1980. *Late date* Bandon, OR, October 27, 1985.

INTERIOR SPRING. Fortine, MT, May 10, 1969; Yakima River delta, WA, June 8, 1987.

INTERIOR FALL. *Adult* Rupert, ID, July 7, 1919. *Juvenile early date* Summer Lake, OR, August 22, 1987. *Late date* Soap Lake, WA, October 2, 1983.

Habitat and Behavior

Most individuals of this species seen in our region have been on mud flats, including their freshwater equivalents on receding lake shores; one bird was reported from a brackish slough. Typical godwits, they probe for a living and travel in flocks, but in our area occurrences are likely to be of single birds. Most of the birds seen in the Northwest have been by themselves, but at least one was with a Marbled Godwit.

Structure

Size at low end of large. Bill long to very long and slightly upcurved. Females larger than males, bill averaging 13 mm longer but with much overlap (Table 26). Legs medium to long. Wings extend well beyond tail tip, primary projection long in adults and juveniles. Tail square, varying to double-notched.

Plumage

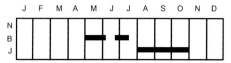

Breeding. Bill black with up to 60 percent of base of lower or both mandibles pink; in many individuals upper edge of upper mandible remains dark. Legs black. Sexual dimorphism marked, breeding males darker than females both above and below. In males crown dark brown with fine pale streaks, rest of head pale with finely brown-streaked cheeks. Conspicuous white supercilium, dark eye stripe. Upperparts blackish, mantle feathers buff-fringed and scapulars and tertials notched with reddish brown to gray-brown. Some individuals much more brightly marked than others, dullest ones almost solid black above. Coverts gray-brown with darker brown streaks. Foreneck chestnut, finely striped with brown; breast and belly dark chestnut with abundant but scattered black bars. Undertail coverts chestnut, heavily barred with black and tipped with white.

In females upperparts more mottled than in males, tertials often plain rather than brightly marked; color of underparts paler, approaching that of dowitcher or Marbled Godwit, often intermixed with white but also barred as in males.

Lower back black; uppertail coverts white,

Breeding male Hudsonian Godwit. Small godwit with long wings and heavy barring in this plumage; dark brown upperparts and rufous underparts indicate male. Churchill, MB, 14 June 1982, Dennis Paulson

longer ones black. Tail black with narrow white tip, outer rectrices white-based. Wing brown above, wing stripe on midwing formed from white tips of outer greater secondary coverts and inner greater primary coverts and white marks on outer web of inner four primaries. Flight feathers and primary coverts darker than secondary coverts. From below wing blackish except for broad white stripe along it formed from largely white central secondary coverts, increasingly narrower white tips on outer secondary and inner primary coverts, and white marks on inner primaries. Axillars blackish.

Molt into breeding plumage begins during March, most birds in full plumage by end of April. Body molt again in migrants during August and September, traces of breeding plumage in some individuals into October.

Nonbreeding. Drab, light gray-brown on upperparts and breast; belly white. White supercilium and dark loral stripe fairly conspicuous. Fine dark streaks on feathers of upperparts produce streaked effect only at close range. So far no records of birds in this plumage in Northwest.

Juvenal. Averages less pale color at bill base than adult, typically only basal third of lower mandible. Browner than nonbreeding adult; upperparts darker, underparts washed with buff. Some mantle feathers and scapulars fringed with buff, some scapulars and longer tertials prominently notched with same color.

Immature. Like nonbreeding adult but with some retained darker coverts. Oversummers at least in some wintering areas in worn nonbreeding plumage with very worn primaries. Some individuals acquire partial breeding plumage.

Identification

All Plumages. The *long, pink-based bill* identifies this as a godwit, and only birds in nonbreeding or juvenal plumage with their bills hidden might be mistaken for any other kind of shorebird. See BLACK-TAILED GODWIT for comparisons.

Breeding Plumage. Breeding adults are *rich, dark reddish below*, darker and obviously smaller than MARBLED GODWIT. They differ from the similarly colored, slightly larger, and equally rare BAR-TAILED GODWIT in the same plumage by being *either fairly heavily barred beneath*

(females) or *very dark, white-spotted above* (males). BAR-TAILEDS show sparse if any ventral barring and are medium brown above, evenly striped and mottled with darker brown. In this plumage HUDSONIAN has *basal half of bill pink*, BAR-TAILED often with bill mostly black (pink no more than one-third of length and often confined to lower mandible).

Nonbreeding and Juvenal Plumages. Nonbreeding birds are *plain-backed* in HUDSONIAN (faintly scalloped in juveniles), heavily patterned with stripes and bars in BAR-TAILED and MARBLED. Plainest winter BAR-TAILED can be almost as plain as HUDSONIAN but usually have at least faintly indicated stripes. HUDSONIAN might be mistaken for WILLET but for its pink-based bill that appears slightly upcurved. The GODWIT's bill is conspicuously longer, over one-and-a-half times the head length, while that of the WILLET is only a bit more than the head length.

In Flight (Figure 8)

A HUDSONIAN GODWIT in any plumage is distinctive in flight, with its *white-based, black-tipped tail, conspicuous wing stripes, and*

Breeding male Bar-tailed Godwit. Godwit with relatively long wings and short legs; barred tail visible above wingtip; dark upperparts, richly colored underparts, and relatively short bill indicate male. Wales, AK, June 1977, Jim Erckmann

blackish underwings. It is most like a WILLET because of the flashiness of its flight pattern, but a WILLET has much wider white wing stripes, vividly bordered by black alulae and primaries above and occupying a third of the wing surface below. The underwing of a HUDSONIAN may have some white in it, but it is mostly black. The godwit's *very long bill* is also distinctive. See the similar BLACK-TAILED GODWIT, thus far unreported from the region.

Voice

Mostly silent in migration, a flushed bird might give the flight call *ta-wit*, like that of a MARBLED GODWIT but higher pitched.

Further Questions

Further information about Alaska breeding populations of this species and their migration pathways would help to clarify its status in the Northwest.

Photos

Bull and Farrand (1977: Pl. 232), Terres (1980: 793), Johnsgard (1981: color Pl. 56), and Farrand (1983: 365) show breeding males, Farrand (1983: 363; 1988a: 153) breeding females. The "winter plumage" bird in Farrand (1983: 365) is a juvenile. Armstrong (1983: 117) and Farrand (1983: 365) show upper- and underwing surfaces of a juvenile and a breeding male, respectively.

References

Baker 1977 (summer diet), Grieve 1987 (identification), Hagar 1966 (breeding biology), Hagar 1983 (migration), Sutton 1968 (sexual dimorphism), Williamson and Smith 1964 (breeding in Alaska).

BAR-TAILED GODWIT *Limosa lapponica*

Anticipation of finding this rare species adds to the excitement of scrutinizing flocks of large shorebirds. In large numbers the Bar-tailed Godwit crosses the Pacific between Alaska and the South Seas, liable to easterly displacement by ocean storms. In spring it is distinctively reddish or whitish below, but fall juveniles must be carefully distinguished from the much more common Marbled Godwit.

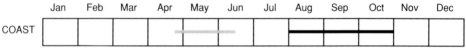

Distribution

The Bar-tailed Godwit breeds on arctic tundra from Alaska west to northern Europe and winters on Old World coasts from New Zealand and other Pacific islands to the Near East, southern Europe, and much of Africa.

Northwest Status

This Eurasian species has a foothold in Alaska, where it breeds widely. These Alaska populations winter in the Old World, but the thousands that have been seen at staging grounds on the Alaska Peninsula must head south across the Pacific rather than follow the roundabout route via the Bering Strait. Therefore it is not surprising that a few of them find their way to the American coast farther south, and they are rare migrants in the Northwest.

Perhaps because of improved observer abilities and coverage, records have proliferated in the last decade, after the second report from the region in 1972 (first in 1931). Now observations average about two per year on the coast, with sightings three times as frequent in fall as in spring. Coastal records through 1990 total 43 (almost neck and neck with the Hudsonian Godwit), with 9 from British Columbia, 22 from Washington, and 12 from Oregon. Most records (58 percent) are outer-coastal, with secondary concentrations at Dungeness (19 percent) and the Vancouver-Victoria area and one at Tacoma. This distribution pattern is different from that of the Hudsonian Godwit, records of which are concentrated in protected waters (although none at Dungeness!) rather than on the outer coast.

Migrants have been seen primarily in May and

early June in spring and in late August, September, and early October in fall. Spring birds have been adults in breeding plumage, except for two in "winter" plumage on April 25 and June 1 in Oregon (these latter were more likely subadults or females in dull breeding plumage). Five of eight August birds for which age was determined were breeding-plumaged adults, three of them females. Five late-August and September records in 1985-89 also involved birds in "winter" plumage, one of which was present until October 11. Most September and October birds for which age was specified were juveniles. Records are of single or paired birds except for groups of six on the Washington coast and four on the Oregon coast.

COAST SPRING. *Early date* Yaquina Bay, OR, April 25, 1980. *High counts* 4 at Bandon, OR, May 14, 1988; 6 at Leadbetter Point, WA, June 8, 1974. *Late dates* Dungeness, WA, June 10, 1980; Willapa Bay, WA, June 10, 1983.

COAST FALL. *Early date* Dungeness, WA, August 1, 1979. *Juvenile early date* Delta, BC, August 21, 1987. *Late date* Colebrook, BC, October 31, 1931.

Habitat and Behavior

Most Northwest records come from ocean beaches and estuaries. This is a typical godwit, foraging by moving along steadily on wet mud or in shallow water with bill probing the substrate. The females with their longer bills often feed in deeper water than the males. Bar-tailed Godwits form large flocks in migration and winter, but the birds on Northwest beaches are likely to be single. Where common, they spread out to forage more than is typical of Marbled Godwits. Individuals have been seen with Marbled Godwits on a number of occasions, by themselves on others.

Structure

Size large. Bill long to very long, slightly upcurved. Bill of adult females averages about 17 mm longer than that of adult males; bill in juveniles averages slightly shorter than in adults (Table 26). Sexual dimorphism in bill length greater than in other godwits, measurements of adults not overlapping. Legs medium. Wings extend beyond tail tip, primary projection long in adults and juveniles. Tail approximately square.

Plumage

Breeding. Bill mostly blackish. In breeding adults pink color typically restricted to basal fifth, usually primarily lower mandible (apparently less in Asian/Alaskan birds than in European subspecies). Legs black. Breeding-plumaged male dark brown above with buff fringes and notches on many feathers, most extensive on coverts. Unmarked rich chestnut below with some white on lower belly. Supercilium pale rufous, conspicuous eye stripe dark brown. Undertail coverts and often sides with black chevrons.

Breeding-plumaged female much duller; upperparts much as in male but pale markings more extensive, producing striped look like that of nonbreeding adults and juveniles. Underparts vary from slightly paler than that of male to almost entirely white with only scattered rufous feathers. Supercilium whitish. August birds that were either breeding-plumaged females or subadults had bills colored as nonbreeders and juveniles.

Lower back brown; uppertail coverts vary from gray-brown barred with brown to white, barred with black. Tail whitish with fairly wide black bars. Wing brown above, white tips of outer greater secondary and inner greater primary coverts forming inconspicuous wing stripe. Darker primary coverts contrast with rest of wing. Underwing coverts and axillars prominently barred brown and white; flight feathers light gray-brown, inner primaries speckled with white.

Nonbreeding. Pale color at bill base more extensive than in breeding adult, occupying about basal 40 percent of both mandibles. Rather plain gray-brown above with darker brown feather centers producing overall striped look. Conspicuous white supercilium and dark eye stripe. Mostly white beneath, breast grayer and with scattered brown streaks. Molt into this plumage takes place during August and September.

Juvenal. Bill color as in nonbreeding adult. More heavily marked than nonbreeding adult.

Crown and mantle brown with narrow buff fringes; scapulars and tertials marked with buff to gray-brown notches, coverts predominantly buff with dark brown streaks. White supercilium conspicuous, dark eye stripe may be more conspicuous in front of or behind eye. Neck and breast finely streaked with brown and washed with buff; belly white, sides sparsely streaked and barred with brown.

Immature. Plumage usually as in nonbreeding adult (or dull breeding female) but primaries very worn or in molt in summer. First-winter birds may retain many worn juvenile tertials and coverts. First-summer birds may be in only partial breeding plumage.

Subspecies

Presumably all of our Bar-tailed Godwits are of the subspecies *L. l. baueri*, breeding in Alaska and Siberia. This form, slightly larger than the European *L. l. lapponica*, can be distinguished from it by its less contrasty flight pattern. Typically in *baueri* the lower back is brown and the rump and tail are whitish, heavily barred with brown, so the tail looks only slightly paler than the rest of the upperparts, while in *lapponica* the lower back and rump are unmarked white, contrasting with the barred tail. The difference is comparable to that between Siberian and European Whimbrels. European Bar-tailed Godwits, unlikely to occur in our region, have a flight pattern with white more extensive than a Greater Yellowlegs but less than a Spotted Redshank, while Siberian/Alaskan birds have a flight pattern more like that of a Red Knot. The underwings and axillars of *baueri* are more heavily barred (regularly barred brown on white ground color) than those of *lapponica* (white with scattered bars if any), thus looking darker in the field.

Identification

All Plumages. This species is a typical godwit, *large with a very long, slightly upcurved, pink-based bill*, but it differs proportionally from the still larger MARBLED GODWIT in a number of ways.

• It is relatively *shorter-legged*, although this would only be obvious with direct comparison.

• It is *longer-winged*, being a long-distance migrant, and this difference is important, because

the wings in BAR-TAILED project about an inch beyond the tail in resting birds and not more than a half-inch in MARBLED.

• The *primary projections are longer* in BAR-TAILED, as long as the distance from bill base to eye in BAR-TAILED and distinctly shorter than that in MARBLED.

• The BAR-TAILED'S bill is slightly shorter, sex for sex, but there is so much overlap that this difference has no value as a field mark.

Breeding Plumage. In addition to proportions, coloration differences allow separation of BAR-TAILED from MARBLED. In breeding plumage BAR-TAILED is:

• *variable*, from chestnut (males) to cinnamon to mixed rufous and white or entirely whitish below (females), only a small percentage of individuals uniformly cinnamon like MARBLED;

• *unbarred* or with scattered short chevron-shaped bars or even spots and/or streaks below, unlike heavily barred underparts of MARBLED (but immature MARBLED are unbarred in spring and summer);

• *often white-bellied*, unlike uniform underparts of MARBLED.

Some individuals have *plain or striped tertials* (always barred in MARBLED). See also HUDSONIAN GODWIT.

Nonbreeding Plumage. In nonbreeding plumage BAR-TAILED is *plain gray-brown above and whitish below*, unlike MARBLED GODWIT, which is very reddish in all plumages. In this plumage it might be mistaken for a nonbreeding HUDSONIAN GODWIT, but BAR-TAILED always looks *striped on upperparts and breast*, HUDSONIAN essentially plain.

Juvenal Plumage. It is in this plumage that BAR-TAILED is most similar to MARBLED in its juvenal and nonbreeding plumages, and the following points should be checked.

• BAR-TAILED is *two-toned and paler beneath*, with buffy breast and whitish belly (entirely buffy cinnamon in MARBLED).

• BAR-TAILED has *rump and tail whitish, barred* (reddish, barred in MARBLED).

• BAR-TAILED has *pale supercilium and dark eye stripe extending conspicuously behind eye* (both fade out quickly in MARBLED).

MARBLED GODWITS can be worn enough by early spring to look paler than usual below, but during that period BAR-TAILED should be in grayer-backed nonbreeding plumage. In addition to these characters, BAR-TAILED has primaries whitish on their inner webs, MARBLED reddish, and the differences can often be seen when an individual stretches or otherwise exposes its wings.

Juvenile Bar-tailed Godwit. As adult, also distinguished from Marbled by paler breast, whitish belly; buff-fringed coverts and buff-notched tertials diagnostic of plumage. Victoria, BC, 20 September 1984, Tim Zurowski

Breeding Marbled Godwit. Godwit with relatively short wings and long legs, pale areas washed with cinnamon-buff; heavily barred underparts diagnostic of plumage. East of Newbrock, AB, 14 July 1975, Dennis Paulson

Juvenile Marbled Godwit. Shape and color diagnostic; unbarred underparts and lightly marked coverts indicate juvenile; bill with less black than breeding adult. San Diego, CA, November 1988, Jo Anne Rosen

In Flight (Figure 8)

The large size and *very long bill* proclaim it a godwit. A juvenile of this species in flight would look much like a MARBLED GODWIT, but it can be distinguished by its contrasts, the whitish belly against the buffy breast and the paler rump and tail against the brown back. A few birds show no contrast in the rump and tail, however. In flight BAR-TAILED shows *brown wings with faint wing stripe*, unlike the very reddish wings of MAR-BLED. See above under Subspecies for further discussion. A well-seen or photographed bird in flight should show an additional field character: because of its *shorter legs* the feet project beyond the tail tip to only about half the length of the toes, or about a half-inch, while in MARBLED they project just about to the full length of the toes, about an inch.

Voice

Usually silent in migration, the species has several flight calls, including a sharp *kip* or *kip kip* and a lower-pitched *kurruck* comparable to the "godwit" call of the MARBLED.

Further Questions

Careful determination of age in fall birds remains of importance, as it will be interesting to compare the frequency of juveniles to adults in rare as well as common species for a thorough understanding of shorebird migration strategies.

Notes

A published record from Neah Bay, July 2, 1974, is not accepted, as photographs of it are not definitive.

Contrary to Farrand (1983: 366) and National Geographic Society (1987: 110), bill length in Bar-tailed and Marbled Godwits overlaps greatly. The range for Asian Bar-tailed is 72-119 mm, for Marbled 82-130 mm. Also, Marbled Godwit is not "much larger" than Bar-tailed (Farrand 1983: 366); it takes some effort to distinguish a size difference when they are seen together in the Northwest.

Some Bar-tailed Godwits in Japan and Australia appear quite white-rumped in flight, and it appears that *baueri* varies more in rump color than is evident from literature descriptions and illustrations; compare the painting in Hayman et al. (1986: 135) with the flight photographs of *baueri* in Pringle (1987: 311) and *lapponica* in Chandler (1989: 201). This seems a more reasonable assumption than the presence of *lapponica* in these flocks so far away from the normal range of that subspecies.

Photos

Photographs in Armstrong (1983: 118) and Farrand (1983: 367) show the sexual dimorphism well. The "winter" bird in Keith and Gooders (1980: Pl. 108) is not typical of that plumage and may be a fresh breeding-plumaged female. The "Bar-tailed Godwits" in Terres (1980: 793) are mostly breeding male and female Black-tailed Godwits; the long-billed birds in the background are Bar-tailed. Two birds on the near edge of the flock are in nonbreeding plumage, a Black-tailed to the left and a Bar-tailed to the right.

References

Conover 1926 (breeding biology), Goss-Custard et al. 1977 (diet), Hindwood and Hoskin 1954 (winter ecology), Smith and Evans 1973 (winter feeding ecology and behavior).

MARBLED GODWIT *Limosa fedoa*

Marbled Godwits evoke the beaches of southern California, where they run unafraid among the bathers. Bills buried to the hilt like big dowitchers, roosting in clumps of Whimbrels, or stitching the air in rufous flight, they are an infrequent but welcome sight on Northwest beaches. When not hiding their very long and slightly upcurved two-toned bills in mud or scapulars, these big reddish shorebirds are unmistakable.

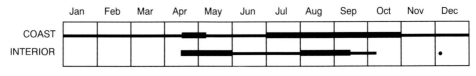

	Jan	Feb	Mar	Apr	May	Jun	Jul	Aug	Sep	Oct	Nov	Dec
COAST												
INTERIOR												

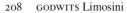

Distribution

The Marbled Godwit is almost confined as a breeding species to the moist grassland of the northern Great Plains of North America, moving from there to the coasts of southern United States and Mexico for the winter. There are also isolated tundra-breeding populations in southwestern Alaska and at James Bay.

Northwest Status

Small to large flocks of this species migrate along the outer coast, with occurrences primarily in April and early May and from July into October. Many more, presumably juveniles, are seen in September than earlier in the fall. The large storm-blown flock at Ashland was quite unusual for a Willamette Valley locality, and there are only a few other records for that valley. North and east of Grays Harbor the species is generally rare, with no more than a few seen each year in the intensely scrutinized Vancouver and Victoria regions, and records are slightly more frequent in spring than in fall.

A few birds have wintered as far north as Vancouver, but most continue south into California or Mexico. Only in Willapa and Coos bays are there wintering flocks in this region, with counts from 70 to 270 birds in recent years. A mixed wintering flock of Marbled Godwits and Long-billed Curlews is now traditional at Tokeland, and the godwits remain in numbers through mid April. Some spring migrants, especially those in late May and June, may be immature birds that did not move to their breeding grounds (most recently observed late-spring godwits appeared to be in nonbreeding plumage). Small flocks summer in a few localities, apparently all first-summer birds and including both sexes.

The Marbled Godwit has bred just outside the Northwest region in Montana and should be sought in grassy valleys west of the Continental Divide in that state. In the interior it is a widespread migrant from late April to mid May and again from early July to mid September. In that subregion it is rare in British Columbia and uncommon in Washington but more frequent in the east and south, nearer to its primary migration route. A bird at Lower Klamath Refuge in December was extraordinarily late if not wintering.

Few concentrations of more than two dozen birds have been reported in the interior, but counts have reached 150 in spring and 300 in fall. A high count on July 10 certainly involved adults, but the August 10 count of 300 could not be assigned to age category. Common in the region at the turn of the century, this species became rare due to hunting pressure during the early part of the 1900s, but it has increased here as elsewhere on the continent in recent years. With its grassland breeding habitat becoming more limited, its populations may never reach former numbers.

COAST SPRING. *High counts* 145 at Ashland, OR, April 25-26, 1981; 250 at Bandon, OR, April 27, 1988; 82 at Blaine, WA, May 6, 1960; 236 at Ocean Shores, WA, April 26, 1986, and 600 there, April 28, 1991.

COAST SUMMER. Leadbetter Point, WA, June 16, 1978 (6).

COAST FALL. *High counts* 175 at Willapa Bay, WA, October 19, 1979; 137 at Willapa Bay, WA, September 9, 1980; 270 at Tokeland, WA, November 4, 1990 (probably wintering); 100+ at Bandon, OR, September 10-23, 1978; 100+ at Yaquina Bay, OR, November 6, 1978.

COAST WINTER. *Northernmost* Victoria, BC, October 28, 1978, to April 20, 1979 (2).

INTERIOR SPRING. *Early dates* Ravalli Refuge, MT, April 23, 1969; Ninepipe Refuge, MT, April 23, 1978; Malheur Refuge, OR, April 21, 1978, February 28, 1991. *High counts* 150 at Ninepipe Refuge, MT, spring 1963; 100 at Ninepipe Refuge, MT, April 23, 1978*; 70 at Deer Flat Refuge, ID, April 25, 1980; 73 at Joseph, OR, May 15, 1982. *Late dates* Malheur Refuge, OR, June 1, 1965; Potholes Reservoir, WA, June 3, 1951.

INTERIOR FALL. *Early date* American Falls Reservoir, ID, June 18, 1980. *Juvenile early date* Willow Creek, WA, August 21, 1940#. *Northernmost records* Okanagan Landing, BC, August 7, 1910; Columbia Lake, BC, August 31, 1941. *High counts* 110 at Malheur Refuge, OR, July 10, 1973; 300 at American Falls Reservoir, ID, August 10, 1976. *Late dates* Malheur Refuge, OR, October 6, 1972; Lower Klamath Refuge, OR, December 3, 1978.

Habitat and Behavior

Marbled Godwits are usually seen in the Northwest on coastal mud flats or, when in the interior, on the muddy shores of big, drying-up reservoirs. They are most likely to be seen in protected estuaries, and they tend to be more localized than other large shorebirds, perhaps because of their strong predilection for flocking. In recent years, large numbers have fed on the golf course at Ocean Shores in spring.

Godwits feed by walking slowly and probing steadily, a very different gait from the purposeful striding and occasional picking or probing of a curlew. They also remain in flocks as they feed, unlike the territorial curlews, and at a distance look more like overgrown dowitchers than they do like similar-sized Whimbrels. The bill is plunged to its entire length into the mud, even the head submerged repeatedly. They seek out the company of Whimbrels and Long-billed Curlews at high-tide roosts and often fly with them.

Structure

Size large. Bill long to very long, slightly upcurved. Females larger than males, bills average 19 mm longer but much overlap (Table 26). Unlike situation in curlews, bills of juveniles grow quickly to about same length as those of adults, presumably to facilitate deep probing. Legs medium to long. Wings extend just beyond tail tip, primary projection short in breeding adults and moderate in juveniles. Tail squarish, central pair of rectrices projecting slightly.

Plumage

Breeding. Bill black at tip and bright pink on basal half, legs dark brownish to black. Looks brown above and buff below from a distance. Crown fairly dark brown, face and neck pale buff with brown streaks. Pale anterior supercilium and dark loral stripe. Mantle feathers, scapulars, and tertials dark brown, notched with buff (tertials barred black and buff at most extreme). Coverts buff, striped and barred with dark brown. Underparts rich buff, breast and sides finely but heavily barred with brown. Males average more heavily barred than females.

Lower back, uppertail coverts, and tail light reddish brown, barred with brown. Wing rufous, in finer detail secondaries and inner primaries rufous, faintly marked with brown; outer primaries and especially greater primary coverts dark brown, contrasting with rest of wing. Un-

derwing entirely rufous, axillars rufous with sparse brown bars; primary tips brown.

Adults can look very dark above in late summer, when pale markings have worn off.

Nonbreeding. Seasonal plumage change less well-marked than in other godwits but still more substantial than in similarly brown curlews. Nonbreeding adults essentially like breeding adults but unbarred below or with sparse barring on sides.

Juvenal. Like nonbreeding adult but virtually lacks breast streaks and no trace of bars. Coverts less heavily marked than those of adult, with poorly developed barring; typically mostly buff with brown central wedge or irregular linear markings. From a distance coverts look paler and plainer than back in juvenile, similar to back in adult. Axillars rufous, without bars. Tertial and primary wear diagnostic in late summer, but some individuals probably cannot be aged after adult's autumn molt.

Immature. Takes at least two years to mature. First-winter birds show juvenile coverts. First-summer plumage distinguished from adult by lack of bars on underparts, very worn primaries in spring and early summer, and flight-feather molt in midsummer.

Subspecies

Two subspecies occur in the Northwest, although almost nothing is known of their distribution and relative abundance, as so few specimens have been collected in the region. *L. f. fedoa* breeds in the northern Great Plains and winters from central California south on the Pacific coast. It may furnish the inland records of Marbled Godwits from the Northwest. *L. f. beringiae* breeds locally on the Alaska Peninsula and is probably the form that winters in the Northwest and occurs in some numbers in migration on the Washington coast. The two differ in average measurements of bill, wing, and tarsus but can be distinguished only in the hand. Although Alaska birds average shorter in all these measurements, they appear to be slightly heavier than interior birds. The great size variation among birds in some migrating flocks may indicate the presence of both subspecies.

Identification

All Plumages. Godwits are *big shorebirds with very long, straight to slightly upcurved, pink-based bills*. This species is *cinnamon-brown all over*, the back heavily marked and the underparts uniform except for the barring in breeding adults. HUDSONIAN GODWITS, very rare, are either brown above and chestnut below or gray above and white below. The almost equally rare BAR-TAILED GODWIT is much more similar to the MARBLED, especially the juveniles (see that species). Adult BAR-TAILED are rarely as uniformly colored as MARBLED, usually either darker reddish or whitish below.

MARBLED GODWITS often associate with similar-sized WHIMBRELS, from which they can be distinguished by their richer reddish coloration and evenly colored underparts (WHIMBRELS are brown with whitish belly), even if bill shape cannot be seen. LONG-BILLED CURLEWS are even more like MARBLED GODWITS in color but considerably larger, usually looking longer-necked. Feeding behavior is diagnostic at a distance; any large shorebirds feeding in a group are GODWITS rather than CURLEWS. The only other large, plain shorebirds are WILLETS, which are gray above and white below.

In Flight (Figure 8)

MARBLED GODWITS can be recognized by their *very long, straight bill* and their *entirely reddish* coloration in flight, both brighter and paler-looking than WHIMBRELS, with which they commonly fly. Look for bill shape, visible at a great distance, and bill length, conspicuously longer in GODWIT. LONG-BILLED CURLEW is colored very similarly but larger; GODWIT's bill, between LONG-BILLED and WHIMBREL in length, will have to be seen.

Voice

The flight call is usually two-syllabled, something like *kerreck* and sounding enough like "godwit" to be memorable, quite different from the *curlee* of the LONG-BILLED CURLEW and the staccato whistles of the WHIMBREL.

Further Questions

It would be worthwhile to determine migration timing of adults versus juveniles in fall and to know which age stage(s) spend the winter in our region. If adults, they should be molting their wings some time during the fall; if juveniles, no wing molt would occur during that time. Late spring migrants, which in late May or early June should be on their nesting grounds if they were intending to breed, should be scrutinized; first-year birds would have very worn primaries. Only by measurements of birds of known sex in the hand will we be able to determine the breeding populations from which Northwest godwits originate, but the relatively longer tarsus of interior birds might be manifested in more of the foot showing beyond the tail tip. Flight photos of individuals from both populations would be of interest.

Photos

An excellent photograph of a Marbled Godwit in flight is in Terres (1980: 793).

References

Gibson and Kessel 1989 (subspecies), Higgins et al. 1979 (breeding ecology), Kelly and Cogswell 1979 (winter movements and habitat use), Ryan et al. 1984 (habitat selection).

TURNSTONES Arenariini

This group comprises two very distinctive species, both of which are common in the Northwest (Table 27). As similar as they are, it is interesting that the Ruddy is a catholic and cosmopolitan bird, migrating to most coastal habitats, well into the southern hemisphere and all across the oceans, while the Black is restricted to North American Pacific rocky shores.

Turnstones are medium-small, short-legged sandpipers with hard, short, pointed bills that are used as pry bars as well as forceps. They are sexually dimorphic in breeding plumage, more so in the Ruddy, and plumage change between seasons is moderate. The two species are almost identical in habits, both breeding on arctic tundra and wintering on the coast. They are as similar in their aggressiveness, chattering calls, and flashy flight patterns as they are in their quick-stepping and rapid-pecking foraging behavior.

Table 27. Turnstone Distribution and Habitat Preference

Species	Hemisphere	Breeding Range	Breeding Habitat	Winter Range	Winter Habitat
Ruddy	both	arctic	tundra	N/equatorial/S	beach/bay/rocks
Black	W	arctic	tundra	N	rocks

RUDDY TURNSTONE *Arenaria interpres*

Shorebird success stories, these "seashore starlings" are at home on palm-fringed South Seas beaches and winter-wave-washed New England rocks as well as Northwest beaches and mud flats. Chattering like windup toys, they dash through shorebird flocks on their way to sundry foraging opportunities, poking wedge-shaped bills into every nook and cranny. Cobbles and seaweed fly into the air and slower-moving sandpipers jump as they pass. Their frenetic activity, along with calico flight pattern and bright spring colors, place them near the top of the list of favored waders for many of us.

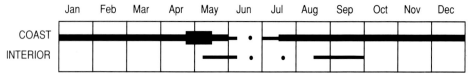

Distribution

The Ruddy Turnstone is an arctic tundra breeder all the way across both hemispheres. It is similarly wide-ranging in winter, throughout the world on temperate and tropical coasts.

Northwest Status

These birds are common spring and less common fall migrants, most frequent on the outer coast but also regular on protected shores. Spring migration begins late in April and peaks in early May, tapering off to the end of that month. Small flocks are the rule, with occasional concentrations of up to a hundred or more. A few birds seen in June were thought to be summering.

In fall flocks are smaller, distribution slightly amplified as young birds wander farther from the main migration routes. Returning adults appear by mid July and peak in late July and early August. High counts are at Bandon, where the species is very numerous. There is a hint of a hiatus before the juveniles appear in numbers later in August, although breeding-plumaged adults and juveniles were collected together at Chatham Island, August 22, 1964, and Ocean Shores, August 20, 1987. Juveniles are found in some numbers through most of September.

Each age category moves through in fall about as rapidly as do adults in spring, the birds seemingly in a hurry to get to the wintering grounds. Occasional individuals or small groups winter on our coasts, sporadically in British Columbia (the high count at Mitlenatch Island is astounding, however) and regularly in Washington and Oregon. The only winter specimens examined include an adult from Sidney Island, November 19, 1980, and an immature from Westport, March 20, 1954.

This species is rare anywhere in the interior, with scattered records throughout and repeated records only at Malheur Refuge. There are few records in spring (two from Idaho, nine from Oregon, and one from Washington), mostly in late May, and even fewer of adults in fall. Juveniles are recorded more regularly, from mid August to mid September.

COAST SPRING. *High counts* 490 at North Bay, WA, May 9-10, 1981; 200 at Ocean Shores, WA, May 13, 1984. *Late dates* Ocean Shores, WA, June 7, 1980; Iona Island, BC, June 1-5, 1989.

COAST SUMMER. Leadbetter Point, WA, June 22, 1980 (4); Ocean Shores, WA, June 17-18, 1988.

COAST FALL. *Early date* Victoria, BC, July 1, 1979. *Adult high counts* 57 at Sidney Island, BC, August 7, 1987; 60 at Dungeness, WA, August 2, 1988; 200 at Bandon, OR, July 18-19, 1981; 179 at Bandon, OR, July 23, 1987. *Juvenile early date* Ocean Shores, WA, August 4, 1979. *Juvenile high count* 150 at Leadbetter Point, WA, August 31, 1979.

COAST WINTER. *High counts* 40 to 50 at Mitlenatch Island, BC, December 19-22, 1965; 8 at Coupeville, WA, January 17, 1988*; 25 at Seaside, OR, January 14, 1967.

INTERIOR SPRING. *Early date and high count* 3 at Malheur Refuge, OR, May 11, 1985. *Late dates* Hart Lake, OR, June 4, 1986; Cow Lake, WA, May 26, 1957.

INTERIOR FALL. *Adults* Camas Refuge, ID, July 19, 1961; Osgood, ID, June 21, 1977; American Falls Reservoir, ID, July 23, 1987. *High counts* 4 at Columbia Refuge, WA, September 1-6, 1988; 5 at Malheur Refuge, OR, September 2, 1983. *Late date* Springfield, ID, September 29, 1984.

Habitat and Behavior

This species shows a breadth of habitat preference as great as any shorebird. Small numbers are regularly seen with Black Turnstones and Surfbirds on rocky shores and jetties, with Sander-

lings and Dunlins on the open ocean beach, and even with Pectoral Sandpipers in *Salicornia* marshes. More typically Ruddies inhabit mud flats, especially those mixed with cobble or rocks. Very often they occur with Dunlins and Red Knots in spring, an eye-filling commingle-ment of rich rufous, black, and white.

Ruddy Turnstones feed by picking and are active foragers, often running through flocks of slower-feeding, probing sandpipers. They constantly live up to their names, turning shells, pebbles, leaves, and any other beach debris (rocks larger than the head, lighter items at times bigger than the bird!) to expose the invertebrates thereunder. Bivalves and barnacles are opened by vigorous pecks. The bill is also used as a shovel, pushed into soft sand or mud to excavate invertebrates just under the surface. This species is among the most varied of shorebirds in many of its attributes, with a catholic taste in food, including carrion and the eggs of colonially nesting terns and other seabirds. No egg-eating has been noted in the Northwest, perhaps for the simple reason that we have no coastal colonies of small terns.

Structure

Size medium-small. Bill short, hard, and tapered to sharply pointed tip; upper edge straight and lower edge slightly upcurved. Legs short. Wings fall slightly short of or extend to tail tip, primary projection moderate in adults and juveniles. Some median coverts quite long in adults, draped over coverts. Tail rounded.

Plumage

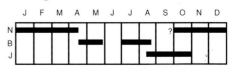

Breeding (Figure 35). Bill black, legs usually bright orange in adults but varying in some individuals (in winter only?) to duller yellow-orange or tinged with brownish. Adults in breeding plumage easily recognized by bright rufous and black backs and solid black breast markings. Males generally brighter than females and sexes usually easily distinguished, although they overlap in intensity of many individual markings. In

RUDDY TURNSTONE

br male

juv

br female

BLACK TURNSTONE

br

juv

Fig. 35. Turnstones. Ruddy with mostly rufous or brown upperparts, white throat, black breast indented on sides and rear, paler legs. Breeding male brighter with whiter head, breeding female more variable, generally duller with darker crown and nape. Juvenile with prominent buff fringes on scapulars and coverts, otherwise like nonbreeding adult. Black with blackish upperparts, entirely black throat and breast, darker legs. Breeding adult with blacker back than nonbreeding adult and juvenile.

males, white to light brown crown, heavily striped with black. Complex head and neck markings vivid black and white, hindneck mostly white. Mantle with rufous central stripe and black edges; upper scapulars mostly rufous and lower scapulars mostly black, tertials mostly rufous with irregular black markings. Retained coverts gray-brown and black, mixed with new rufous ones. Throat white, breast black.

Lower back white; anterior uppertail coverts black, posterior ones white. Tail white at base, brown in middle, and black toward end, tipped with white. Outermost rectrices edged with

white. Wing conspicuously patterned from above, brown with white wing stripe formed from broad tips of greater secondary coverts and markings on outer webs of five inner primaries. White humerals and some white inner lesser secondary coverts also contrast with other rufous or brown coverts. Turnstones and avocets only North American shorebirds in which humeral feathers stand out as differently colored from feathers around them. Underwing and axillars white, tips of flight feathers light gray-brown.

In females crown and hindneck brown, streaked with black; distinct from contrasty white-naped males. Upperpart pattern more irregular and mottled, rufous and black in smaller patches, giving overall brown tone. As in *Pluvialis* plovers, male Ruddy Turnstones more likely than females to molt coverts and tertials in spring, replacing brown winter feathers with rufous. Molt begins early in spring, all adults in full breeding plumage by their appearance in Northwest. Apparently pass through in autumn without molting, perhaps indicating long-distance migration, although some rufous tips are worn off by then, giving them overall darker appearance than spring adults.

Nonbreeding (Figure 36). Upperparts relatively plain and dark, most individuals showing scarcely a hint of ruddy. Head mostly brown, with faint indications of breeding-plumage pattern. Upperparts brown to blackish; in some birds most feathers with broad gray-brown fringes, in others with dull rufous-brown fringes. Breast markings mixed brown and black, enclosing pale area on each side of breast. Coverts dull rufous or gray-brown with darker central streak.

Juvenal (Figure 35). Much duller than breeding-plumaged adult and typically duller but neater looking than nonbreeding-plumaged adult. Medium to dark brown above with scapulars narrowly buff- or whitish-fringed and coverts broadly buff-fringed. Scapulars smaller than those of adults, allowing more coverts to be seen.

Immature. Pale fringes of juvenile gradually wear away, producing first-winter plumage darker and more uniform than nonbreeding adult. Most immatures spend first summer on wintering grounds; any moving into this region would be recognizable by very worn primaries.

Subspecies

Birds migrating through the Northwest represent *A. i. interpres*, which breeds in Eurasia and Alaska and winters on both sides of the Pacific. It is this subspecies that migrates to the Pacific islands, rather than Canadian *morinella*, as claimed by Hayman et al. (1986). Siberian and Alaskan *interpres* typically migrate south across the Pacific Ocean in fall and up the Asian Pacific coast in spring, and it is of interest that there is also an American Pacific coast migration of this form. *A. i. morinella* breeds in the Canadian Arctic and migrates primarily down the Atlantic coast. Specimens of this subspecies have been cited from the Pacific coast, but it must be much less common than *interpres* here. The Northwest interior is between the migratory routes of the two subspecies, as Ruddy Turnstones are much rarer there than on the Great Plains.

In breeding plumage *interpres* shows more black above than does *morinella*. The subspecies are not easily distinguishable in the field, and distinction is made even more difficult because many Alaskan birds are intermediate between *morinella* and typical *interpres* of Eurasia. Nevertheless, birds in the hand are distinctly different. In nonbreeding plumage *interpres* looks blackish, with little of the rufous pattern of *morinella*. Juveniles of *interpres* are duller and darker than those of *morinella*, with narrower and grayer fringes on the feathers of the upperparts. In both of these plumages the coloration of the upperparts of *interpres* is surprisingly like that of a Black Turnstone.

Identification

All Plumages. The two species of turnstones can be recognized as such by being sandpipers with *short, wedge-shaped bills*. The RUDDY additionally is characterized by its *white throat, bilobed, dark breast markings, and short, bright orange legs*. It is very slightly smaller and less robust than the BLACK.

Breeding Plumage. In breeding plumage the combination of *rufous-and-black back and black breast markings* are distinctive. DUNLINS are only superficially similar, instantly recognizable by their much longer bills and black patch on belly rather than breast.

Nonbreeding and Juvenal Plumages. Juvenile RUDDIES, much duller than adults, stand a real chance of being mistaken for BLACK TURNSTONES in the same plumage, but the pattern of the breast markings is diagnostic, with *breast markings bilobed* in RUDDIES and cut straight across in BLACKS. In addition, the RUDDY's *white throat and pale lateral breast markings* allow easy distinction. The back color in a RUDDY is dark but distinctly brown, rather than blackish as in a BLACK TURNSTONE. Some BLACK TURNSTONES have brighter legs than indicated by most field guides, causing confusion when leg color is used as a primary field mark.

In Flight (Figure 11)

In any plumage turnstones in flight are more vividly patterned than other shorebirds. The strong white wing stripe is complemented by a narrow white stripe across the wing base formed by feathers underlying the scapulars and exposed only in flight, and the white back and black rump are followed by a white-based, black-tipped tail, creating a flickering *calico flight pattern*.

The RUDDY always shows *brown scapulars and coverts* in flight, while the BLACK looks entirely black and white, but their pattern is essentially identical, and they are difficult to distinguish at a distance. Look also for the RUDDY's *light head markings*, very obvious in breeding adults.

Voice

The RUDDY's loud flight call is a series of notes rolled into a low, semi-musical chatter, unlike any other shorebird. Birds at times reduce the chatter to a few notes or even a single note. They may chatter continually when feeding, perhaps to keep other birds out of their way rather than for "positive" communication.

Further Questions

Sex-ratio counts should be taken in both spring and fall to determine whether there is differential timing in the migration of the sexes. Also, we know very little about which age stage winters in the region. Additional work with specimens and captured birds and a banding program will be necessary to determine what breeding population our migrants come from—*interpres* from Alaska, *morinella* from Canada, or both. With adults so much more common in spring than fall, are spring birds coming from wintering populations on South Pacific islands as well as the American Pacific coast?

Notes

A published record of "over 100 between Seattle and Bainbridge Island," October 8, 1954, is an unusually large number, especially so late in the fall, if a valid record.

The nonbreeding-plumaged individual illustrated in Hayman et al. (1986: 161) is much brighter above than is typical of birds in this plumage (the accompanying text states "lacking all chestnut on upperparts").

Photos

The breeding-plumaged Ruddy Turnstones illustrated in Bull and Farrand (1977: Pl. 241), Udvardy (1977: Pls. 185-86), Keith and Gooders (1980: Pl. 124), Armstrong (1983: 131), and Farrand (1983: 369; 1988a: 157; 1988b: 161) are all males. Even with our present knowledge of dimorphism in this and other species of shorebirds, females are not making it into the field guides; thus half the birds seen each spring are not as pictured! Hammond and Everett (1980: 111), however, show a breeding female, and the "winter plumage" bird in Bull and Farrand (1977: Pl. 196) appears to be one also. The "male in nuptial plumage" in Johnsgard (1981: black and white Pl. 13) is probably a female (compare it with the plates immediately preceding and following it). Most of the birds in Hosking and Hale (1983: 17) are in fresh nonbreeding plumage, some with a trace of breeding. A comparison of molting adults and juveniles is afforded in Hosking and Hale (1983: 174). The "first winter" bird in Keith and Gooders (1980: Pl. 99) appears to be a juvenile.

References

Bergman 1946 (general biology), Beven and England 1977 (general ecology), Branson et al. 1979 (measurements and molt in England), Brearey and Hildén 1985 (nesting association with gulls and terns), Browning 1974 (Northwest subspecies), Burger et al. 1979 (aggressive behavior in migration), Ferns 1978 (breeding plumage variation), Fleischer 1983 (winter flocking and foraging), Groves 1978 (age-related foraging behavior), Harris 1979 (feeding ecology in winter), Johnson 1973 (reproductive condition of summering birds), Johnson 1977 (plumage and molt of summering birds), Johnson 1979 (summary of 1973/1977 papers), Jones 1975 (winter diet and feeding behavior), Loftin 1962 (summering), MacDonald and Parmelee 1962 (summer feeding behavior), McKee 1982 (winter feeding behavior), Metcalfe 1986 (winter flocking), Metcalfe and Furness 1985 (population biology), Metcalfe and Furness 1987 (ag-

Breeding male Ruddy Turnstone. Turnstone with brilliant rufous, black, and white upperparts and orange legs; clean pattern and mostly white head diagnostic of breeding male. Arctic National Wildlife Refuge, AK, June 1976, Jim Erckmann

Juvenile Ruddy Turnstone. Orange legs and white-centered, bilobed breast pattern diagnostic of species; buff-fringed coverts indicate juvenile. Port Mahon, DE, 22 September 1984, Dennis Paulson

Nonbreeding Black Turnstone. Black and white turnstone, plainer and browner in this plumage; fresh plumage in March indicates adult (first-year bird more worn). Bandon, OR, March 1984, Tom Crabtree

gressive behavior), Nettleship 1973 (breeding-biology), Paget-Wilkes 1922 (breeding biology), Parkes et al. 1971 (egg-eating), Robertson and Dennison 1979 (winter feeding ecology), Thompson 1974 (migration across Pacific), Vuolanto 1968 (breeding biology), Whitfield 1986 (breeding plumage variation), Whitfield 1988 (nonbreeding plumage variation).

BLACK TURNSTONE *Arenaria melanocephala*

Busily scrutinizing the substrate or pecking with pointed barnacle bill, this common species is darker than its rock-climbing associates, in all plumages blackish with white belly. Chattering shrilly when startled into flight, a flock flickers black and white like the surf before settling again to disappear on the dark rocks.

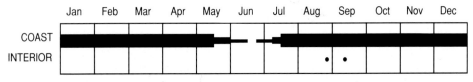

Distribution

Black Turnstones breed on coastal tundra of western Alaska and winter along the Pacific coast from southeastern Alaska to northern Mexico.

Northwest Status

This species is common from mid July to mid May wherever there are rocky shores or jetties on both outer and protected coasts. Two mid June records may indicate occasional summering. Juveniles arrive in August and become an increasing part of the population but never outnumber adults. Both age classes winter, although 48 of 55 winter specimens examined from the Northwest are adults, including 23 of 26 in one series from the tip of the Olympic Peninsula. With flock sizes ranging up to 100 and local concentrations to 200 almost any time during their occurrence here, migration peaks are not especially evident, but numbers build up in the Fraser River delta from September to November. The awesome 3,560 on the Comox Christmas Bird Count on December 19, 1982, is one of the largest recorded concentrations of the species.

The species is a confirmed coast dweller, with but four Willamette Valley records: Wapato Lake, November 12, 1913; Coburg, December 19, 1974; Medford, April 30, 1984; and Ankeny Refuge, May 13, 1986. Presumably all four birds were in transit. There are only two records from the interior subregion, probably juveniles from the dates.

Christmas Bird Count data indicate a decrease in regional wintering populations in recent years. Nine of 19 possible highest counts were obtained during the 1974-78 period and 9 lowest counts during the 1984-88 period.

COAST SPRING. *High counts* 450 flying past north jetty of Columbia River, WA, May 11, 1968; about 500 at Pacific Rim National Park, BC, April 25, 1972. *Late date* Ocean Shores, WA, June 9, 1989#.

COAST SUMMER. Ocean Shores, WA, June 11, 1977; Sidney, BC, June 15, 1978.

CHRISTMAS BIRD COUNTS	Five-Year Averages		
	74-78	79-83	84-88
Campbell River, BC	15	16	13
Comox, BC	82	1,220	156
Deep Bay, BC	251	224	176
Duncan, BC	36	34	20
Ladner, BC	26	18	33
Nanaimo, BC	66	131	26
Pender Islands, BC	5	9	66
Vancouver, BC	153	69	45
Victoria, BC	133	122	170
White Rock, BC	4	10	12
Bellingham, WA	17	21	16
Grays Harbor, WA	89	29	16
Leadbetter Point, WA	38	38	27
Seattle, WA	42	15	22
Sequim-Dungeness, WA	20	34	26
Tacoma, WA	22	29	11
Coos Bay, OR	139	92	97
Tillamook Bay, OR	210	128	146
Yaquina Bay, OR	133	71	84

COAST FALL. *Early dates* Victoria, BC, June 26, 1979; Smith Island, WA, June 26, 1988. *Juvenile early date* Ocean Shores, WA, August 6, 1982*. *High counts* 1,000 at Port Hardy, BC, September 15, 1939; 400 at Penn Cove, WA, October 29, 1989*.

INTERIOR FALL. Ochoco Reservoir, OR, September 8, 1985; Lake McDonald, MT, August 28, 1957.

Habitat and Behavior

Black Turnstones prefer but are not restricted to rocky habitats, including isolated rock outcrops and jetties. At high tide I have seen them feeding a short way up rocky streams emptying into the Strait of Juan de Fuca. They also occur in small numbers on sand beaches and mud flats, often with Ruddy Turnstones. Birds are seen in these atypical habitats both spring and fall and include adults and juveniles. They are visual feeders, moving steadily and slowly over the rocks and rapidly removing their superabundant prey while foraging. Acorn barnacles predominate in the diet, and the birds acquire them either by inserting the closed bill within the shell opening and gaping or levering the shell open or by hammering at the shell, breaking it up. They swallow only the soft parts, while Surfbirds pull the whole barnacle from the rock. Limpets are also commonly taken, and they are pried loose with the sharply pointed bill that functions somewhat like that of an oystercatcher.

Occurring largely on solid rocks, the Black Turnstone, unlike its cobble-foraging ruddy relative, scarcely has the chance to live up to its name. However, I once watched eight birds on a huge pile of oyster shells flipping over shells as large as themselves to get particles remaining after the oysters had been shucked. This may be common behavior in turnstones, as my field companion had seen both species exercising their neck muscles at this spot on a previous occasion. Both turnstones also plow through algal clumps where they are present, and "turnweed" would have been an appropriate name for either species.

Flock size is usually small, from a few birds to a few dozen, but may attain the low hundreds where concentrated. Individuals feed close together on both rocks and softer substrates, although they are aggressive to one another and to

Surfbirds at times. They roost in loose groups on the higher rocks adjacent to their feeding areas.

Structure

Size medium-small. Bill short, hard and finely pointed; legs short. Wings reach tail tip or fall just short of it, primary projection short to moderate in adults and juveniles. Tail rounded.

Plumage

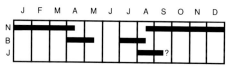

Breeding (Figure 35). Bill black, legs typically dull pinkish brown or orange-brown. Males crisply black on upperparts and breast with fine white streaks on crown and cheeks, large white spot between bill and eye, hint of narrow white supercilium, and white flecks on breast. Females duller, brown-tinged and with less white on head.

Lower back white; anterior uppertail coverts black, posterior ones white. Tail black, with white at base and extreme tip. Wing complexly patterned, blackish above with wing stripe formed from broad white tips of greater secondary coverts and white bases of secondaries and five inner primaries. White humerals and some white inner lesser secondary coverts also contrast with other blackish coverts. Underwing and axillars white, tips of flight feathers light gray-brown.

Body molt in adults in April and July to August, with wing molt throughout August and into September. At that time adults look scruffy, with whitish flight-feather bases showing where uncovered by missing feathers.

Nonbreeding (Figures 35, 36). Nonbreeding adult browner than breeding adult, lacking white head and breast markings and thus uniformly colored on upperparts and breast. Some scapulars and tertials and greater coverts narrowly fringed with white.

Juvenal. Almost identical to nonbreeding adults but in early fall easily distinguished from disheveled molting adults by pristine appearance. In flight juvenile wings intact, those of adults with prominent gaps. Some indication that first-year birds have wider white wing stripes than

WANDERING TATTLER

ROCK SANDPIPER

SURFBIRD

BLACK TURNSTONE

Fig. 36. Nonbreeding rock shorebirds. Black Turnstone stands out by darker coloration, breast/belly contrast particularly strong. Surfbird and Rock Sandpiper both paler with spotted sides, former conspicuously larger and shorter-billed, latter smaller and longer-billed. Wandering Tattler with long bill and long wings, gray with unspotted sides.

adults, but this needs further confirmation. Adult and immature shorebirds usually have same wing patterns, but see Willet also.

Immature. By midwinter adults in fresh plumage with broad white tail tips, first-winter birds beginning to wear substantially and a bit paler than adults with white tail tips worn off (easy to see in specimens, difficult in field). Differences persist into spring and are even visible on returning birds in late July. White tail tip distinguishing adult from immature should be fairly evident in flight. Most birds probably attain adult plumage in first summer and migrate to breeding grounds.

Identification

All Plumages. With typical turnstone characteristics of *short, wedge-shaped bill and short legs*, this species is essentially identical to the

RUDDY in size, shape, and actions but differs by its *uniformly dark brown to blackish coloration, brown to orange-brown legs, and entirely dark breast meeting white belly in a straight line*. There is no trace of the RUDDY's white throat or lateral breast patches. Even in dull juveniles with orange-brown legs, which look a lot like RUD-DIES, the entirely dark breast and breast-belly demarcation are diagnostic.

Nonbreeding and Juvenal Plumages. BLACK TURNSTONES occur much of the time with SURF-BIRDS and ROCK SANDPIPERS, both similarly colored for camouflage against dark rocks. Breeding-plumaged adults of the three are quite distinct, but in nonbreeding and juvenal plumages they all present a similar appearance—dark all over except for white belly. The other two species are both paler, a medium gray rather than

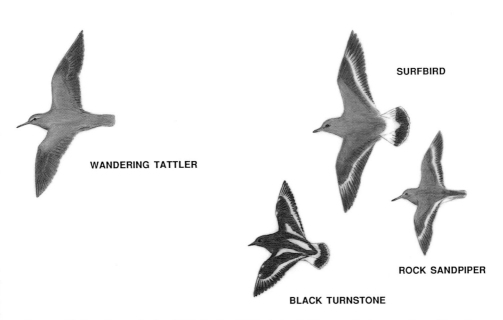

WANDERING TATTLER

SURFBIRD

ROCK SANDPIPER

BLACK TURNSTONE

Fig. 37. Nonbreeding rock shorebirds in flight. Wandering Tattler unmarked gray, solitary flier; others conspicuously marked, usually in flocks. Rock Sandpiper with conspicuous wing stripes, Surfbird also with white rump and tail base, Black Turnstone with black-and-white striped effect.

the turnstone's *overall blackish coloration*. The contrast between breast and belly is stronger in the turnstone because of its darkness. In addition, the larger SURFBIRD has bright yellow legs, the smaller ROCK SANDPIPER a conspicuously longer bill.

In Flight (Figures 11, 37)

The BLACK TURNSTONE's flight pattern is like that of a zebra, with *alternating black-and-white stripes and patches*. Only the RUDDY TURN-STONE duplicates this pattern, and it appears browner and paler-headed except in poorest light or at a distance. SURFBIRDS are almost as showy in flight, but attention is usually focused on their white tail base, while the turnstone is conspicuously patterned all over. ROCK SANDPIPERS are dull in comparison, with typical *Calidris* flight pattern.

Voice

Both turnstones are highly vocal, but the BLACK's shrill, high-pitched chatter is easily distinguished

from the RUDDY's lower, slower, and more melodic series of notes. The BLACK TURNSTONE's call might most likely be mistaken for that of a BELTED KINGFISHER. The two occur together in some areas, and one of the KINGFISHER's rattling calls is surprisingly similar to the TURNSTONE's call.

Further Questions

The difference in width of the white wing stripe between the two age classes needs further confirmation. Good photographs in flight would probably show the differences, especially if both age classes were known to be in the same flock.

It would be of interest to document how frequently Blacks and Ruddies turn over items while feeding *in the same habitat*, and what other differences there might be in their foraging methods, if any. Also, there is little published information about differences in foraging among any of the rock-inhabiting species. And why are Black Turnstones aggressive to one another at some times and not at others?

Notes

Leg coloration in this species has been depicted as gray or black in older field guides and as yellowish in Peterson (1990: 143), when in fact it can be almost as orange as the dullest Ruddy. Contrary to statements in Hayman et al. (1986: 161, 340), the basic coloration of nonbreeding adults and juveniles is identical.

Photos

Armstrong (1983: 132) and Farrand (1983: 371; 1988b: 160) show pale-legged individuals.

References

Conover 1926 (breeding biology), Gill et al. 1983 (breeding and winter site fidelity).

CALIDRIDINES Calidridini

This group includes 24 species, all of which have been recorded from North America and all of which are discussed here (Tables 28 and 29). Only about half of them are regular visitors to the Northwest, however.

Although the tringines are the classical sandpipers, most lay persons think of calidridines such as the Sanderling when they hear the word "sandpiper." As with the Tringini, this tribe has never had a common name, but the technical term "calidridines" (incorrectly written "calidrines" and "calidrids" in some publications) has been in use among shorebirders for some time now. "Beachpipers" would be an appropriate name for the tribe for anyone unwilling to use the technical term. The genus *Calidris* is the primary component of the tribe, but several closely related genera that differ in mating habits or bill structure are included in the group. Calidridines are very small to medium sandpipers with short to long bills and short to medium legs. Appendix 3 compares calidridine proportions, and it can be seen that both relative bill length and relative leg length vary by a factor of two.

Bills vary from short and stubby in the rock-feeding Surfbird and short and fine in the grassland Buff-breasted Sandpiper to rather long and slightly drooped in the deep-probing Dunlin, Curlew Sandpiper, and Stilt Sandpiper. Broad-billed and Spoonbill sandpipers show their own specializations. Legs are shortest in rock-feeding Surfbird and Rock Sandpiper, longest in wading Stilt Sandpiper and running Ruff.

All typically run or walk over sand or mud substrates, regularly picking or probing, the longer-billed species often feeding in water as well. Breeding habitat is tundra or open muskeg in boreal forest, wintering habitat typically beaches but freshwater marshes in some species and both beaches and marshes in others. There is a single upland species, the Buff-breasted Sandpiper. Because of substantial differences between breeding and nonbreeding habitats, most members of this group show considerable plumage change. Calidridines are usually brightly marked above and, in a few species, richly colored below in breeding plumage, sand- and mud-colored above and mostly white below in nonbreeding plumage.

Calidridines are arctic and subarctic breeders, some of them migrating relatively short distances or even remaining resident at high latitudes (Rock and Purple sandpipers), others going all the way to the other end of the earth in their respective hemispheres.

Sandpipers of this group, presumably because of their small size, form the largest flocks of any shorebirds, whether feeding, roosting, or in flight. A carpet of roosting Dunlins or Western Sandpipers furnishes amazement at both the numbers and the efficiency of packing, and a twisting, turning flock of hundreds or thousands of them in flight, transmogrified again and again, is one of nature's unforgettable sights.

References

Blomqvist 1983a (bibliography of *Calidris* and Broad-billed Sandpiper).

Table 28. Distribution and Habitat Preference of Larger Calidridines

Species	Hemisphere	Breeding Range	Breeding Habitat	Winter Range	Winter Habitat
Surfbird	W	arctic	dry tundra	N/equatorial/S	rocks
Great Knot	E	arctic	dry tundra	equatorial/S	bay
Red Knot	both	arctic	dry tundra	N/equatorial/S	bay/beach
Sanderling	both	arctic	tundra	N/equatorial/S	beach
White-rumped Sandpiper	W	arctic	tundra	S	bay/marsh
Baird's Sandpiper	W	arctic	dry tundra	equatorial/S	grassland
Pectoral Sandpiper	both	arctic	wet tundra	equatorial/S	marsh
Sharp-tailed Sandpiper	E	arctic	wet tundra	equatorial/S	marsh/bay
Purple Sandpiper	Atlantic	arctic	tundra	N	rocks
Rock Sandpiper	Pacific	arctic	tundra	N	rocks
Dunlin	both	arctic	tundra	N/equatorial	bay/beach
Curlew Sandpiper	E	arctic	tundra	equatorial/S	bay/marsh
Stilt Sandpiper	W	arctic	tundra	equatorial/S	marsh
Broad-billed Sandpiper	E	arctic	tundra	equatorial	bay
Buff-breasted Sandpiper	W	arctic	tundra	S	grassland
Ruff	E	arctic/sub./temp.	tundra/marsh	N/equatorial/S	marsh

SURFBIRD *Aphriza virgata*

This species stands out in the gang of four—the inseparable four species of rock shorebirds—as the large, gray sandpiper with a plover bill. Its plain gray back and white, black-tipped tail provide visual stability in flickering flocks of Black Turnstones. The fancy breeding plumage is seen briefly, the rufous back splotches of spring fading to white by midsummer.

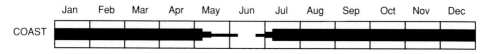

Distribution

The Surfbird is poorly known as a breeding species, restricted to alpine tundra of central Alaska and the Yukon. In winter it is found on rocky coasts from southeastern Alaska sparingly to Chile.

Northwest Status

A winter visitor and migrant on rocky coasts, this species is typically present from early July well into May; juveniles arrive a month after adults. Detectable migration occurs in British Columbia from mid April to early May and from the end of June to mid September. Both adults and immatures winter in the Northwest, the adults apparently more common.

The Surfbird is more restricted in distribution and generally less common than the Black Turnstone. Turnstones are common and widespread along both outer coast and protected shores of the region, but Surfbirds occur only locally in the latter areas. For example, of 17 Christmas Bird Count localities that regularly support midwinter populations of Black Turnstones in protected waters of Washington and British Columbia, Surfbirds are reported from only four. These four localities are widespread, however, extending from Roberts Creek on the mainland and Nanaimo and Victoria on Vancouver Island all the way down to Restoration Point in Puget Sound.

On the north shore of the Olympic Peninsula few Surfbirds are seen in the frequent Black Turnstone flocks. On the outer coast they are more widespread but again less common than turnstones. Curiously, the reverse is consistently true at the jetty at Ocean Shores. Small flocks are the rule, with only occasionally up to a hundred birds or more at one place and time. However, flocks of thousands of birds have been seen during spring and fall in Barkley Sound, which must be an important staging area for further migration. Large numbers of birds have been found wintering in the same area, with peak counts in the hundreds.

Common as this species is on the coast, it has not yet been detected in the interior of the region; the Rock Sandpiper is the only other common coastal shorebird sharing this distinction. As there are a number of Surfbird records from interior California and even a few from central and eastern North America, an interior record from the Northwest would not be earthshaking.

COAST SPRING. *High count* 4,500+ at Turtle Island, BC, April 25, 1972 (with 500 Black Turnstones). *Late date* Ocean Shores, WA, June 4, 1983.

COAST FALL. *Early date* Victoria, BC, June 28, 1981 (9). *Juvenile early date* Ocean Shores, WA, July 26, 1979. *High counts* 1,000 at Fleming Island, BC, August 10, 1967; 2,000+ at Tzartus Island, BC, August 14, 1968; 148 at Ocean Shores, WA, July 21, 1982*.

COAST WINTER. *High counts* 450 at Tofino, BC, November 17, 1980; 450 at Tofino, BC, November 12, 1982; 130 at Kalaloch, WA, March 1, 1969; 220 at Tillamook Bay, OR, December 15, 1973.

CHRISTMAS BIRD COUNTS	Five-Year Averages		
	74-78	79-83	84-88
Nanaimo, BC	121	83	3
Pender Islands, BC			73
Victoria, BC	11	6	12
Grays Harbor, WA	94	44	55
Seattle, WA	18	15	22
Coos Bay, OR	45	22	59
Tillamook Bay, OR	88	73	140
Yaquina Bay, OR	111	59	82

Habitat and Behavior

Surfbirds are only rarely seen away from their preferred rocky shores and jetties. The occasional individuals or small flocks that appear on sand beaches (especially near rocks) or mud flats usually do not linger very long, except at Bandon, where they and other rock shorebirds regularly forage on mud. They are almost obligate associates of Black Turnstones, rarely found in any place lacking the smaller species. These two species and the Rock Sandpiper commonly mingle in mixed-species flocks both while foraging and in flight, but I have seen flocks that consisted solely of Surfbirds even when turnstones were present. They roost on dry rocks just above the splash zone or fly short distances to offshore pinnacles.

Surfbirds forage in tight flocks but are at times aggressive to other birds in close proximity. They feed by picking invertebrates such as snails from intertidal rocks, but their primary feeding mode involves plucking off young mussels, barnacles, and limpets with a sideways tug of the head; they later regurgitate the shells. Note the scars of removed barnacles all over the rocks where these birds have been feeding. This is the only sandpiper with a plover-shaped bill adapted to pull prey loose from the substrate.

Structure

Size medium. Bill short, straight and blunt-tipped, amazingly ploverlike for a sandpiper. Bill shape remarkable example of convergent evolution (plovers also pick), as Surfbird's ancestor probably a knotlike bird. Legs short, feet heavy for shorebird. Wings reach tail tip or slightly beyond, primary projection moderate to long in adults and juveniles. Tail square.

Plumage

Breeding. Bill black with base of lower mandible yellow, legs pale yellow with tinge of greenish. Head and neck white, heavily striped with black, crown stripes wider. Mantle feathers black, fringed with white. Scapulars vary greatly, from gray-brown (retained from nonbreeding plumage) to black with extensive gray tips to entirely black with or without rufous notches to largely rufous with black subterminal blotches; most have white tips. Overall effect one of irregular rufous blotches, although some birds show no rufous at all. Coverts gray-brown (old) or black (new), many with white fringes. Underparts white with dense black chevrons, those on breast U-shaped and those on sides and much of belly heart-shaped.

Lower back black, uppertail coverts white. Tail black with white base and narrow white tip. Wing dark gray above; conspicuous wing stripe formed from broad white tips of greater secondary co-

Breeding Surfbird. In late summer, plumage very worn and scapulars extremely faded. Brooks Peninsula, BC, July 1987, Tim Zurowski

Breeding Surfbird. Rock-frequenting sandpiper with ploverlike bill and yellow legs; heavy black spotting on breast and sides diagnostic of plumage; unworn plumage and extensively rufous scapulars in spring. San Diego, CA, April 1976, J. R. Jehl, Jr.

Juvenile Surfbird. Fine fringes on upperparts and breast diagnostic of plumage. Ocean Shores, WA, late August 1984, Urban Olsson

verts, narrower white tips of greater primary coverts, and white basal markings on outer webs of inner primaries and inner webs of outer primaries. Underwing white except gray flight-feather tips, light gray-brown primary coverts, and dark gray marginal coverts. Axillars white.

Bright rufous faded to pale buff or even white in returning adults in July, striking example of plumage fading between spring and fall. White tips to scapulars also worn off, these changes producing contrasty black-and-white look above. Autumn body molt primarily in August.

Nonbreeding (Figure 36). Upperparts and breast gray, belly and restricted throat patch white. Plainness relieved only by white fringes on some coverts and a series of conspicuous and linearly arranged spots along sides, in some individuals extending across belly.

Juvenal. Similar in overall coloration to nonbreeding adult, although slightly paler. Head streaked and breast finely barred, in contrast with more uniform coloration of adult. Scapulars with faint, narrow, dark subterminal bars and white tips; coverts more conspicuously scalloped with dark subterminal and white terminal fringes.

Immature. By October adults and juveniles indistinguishable in field, although at close range slightly more worn look of juvenile's primaries might be apparent in comparison. Juveniles with more evidently white-tipped coverts than adults, difficult to detect in field. Unlike Black Turnstone, juvenile's white tail tip does not seem to wear off during winter. Very few birds in summer south of breeding range, so most birds probably mature and breed in first year.

Identification

All Plumages. SURFBIRDS are distinctive in any plumage as medium-sized sandpipers (with sandpiper feeding actions) with *short, ploverlike bill and short, yellow legs.*

Breeding Plumage. In breeding plumage, with rusty-marked back and vividly spotted sides, SURFBIRDS are colored like no other shorebird except the smaller ROCK SANDPIPER with which they regularly associate. The SURFBIRD's larger size, shorter bill, and lack of dark ear coverts and breast blotch allow easy distinction. A breeding-plumaged GREAT KNOT with bill hidden might be mistaken for a SURFBIRD.

Nonbreeding Plumage. BLACK TURNSTONES and ROCK SANDPIPERS are the only species from which SURFBIRDS must usually be distinguished. In nonbreeding and juvenal plumage they differ from both by being *larger*; from the BLACK TURNSTONE, by being *paler and yellow-legged, with spotted sides*; and from the ROCK SANDPIPER, by being *considerably larger and short-billed.*

Juvenal Plumage. Juvenile SURFBIRDS are patterned like juvenile RED KNOTS but are darker. The short bill of the SURFBIRD is an obvious mark of discrimination.

In Flight (Figures 11, 37)

The SURFBIRD's flight pattern is unique; no other Northwest shorebird is *conspicuously stripe-winged and band-tailed* except for the much larger HUDSONIAN GODWIT. BLACK TURNSTONES are even more vividly patterned, black and white on the back as well as the wings and tail, and ROCK SANDPIPERS are drab, with narrower wing stripes and no tail pattern. SURFBIRDS stand out in mixed flocks even as silhouettes by their *larger size.*

Voice

SURFBIRDS are quiet birds, their vocalizations apparent only at close range. Soft, single, high-pitched notes are heard regularly from flocks while they are feeding, less often in flight. Three juveniles flushed at very close range called with a turnstonelike low chatter—*chut* or *ka-chut.*

Further Questions

Many shorebirds of Surfbird size do not breed until two years of age, yet there seem to be few if any records of summering Surfbirds in our region. Regular June field trips to areas where Surfbirds winter would be worthwhile, with attempts to recheck any birds seen in late May or early June to see if they are indeed summering; photographs and plumage descriptions would help.

Surfbirds should benefit by association with the noisy and alert Black Turnstones, which make up for the silence of the other two members of their winter flocks. If Surfbirds did flock with turnstones for the warning system provided by that species, then they should rarely be found away from them, and it would be of interest to document how often, where, and when they are found separately. In addition, we would predict that Surfbirds away from turnstones would be more vigilant, feed at a slower rate, spend more time with heads up, be more easily flushed, and be more likely to call when flushed.

Notes

The illustrations of nonbreeding plumage in Hayman et al. (1986: 161) are much too dark.

Photos

Farrand (1983: 373) shows a bird in fresh breed-ing plumage, Udvardy (1977: Pl. 222) one in worn breeding plumage. The birds in flight in "winter plumage" in that book (Udvardy 1977: Pl. 227) are also in worn breeding plumage. One of the "Surf Birds" shown in Hosking and Hale (1983: 59) is a Black Turnstone.

References

Dixon 1927 (breeding biology), Jehl 1968 (taxonomic relationships), Marsh 1986 (predation on mussels), Miller et al. 1987 (breeding vocalizations), Navarro et al. 1989 (diet).

GREAT KNOT *Calidris tenuirostris*

From book descriptions, the three plumages of this Siberian accidental recall the same plumages of three common Northwest species—breeding Surfbird, nonbreeding Red Knot, and (more distantly) juvenile Pectoral Sandpiper. Both larger and more extended than the other knot, with a longer and slightly droopier bill, the Great Knot will present its own impression when actually encountered.

	Jan	Feb	Mar	Apr	May	Jun	Jul	Aug	Sep	Oct	Nov	Dec
COAST									▬ •			

Distribution

The Great Knot breeds in alpine tundra in northeastern Siberia and winters primarily on the north coast of Australia, much more sparingly west to Pakistan and north to southern China. It is common in few areas and apparently migrates in long-distance flights, probably mostly over the western Pacific.

Northwest Status

This species is known in North America from only a handful of late-spring records around the Bering Sea and should be considered unlikely to occur in this region, especially in autumn. Nevertheless, there are two fall records from the Northwest documented by photographs, an apparent nonbreeding adult and a juvenile, and two additional sight records (see under Notes).

COAST FALL. Yaquina Bay, OR, September 28, 1978; Bandon, OR, September 1-19, 1990.

Habitat and Behavior

This largest *Calidris* forages primarily on intertidal mud and sandy mud. The two knots feed in much the same fashion and often forage together on their wintering grounds. Great Knots are somewhat more active than Red Knots and apparently rely more heavily on visual searching for prey, although they also probe like the other species. They occur in quite large flocks where common, often associating with Bar-tailed Godwits, but single birds would be expected in the Northwest. The Oregon adult was with a flock of Black-bellied Plovers, while the juvenile foraged on mudflats and roosted on rock jetties.

Structure

Size medium. Bill medium, fairly slender and often slightly drooped at tip. Legs short. Wings project beyond tail tip, primary projection long in adults and juveniles. Tail square.

Plumages

Breeding. Bill black, legs gray. Head and neck grayish brown, crown and hindneck heavily streaked with black. Mantle feathers, coverts, and tertials blackish with white fringes; scapulars vary from gray-brown to rufous and black with white tips. White scapular tips wear off, but, unlike Surfbird, rufous markings have not faded much in worn late-summer birds. Breast blackish (solid or spotted), belly white; sides marked with black spots and chevrons. Undertail coverts sparsely black-spotted. Lower back gray-brown, feathers with darker tips and pale fringes. Uppertail coverts white, usually with brown subterminal spots. Rectrices gray-brown. Wing brown above, white wing stripe formed by tips of primary and secondary coverts. Underwing pale gray-brown, greater coverts tipped white; lesser and median coverts and axillars white.

Breeding plumage attained in April (even late March) and lost in late August and September, although many birds retain black spotting on breast into early October.

Nonbreeding. Legs greenish to greenish yellow. Upperparts gray-brown (tertials typically darker brown), most feathers with fine dark shaft streaks and fine white fringes. White supercilium variable, may be almost complete or lacking either before or behind eye; usually does not reach bill base. Brown loral stripe not conspicuous, postocular stripe even less so. Breast and sides irregularly streaked with brown, breast also with scattered larger, darker spots; belly white.

Juvenal. Legs as in nonbreeding adult. Overall color and head pattern in lighter birds as in nonbreeding adult, but mantle feathers, scapulars, coverts, and tertials with heavy dark brown central stripes, creating darker, more patterned effect. In darkest birds, head and neck more darkly streaked and mantle feathers and scapulars blackish brown with white fringes and notches. Unlike nonbreeding adult, supercilium complete in some individuals. Breast suffused with buff and finely and evenly streaked or spotted with brown, sides similarly but more faintly marked; belly white.

Immature. As nonbreeding adult, perhaps recognizable by retained darker tertials until at least midwinter. Some first-summer birds attain partial breeding plumage. Most birds apparently spend first summer on wintering grounds.

Identification

All Plumages. The GREAT KNOT looks much like the RED KNOT but is larger in all dimensions, although still considerably smaller than a BLACK-BELLIED PLOVER. Typically it appears more slender and more pointed behind than the RED KNOT, although this is best judged in comparison. The *bill is relatively longer* than that of a RED KNOT, and it looks more slender, probably because of its greater length; the *tip may droop very slightly*, a good field mark. The posture is typically *more upright, with longer neck* than in the other knot, a good clue to examine such a bird more closely.

Breeding Plumage. GREAT KNOTS in breeding plumage—about the bulk of SURFBIRDS—are colored rather like them in the same plumage, with heavily *dark-spotted upperparts, rufous-marked scapulars, and blackish breast*; their *long bill* distinguishes them. They are also longer-necked and longer-legged and feed in entirely different habitats. Otherwise, they look like no other shorebird.

Nonbreeding Plumage. As might be expected, as they feed in quite different habitats, nonbreeding adults and juveniles are very differently colored from SURFBIRDS. Nonbreeding adults are gray all over like RED KNOTS in similar plumage, but even in this plumage they are more *distinctly striped on the crown and back and usually distinctly spotted on the breast*. In birds in both juvenal and nonbreeding plumage, the white supercilium is relatively poorly defined, in many birds absent or extending only about halfway from the eye to the bill; the RED KNOT typically shows a full supercilium from the bill base to well behind the eye.

Juvenal Plumage. Birds in this plumage have *heavily and evenly dark-striped upperparts and heavily dark-spotted breast*. The effect recalls a PECTORAL SANDPIPER, although the ground color is gray-brown rather than brown and the markings more contrasty, much heavier than those of RED KNOTS in any plumage. Darker birds are marked like SANDERLINGS on the upperparts but are much larger with a spotted breast.

In Flight

In flight the GREAT KNOT is as similar to the RED KNOT as to any other shorebird, although slightly larger; being long-winged, it looks large enough also to warrant comparison with BLACK-BELLIED PLOVER. Unlike either of these species, it appears distinctly *white-rumped*, while the RED KNOT

Juvenile Great Knot. Large calidridine with long wings and heavy, medium-length bill; blackish-streaked back, black-mottled breast, whitish-fringed coverts and tertials with dark subterminal markings, and pale legs diagnostic of plumage and species. Kyushu, Japan, mid September 1984, Urban Olsson

Juvenile Red Knot. Large calidridine with chunky shape, long wings, and head-length bill; pale upperparts with fine dark subterminal and whitish terminal fringes diagnostic of plumage and species. Crockett Lake, WA, 10 September 1989, Harold Christenson

appears pale-tailed with spotted and barred rump and the BLACK-BELLIED distinctly white-tailed. The *conspicuous wing stripe* is only slightly less so than that of the RED KNOT.

Voice

Like its smaller relative, the GREAT KNOT is not very vocal, rarely calling when flushed; its calls are low single or double whistles.

Notes

Two additional unpublished sight records from the Northwest are not supported by photographs. One from La Push, September 6, 1979, was re-

ported to be in full breeding plumage, odd for that date, but some individuals of this species may retain that plumage fairly late. Another from Boundary Bay, May 13, 1987, is even more peculiar, as it was apparently in nonbreeding plumage, and the first-year birds that exhibit that plumage in May are normally nonmigratory.

References

Hindwood and Hoskin 1954 (winter ecology), Jehl 1968 (photo of wing and tail), Lethaby and Gilligan 1991 (Oregon record), Marchant 1986 (identification and habits), Myers et al. 1982 (general biology), Portenko 1933 (life history).

RED KNOT *Calidris canutus*

Hundreds to thousands of rich rufous breasts dazzle the eye at just a few Northwest beaches each spring when these sandpipers stop briefly on their way to Alaska. Other than at these knot hot spots, they are widespread but uncommon in the region. Often with either Black-bellied Plovers or Short-billed Dowitchers, these short-billed, big calidridines, whether red-breasted adults or delicately scalloped juveniles, stand out in their crowd.

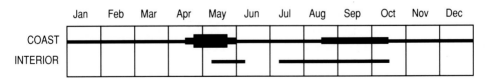

Distribution

Red Knots breed on high-arctic tundra very locally all around the world. Each population has its own far-flung coastal wintering ground, primarily in Florida and southern South America in the New World and ranging from western Europe south along the west coast of Africa and all around Australia in the Old World.

Northwest Status

These birds are primarily coastal migrants, considerably more common in spring than in fall. Breeding high in the Arctic, they are late spring migrants, seldom seen before mid April and usually present through May. The migration peak seems to coincide with or lag slightly behind those of the abundant species of the same habitats: Western Sandpiper, Dunlin, and Short-billed Dowitcher.

Knots can be very abundant in Grays Harbor at the peak of migration, with a recorded peak of more than 6,000 birds on April 27, 1981. Numbers remained in the thousands at least through May 7 of that year. The birds were locally distributed, with concentrations at Bottle Beach on the south side of the harbor and Point New and Bowerman Basin on the north side. The largest number reported at Leadbetter Point in one day during a recent year's census there was 100 birds, but flocks of thousands have been reported in past years in Willapa Bay. Thus Grays Harbor and Willapa Bay are major staging areas for this species in spring migration. It is common nowhere else on the Pacific coast south of southern Alaska, and the source of these huge spring flocks is not known. They greatly exceed the thousand or so birds known to winter in California.

The species is uncommon anywhere on the coasts of British Columbia and Oregon, with high counts rarely reaching 100. In most springs a dozen or two birds represent Oregon's peak count. There are few records from the Puget Sound region, most of them in spring when the largest numbers pass through the region. The species has been recorded only once from the heavily scrutinized Seattle area, an August adult. Eleven at Medford on May 11, 1984, was an outstanding number for the Willamette Valley, where there are also few records. Occasionally a few individuals summer at Grays Harbor (20 once) or Willapa Bay (26 on one occasion, 38 on another), with very few records of summering knots elsewhere in the region. These flocks, in nonbreeding plumage, are probably all first-year birds; two collected were males.

The fall migration route of the large numbers of Red Knots that pass up our coast in spring remains a mystery. In fall only small numbers of them pass through the Northwest, a few adults in July and August and juveniles in August and September. An adult and a juvenile were collected together at Seal Rocks, Oregon, on August 19, 1914, and single adults in small flocks of juveniles were seen at Westport on August 22, 1986*, and at Ocean Shores on August 25, 1988*. Records from Pacific Rim National Park are surprisingly concentrated during the adult migration, but most of the knots seen elsewhere on the coast in fall are juveniles, usually not exceeding a dozen or two birds. Red Knots have been seen in very small numbers with Dunlin flocks that had just arrived on the ocean beach at Ocean Shores in mid October, presumably birds that lingered somewhere to the north until then. One group of eight on October 18, 1986*, were adults.

Still smaller numbers regularly remain in the region into mid December, perhaps late migrants. There are few midwinter records, but single birds have been seen several times in winter in southern British Columbia, there are winter records from Washington every few years, and 22 birds wintered at Coos Bay in 1976-77. To the south moderate numbers of Red Knots winter locally along the California coast. One winter specimen from the Northwest (Tokeland, December 17, 1960) is an immature, as was a bird photographed at Ocean Shores on December 19, 1976; another winter specimen (Iona Island, March 4, 1950) is an adult.

The Red Knot is rare in the interior of the region, with records of one or two at a time from British Columbia, Washington, Oregon, and Idaho. There are about twice as many records of fall juveniles (August and September) as spring adults (May) and very few records of fall adults.

COAST SPRING. *High counts* 120 at Tillamook, OR, May 10, 1964; 143 at Tillamook, OR, May

10, 1976; 6,100 at Grays Harbor, WA, April 27, 1981; 40 at Tofino, BC, May 8, 1990.

COAST FALL. *Adult high count* 36 at Long Beach, BC, July 12, 1972. *Juvenile early date* Seal Rocks, OR, August 19, 1914. *Juvenile high counts* 18 at Saanich, BC, September 18, 1955; 65 at Leadbetter Point, WA, August 26, 1979; 100 at Ocean Shores, WA, September 7, 1982; 26 at Tillamook Bay, OR, August 29, 1982.

COAST WINTER. *Northernmost* Iona Island, BC, March 4, 1950#; Chatham Island, BC, January 6, 1959; Crescent Beach, BC, February 8-15 and December 14-16, 1975; Victoria, BC, February 12, 1975, and December 19, 1978.

INTERIOR SPRING. *Early date* Nampa, ID, May 4, 1982. *Late date* Malheur Refuge, OR, June 6, 1977.

INTERIOR FALL. *Adults* Springfield, ID, July 31 to August 2, 1986, July 11, 1987. *Juvenile early date* Reardan, WA, August 18, 1973. *Late date* Walla Walla River delta, WA, October 10, 1987.

Habitat and Behavior

Red Knots are birds of open mud flats and, less commonly, ocean-front sand beaches. They frequently associate with Short-billed Dowitchers, Dunlins, Ruddy Turnstones, and Black-bellied Plovers in these habitats. Occasionally birds associate with turnstones and Surfbirds in rocky habitats. They pack as tightly while feeding as any of our sandpipers, typically remaining in cohesive flocks, even of hundreds of birds.

Although a Red Knot's bill appears short for the size of the bird, it is slender, and knots feed primarily by probing in mud and sand, not very differently from the Dunlins and Short-billed Dowitchers with which they feed at times. They alternate a short series of probes in mud or sand and a quick run of a foot or so. They also capture prey by pecking but use this feeding method less frequently than probing in our region.

Structure

Size at large end of medium-small. Bill short to medium, straight and evenly tapered to moderately fine tip, legs short. Wings usually extend just beyond tail tip, primary projection long in adults and juveniles. Tail square.

Plumage

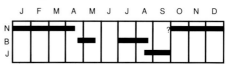

Breeding. Bill black, legs dark olive to dark gray or almost black. Overall appearance mottled above, bright rufous below. Crown, cheeks, and hindneck narrowly but heavily striped with black. Mantle feathers black with gray-brown fringes. Scapulars and tertials range from gray-brown (old) to variably marked; many are mostly black or black with large, paired subterminal rufous spots and white tips. Coverts gray-brown except for a few marked like brighter scapulars. Underparts bright rufous, white on belly varies from extensive to almost absent. Posterior sides and undertail coverts sparsely barred with black. Males average more brightly marked than females, with more extensive rufous above; females also with more white below and more likely to retain few barred winter feathers on breast. These are average differences, perhaps with total overlap.

Lower back and uppertail coverts barred black and white, with scattered rufous. Tail gray; some birds with central rectrices dark brown, barred with rufous near tip. Wing gray-brown above, white wing stripe formed from narrow tips of both greater secondary coverts and inner greater primary coverts. Flight feathers and especially primary coverts darker than rest of wing. Underwing light gray-brown; greater coverts faintly patterned with gray-brown, marginal secondary coverts barred with dark brown. Axillars white with narrow brown bars; only calidridine with barred axillars.

Autumn and even late-spring adults can have largely blackish crowns and mantles as their pale fringes wear off.

Nonbreeding. Legs greenish, paler than in breeding adults. Upperparts and breast light gray-brown with faintly darker streaks, underparts white with fine brown wavy bars on breast and sides. White supercilium and gray-brown loral stripe moderately conspicuous, ear coverts also dusky. Most adults passing through region in July appear to be in full breeding plumage, but body molt begins in some populations on breeding grounds; largely complete by end of September.

Juvenal. Legs greenish, may be paler than in adults. Coloration overall similar to that of non-breeding adults, but breast more likely to be less conspicuously marked and with fine brown streaks and dots instead of bars. Scapulars, coverts, and tertials with narrow, dark subterminal fringe and white or buff terminal fringe. In younger juveniles, entire underparts may be washed with buff, persisting even into September in some individuals.

Immature. Retention of finely patterned juvenile coverts by first-winter birds facilitates their identification at close range. Pale and worn-looking by spring, some molt partway into breeding plumage, with much rufous below, but most remain in nonbreeding plumage during first summer. First-summer birds look like nonbreeding adults but have very worn or molting primaries. Only a few of them have been seen in the region, but most or all birds among 38 at Leadbetter Point on June 4, 1979, were birds of this sort. Oversummering common in parts of winter range.

Subspecies

Birds on the Pacific coast of North America have been difficult to allocate to subspecies, and they are surprisingly variable if all from the same population. Some workers consider them to be *C. c. rogersi*, which breeds in northeast Siberia and perhaps in Alaska and winters primarily in Australia and New Zealand but perhaps also on the Pacific coast of South America. Other authorities synonymize *rogersi* with *C. c. canutus* of central Siberia, which winters primarily in Africa. As a third alternative, Pacific coast populations have been considered intermediate between *canutus* and *C. c. rufa* of central Canada, which winters in southeastern North America and South America. Finally, recent studies indicate that our birds are probably from a population that breeds on Wrangel Island, Siberia, and is distinct from all named populations. These birds may winter entirely in the New World.

Typical individuals of this population are darker than those of *rufa*, with somewhat less white on the belly. The scapulars and tertials tend to have more dark than light coloration, while those feathers on eastern knots are more extensively pale. Southbound autumn adults, with pale fringes worn off, are particularly dark above, while *rufa* at the same time of year still looks rather pale. There is too much overlap, however, to allow easy separation, even in the hand. Both *rogersi* and *canutus* are even darker than Wrangel Island birds, true in juveniles as well as breeding adults.

Identification

All Plumages. About the size of DOWITCHERS, RED KNOTS are most likely to be mistaken for them. In any plumage, as soon as the *much shorter bill* is seen there is no problem.

Breeding Plumage. In breeding plumage, sleeping or foraging RED KNOTS with bills hidden can be distinguished by their *gray-and-black speckled back, bright, unmarked rufous breast,* and (usually) *white undertail coverts.* DOWITCHERS are much browner on the back, are not as brightly colored on the breast, and look evenly colored from chin to tail, without the very strong contrast between upperparts and underparts that characterizes the KNOT. See also CURLEW SANDPIPER, another rufous species.

Nonbreeding and Juvenal Plumages. In nonbreeding or juvenal plumage, KNOTS have mostly *white breast and sides,* DOWITCHERS gray (again of value for identifying birds with bills hidden). In these plumages they are as likely to be mistaken for a BLACK-BELLIED PLOVER but are smaller with slender, sandpiper bill and sandpiper foraging mode. Although similar in shape and relative bill length, KNOTS are much larger than SANDERLINGS and never so pale as nonbreeding adults of that species.

The juvenile's *finely dark-scalloped upperparts* are shared only by SURFBIRDS and TEMMINCK'S STINTS.

In Flight (Figure 11)

RED KNOTS fall in a small group of shorebirds that are *stripe-winged and white-tailed.* The only other member of the group in our region is the BLACK-BELLIED PLOVER, which looks similar at a distance. KNOTS have a less conspicuous wing stripe and less contrasty white tail than the larger species. The KNOT'S rump is heavily barred, the tail plain gray rather than white with bars as in the BLACK-BELLIED and most other white-tailed species. As soon as the underside of a bird in question is seen it is easy, as the KNOT lacks the BLACK-BELLIED's black axillars.

Voice

RED KNOTS are quiet shorebirds, typically not calling when flushed. Their flight call is a soft, melodious *cur-ret*, sounding as much like that of a PACIFIC GOLDEN-PLOVER as any other regional shorebird. It is most likely to be given by single birds in flight, probably trying to locate other knots.

Further Questions

At this point we cannot be sure where the thousands of knots that appear at Grays Harbor and the Copper River delta of Alaska each spring are coming from, or where they are going, or the route of their fall migration. Wrangel Island may be their breeding destination, but they are not known to winter in large numbers anywhere on the American Pacific coast. Australia, the wintering grounds of most Siberian knots (*rogersi*), seems unlikely to be the source of Northwest migrants. Long over-water flights to wintering grounds may explain the scarcity of Red Knots in the Northwest in autumn, but this is purely speculative. The movements of Red Knots around the edges of the Atlantic Ocean are now reasonably well understood, and a major program to capture and mark Pacific Ocean knots needs to be undertaken.

Notes

Breeding-plumaged adults have dark gray legs rather than greenish as stated and shown by Hayman et al. (1986: 182-83) and most other guides, although Farrand (1983: 372) is correct.

Photos

The knots in fresh breeding plumage in Farrand (1983: 373; 1988a: 178; 1988b: 172) are of the subspecies *rufa*, as is a more worn one in Udvardy (1977: Pl. 223). The "adult winter" bird in Chandler (1989: 88) may be in second-winter plumage, with both body and wing molt much in advance of the adult bird to its right. The illustration in Hosking and Hale (1983: 126) affords a good comparison of juveniles (some molting into first-winter plumage) and nonbreeding adults, with a single breeding-plumaged bird that has already molted its head feathers. The bird in nonbreeding plumage in Farrand (1983: 373) is unusually darkly streaked on the breast, reminiscent of a Great Knot. The juvenile in Keith and Gooders (1980: Pl. 97) is more conspicuously white fringed than is typical of American knots. The "immatures" in Udvardy (1977: Pl. 230) are adults in nonbreeding plumage; the "winter" bird in Hammond and Everett (1980: 120) is a juvenile.

References

Barter et al. 1988 (Pacific Ocean subspecies), Barter et al. 1989 (biometrics and molt in Australia), Burger et al. 1979 (aggressive behavior in migration), Conover 1943 (subspecies distinctions), Flint 1972 (breeding biology), Goss-Custard 1970 (winter foraging and spacing behavior), Goss-Custard et al. 1977 (diet), Harrington 1983 (Atlantic migration), Harrington and Leddy 1982 (association of individuals in winter flocks), Harrington et al. 1988 (winter site fidelity and survival), Hindwood and Hoskin 1954 (winter ecology), Hobson 1972 (breeding biology), Nettleship 1974 (breeding biology), Prater 1972 (winter food and feeding habits).

SANDERLING *Calidris alba*

Sanderlings seem mechanical toys, on wheels rather than feet as they zigzag up and down the beach with each wave. When finished feeding, one bird after another collects in dense roosting flocks, seemingly asleep but ready to hop one-legged away from any disturbance. Pale adults and spangled juveniles are familiar beach occupants, but the bright orangey birds of spring are too briefly present, gone in a flurry of white wing stripes and sharp calls.

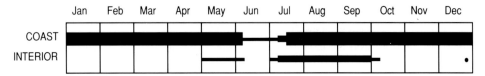

	Jan	Feb	Mar	Apr	May	Jun	Jul	Aug	Sep	Oct	Nov	Dec
COAST												
INTERIOR												

Distribution

Sanderlings are high-arctic tundra breeders, scattered around the top of the globe. In winter they spread out over much of the world's coasts, from north-temperate latitudes through the Tropics to the ends of the southern continents. A comparison of the Sanderling's breeding and winter ranges is a lesson in shorebird biology in itself. Thus far this is the only shorebird species in which the same color-banded individual has been observed on both Pacific (spring) and Atlantic (fall) coasts of North America during its annual cycle, an oval migration path even wider than is followed by other shorebirds.

Northwest Status

Primarily a bird of sand beaches of the outer coast, this species is abundant there throughout the winter. The first fall migrants begin to return in small numbers by mid July, and the highest count of autumn 1979 at Ocean Shores was 1,000 on July 26; subsequent peaks of over 500 birds occurred during early September and late October. Interestingly, the September peak consisted primarily of adults rather than a wave of juveniles as would be expected in most shorebirds. Leadbetter Point, to the south, had peak numbers of migrants through most of September and October in 1978. At Pacific Rim National Park, migratory movements peak in August and September.

The north coast of Oregon and the south coast of Washington support the largest winter concentrations of Sanderlings anywhere in North America, with Christmas Bird Count totals of thousands of birds (over 5,000 at times) at several localities in both states. Wintering populations diminish north of Grays Harbor, with highest counts in the low hundreds on the west coast of Vancouver Island and in the Vancouver area. Both adults and immatures winter in the Northwest, the former more common on the outer coast and the latter in protected waters. Sanderlings shift their wintering grounds from year to year for

reasons not entirely understood, and the highest Christmas Bird Count for the species anywhere in the Northwest in 1986 was 799.

Spring migration counts are also higher in our region than elsewhere on the continent, and more than 20,000 birds at once have been seen in late April and May on southern Washington and northern Oregon beaches. The large flocks avoid the west coast of Vancouver Island, with spring peak numbers there in the low hundreds. Probably because it nests in the high Arctic, the Sanderling remains common late in the spring (note the May 25 peak count), and 1,350 birds were still present at Ocean Shores on June 1, 1980*. Many Sanderlings spend their first summer on their wintering grounds, and a few such birds can be found in the Northwest through June and early July, including small groups at Leadbetter Point. However, a search for summering birds on the coast in 1989 proved fruitless.

Much smaller numbers winter on and migrate along the protected shores of the region wherever there are open beaches. Numbers of Sanderlings usually peak on protected waters, as for example around Vancouver, from August to October while juveniles are on the move. Although juveniles are usually much less common than adults on ocean beaches, two-thirds of 407 birds closely scrutinized at Ocean Shores on October 18, 1986*, were juveniles.

Sanderlings are rare to uncommon migrants throughout the interior, where groups of a few birds are seen, foraging with more common species such as Western Sandpipers. Only scattered individuals are seen in spring, usually in May. Larger numbers of both adults and juveniles, including small flocks, are seen from July through September. There is only a single interior winter record.

Christmas Bird Count data indicate an increase in regional wintering populations in recent years. Eight of 13 possible lowest counts were obtained during the 1974-78 period and 7 highest counts during the 1984-88 period. This stands in contrast with recent evidence for a substantial decline in Sanderling numbers on the east coast of North America. Pacific coast populations have fluctuated in recent years but apparently with no overall decline.

COAST SPRING. *High counts* 20,000 at Sunset Beach, OR, May 21, 1977; 30,000 at south jetty of Columbia River (3 miles of beach), OR, May 25, 1978.

COAST FALL. *Juvenile early dates* Comox, BC, August 17, 1931#; Swantown, WA, August 17, 1985*. *High counts* 600 at Long Beach, BC, September 5, 1982; 500 at Long Beach, BC, September 26, 1985; 4,000+at Leadbetter Point, WA, September 24 to October 9, 1978; 1,000 at Ocean Shores, WA, July 26, 1979*.

COAST WINTER. *High counts* 450 at Chesterman Beach, BC, November 15, 1982; 3,253 at Grays Harbor, WA, December 19, 1981; 3,812 at Grays Harbor, WA, December 21, 1985; 5,853 at Columbia River estuary, OR, December 18, 1983; 3,166 at Florence, OR, December 28, 1984; 5,238 at Coos Bay, OR, December 22, 1985.

INTERIOR SPRING. *Early dates* Lewiston, ID, May 5, 1984 ; Malheur Refuge, OR, April 24, 1986. *High count* 150 at Creston, BC, May 21, 1956. *Late dates* Malheur Refuge, OR, June 8, 1983; Swan Lake, BC, June 7, 1951.

INTERIOR FALL. *Early date* Chewelah, WA, July 6, 1982. *Adult high count* 75 at Okanagan Landing, BC, August 2, 1908. *Juvenile early date* Turnbull Refuge, WA, August 26, 1955#.

CHRISTMAS BIRD COUNTS	Five-Year Averages		
	74-78	79-83	84-88
Comox, BC	50	71	98
Ladner, BC	95	220	192
Vancouver, BC	145	162	186
Victoria, BC	14	19	50
Bellingham, WA	15	50	67
Grays Harbor, WA	2,770	1,390	1,864
Kitsap County, WA	118	192	118
Leadbetter Point, WA	314	550	640
Seattle, WA	50	46	47
Sequim-Dungeness, WA	48	235	656
Coos Bay, OR	591	285	1,842
Tillamook Bay, OR	1,073	687	932
Yaquina Bay, OR	531	485	514

Juvenile high count 80 at Potholes Reservoir, WA, September 21-22, 1980. *Late date* Stratford, WA, October 11, 1970.

INTERIOR WINTER. Okanagan Landing, BC, December 29, 1932.

Habitat and Behavior

Although Sanderlings are quintessential birds of the broad sandy beaches of the ocean front, small flocks of them can also be found in winter on mud flats, gravelly beaches, and rocky shores. They are one of two shorebirds (Black Turnstone the other) that regularly forage on floating kelp beds. Oddly, hundreds of Sanderlings often feed in a brackish marsh at Swantown, when high tides drive them away from the nearby beach, and near Portland they feed in wet, grassy fields on occasion. Most freshwater occurrences of Sanderlings, however, are during migration. In the interior they are most likely to be seen at large reservoirs. In this species more than most shorebirds, there is a clear association in autumn of adults with optimal habitat (open ocean beach) and juveniles with all others.

Sanderlings are the classical in-and-out-with-the-waves sandpiper, but other small sandpipers, particularly Dunlins and Western Sandpipers, feed along with them at times, so identification cannot be based only on this behavior. Sanderlings are the birds that make the band of probe holes along Northwest ocean beaches. As if a million golf tees had been driven into and then removed from the sand, this band may stretch for miles along some beaches, and just a few hundred birds can make many thousands of holes in a few hours of foraging. This is impressive visual evidence of the sandpiper tactile-foraging strategy.

Sanderlings eat a variety of small invertebrates, particularly amphipod crustaceans, farther south on the Pacific coast. One study on the Washington coast found young razor clams to be major prey items. Like Ruddy Turnstones, Sanderlings eat carrion wherever they find it. Sanderling feeding flocks range from a few to dozens of birds loosely spread along the beach but coming together when they flush. Usually they are tolerant of one another, but individual Sanderlings often set up feeding territories, which they defend vigorously against other Sanderlings (running right past other species such as Dunlins to do so).

Roosts are typically on the upper beach or even just above the waves on mid beach, and they are made up of dense flocks of up to hundreds of birds. The flocks can be pure Sanderlings or well mixed with Dunlins and Western Sandpipers, the latter especially during spring migration. Roosting Sanderlings provide perpetual opportunities for "poor one-legged sandpiper" comments by naïve observers. They virtually always roost on one leg, as do most shorebirds, an adaptation for conserving heat (as is the tucking of the bill into the scapulars), and they are surprisingly persistent in remaining one-legged. As a roosting Sanderling is slowly approached, it will hop rapidly away on that single leg, convincing evidence that it is indeed crippled. When an entire flock does the same, seeming to defy laws of gravity and momentum with impunity, concern is quickly replaced by amusement. No wonder truly one-legged shorebirds can forage, albeit awkwardly, to survive.

Nonbreeding Sanderling. Beach-running calidridine with short bill and black legs; pale upperparts and snow white underparts characteristic of plumage; note lack of hind toe. Port Isabel, TX, 25 January 1991, Dennis Paulson

Structure

Size small. Bill short and straight, legs short. Only sandpiper that lacks hind toe, presumably an adaptation to rapid running. Wings extend slightly beyond tail tip, primary projection short to moderate in adults and moderate to long in juveniles. Tail double-notched, central rectrices pointed and extending beyond others.

Plumage

Breeding. Bill and legs black. Overall impression is of an orange-rufous bird. Males average slightly brighter than females, not sufficient for differentiation other than in some pairs on breeding grounds. Head and hindneck gray-brown to rufous, striped with brownish black. Mantle feathers black with rufous fringes. Scapulars and tertials gray-brown (either new or retained) or black with rufous notches and edges and white tips, coverts gray-brown. Foreneck and breast rufous, heavily spotted with brown; rest of underparts immaculate white.

Lower back gray-brown; central uppertail co-

verts same color, often barred black and rufous. Central rectrices blackish, outer ones light gray-brown. Wing brown above; conspicuous white wing stripe formed from broad tips of greater secondary coverts, narrow tips of greater primary coverts, bases of secondaries, extensive markings on outer webs of five or six inner primaries, and pale inner webs of outer primaries. Underwing and axillars white, tips of flight feathers light gray-brown.

Attains breeding plumage later than any other Northwest shorebird, not beginning molt until mid to late April and most finishing it by mid or late May. A third of 1,350 birds at Ocean Shores on June 1, 1980, still had not completed molt. More unusual, one in full breeding plumage at Swantown on April 13, 1986, in large flock of nonbreeding-plumaged birds. In spring, back typically molts before head and breast.

At other end of season, birds arrive in breeding plumage, but many already in molt by late July, more rapidly than most sandpipers. Many in full nonbreeding plumage by early August, virtually all by end of that month. Flight feathers molted mostly during September, greater separation of body and wing molt than typical of shorebirds in this area.

Nonbreeding (Figure 15). Above very pale gray-brown and below entirely white, uniformly

Juvenile Sanderling. Spangled upperparts diagnostic of plumage; by this date, buff on breast faded away and pale scapulars of first-winter plumage appearing. Ocean Shores, WA, 5 September 1983, Dennis Paulson

pale appearance broken by dark brown eye, black bill and legs, and black mark at wrist (lesser wing coverts) that is most apparent in aggressive birds. Only other marking of note, fairly distinct white supercilium.

Juvenal. Similar to nonbreeding adult but heavily marked above, crown and mantle striped with black and white-notched and -fringed black scapulars producing spangled or checkered pattern. Tertials darker than in adults, darkening to dark brown at tip and conspicuously fringed and, in some individuals, notched with white. White supercilium contrasts more with crown and eye stripe than in adult. Breast streaked on sides and may be washed with pale buff. Molting adults with black summer feathers scattered on upperparts (visible into September) can be confusing; however, pattern never regular like that of juveniles. Juvenal plumage retained through September, rapidly molted into first-winter plumage in October.

Immature. First-winter plumage visibly distinct from that of adults by darkly patterned juvenile tertials retained throughout winter. Close look necessary to distinguish these darker, paler-edged feathers (some may show notching) from uniform gray tertials of adults. A few dark scapulars also retained through winter by some birds.

Identification

All Plumages. SANDERLINGS at the ocean are often distinguished by their *rapid running*, but other small sandpipers may perform the same behavior at times, and SANDERLINGS on beaches with no wave action forage like other sandpipers.

Breeding Plumage. Breeding-plumaged adults are relatively unfamiliar, as they are seen here for a shorter time than those of most shorebirds. They are distinctive in their *mottled orangey breast and upperparts*. Juvenile BAIRD'S SANDPIPERS are perhaps most like these birds, and the two species may forage similarly on dry parts of sand beaches. However, the BAIRD'S coloration is more subdued, that of the SANDERLING bright and often blotchy. SANDERLINGS never show the scalloped effect of BAIRD'S and are not so long-winged and pointy looking at the rear. Finally, by the time juvenile BAIRD'S appear in the Northwest (early August), there should be few if any SANDERLINGS in full breeding plumage.

In some adult SANDERLINGS in breeding plumage in July, the back looks dark brown with buffy edgings, the breast a clearer, brighter buff, and these birds could easily be mistaken for RUFOUS-NECKED STINTS, a much smaller species. If there is real doubt about a close-range bird, note that the SANDERLING *lacks a hind toe.*

Nonbreeding Plumage. In nonbreeding plumage, which we see in the Northwest from late July through May, SANDERLINGS are easily distinguished from other small sandpipers by their *overall pale coloration,* light gray-brown above and pure white below. DUNLINS are brown-backed and brownish-breasted. RED KNOTS and WESTERN SANDPIPERS are light-breasted and halfway between DUNLINS and SANDERLINGS in back color but are considerably larger or smaller than SANDERLINGS respectively. The SANDERLING'S *bill is straight and fairly short* in comparison with that of a DUNLIN or WESTERN SANDPIPER, although the rarer SEMIPALMATED SANDPIPER approximates its proportions.

Juvenal Plumage. Juveniles are still obviously SANDERLINGS, although darker and more patterned than the nonbreeding adults. There is no other shorebird that is *pure white beneath and light gray-brown, spangled with black above.*

In Flight (Figure 12)

Nonbreeding SANDERLINGS are *paler* than any other regional shorebird, with *conspicuous white wing stripes and striped tail.* Closest to them are SNOWY PLOVERS, which are distinctly smaller and less conspicuously wing-striped, and their short plover bill can be seen under most conditions. Also similar is a nonbreeding RED PHALAROPE, pale (not so pale as an adult SANDERLING but at a distance more like a juvenile) and with fairly conspicuous wing stripe. Look for the dark eye patch and entirely dark tail in the phalarope. Also, the SANDERLING'S *wing tip is much darker than the wing base,* a condition that increases the conspicuousness of the wing stripe.

Further Questions

Sanderlings represent perhaps the most obvious case of juvenile and adult separation on feeding grounds, in particular because the two plumage types occur synchronously and can be so easily separated during the fall, and it would be of interest to gather more evidence of this. What percentage of birds are juveniles at various localities in September and October (this is of special interest for all birds away from the outer coast)? When adults and juveniles are present, do they feed together or apart? What are the dominance relationships between them? What percentage of the time do individuals of each age class forage or roost (fortunately these two endeavors take up just about all the time of a shorebird and can be easily distinguished)?

Studies of feeding territoriality in Sanderlings have been conducted in California, and it would be worthwhile to learn whether this phenomenon occurs under similar circumstances in the Northwest. In many bird species, slightly opened wings signal aggression. Are the Sanderling's black lesser wing coverts significant in aggressive displays?

The evidence left by Sanderling probes on the beach has to reward detailed study by the student of shorebird foraging. Why are the holes patchily distributed at times? What is the success rate (if successful probes can be distinguished)? The simplest question, involving the identity of what they are removing from the holes, has yet to be answered.

Many Sanderlings were color-banded on the Pacific coast of North and South America in the 1980s, and it would be of value to continue to watch for these birds (Sanderlings have lived as long as a dozen years) and to report any details about color-banded birds to The Sanderling Project, Bodega Marine Laboratory, P. O. Box 247, Bodega Bay, CA 94923. Snowy Plovers and other shorebirds are also being banded by the same and other researchers.

Notes

The black bend of the wing, emphasized as a distinctive mark in some field guides, is often not visible in relaxed Sanderlings. The birds illustrated in National Geographic Society (1987: 129) are too small relative to the other birds on the plate; Sanderlings are just about the size of Dunlins. The juvenile in Peterson (1990: 141) should not be labeled "immature winter;" the juvenal plumage is held only until October.

Photos

Keith and Gooders (1980: Pl. 92) show a bird well along in its molt into first-winter plumage. The "first-winter" bird in Chandler (1989: 91) is probably an adult, as persistent first-winter tertials are typically darker and usually notched. The "winter-plumage adult Sanderling" in Farrand (1988a: 179) is a dull juvenile Semipalmated Sandpiper.

References

Burger and Gochfeld 1991 (human influence on foraging behavior), Burger et al. 1979 (aggressive behavior in migration), Connors et al. 1981 (interhabitat movements), Dwight 1900 (molt and plumages), Evans et al. 1980 (population ecology, food, and foraging), Gerritsen and Meiboom 1986 (tactile foraging), Maron and Myers 1984 (sex distinction), Maron and Myers 1985 (winter feeding and activity patterns), Myers 1980a (population ecology, habitat use, territoriality, and foraging behavior), Myers 1981 (winter distribution of age and sex classes), Myers 1983b (flock associations), Myers et al. 1979 (winter territoriality), Myers et al. 1980 (winter foraging behavior), Myers et al. 1981 (winter territory size), Myers et al. 1984 (Pacific coast distribution), Myers et al. 1990 (migratory routes), Parmelee 1970 (breeding biology), Parmelee and Payne 1973 (mating system), Pienkowski and Green 1976 (breeding biology), Royal 1939 (diet on Washington coast), Silliman et al. 1977 (winter foraging behavior), Spaans 1980 (size variation and molt), Summers et al. 1987 (population biology), Wood 1987 (sex distinction).

STINTS

Very small sandpipers have traditionally been called stints in Eurasia and peeps in North America. Here I am using the term "stint" for the seven species in this group, as have three excellent recent treatments of stint identification (Grant 1984, Veit and Jonsson 1984, Colston and Burton 1988). All three accounts perhaps overemphasize some of the minor (and average) differences among the species, but all three discuss variation and urge caution accordingly. The seven species of stints include three essentially restricted to the New World and four to the Old World. There are three species pairs and a single isolated species (Table 29). The Spoonbill Sandpiper is in essence a stint with an odd bill, but I have left it out of the group for this discussion.

Stint Identification. With the recent flurry of articles on stint identification, the literature abounds with field marks to distinguish stints from one another. It is relatively easy to distinguish three groups of stints—black-legged, yellow-legged, and Temminck's—but within the first two groups, in particular the black-legged species, matters become more difficult. The majority of characteristics represent average differences that would allow distinction of many if not

Table 29. Stint Distribution and Habitat Preference

Species	Hemisphere	Breeding Range	Breeding Habitat	Winter Range	Winter Habitat
Semipalmated Sandpiper	W	arctic	tundra	equatorial	bay
Western Sandpiper	W	arctic	tundra	N/equatorial	bay
Rufous-necked Stint	E	arctic	tundra	equatorial/S	bay/marsh
Little Stint	E	arctic	tundra	equatorial/S	bay/marsh
Temminck's Stint	E	arctic	tundra	equatorial	marsh
Long-toed Stint	E	subarctic	taiga	equatorial	marsh
Least Sandpiper	W	subarctic	taiga	N/equatorial	marsh
Spoonbill Sandpiper	E	arctic	tundra	equatorial	bay

most individuals in a mixed flock of two of the similar species (which rarely happens) but would not guarantee identification of all single individuals. And, after the relatively easy separation of Least and Western sandpipers and the less easy but generally possible separation of Western and Semipalmated sandpipers, Northwest stint identification concerns the picking out of single individuals of *very rare* species. Even distinguishing Semipalmated from Western is not always certain, and the variation in the common species must be thoroughly understood before the accidental ones can be picked out in the field.

Familiarity Bias. See the Introduction for a discussion of this phenomenon, which may be acute with regard to stint identification. Note the "gray morph" of the Little Stint shown in Veit and Jonsson (1984: Fig. 15), so different from our usual concept of the species, and note the variation in Semipalmated Sandpiper juveniles illustrated in the same source. These are the two species that the authors know very well, and they recognize the great variation in them. Probably all of the stints vary similarly, but who has noted this for Long-toed and Rufous-necked stints?

Size and Shape. All stints are *very small* sandpipers, but the largest (female Western Sandpiper at 29 grams on breeding grounds) is about one-third larger than the smallest (male Least Sandpiper at 20 grams on breeding grounds). Much has been made of shape differences in some of the species; for example, the Semipalmated Sandpiper has been called "chunkier" or "plumper" than the Little Stint (Oddie and Marr 1981, Veit

and Jonsson 1984). Yet the two species weigh about the same, averaging around 25-26 grams in summer and 21-22 grams in winter. From their measurements (Appendix 3), there seems no reason for a Semipalmated to look different in any way in shape or size from a Little, and I have not been able to see that difference in numerous photographs of the two species (not having had the chance to compare them in the field). Either can look plump or not, depending on how it is standing. A Little Stint photographed in a flock of Semipalmated Sandpipers in Delaware (Rowlett 1980) is shaped and sized exactly like the birds around it. Direct comparison of Little and Semipalmated skeletons indicated no difference at all in size of the major skeletal elements. Perhaps the slightly thicker bill and slightly shorter wings make the Semipalmated appear plump.

Just as in other *Calidris* sandpipers, the stints vary slightly in shape because of their relative wing and tail lengths. Two species stand out as looking more pointed behind than the others: Temminck's because of its long tail and Rufous-necked because of its long wings and tail (and perhaps because of its slightly shorter legs). In Temminck's Stint the tail appears always to project well beyond the wing tips, whereas this happens rarely in other stints, apparently more likely in Least or Long-toed than in the black-legged species (Table 30). The wing tips commonly project beyond the tail in Western, Rufous-necked, and Little, but can do so in any species other than Temminck's. Because Rufous-necked has a longer tail as well as longer wings, its wing

Table 30. Stint Attributes

Species	Weight	Toes	Leg Color	Wing Projection	Primary Projection	Bill Length	Bill Shape
Semipalmated	26	webbed	black	+	+	short	thicker
Western	27	"	"	+	+	long	drooped
Rufous-necked	26	unwebbed	"	+	++	short	thinner
Little	25	"	"	+	+	"	"
Temminck's	24	"	yellow	–	+	"	"
Long-toed	24	"	"	o	o	"	"
Least	21	"	"	o	o	"	"
Spoonbill	28	"	black	+	++	long	spatulate

tips do not project beyond its tail tip much more than in Little. Shape varies enough depending on how the bird is standing (for example, as birds bend down to feed, their wings may be drawn forward) that it should be used as an indicator rather than a definitive field mark.

Primary Projection. Stints vary considerably in their primary projections, which should be judged only on juveniles or fresh-plumaged winter birds. There is very little primary projection in Least and Long-toed, the tertials almost completely covering the primary tips. In Temminck's the tips of one or two primaries typically show beyond the tertials. In Semipalmated and Western, two to three primaries show, the third only slightly, and in Little and Rufous-necked, more of the third primary shows on average. Although the difference between yellow-legged and black-legged groups is pronounced, that between the two black-legged groups is not so pronounced. In studying many photographs, I have not consistently been able to distinguish with certainty Rufous-necked and Little stints from Semipalmated Sandpiper by this characteristic.

Bill Size and Shape. The emphasis in stint-identification papers on differences such as bill size and thickness can be misleading when accepted as gospel. Although there are modes, the birds simply vary in bill shape just as they do in coloration. Figure 38 shows typical stint bills and their range in length. Some Semipalmated Sandpipers look as if their bill is finely pointed, for example, no different in any way from the "finer-billed" Rufous-necked and Little stints. Western, with its long-billed females, stands out from the other species, but the shortest-billed juvenile male Westerns overlap with other species such as Semipalmated. Similarly, the very fine-billed Temminck's overlaps greatly with other species such as Little and Long-toed and somewhat with the still larger-billed species.

Leg Color. Stints come in two basic leg colors: *black* (varying to olive-gray in juveniles), which includes Semipalmated and Western sandpipers and Rufous-necked and Little stints; and *yellow* (varying from dull to bright yellow or even yellow-orange or yellow-green), which includes Temminck's and Long-toed stints and Least Sandpipers. This characteristic is intended more for close-range checking than for distant identi-

fication of the common species, but it would be essential to note for rare species.

Bear in mind that yellow-legged species can easily darken their legs with wet mud and black-legged species lighten theirs with dried mud (thus the first steps of stint identification might be capture and leg-washing). British observers have reported single Temminck's and Long-toed stints with black and dark brown legs, respectively, and a Little Stint with yellow legs, and enough close observation on this side of the Atlantic ought to add anomalous leg colors to what we know about American stints.

Toe Webbing. Most stints have unwebbed toes, but two species (Semipalmated and Western) have small webs at the base of their toes, better developed between the outer and middle ones. This characteristic might reflect their basically coastal mud-flat distribution in winter, while the other stints are common in interior habitats as well. Perhaps Little and Rufous-necked stints, while common on mud flats, have not been tied to them sufficiently to evolve similar webs, or perhaps the mud is sandier where they live. The other three stints do not typically forage on such soft substrates.

Toe webbing can be very difficult to see in the field, and on several occasions stints have been collected because they were thought to lack webs and thus to be one of the rare species, only to show clearly in the hand the "missing" webs.

Plumages. Even more than with many other shorebirds, it is mandatory to determine the plumage stage (breeding, nonbreeding, juvenile, or molting between two stages) of a stint to facilitate its identification. First-winter stints look about like nonbreeding adults, but many first-winter birds molt relatively little in spring. Some of them remain on their wintering grounds, but others migrate north and thus produce individuals in spring migration and on the breeding grounds not very different from winter-plumaged birds. Thus nonbreeding-plumaged birds could be seen at any time of year, although most likely from September through March.

Only Western and Least sandpipers typically molt into nonbreeding plumage in our region. All of the Eurasian stints and the Semipalmated Sandpiper are longer-distance migrants, wintering to the south of our latitudes, and they should

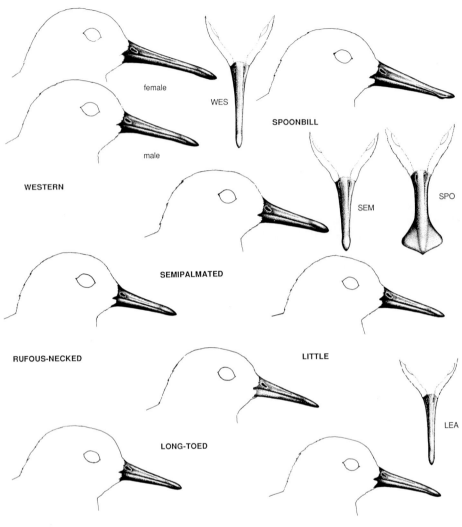

female

WES

WESTERN

male

SPOONBILL

SEM

SPO

SEMIPALMATED

RUFOUS-NECKED

LITTLE

LONG-TOED

LEA

TEMMINCK'S

LEAST

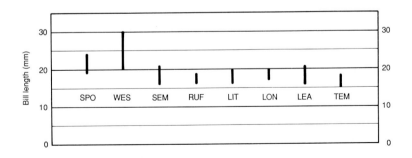

typically move through the Northwest in fall while still largely in breeding plumage. This mode is certainly true for the Semipalmated but naturally difficult to determine for the others because of their rarity. There is of course the possibility that some individuals of the Eurasian species may do something not so typical, paralleling their unusual presence on this side of the Pacific.

Breeding Plumage. Breeding-plumaged birds should be present from April through July and at times as late as September, with body molt in adults occurring primarily in March/April and August/September. Individuals in fresh breeding plumage have feathers of back and wings broadly edged with light brown, bright rufous, buff, and white in different combinations in different species. This is how they are seen in spring, with the primary source of variation being the amount of retained nonbreeding plumage. However, some newly molted feathers, mixed with bright breeding plumage, are nevertheless drab. These feathers may be grown at the beginning of the molt before hormones signal "time to be bright."

As in other shorebirds, the pale edges and fringes become narrower with wear, producing birds that are darker and darker as summer progresses. Wear and tear must be maximal on the breeding grounds, as returning birds in early July are amazingly different from the same birds that passed through two months earlier in May. Western and Semipalmated sandpipers may end up largely black and gray, Least Sandpipers largely black and dark brown. Most of these birds move rapidly through the Northwest, but in August Leasts and Westerns begin a fairly rapid body molt, thus becoming paler and paler as new nonbreeding feathers come in on the upperparts. Leasts begin this molt slightly earlier than Westerns.

Fig. 38. Stint and Spoonbill Sandpiper bills. Illustrations intended to typify each species; overall similarity in shape points out difficulty in use as field mark. Semipalmated a bit thicker, with tip slightly expanded from above, but variable. Western only species with substantial sexual dimorphism in bill length, allows sexing of many individuals. Spoonbill "spoon" not easily seen from side. Bill-length measurements (below) show substantial overlap among all but two longer-billed species. Extent of variation probably greater than indicated in Rufous-necked and Long-toed.

erns. By the first week in September, stints, both migrants and intended winterers, can be seen in full nonbreeding plumage.

Nonbreeding Plumage. Because this is a dull plumage to begin with, wear and fading affect it very little, and wintering Leasts and Westerns look about the same in March as they do in September. Some of the vagrant Siberian stints that hit our shores in autumn may molt at least partially into this plumage while in the Northwest, and a few adult Rufous-necked Stints seen in August were well along in body molt.

Juvenal Plumage. As in other shorebirds, juvenile stints are most brightly marked just when they assume this plumage; probably all have buffy breasts at that time. The buff fades rapidly as they move south, although the upperparts generally remain brightly marked while they are in this region. Nevertheless, pale fringes slowly diminish in width, the latest juveniles not quite so vivid as the earliest. Juveniles should be present from July through September, with body molt into first-winter plumage during September and October. Juvenile Westerns and Leasts, perhaps many of them birds that intend to winter in the region, begin to show considerably plainer backs as October progresses.

Immature Plumage. Many stints breed in their first year, but in at least some species, other individuals do not breed until their second year. This may be the rule in Rufous-necked and is the case in at least some Semipalmated and Western. Most of these birds, in a first-summer plumage much like that of nonbreeding adults, probably remain on their wintering grounds, primarily to the south of our region. Some birds seen in spring in less than full breeding plumage may be one-year-olds.

Vocalizations. As in other shorebirds, flight calls of stints are species-specific. Nevertheless, there is variation within each species, some species apparently have more than one typical call, and no one writing about stints to date (including this author) has heard enough individuals of all the species to compare them effectively. Literature descriptions are confusing, especially for species such as the Little Stint, which has been described by so many authors. Each of the three American species, which I know well, has one typical flight call, with variation relatively rare but promoting confusion, and I assume the same is true for the four Eurasian species. Thus vocalizations are not enough to confirm the identification of a suspected rare species but of course are of value in its assessment. Certainly tape recordings of such individuals would be of great value along with photographs.

Eurasian Stints in the Northwest. There are numerous (five to ten for each species) sight records of these stints on file from the region, primarily from an era when their occurrence was predicted but the information available about them was not adequate to identify them

definitively. Some of the records may be valid, while others are certainly not, based on their written descriptions, and I am choosing to be conservative in my assessment of the occurrence of Eurasian stints in the Northwest by accepting only records that are documented by photographs. Such records can be reassessed at any time in light of new information.

References

Colston and Burton 1988, Jonsson and Grant 1984, Veit and Jonsson 1984, Wallace 1974 (all identification).

SEMIPALMATED SANDPIPER *Calidris pusilla*

Outliers of the immense population of this eastern species move through the Northwest each spring and fall, rewarding diligent stint scrutiny. Pickers rather than probers, their stubby bills and subdued patterns allow distinction from the Western and Least sandpipers with which they associate.

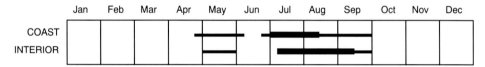

Distribution

The Semipalmated Sandpiper breeds on tundra throughout the North American Arctic from western Alaska to the Atlantic. Migrants are abundant on both Atlantic coastal and interior routes, but different populations have different migration strategies. Both central and eastern Canadian-breeding birds move southeast toward the Atlantic coast in fall and then follow an offshore route to northern South America. Eastern birds come back up the Atlantic coast, central ones across the Gulf of Mexico and up the middle of the continent. Alaskan birds apparently move both south and north across mid continent. Juveniles wander more widely to both the Pacific coast and the Southeast. Wintering occurs primarily in northern South America and around the Caribbean, with small numbers north to Florida.

Northwest Status

The Semipalmated is seen only in migration in the Northwest, with small numbers passing through the coastal subregion in spring from late April through May. The only substantial numbers reported are from Browning Inlet at the north end

of Vancouver Island, where up to 40 birds per day were seen in early May 1969. Numbers of this magnitude or greater need additional confirmation, as no other spring counts from the region have involved more than a few birds.

Adults return in late June and are present to early August. Juveniles occur from late July to mid September. Many Alaska-breeding birds must pass down the Canadian Pacific coast in fall, as the species is regular and at times almost common around Vancouver. Dozens have been seen at the Iona Island sewage ponds in a day, with high counts in the hundreds at times, mostly adults and substantial numbers compared with the few birds anywhere else in the region. Small numbers are seen regularly on southern Vancouver Island, although a single count of 200 is extraordinary. These coastal British Columbia birds almost surely head inland immediately as they move south, as the species is always uncommon farther south in the Pacific states. Perhaps they accompany Western Sandpipers, some of which are known from banding records to fly from Vancouver directly to Kansas.

A few Semipalmateds are regularly seen at

freshwater and brackish ponds in the Puget Sound area, with maxima of both adults and juveniles reached in July. They are scarce on the coast of Washington, but in Oregon they are regular on the coast and quite uncommon anywhere else (for example, few records from the Portland area), an odd discrepancy. Almost all Oregon sightings are of juveniles. Although there are alleged winter records of this species from the Northwest, none is adequately documented, and the species otherwise winters primarily south of the United States.

The Semipalmated Sandpiper is a rare species anywhere in the interior in spring, with a few records each for Oregon, Washington, and southern British Columbia (only two for the well-studied Okanagan region), one for northern Idaho, and two for western Montana.

In fall it is a regular migrant through the interior, with small numbers of adults passing through in mid July and larger numbers of juveniles from late July to mid September. It is more common toward the eastern part of the region, and comparison with the Western Sandpiper is of interest. In eastern Oregon, Western is locally common to abundant and Semipalmated scarce. In eastern Washington and Idaho, Western is common, Semipalmated regular in small numbers. In south-central British Columbia, both seem present in about equal small numbers, but at Fortine, Montana, Semipalmated is more common. The change becomes complete on the Great Plains east of our region, where Semipalmated is abundant, Western relatively rare.

COAST SPRING. *Early dates* Medford, OR, April 14, 1964; Saanich, BC, April 23, 1984. *Late dates* Ocean Shores, WA, June 1, 1980; Iona Island, BC, June 7, 1975.

COAST FALL. *Early dates* Iona Island, BC, June 20, 1989; Tillamook, OR, June 29, 1986. *Juvenile early date* Ocean Shores, WA, July 12, 1980#. *Adult high counts* 200 at Witty's Lagoon, BC, July 12, 1981; 200 at Iona Island, BC, July 18, 1986; 125 at Iona Island, BC, July 18, 1987; 25 at Protection Island, WA, July 21, 1968. *Juvenile high counts* 168 at Iona Island, BC, August 1, 1988; 10 at Seattle, WA, July 30, 1980; 15 at Tillamook, OR, August 18, 1985. *Late dates* Ocean Shores, WA, September 28, 1977; Manzanita, OR, September 25, 1987.

INTERIOR SPRING. *Early dates* Boise, ID, May 3, 1981; Hatfield Lake, OR, May 3, 1987. *High count* 6 at Cow Lake, WA, May 10, 1958. *Late date* Moscow, ID, May 29, 1953.

INTERIOR FALL. *Early date* Bend, OR, July 8, 1984. *Juvenile early dates* Okanagan Landing, BC, July 28, 1915#; Potlatch, ID, July 28, 1951#. High counts 60 at Vernon, BC, August 3, 1911; 230 at Sirdar, BC, August 28, 1947 (possibly Westerns); 30 at Mann's Lake, ID, August 9, 1990; 60 at Fortine, MT, August 18, 1987*. *Late dates* Okanagan, BC, September 16, 1916#; Malheur Refuge, OR, September 25, 1983 (sight record); Sandpoint, ID, September 28, 1950 (sight record).

Habitat and Behavior

Semipalmated Sandpipers are typically birds of interior freshwater ponds in the Northwest, even in coastal areas, but on the Oregon coast they are usually seen on mud flats of coastal bays, and a few have been seen on exposed ocean beaches. On the Atlantic coast, where they are the most abundant sandpipers, they occur in virtually all habitats. They are often seen with Western Sandpipers on the coast and with Western and Least sandpipers in fresh water.

As might be guessed because of their short bills, Semipalmated Sandpipers forage by picking prey from the surface in almost ploverlike fashion, running and stopping jerkily. They can often be distinguished by this foraging method, as they move through flocks of methodically probing Western Sandpipers at an obviously faster rate. This is only a hint and cannot be used as a definitive field mark, as many shorebirds at times utilize atypical foraging methods, especially when in habitats unusual for them. Note also that the Semipalmated often feeds shoreward from the deeper-probing Western.

Structure (Figure 38)

Size very small. Bill short, straight, and blunt-tipped; legs short. Slightly developed webs between toes, outside one larger. Wings project to or just beyond tail tip, primary projection short in adults and moderate in juveniles. Tail double-notched, central rectrices pointed and projecting.

Plumage

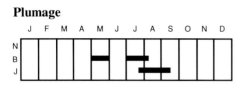

Breeding. Bill and legs black, latter may be tinged gray or greenish. Typical birds more uniformly colored above than most stints, with entirely gray-brown upperparts, regularly or irregularly mottled with dark brown or black where dark feather centers of mantle, scapulars, and tertials show; scapulars with or without paler outer edges. In some individuals, basic color of crown, ear coverts, mantle, scapulars, and tertials slightly more reddish brown. Fairly conspicuous supercilium and throat white, narrow loral stripe brown. Breast gray-brown, finely striped and spotted with dark brown; sides often with a few dark streaks.

Lower back, central uppertail coverts, and central pair of rectrices black; outer uppertail coverts white, rest of tail gray. Wing brown above, with narrow white wing stripe formed by moderately broad tips of greater secondary coverts and quite narrow tips of greater primary coverts. Linear white markings on outer webs of inner primaries and paler inner webs of flight feathers add to distinctness of stripe. Underwing and axillars white, gray-brown on flight-feather tips and greater primary coverts and darker brown on marginal coverts.

Birds that have not entirely molted show considerable plain gray-brown color above, spring birds sometimes differing from those in nonbreeding plumage by no more than scattered dark feathers; these may be first-summer birds. Returning adults in early autumn blacker above and more contrasty, as some pale feather edges have faded to white and others have worn off.

Nonbreeding. Virtually identical to Western Sandpiper, plain gray-brown above and with lightly streaked breast (see below for minor differences). Adults migrate through our region rapidly in autumn with relatively little molt; few in molt show many plain winter feathers on upperparts and some reduction of breast striping. Full nonbreeding plumage rarely if ever seen in Northwest.

Juvenal. Legs olive in young juveniles, becoming darker with age. A variable plumage, seemingly more so than in juvenile Westerns. Typical bird medium brown above, most feathers fringed or edged on outside with pale buff and tipped with whitish. In some individuals feather centers darker, making fringes more conspicuous and forming distinct scaly pattern. Some birds so pale as to simulate nonbreeding plumage, this confusion dispelled at close range when still paler feather edges seen. In many birds overall tones darker, edging not so contrasty, and mantle feathers, scapulars, and tertials may all have dull rufous fringes, in brightest birds almost as highly colored as juvenile Western. All have gray-brown coverts with buff fringes. White mantle and scapular lines typical of juvenile stints usually absent in this species or, if present, narrow and indistinct. Throat and supercilium white, loral stripe and ear coverts brown. Breast suffused with light tan and streaked with brown on each side.

Juveniles regularly appear in region by late July, surprisingly early for arctic-breeding species. Autumn molt usually delayed until wintering grounds, primarily south of United States; unmolted juveniles in Southeast as late as early November.

Immature. Many juveniles molt outermost primaries on wintering grounds. Some return north to breed in first summer, while others do not, oversummering in nonbreeding plumage.

Identification

All Plumages. Only two other regularly occurring sandpipers are small enough to be confused with this one. LEASTS should be immediately separable by their overall browner coloration in any plumage, especially the darker breast, although some breeding SEMIPALMATED may approach the condition typical of LEAST. The SEMIPALMATED's *dark legs* always allow distinction from LEAST unless the latter's yellow legs are covered with dark mud. The bills of the two species are about the same length but on the average are shaped differently, the SEMIPALMATED a bit blunter and less likely to be slightly curved.

SEMIPALMATED and WESTERN are essentially identical in size and shape, WESTERN averaging very slightly larger. SEMIPALMATED can be distinguished in any plumage by its *shorter, stubbier bill*. In the Northwest the two species barely overlap in bill length, but note that SEMIPALMATEDS from eastern North America have slightly longer bills and overlap a bit more with WESTERNS. SEMIPALMATED typically has a distinctly bulbous-tipped bill, therefore blunter-looking than WESTERN, but a longer-billed SEMIPALMATED may have a somewhat finer bill tip.

Breeding Plumage. Breeding-plumaged SEMIPALMATED and WESTERN SANDPIPERS, whether fresh or worn, can be easily distinguished. SEMIPALMATED usually has *little or no rufous* coloration above. Although individuals can show reddish on any of the feathers of the upperparts, it is never as bright or extensive as the WESTERN'S copper tones. Adults, especially worn fall birds, have *much paler breasts and less heavily marked sides* than WESTERNS, in which the breast streaks are more conspicuous and the sides are heavily marked with chevrons as well as streaks.

Nonbreeding Plumage. Plumage differences (breast streaking, head streaking) between SEMIPALMATED and WESTERN in nonbreeding plumage are no more than average ones, with sufficient variation to complicate any assessment of individual birds. In this plumage the bill remains the best way to tell them apart.

Juveniles that retain their rufous scapulars into late autumn can easily be identified as WESTERNS, but nonbreeding adult WESTERNS first appear by late August, and, because most of the late and wintering WESTERNS are young males, with the shortest bills of the species, there is a real chance for confusion. Plain, short-billed birds seen among brightly colored, longer-billed birds will very likely *not* be SEMIPALMATEDS, and probably all the late-autumn and winter sight records of SEMIPALMATEDS from the Northwest have been these male WESTERNS.

Juvenal Plumage. All juvenile WESTERN SANDPIPERS seem stamped from a single mold, and most SEMIPALMATEDS differ from them in having *little or no rufous above* as in breeding birds. Nevertheless, the most brightly marked SEMIPALMATEDS are about as rufous above as WESTERNS. Those that are richer reddish brown above will usually be that way on the crown and mantle as well as elsewhere, whereas WESTERNS

typically have their rufous most prominent on the bases of the upper scapulars and tertials. SEMIPALMATEDS that have little rufous but instead show dull medium-brown to almost black feather centers above, with conspicuous light edges, look more like miniature juvenile SANDERLINGS when they are more vivid or miniature juvenile BAIRD'S when they are more subdued. These birds will never be mistaken for WESTERNS.

SEMIPALMATEDS also typically are buffier on the breast (particularly earlier in the season) and more heavily streaked on the breast sides than WESTERNS, with a slightly more contrasty head pattern caused by slightly darker crown and ear coverts. Additionally, there are a number of minor average differences that I consider too variable to discuss here. By themselves they are not definitive, but in the aggregate they might be of value in identifying a puzzling individual. These differences are enumerated by Veit and Jonsson (1984) and Grant (1986).

In Flight

This species looks about like the WESTERN SANDPIPER in flight.

Voice

The flight call of the SEMIPALMATED is distinctive, a short *chert* or *chut* that may be doubled, distinctly lower and shorter than the high, rolled call of the LEAST or the single squeaky syllable of the WESTERN.

Further Questions

The variation in juvenal plumage in this species is unusually great, and some description of plumage should accompany all sightings of juveniles, especially if comparison is possible when several individuals are present. Also, note should be taken of unusual foraging methods, for example, Semipalmated moving slowly and probing as is typical of Western. In general, notes on comparative foraging of birds of different species in the same habitat should provide additional knowledge of shorebird biology and be of great value for future field identifiers.

Notes

A record of 100 at Crescent Beach, British Columbia, May 8, 1948, is considered doubtful, as are reports at the same locality of six on January 4, 1941, and 10 on December 16, 1941. An unusually early spring record from Drayton Harbor, April 7, 1962, is also not adequately documented to be acceptable. Records after September—for example, in British Columbia at Long Beach, October 5, 1982, and Victoria, October 14, 1975—are difficult to interpret but may be valid.

The comparison between male and female bills in Farrand (1983: 376) could be used to represent extreme variation, as the differences are too great to be average ones.

Photos

The degree of variation in juveniles can be seen by examining photographs herein and in Bull and Farrand (1977: Pl. 221), Armstrong (1983: 144), Farrand (1983: 377; 1988a: 179, 182 [called a Sanderling]; 1988b: 168), Veit and Jonsson (1984: 856-58), and Chandler (1989: 16, 93). The "Semipalmated Sandpiper winter plumage" in Udvardy (1977: Pl. 228) is a Sanderling in first-winter plumage. The "winter plumage" Semipalmated in Farrand (1983: 379; 1988b: 168) has a bill long enough to be a male Western, although it is within the range of eastern Semipalmated females. As it is in the worn nonbreeding plumage of mid or late winter, it is surely a Western if photographed anywhere north of southern Florida. Compare it with the fresh-plumaged Western at the bottom of the same page.

References

Ashkenazie and Safriel 1979a (breeding biology), Ashkenazie and Safriel 1979b (time-energy budgets), Ashmole 1970 (winter foraging behavior), Baker 1977 (summer diet), Baker and Baker 1973 (summer and winter foraging behavior), Buck et al. 1966 (identification), Burger et al. 1979 (aggressive behavior in migration), Cartar 1984 (size comparisons with Western Sandpiper), Dunn et al. 1988 (fall migration and fat deposition), Evanich 1989 (status in Oregon), Grant 1981 (identification), Grant 1986 (identification), Gratto 1988 (site tenacity and breeding age), Gratto and Morrison 1981 (postjuvenal wing molt), Gratto et al. 1981 (breeding success), Gratto et al. 1985 (site tenacity and mate fidelity), Harrington 1982 (bill-length variation and habitat use in migration), Harrington and Groves 1977 (aggressive behavior in migration), Harrington and Morrison 1979 (migration), Harrington and Taylor 1982 (sex distinction), Holmes and Pitelka 1968 (summer diet), Jackson 1918 (molt), Lank 1979 (migration and marking techniques), Lank 1989 (migratory departure cues), Miller 1983c (aerial displays and vocalizations), Norton 1972 (incubation schedules), Norton and Safriel 1971 (homing of breeding birds), Oddie and Marr 1981 (distinction from Little Stint), Page and Bradstreet 1968 (fall migration), Page and Middleton 1972 (fat deposition during migration), Page and Salvadori 1969 (fall weights), Paulson 1983 (juvenile fledging and migration), Phillips 1975 (distribution and identification), Stevenson 1975 (identification), Wallace 1979 (identification), Weber 1981 (status in Washington and northern Idaho), Wilson 1990 (foraging ecology).

WESTERN SANDPIPER *Calidris mauri*

It is difficult to imagine the numbers of these little birds that pile up on our coasts in late April and again in early July on their rapid way to and from the Arctic. Ebbing and flowing with the tides, they are at their most spectacular when a raptor flushes them and flocks coalesce over the mud flats into pulsating macro-organisms. Know the Western's black legs, its longish, slightly droopy bill, and its rufous back and head markings well to be able to identify any small sandpiper that deviates from it.

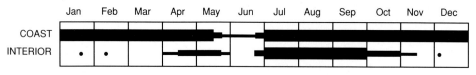

Distribution

The Western Sandpiper is an Alaska breeder, occurring primarily on wet tundra of the western coastal plain of that state—a surprisingly small breeding range for such an abundant bird—and extending into far northeastern Siberia. The winter range is from southern United States, on both Atlantic/Gulf and Pacific coasts, south to northern South America. Migrants are much more abundant in fall than in spring in the interior and East, and the spring route of eastern-wintering birds is entirely open to conjecture, perhaps a single long flight.

Northwest Status

During migration this species is by far the most abundant shorebird in the Northwest. Spring migrants begin to arrive in mid April and peak near the end of that month, when hundreds of thousands are present in Grays Harbor and tens of thousands in many other estuaries on both outer coast and protected waters. The highest count in Grays Harbor was just over a half-million on April 25, 1981, and it was thought that there were many more present on the preceding two days. Birds are common as far inland as the Willamette Valley.

Near Vancouver spring migrants remain for an average of about three days on their way north. At Grays Harbor numbers drop off sharply in early May, by mid month only hundreds are present at favored localities, and by the end of the month they are gone. At the Fraser River delta, many birds are still present through mid May. The last week of May and the first three weeks of June provide few birds of this species at Grays Harbor, perhaps all of them injured or otherwise unable to migrate north. Western Sandpipers oversummer south of their breeding range in other areas, however, and small groups have been seen at Leadbetter Point through June.

Fall migrants return with a rush in the last week of June, reaching the tens of thousands by the first week of July. Adults in full breeding plumage make up these flocks, and after the first torrent they keep coming, although in much reduced numbers, through July and much of August. Juveniles begin to arrive early in August (rarely as early as mid July), and by late August they greatly outnumber adults. Juvenile peaks have

been recorded throughout August and early September. Birds in this age class are more widespread than adults, for example occurring by the thousands in inlets in southern Puget Sound where adults are relatively scarce.

They remain common through September, and their numbers are often augmented by substantial groups of birds that arrive with the big Dunlin flocks in October. After that month, wintering populations can be found in most coastal areas, often associated with groups of Dunlins but usually in the minority. Grays Harbor, Willapa Bay, and Coos Bay support the largest numbers, with occasional counts in the thousands. Fewer winter in other Oregon estuaries and still fewer in protected coastal localities of Washington and British Columbia, north to Victoria and Vancouver. The Willamette Valley supports small flocks south to Eugene.

Males migrate slightly before females in spring, predominating in the earlier flocks. In fall this is reversed, females more numerous among first-arriving birds. Wintering birds in Grays Harbor are primarily if not all first-year males. Both sexes, but mostly males, have been evident in small wintering flocks farther north. As primary molt occurs in some adults still in the region in early September, that age stage probably winters also.

Western Sandpipers are uncommon spring (late April to early May) and at least locally common fall (July through September) migrants throughout the interior of the region, decreasing toward the east. They are rare in spring and locally common in fall in Idaho, and much more numerous in fall than in spring in western Montana, where they are uncommon at any time.

High counts in spring involve no more than a few birds at most localities. Fall counts are much higher, with peaks in southern Oregon in the thousands. The late September record of 23,000 at Malheur Refuge is unusual for the large numbers so late in the fall, and nowhere else in the region do numbers even approach this total. There are a few winter records from the interior, the only January record during a mild winter.

COAST SPRING. *High counts* 5,000 at Tillamook, OR, May 8, 1976; 5,000 at Florence, OR, April 28, 1984; 1,500 at Medford, OR, April 30, 1984; 520,000 at Grays Harbor, WA, April 25,

Juvenile Semipalmated Sandpiper. Bright extreme of this plumage, with rufous crown and upper scapulars; note very plain coverts and faint indication of mantle line. Jamaica Bay Wildlife Refuge, NY, August 1984, Urban Olsson

Breeding Semipalmated Sandpiper. Stint with black legs and short, thick bill; dullest extreme of breeding plumage, worn and with few scapulars replaced by non-breeding plumage; note toe webbing. Maine, August 1983, Joseph Van Os

Juvenile Semipalmated Sandpiper. Duller individual, with scalloped appearance like Baird's Sandpiper. Seattle, WA, 31 July 1983, Jim Erckmann

Breeding Western Sandpiper. Long-billed, black-legged stint with bright rufous crown, cheeks, and scapulars and black side chevrons in this plumage. Ocean Shores, WA, 22 April 1984, Richard Droker

Juvenile Western Sandpiper. Especially bright fresh-plumaged juvenile, with rufous wash over entire upperparts; long bill indicates female. Seattle, WA, 4 August 1983, Jim Erckmann

Juvenile Western Sandpiper. More typical juvenile (also a month later), with rufous confined to upper scapulars; note forking of mantle feathers (just above upper scapulars) and separation of upper and lower scapulars by a tertial. Crockett Lake, WA, 10 September 1989, Harold Christenson

1981; 100,000 at Roberts Bank, BC, April 25, 1979; 60,000 at Roberts Bank, BC, April 25, 1982.

COAST FALL. *Adult high counts* 100,000 at Iona Island, BC, July 1, 1979; 10,000 at Blaine, WA, July 6, 1968; 30,000 at Grays Harbor, WA, July 10, 1979; 26,500 at Grays Harbor, WA, July 21, 1982; 15,000 at Tillamook, OR, August 2, 1983 (may include juveniles). *Juvenile early date* Iona Island, BC, July 18, 1983*. *Juvenile high counts* 35,000 near Tofino, BC, September 4, 1977; 20,000 at Iona Island, BC, September 6, 1982; 20,000 at Ocean Shores, WA, September 13, 1987; 80,000 at Tillamook and Netarts bays, August 25-26, 1979.

COAST WINTER. *High counts* 2,500 on southwest coast, WA, January 15-16, 1983; 2,000 at Coos Bay, OR, December 16, 1979.

INTERIOR SPRING. *Early dates* Malheur Refuge, OR, April 2, 1966, February 14, 1990 (7); Potholes, WA, April 2. *High counts* 1,120 at Summer Lake, OR, April 21-22, 1987; 1,029 at Summer Lake, OR, May 1, 1987; 75 at Banks Lake, WA, spring 1963. *Late date* Haynes Point, BC, May 24, 1966.

INTERIOR FALL. *Early dates* Potholes, WA, June 28, 1954; Malheur Refuge, OR, June 21, 1988. *Adult high count* 8,000 at Stinking Lake, OR, July 9, 1975. *Juvenile early date* ID, July 28#. *Juvenile high counts* 500 at Salmon Arm, BC, September 6, 1970; 300+ at Soap Lake, WA, August 9, 1958; 200 at Reardan, WA, August 14, 1965; 23,000 at Malheur Refuge, OR, late September 1979; 9,200 at Deer Flat Refuge, ID, August 25, 1990. *Late date* Malheur Refuge, OR, November 10, 1968.

INTERIOR WINTER. Vantage, WA, December 5, 1980 (small flock); Springfield, ID, January 19, 1975.

CHRISTMAS BIRD COUNTS	Five-Year Averages		
	74-78	79-83	84-88
Grays Harbor, WA	863	310	356
Leadbetter Point, WA	151	755	156
Coos Bay, OR	771	685	676
Tillamook Bay, OR	244	121	65
Yaquina Bay, OR	119	56	70

Habitat and Behavior

Westerns are mud-flat birds, but they occur in smaller numbers in most shore environments in the Northwest, for example in the big flocks of shorebirds on the outer ocean beach during the height of migration. They invade tangled salt marshes much less frequently than do Least Sandpipers, although the two feed together regularly at the upper edges of mud flats and in virtually all freshwater sites.

Westerns, with their longer bills, feed by probing in deeper water or farther out on mud flats than do Least and Semipalmated. Where a series of environments are provided, these species separate out spatially, but in more limited areas, for example tiny freshwater ponds, they may overlap entirely in habitat (including water depth) and feeding mode. When on the ocean beach with Sanderlings and Dunlins, Westerns are more likely to be up on the moist sand while the larger species are feeding at the edge of the waves.

Western Sandpipers and Dunlins form the largest flocks of any Northwest shorebirds, individual flocks numbering up to the thousands both at roosts and in flight. When feeding, Westerns spread out considerably and at some times show substantial aggression among flock members; at other times, dozens feed close by one another with no sign of antagonism. In some flocks it is obvious that only a few individuals are persistently aggressive.

Typical of shorebirds, they may fly many miles between roosting and feeding sites, flocks of all sizes sweeping low over water and beaches at high speed, often mixed with Dunlins and Short-billed Dowitchers. The sound of their wings precedes their occasional flight calls as they pass close enough to the onlooker at times to provoke apprehension. Other flocks, perhaps passing over headlands, attain hundreds of feet in the air. These back-and-forth flights are as spectacular in their speed and directness as the whirling predator-alarm flights are in their grace and rhythm.

Structure (Figures 38, 39, 44)

Size very small. Bill short, but longer than typical stint bill. Only stint (and one of relatively few sandpipers) in which there is a chance to distinguish sexes in field. Male bills distinctly shorter than those of females, averaging slightly less than

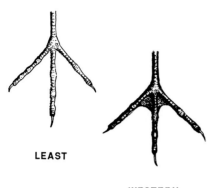

LEAST

WESTERN

Fig. 39. Western and Least Sandpiper feet. Western with webs between front toes, Least without; Least also with relatively long toes for its size.

an inch in length, while females average slightly more than an inch. Average difference only 4 mm, but many birds can be sexed by this character, especially when both sexes together. Legs short, with webbing between toes exactly as in Semipalmated. Wings project to tail tip or slightly beyond, primary projection short in adults and short to moderate in juveniles. Tail double-notched, central rectrices projecting.

Plumage

Breeding. Bill and feet black. In fresh breeding plumage, acquired in March and April, marked above with rich reddish. Crown feathers and some mantle feathers fringed with rufous; lores brown, ear coverts rufous-tinged. Scapulars and tertials broadly fringed with rufous and tipped or edged with white, some marked with rufous basally as well. Coverts plain gray-brown. Underparts white; breast light gray-brown marked with black streaks, sides marked extensively with black streaks and chevrons.

Lower back, central uppertail coverts, and central pair of rectrices black; outer uppertail coverts white, rest of tail gray. Wing brown above, with narrow white wing stripe formed by moderately broad tips of greater secondary coverts and quite

narrow tips of greater primary coverts. Linear white markings on outer webs of inner primaries and paler inner webs of flight feathers add to distinctness of stripe. Underwing and axillars white, gray-brown on flight-feather tips and greater primary coverts and darker brown on marginal coverts.

Some birds show scarcely any rufous in spring, although whether due to immaturity, late molting, or individual variation is not known. Returning adults in fall grayer and faded; black on back more prominent and patchier, rufous also more prominent by exposure of more of scapular and tertial bases. Body molt begins in August in adults that intend to winter in region.

Nonbreeding. Very plain gray-brown above, feathers with fine dark streaks; white below, with fine breast streaks better developed than in juveniles. Supercilium white; loral stripe brown, ear coverts as upperparts. Few adults in this plumage mixed with bright juveniles in September.

Juvenal. Differs from breeding adult by lack of rufous tones on head. Head pattern like that of nonbreeding adult but darker and slightly more distinct. Crown and hindneck gray-brown with brown stripes. Mantle feathers fringed with gray-brown or rufous, scapulars and tertials with rufous and white. Coverts gray-brown with buff fringes. Underparts largely unmarked, with streaks on sides of breast and faint buffy wash across it. Typically a white mantle line, more rarely a white scapular line.

Immature. Gray upperpart feathers of first-winter plumage begin to appear at end of August, but rusty scapulars persist into October in most individuals. Birds that remain for winter become indistinguishable from nonbreeding adults in field. No primary molt occurs in first winter, perhaps because this species winters farther north than most other stints. Majority of individuals, but definitely not all, migrate north to breed in first summer. Summering birds rare anywhere on Pacific coast of North America but may be more regular on south Atlantic and Gulf coasts and in South America.

Identification

All Plumages. By far the most abundant Northwest shorebird, the WESTERN serves as the baseline for identification of all other stints. Basic characteristics in all plumages include *small size* and a *longish bill* that appears to droop slightly at the tip, especially in females (the bill length of this species scarcely overlaps that of any of the other stints). Although the largest of the stints, it is still a small bird, the next-largest *Calidris*—BAIRD'S and WHITE-RUMPED—about two-thirds again heavier.

See SEMIPALMATED and LEAST SANDPIPERS for additional comments on distinguishing these common species, and the other stints, all of which are extremely rare, for their characteristics. The WESTERN SANDPIPER is a quite distinctive stint but variable enough, especially in worn breeding plumage, that at least short-billed males might generate some confusion.

Breeding Plumage. In breeding adults the *rufous-based upper scapulars and rufous head markings* provide the best field marks. These reddish patches against the otherwise basic gray of these birds give them a contrasty look different from the overall blackish to brown to reddish brown LEAST SANDPIPERS and the overall dull to mottled to buffy brown SEMIPALMATED SANDPIPERS in this plumage. In addition, the *prominently streaked sides* provide distinction from all other stints, and the breast is more heavily marked than that of the SEMIPALMATED.

Nonbreeding Plumage. Perhaps the only other species worth comparing with the WESTERN is the DUNLIN in nonbreeding plumage. DUNLINS are about twice as heavy as WESTERNS, and the latter look *conspicuously smaller and shorter-billed* when mixed in with the big DUNLIN flocks. In addition, they are much *paler-breasted*, looking all white beneath or with a pale wash across the breast, while DUNLINS look brown-breasted. The WESTERN is the only black-legged stint likely to be seen in this plumage in the Northwest, but see the other species (SEMIPALMATED, RUFOUS-NECKED, and LITTLE) for differences. WESTERNS are larger and much paler-breasted than nonbreeding-plumaged LEASTS, with which they are seldom seen during the winter.

Juvenal Plumage. As in breeding plumage, the *rich reddish scapulars* furnish a good field mark, especially in contrast with the gray appearance of the rest of the plumage. The much browner LEASTS look quite different, but some juvenile SEMIPALMATEDS approach WESTERNS in

color, with scapulars more reddish than the rest of the upperparts; these birds should be distinguishable by bill length. The buffy wash on the breast is usually paler in WESTERN than in SEMIPALMATED, but the two species overlap.

In Flight (Figure 12)

WESTERNS are easily recognizable in flight because of their *small size and narrow white wing stripe*, with only other stints likely to be mistaken for them. A SEMIPALMATED would look essentially similar unless color and bill length could be seen at close range. LEASTS can be distinguished from WESTERNS by their smaller size (apparent in mixed flocks) and browner upperparts and breast. WESTERNS in migration occur in much larger flocks than LEASTS, which are usually more scattered in the first place. WESTERNS wheel and flash in breathtaking aerial ballet much like the similarly large DUNLIN flocks that occupy the flats during winter; the two may mix freely in spring when both are common.

Voice

The WESTERN'S flight call is a high-pitched *dzheet*. Individuals of this species are much less vocal than the LEAST SANDPIPERS with which they often mingle, but calls will be heard frequently in big flocks of WESTERNS. The call is clearly monosyllabic, in comparison with the LEAST'S rolling almost-trill. The SEMIPALMATED has a shorter, lower-pitched call than the WESTERN.

Further Questions

As female Western Sandpipers desert their mates soon after their eggs hatch, it is likely that many more females than males reach our region in early fall. Alternatively, it may be that most of the early July birds are failed breeders of both sexes. Both sexes are clearly represented among the early migrants, but we need more careful censuses with considerable effort to sort out the sexes at that time. So far only a single study near Vancouver has sampled sex ratios in captured birds during migration. There is also some evidence that juvenile females migrate slightly before juvenile males, a surprising situation if real. Similarly, toward the end of the migration in late September and October, sex ratios should be determined. It appears that most birds present after mid September are males, but this should also be further documented, especially away from Grays Harbor, the only place Western Sandpiper wintering has been studied.

It would be especially interesting to determine whether males and females had exactly similar feeding habits or if there were some spatial segregation between the sexes, say in water depth or distance from the tide's edge. One recent spring observation showed many more males than females in some flocks on the ocean beach, and even the sex ratios in individual feeding flocks would be worthwhile to determine. Finally, the small numbers of Western Sandpipers that are seen between spring and fall migrations should be scrutinized for plumage (are some of them immature?) and injuries (or are they merely unable, for some reason, to finish their northward flight?).

Some Western Sandpipers accompany October-arriving Dunlin flocks, and we should know whether these are adults or immatures. Check birds in these flocks for retained reddish scapular feathers (indicating immatures) and wing molt (indicating adults that have been present for some time).

Photos
The Western Sandpiper arranging "her" eggs in Hosking and Hale (1983: 103) appears to be a male. A juvenile unlabeled as such is in Bull and Farrand (1977: Pl. 222). The adults in Farrand (1983: 379-81) are males, the juvenile probably a female.

References
Ashmole 1970 (winter foraging behavior), Baldassarre and Fischer 1984 (diet and foraging behavior in migration), Brown 1962 (breeding behavior), Buchanan 1988 (status in southern Puget Sound), Butler and Kaiser 1988 (British Columbia populations), Butler et al. 1987 (migration periods, age and sex ratios, and weights in south coastal British Columbia), Cartar 1984 (size comparisons with Semipalmated Sandpiper), Holmes 1971b (breeding biology), Holmes 1972 (breeding biology and annual cycle), Holmes 1973 (breeding behavior), Loftin 1962 (summering), Ouellet et al. 1973 (plumages, identification, and occurrence in eastern Canada), Page and Fearis 1971 (sexual differences), Page et al. 1972 (age and sex composition of winter flocks), Paulson 1983 (juvenile fledging and migration), Senner 1976 (diet in migration), Senner and Martinez 1982 (migration), Senner et al. 1981 (spring migration in Alaska), Stevenson 1975 (identification).

RUFOUS-NECKED STINT *Calidris ruficollis*

This casually occurring stint should be easily identifiable in rufous-headed breeding plumage, but only lengthy study will provide enough details to distinguish juveniles or nonbreeding adults from the much more common Semipalmated Sandpiper. The long, pointed look of the species might focus the observer's attention closely enough to detect its unwebbed toes, but even then the difficulty of distinguishing it from the Little Stint remains.

	Jan	Feb	Mar	Apr	May	Jun	Jul	Aug	Sep	Oct	Nov	Dec
COAST												

Distribution
The Rufous-necked Stint breeds on the Siberian tundra and winters in southeast Asia and Australasia, where it is by far the most common stint.

Northwest Status
Breeding locally in western Alaska and abundant on the Asian Pacific coast, this should be the most likely Siberian stint to occur in the Northwest, and this has been the case. About ten well-documented breeding-plumaged individuals have been recorded since 1978, equally from Iona Island and the Oregon coast (do they skip Washington entirely?) and all in fall migration from June

20 to August 26. Additional records of adults from California, including one in spring, indicate that the species probably occurs annually in very small numbers along the American Pacific coast.

COAST FALL. Iona Island, BC, June 24-25, 1978; Iona Island, BC, July 13-15, 1978; Iona Island, BC, August 25-26, 1978; Iona Island, BC, July 3-4, 1986; Iona Island, BC, June 26 to July 5, 1988; Tillamook, OR, June 20, 1982; Tillamook, OR, July 3, 1982; Tillamook, OR, August 19-26, 1982 (2); Bandon, OR, June 25, 1984.

Habitat and Behavior

Like other black-legged stints, this is a species of mud flats and pond edges, occurring in migration on both fresh and salt water in its Asian and Australian range; wintering is mostly on the coast. It feeds by rapidly moving across the substrate with head down, picking here and there for surface invertebrates, but it also may probe like a Western Sandpiper. Relatively short-billed like a Semipalmated, most of its feeding is similarly above the water line. Feeding birds spread out over the substrate but come together in small flocks when flushed. They often roost scattered in flocks of larger shorebirds. Although in large numbers on the other side of the Pacific, where it is the common small sandpiper, it is likely to occur singly in our region. Northwest birds have invariably associated with Western Sandpipers, as would be expected.

Structure (Figure 38)

Size very small. Bill short, finely pointed, and slightly droopy; legs short. Wings long, reaching tail tip or projecting beyond it, in some individuals well beyond; primary projection moderate in adults and juveniles. Tail double-notched, central rectrices pointed and projecting.

Plumage

Breeding. Bill and legs black. Adults in fresh breeding plumage striking, with most of head, neck, mantle, and upper breast rich rufous. Males average brighter than females, especially on foreneck, but considerable overlap. Rufous of foreparts frosted with white tips early in spring and duller and faded later in summer and, even when fresh, varying in intensity among individu-als. Variation not well understood, but some individuals in spring with much white on throat. Crown, hindneck, and mantle striped with black. Dark loral stripe. Scapulars and tertials vary from gray-brown with black centers to black with broad rufous fringes. Coverts gray-brown. Rufous on upper breast bordered behind with scattered dark brown spots that extend a short way down sides.

Lower back black, scalloped with gray-brown. Central uppertail coverts black, outer ones white; tail gray-brown except for dark brown central rectrices. Wing brown above, narrow white wing stripe formed by moderately broad tips of greater secondary coverts and quite narrow tips of greater primary coverts. Linear white markings on outer webs of inner primaries and paler inner webs of flight feathers add to distinctness of stripe. Underwing and axillars white, gray-brown on flight-feather tips and greater primary coverts and darker brown on marginal coverts.

Breeding plumage acquired in March and April; molt out of it begins in July, with white feathers appearing on throat and breast and gray feathers on back and wings. Nonbreeding plumage fully attained between mid August and mid September, rarely traces of breeding plumage retained into October.

Nonbreeding. Light gray-brown above, feathers with conspicuous dark streaks (rarely dark centers); white below, often with fine streaks at side of breast. White supercilium and dark loral stripe. Most adults migrate south rapidly while still largely in breeding plumage, but later individuals likely to be well along in body molt, and entirely molted individual a possibility.

Juvenal. Legs of young juveniles might be olive, as in related species, but no information exists. Relatively dull plumage but considerable variation, with some individuals distinctly more brightly marked. Crown and mantle gray-brown to dull rufous with dark stripes. Scapulars often brightly marked, edged with rufous and tipped with white. Tertials typically drab, not especially dark-centered, and fringed with gray-brown to dull buff rather than the richer tones of other species. White mantle and scapular lines usually not well-developed. Coverts plain buffy brown. Supercilium typically interrupted by fine streaks above eye. Breast light gray or buffy, with or without fine streaks on sides.

Immature. Indistinguishable from adult after postjuvenal molt. Many individuals molt outer primaries during first winter, majority (higher proportion than in other stints) oversummer in southern hemisphere in worn nonbreeding plumage. Bird collected in southern California on August 17 in this plumage.

Identification

All Plumages. This species is one of the group of stints with *black legs*, all the species of which are of similar size. It is further distinguished from all but LITTLE STINT by its *unwebbed toes*, which are of course difficult to see clearly in most birds. Its wings are longer than in any other stint (it is probably the longest-distance overwater migrant of the group), and, coupled with slightly shorter legs than the average of its size group, it has a "rakish" or pointed look toward the rear. This is an average difference, not obvious in all individuals. Going along with the longer wings is the longest primary projection of the group, shared with LITTLE STINT; three primaries regularly project beyond the tertial tips. The bill is fairly short and fine, between those of SEMIPALMATED and LITTLE, but the three species overlap substantially.

Breeding Plumage. With their *rufous throat and breast*, most breeding-plumaged adults are easy to identify. Birds with white throats, however, could be mistaken for LITTLE STINTS, which show rufous extensively on the sides of the head and are otherwise differently patterned, with more contrasty upperparts. The WESTERN SANDPIPER has a bright rufous cap and dull rufous ear coverts (and is otherwise very gray), but neither LITTLE nor WESTERN normally has rufous on throat or breast. However, be aware of the great similarity of a breeding-plumaged SPOONBILL SANDPIPER.

SANDERLINGS in breeding plumage have been called RUFOUS-NECKED STINTS on more than one occasion, even by experienced observers. SANDERLINGS are a duller orange-rufous on the head, breast, and upperparts, the orange evenly marked with short stripes and spots, while the STINT is more brightly colored, especially on the throat and breast, and *the lower throat is unmarked*, with dots or streaks only at the lower edge of the rufous area. Their typical foraging behavior would be different, but SANDERLINGS do slow down from a run on occasion. They are clearly different in size, although comparisons might be necessary.

Nonbreeding Plumage. Nonbreeding birds look essentially like WESTERN and SEMIPALMATED SANDPIPERS in coloration and could only be distinguished by their *unwebbed toes*. This characteristic should be photographed clearly to document a bird in nonbreeding plumage, which is otherwise so similar to a short-billed male WESTERN, the only black-legged stint typically seen in this plumage in the Northwest. A bird in a flock of WESTERNS might be recognized by its slightly longer-looking wings and shorter legs. See also LITTLE STINT, also almost identical in this plumage.

Juvenal Plumage. Short of a bird in the hand, it may be impossible to distinguish without doubt a juvenile of this species from the similar SEMIPALMATED SANDPIPER and LITTLE STINT, even though it is the most likely among the unlikely stints from Siberia.

As in SEMIPALMATED, individual birds may be strongly rufous above or more subdued. Colored essentially like SEMIPALMATED, they present great difficulties in identification, unless the presence or absence of webbing can be clearly seen. The best characters for separation of the two species involve the crown. In RUFOUS-NECKED the *lateral crown is often pale*, the feathers contrasting with the slightly darker central crown and rather little with those of the hindneck. In SEMIPALMATED the entire crown is distinctly darker and browner than the hindneck, even rufous in some individuals. Furthermore, in RUFOUS-NECKED the *supercilium is typically interrupted above the eye*, while it is typically continuous in SEMIPALMATED. Both of these differences give RUFOUS-NECKED a less capped look than SEMIPALMATED, but, as with so many of these stint characteristics, this is strongly indicative rather than definitive. See LITTLE STINT for distinction from that species.

In Flight

In flight this species looks like the other stints of its size and general coloration range—WESTERN and SEMIPALMATED.

Voice

The usual flight call is a squeak rather similar to that of a WESTERN SANDPIPER, only audible at fairly close range. These two species may not be distinguishable vocally, but the call of the RUFOUS-NECKED is quite distinct from those of the similar-looking SEMIPALMATED SANDPIPER and LITTLE STINT.

Notes

Although it is likely that juveniles occur as regularly as adults do in the Northwest, a number of sight and photographic records of that age class attributed to this species remain equivocal.

Photos

Chandler (1989: 97) shows a nonbreeding adult with an outstretched wing.

References

Alström and Olsson 1989 (identification), Evans 1975 (molt), Melville 1981 (spring measurements, weights, and molt), Myers et al. 1982 (general biology), Paton and Wykes 1978 (molt), Sinclair and Nicholls 1976 (identification), Thomas and Dartnall 1971a (winter feeding behavior), Thomas and Dartnall 1971b (molt), Veit 1988 (identification of California specimen), Wallace 1979 (identification).

Juvenile Rufous-necked Stint. Short-billed, black-legged stint with long primary projection and relatively short legs; brightly colored juvenile, almost as contrasty as a Little Stint but with somewhat duller head and coverts and no obvious mantle and scapular lines; darker central crown and interrupted supercilium perhaps best distinction from Semipalmated if lack of toe webbing not surely evident. Japan, September 1984, Urban Olsson

Juvenile Rufous-necked Stint. Duller individual, easily distinguished from Little Stint by plainness; much more similar to Semipalmated, even in its blunt-tipped bill, but pale lateral crown and interrupted supercilium indicative of species. Japan, September 1984, Urban Olsson

Juvenile Little Stint. Short-billed, black-legged stint much like Rufous-necked but typically brightly patterned, with conspicuous mantle and scapular lines and contrastily patterned coverts. Halland, Sweden, September 1985, Urban Olsson

LITTLE STINT *Calidris minuta*

This least likely of the world's stints to find its way to the Northwest has nevertheless done so. Watch especially for a very brightly patterned short-billed juvenile stint with black legs and unwebbed toes or an even less likely reddish-faced, white-throated adult. A photograph and a phone call to the local rare-bird alert would be in order in either case.

	Jan	Feb	Mar	Apr	May	Jun	Jul	Aug	Sep	Oct	Nov	Dec
COAST												

Distribution

Little Stints breed on tundra from Siberia, perhaps overlapping with Rufous-necked, to northern Scandinavia. Most birds head south and west, wintering in much of Africa, the Near East, and the Indian subcontinent, with only very small numbers recorded to the east of those regions.

Northwest Status

From its more westerly breeding and wintering distribution, this species must be considered the least likely Siberian stint to occur in the Northwest. Nevertheless, two juveniles have been photographed in the region, both in Oregon, and one or more sight records of breeding adults are probably valid (see under Notes). Furthermore, two juveniles were photographed in northern California (September 14-22, 1983, and September 10-21, 1985), and both spring adults and fall juveniles have been repeatedly recorded in Alaska.

COAST FALL. Tillamook, OR, September 7, 1985; Bandon, OR, September 12, 1986.

Habitat and Behavior

This species is similar to Western and Semipalmated sandpipers in occurring in a wide range of freshwater and saltwater habitats, primarily in the open instead of in marsh vegetation. It is probably most common on mud flats in its normal range but might occur anywhere in the Northwest. The Oregon birds were on mud flats with Western and Least sandpipers and Semipalmated Plovers. Little Stints may feed like Least Sandpipers, moving steadily along with head down and picking almost constantly at the surface, but they also move faster, and, detecting prey visually, pick at intervals as is more typical of Semipalmated Sandpipers.

Structure (Figure 38)

Size very small. Bill short, straight to very slightly curved, and finely pointed; legs short. Wings extend to or a bit beyond tail tip, primary projection short in adults and moderate in juveniles. Tail double-notched, central rectrices pointed and projecting.

Plumage

Breeding. Bill and legs black. Breeding-plumaged adults richly colored above, brightest of all stints in this plumage. Most feathers of crown, cheeks, mantle, scapulars, and tertials fringed with bright rufous, interrupted by white supercilium in some individuals. Centers of most feathers of upperparts dark, producing substantial contrast with rufous fringes. Coverts gray-brown, many tertials tipped with this color in duller-colored birds. Usually a distinct pair of white mantle lines. Throat white, contrasting with rufous head; dark streaks and spots on sides of breast, with traces of rufous in same area in some individuals.

Lower back and central uppertail coverts black, outer uppertail coverts white; tail gray-brown except for darker brown central rectrices. Wing brown above, with narrow white wing stripe formed by moderately broad tips of greater secondary coverts and quite narrow tips of greater primary coverts. Linear white markings on outer webs of inner primaries and paler inner webs of flight feathers add to distinctness of stripe. Underwing and axillars white, gray-brown on flight-feather tips and greater primary coverts and darker brown on marginal coverts.

Adults occurring in Northwest should be in this plumage, acquired on wintering grounds in March and April and lost there in September and October.

Nonbreeding. Plain gray-brown above, typically with well-defined darker brown feather centers, and white below, with fine streaks on sides of breast. White supercilium and dark loral stripe round out conspicuous markings, as in other black-legged stints.

Juvenal. Young juveniles with olive legs, older ones with black. Brightly marked, with conspicuous white mantle and scapular lines. Crown and mantle feathers fringed with rufous, in contrast with gray-brown-bordered hindneck feathers. Scapulars and tertials brown, darkening to black toward tips, with bright rufous and white edges. Coverts brown-centered and buff-fringed. Conspicuous white supercilium (often split) and dark loral stripe. Breast varies from white to pale buff, usually with fine streaks on sides. Some individuals with more muted pattern (see under Identification).

Immature. Juveniles molt into first-winter plumage on wintering grounds during October, then molt primaries during midwinter; indistinguishable from adults after that time. Most birds probably mature in first year, as few seen in Africa during breeding season.

Identification

All Plumages. In any plumage, the LITTLE STINT falls with the RUFOUS-NECKED as one of a pair of stints with *black legs and unwebbed toes*. It is almost identical to the SEMIPALMATED SANDPIPER in proportions, the bill very slightly thinner and more pointed, with the RUFOUS-NECKED bill fitting into the scarcely perceptible gap between those two species. The primary projection averages slightly longer than in the two web-footed stints, although not as long as in RUFOUS-NECKED. Although approximately the same in all measurements, the LITTLE has been perceived as slimmer than the SEMIPALMATED when they have been seen together; a bird by itself could not be so distinguished.

Breeding Plumage. Birds in this plumage must be carefully distinguished from RUFOUS-NECKED STINTS in the same plumage.

• Typical breeding-plumaged LITTLE STINTS have *crown and face rufous*, contrasting with the white throat and breast. RUFOUS-NECKED are usually entirely rufous about the head and breast in this plumage, although individuals, especially in early autumn, may be faded enough so the

rufous on the breast is not especially apparent (or some of it may have been lost in molt). Nevertheless, there is usually some rufous behind the white throat of a RUFOUS-NECKED.

• The mantle feathers, scapulars, and tertials are usually dark-centered and brightly rufous-edged; thus the LITTLE has *entire upperparts brightly marked with rufous*, the color more extensive than on any other stint with the exception of RUFOUS-NECKED. Some RUFOUS-NECKED are just as brightly marked above, but these can be distinguished from LITTLE by their reddish breasts.

• In most LITTLES the *tertials are bright and contrasty*, while in most RUFOUS-NECKED they are drab gray-brown, whether new or retained feathers from the previous autumn.

• In LITTLE the *coverts have dark centers and pale edges*, whereas in RUFOUS-NECKED the coverts are usually largely gray, contrasting strongly with the mantle and scapulars. Again, however, occasional RUFOUS-NECKED have coverts as brightly marked as typical LITTLE.

• Finally, LITTLE usually has *conspicuous mantle lines*, poorly developed or lacking in RUFOUS-NECKED and other stints.

Nonbreeding Plumage. Nonbreeding LITTLE STINTS look much like SEMIPALMATED SANDPIPERS or RUFOUS-NECKED STINTS in the same plumage; none of the three is likely to occur in this plumage in our area. There are slight average differences, as outlined by Veit and Jonsson (1984), but the degree of overlap makes it unlikely that a bird can be definitely identified by sight or analysis of photographs. LITTLE typically has dark-centered scapulars that produce a characteristically *blotched look*, the upperparts more patterned than is typical of RUFOUS-NECKED or any of the other black-legged stints. There is enough overlap in this characteristic to suggest caution, and a small percentage of RUFOUS-NECKED STINTS also show blotched upperparts. Bear in mind that a LONG-TOED STINT in this plumage is similarly or even more distinctly blotched.

Juvenal Plumage. Juvenile LITTLE STINTS are among the most contrasty of stints in this plumage, with *dark-centered feathers with rich rufous edges above*, and they have the most *conspicuous white mantle and scapular lines* of any of the juvenile dark-legged stints. Note, however, that some individuals are very dull (Veit and Jonsson 1984: Fig. 15), surprisingly different from the typical bright birds. LITTLE STINTS are unlikely to be confused with WESTERN SANDPIPERS, although the two are somewhat similar in plumage. Some WESTERNS have both mantle and scapular lines well developed and are as richly colored as LITTLES, but the bill-length difference should easily separate them. In addition, WESTERNS typically show rufous only along the upper scapulars (in some continued onto the tertials) and are gray elsewhere, while LITTLES are typically browner on the mantle and crown and more heavily marked on the coverts.

Juvenile SEMIPALMATED SANDPIPERS are usually duller than LITTLE STINTS, but both species vary considerably and overlap in most of the characteristics discussed by Veit and Jonsson (1984). Nevertheless, a juvenile SEMIPALMATED would be unlikely to be as bright overall as a juvenile LITTLE, and it will of course be the more typical, brightly marked LITTLES that are identifiable in any case.

• LITTLES on the average show more conspicuous mantle and scapular lines, but there is overlap, because both dull LITTLES and some SEMIPALMATEDS have poorly defined lines.

• LITTLES are also more likely to show vividly patterned tertials and coverts, while those of SEMIPALMATED are more subdued.

• Only one characteristic is definitive, the *absence of toe webbing* in LITTLE, and that attribute must be carefully observed.

LITTLE and RUFOUS-NECKED juveniles overlap in structural and plumage characteristics to a degree that makes separation of individuals very difficult. Wallace (1979: 273) summed up the problem of distinguishing these two species: "… there is the danger of observers straining for differences and forgetting the quite wide morphological variation in the juvenile plumages of both Red-necked and Little Stints." Nevertheless, at least the extreme individuals should be identifiable.

• A very brightly marked LITTLE is more colorful than virtually any RUFOUS-NECKED, distinguished by *bright rufous edgings on dark-centered dorsal feathers.* In RUFOUS-NECKED, only rarely are all tertials vividly marked dark

brown and rufous, while this is the norm in LITTLE.

• In most juvenile RUFOUS-NECKEDS the coverts are drab, with medium-brown centers and buff edges, while most juvenile LITTLES have the *coverts dark-centered and reddish-edged.* Thus a juvenile LITTLE presents a uniformly spangled appearance, whereas a juvenile RUFOUS-NECKED usually shows a brightly marked back and scapulars contrasting with grayer, duller coverts and tertials.

• The *posterior scapulars have white outer edges* in LITTLE, white tips in RUFOUS-NECKED, which makes the former species appear more striped, the latter more scalloped. This edging is, of course, what produces the white mantle and scapular lines that are better developed in LITTLE than in RUFOUS-NECKED.

• RUFOUS-NECKED STINTS, with their longer wings and tail, tend to look *more pointed toward the rear* than do LITTLES, but it may be hopeless to try to distinguish individual birds by this characteristic.

Finally, bear in mind that juvenile LEASTS are also quite brightly marked above, but their brownish breasts and yellow legs will distinguish them from LITTLES at a second glance.

In Flight

The LITTLE STINT looks like a SEMIPALMATED or WESTERN SANDPIPER in flight, perhaps with a slightly more distinct wing stripe.

Voice

The call is a sharp single note, squeaky but shorter than the WESTERN SANDPIPER'S call. It has been considered reminiscent of the calls of SANDERLING or RED-NECKED PHALAROPE but to my ear is different from either. Some birds may give a trilled note, more similar to calls of other stints. The calls of LITTLE and RUFOUS-NECKED STINTS may aid in distinguishing them if a bird is unquestionably one or the other. The RUFOUS-NECKED typically gives a slightly rolling or trilled call, in contrast with the LITTLE'S single, sharp call. Even here, however, there may be problems, as some descriptions of the calls of the two species indicate overlap.

Notes

Some of the additional records (three adults, July 10-21, and three to four juveniles, July 25 to September 3) from Iona Island and Boundary Bay may be valid, although only one was photographically documented. The photograph—of a juvenile at Iona Island on July 25-27, 1973—is not definitive. An adult from Boundary Bay, July 10, 1988, was well described and is probably valid.

Some characteristics that have been cited in recent literature to distinguish this species from the Rufous-necked Stint are variable enough so that each is not definitive by itself, although they are useful indicators. (1) The tertials in some breeding adults and juveniles of Rufous-necked are as dark-centered and brightly edged as in any Little. (2) The lower scapulars in some juvenile Rufous-necked are as dark-centered as is typical of Little. (3) The wing coverts in some juvenile Rufous-necked are dark-centered as is typical of Little. (4) The color of the breast and the extent of its lateral streaks of juveniles is too variable in all black-legged stints to be definitive. (5) The bill shape and size of Little and Rufous-necked overlap greatly.

Photos

The range in juveniles is shown by Chandler (1989: 99). The "winter" birds in Hammond and Everett (1980: 121) and Keith and Gooders (1980: Pl. 91) are also juveniles.

References

Bengtson and Svensson 1968 (prey choice in migration), Dean 1977 (molt), Jackson 1918 (molt), Lifjeld 1984 (feeding ecology in migration), Middlemiss 1961 (winter habits, weight, and molt), Oddie and Marr 1981 (identification), Pearson 1984 (molt), Pearson et al. 1970 (weight), Sinclair and Nicholls 1976 (identification), Tree 1974 (ageing and sexing).

TEMMINCK'S STINT *Calidris temminckii*

This little sandpiper from Siberia may be the easiest of the stints to identify, in the unlikely event of an encounter with it. The best field mark of Temminck's Stint—its long tail with sparkling white edges—is saved for last, but it is also recognizable by being plainer than any of the other stints in each of its plumages. As mousy as the Least Sandpiper, it is similar in habitat and behavior.

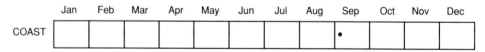

	Jan	Feb	Mar	Apr	May	Jun	Jul	Aug	Sep	Oct	Nov	Dec
COAST									•			

Distribution

Temminck's Stint breeds from Siberia to northern Scandinavia, primarily in moist meadows at low arctic and subarctic latitudes. It winters from the Indian subcontinent through much of Africa.

Northwest Status

This Old World species is a rare migrant, both spring and fall, on western Bering Sea islands and is unlikely to occur in the region except as a vagrant. There is a single well-documented Northwest record, a juvenile photographed in southern British Columbia.

COAST FALL. Reifel Island, BC, September 1-4, 1982.

Habitat and Behavior

Temminck's is usually a bird of fresh water, often small, grassy ponds, but it may occur in marshes or tide pools on the coast. Given its habitat preferences, it should occur with Least rather than Western sandpipers. It occurs singly or in small, loose flocks, but birds on our side of the Pacific would most likely be alone or with other species of stints. The Reifel Island bird foraged with Least and Pectoral sandpipers in a muddy ditch. This species typically feeds by picking very much like a Least Sandpiper, in a slow and steady manner with head down, seemingly crouched because of its short legs.

Juvenile Temminck's Stint. Small, fine-billed, yellow-legged stint with tail projecting well beyond wingtips and eye-ring prominent on plain gray head; subterminal black and terminal white fringes on upperparts diagnostic of plumage and species. Reifel Refuge, BC, 2 September 1982, Jim Erckmann

Structure (Figures 38, 40)

Size very small. Bill short, straight, and finely pointed; legs short. Wings fall considerably short of tail tip, primary projection short in adults and juveniles. Tail long for stint, double-notched and with projecting, pointed central rectrices.

Plumage

Breeding. Bill black; legs pale, varying from yellowish to greenish, occasionally dark. Breeding plumage variable, probably because many individuals replace only part of nonbreeding plumage in spring. Gray-brown and rather plain above, with scattered irregular markings of black, buff, and rufous. These markings most distinct on scapulars, where some may be chevron-shaped. White eye-ring conspicuous against dull facial pattern. Breast washed with same color as back and finely streaked, looking plain from a distance.

Lower back, central uppertail coverts, and central rectrices brown; outer uppertail coverts and outer rectrices white. Wing brown above, with narrow white wing stripe formed by moderately broad tips of greater secondary coverts and quite narrow tips of greater primary coverts. Linear white markings on outer webs of inner primaries and paler inner webs of flight feathers add to distinctness of stripe. Underwing and axillars white, gray-brown on flight-feather tips and greater primary coverts and darker brown on marginal coverts.

Autumn adults quite dull; reddish markings of spring subdued and mostly restricted to scapulars, breast with small spots. This plumage retained at least into late August, most body molt occurring during September. Primary molt may begin on breeding grounds and be suspended during migration.

Nonbreeding. Plain gray-brown upperparts, relieved only by faintest of dark streaks. Face plain, narrow white eye-ring conspicuous, gray-brown loral stripe inconspicuous. Breast washed with light gray-brown, unstreaked or with faintly indicated streaks; belly white.

Juvenal. About same color as nonbreeding adult but with most mantle feathers, scapulars, tertials, and coverts clearly fringed with whitish

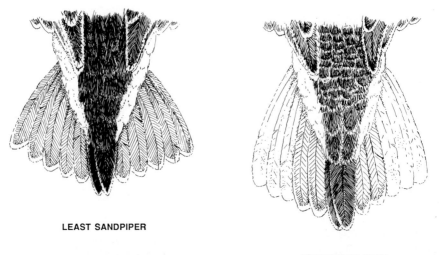

LEAST SANDPIPER

TEMMINCK'S STINT

Fig. 40. Least Sandpiper and Temminck's Stint tails. Least tail typical of most stints and many other calidridines. Temminck's only stint with pure white outer tail feathers, long-tailed compared with other species.

to buff, with narrow brown subterminal fringe. These markings rather like those of juvenile Surfbirds and Red Knots and totally different from dark-centered feathers of other stints. Some with subterminal fringes on scapulars very broad, producing barred effect. Pale fringes may wear off by late September, but scalloped pattern produced by dark fringes should remain until next molt.

Immature. First-winter birds may show retained scalloped coverts. At least outer primaries molted in first winter. Most if not all breed in first summer.

Identification

All Plumages. This stint is distinctively shaped, with *short legs and long tail*, giving it a somewhat attenuated look (like a tiny BAIRD'S SANDPIPER) that could distinguish it even in silhouette. TEMMINCK'S is the shortest-legged stint, but its tarsus averages only a millimeter shorter than that of LEAST, so the difference between the two species is not striking.

The tail can project beyond the wing tips in other stints, presumably depending on how the wings are held, but generally not more than the eye diameter, while in TEMMINCK'S the tail usually projects beyond the wings a distance two to

three times the eye diameter. From a survey of photographs, only a very few individuals of any of the other stints (observed in RUFOUS-NECKED and LEAST) would be likely to show as much tail as TEMMINCK'S, and only the LEAST has the tail commonly extending beyond the wings.

The tail furnishes another characteristic, the *pure white outer tail feathers*. In all stints the outer feathers vary from paler to darker gray, and caution is advised here, as these gray feathers can look white in flight in some lighting conditions. The two central feathers are dark in all species (palest in TEMMINCK'S), the next two pairs are gray in all species, and the outer three pairs are white in TEMMINCK'S, gray in the others (Figure 40). Note that TEMMINCK'S is the only stint with a *brown rump*, the others with black rumps darker than the back color. The contrast between rump and outer tail feathers might not be any greater in this species, yet the whiteness is striking on a bird in flight or one that exposes its tail when preening. The long, fluffy undertail coverts are white in all stints, and they may appear to be the edge of the tail in a perched bird, so caution is advised here also.

TEMMINCK'S is one of three stints with *pale legs*, their color varying from yellowish through greenish to (occasionally) all dark. Even without a tail, a TEMMINCK'S STINT can be distinguished in all plumages by its *plainness*, certainly the easiest of the stints to identify by plumage alone.

Breeding Plumage. Breeding-plumaged birds in fresh plumage are less colorful than other spring stints, but some of the latter on their return trip south may be so worn as to simulate the patchy look of a TEMMINCK'S. The *dark spots on the back are randomly scattered* rather than aligned in patterns as in other stints, and they are irregularly shaped, each one with both front and rear edges jagged. The tertials are never brightly edged as they are in most species, nor are there white lines on the mantle or scapulars. The breast streaks are less well defined than in any other stint in this plumage.

Nonbreeding Plumage. Nonbreeding adults are *entirely plain*, scarcely showing any markings at all above and with an unstreaked dusky breast reminiscent of the breast of a juvenile SPOTTED SANDPIPER but darker. Nonbreeding LEAST SANDPIPERS can be rather plain and are most like TEMMINCK'S of all the stints, but the TEMMINK'S is even plainer, and its *long tail* should cinch the identification if there is some question about plainness.

Juvenal Plumage. Juveniles are pleasingly patterned with *scallops of brown and pale buff or white* bordering most of the back and wing feathers, a very different pattern from the striped and blotched look of other juvenile stints.

In Flight

TEMMINCK'S STINTS often "tower" into the air like little SNIPES, perhaps their long tails an asset for zigzagging ascent. LEAST SANDPIPERS may do the same, and TEMMINCK'S also flies in the same fluttery style as that species or even with wings bowed like a SPOTTED SANDPIPER. With any decent look, the *pure white outer tail feathers* (Figure 40) on TEMMINCK'S should be seen, providing the realization that the pale edges on other stint tails don't sparkle in the sun like this. Other stint tails, even when they look white, show a dark stripe down the middle of the tail (the two central rectrices), but in TEMMINCK'S the white zone stands out from a wider dusky central area (placing this species somewhere between "stripe-tailed" and "white-margined").

Voice

The flight call is a short, musical trill, often given in series. This species is usually vocal when flushed, again recalling SOLITARY SANDPIPER or COMMON SNIPE, which almost invariably announce their presence. However, the British Columbia TEMMINCK'S seemed a bit quieter and less likely to flush than the LEASTS with which it associated.

Notes

There are several additional published sight records, possibly valid but not sufficiently documented, from the Vancouver area (Reifel Island, September 6-7, 1981; Blackie Spit, December 14, 1980) and interior Washington (Frenchman Hills, September 1, 1981).

The illustration of the tail of this species in Johnsgard (1981: 246) is inaccurate, perhaps reversed with that of the Little Stint (but the heads appear to be correct). Published accounts even as recent as Colston and Burton (1988) and Peterson (1990) have been misleading with regard to the relative length of wing and tail in this species.

Photos

The "winter" birds in Hammond and Everett (1980: 121) are actually in fresh breeding plumage.

References

Breiehagen 1989 (breeding biology and mating system), Hildén 1975 (breeding biology and mating system), Hildén 1978 (population biology), Hildén 1979b (territoriality and site fidelity), Kautesk et al. 1983 (occurrence in Northwest), Southern and Lewis 1937 (breeding behavior).

LONG-TOED STINT *Calidris subminuta*

This close relative of the Least Sandpiper is accidental in our region. Do not expect a jacanalike bird, but do look for relatively long legs and toes, upright stance and a look of alertness. Lengthy appraisal will reveal other, more subtle field marks, especially the patterning of the head and mantle.

	Jan	Feb	Mar	Apr	May	Jun	Jul	Aug	Sep	Oct	Nov	Dec
COAST									•			

Distribution

The Long-toed Stint breeds in boreal-forest clearings in subarctic Siberia, wintering in southeast Asia and, less commonly, Australasia.

Northwest Status

There is a single unequivocal record of this species from the Northwest, a juvenile photographed on the Oregon coast in 1981; another was present on the central California coast from August 29 to September 2, 1988. The species is otherwise known from North America as a rare spring and fall migrant through the western Bering Sea islands.

COAST FALL. South jetty of Columbia River, OR, September 5-12, 1981.

Habitat and Behavior

Like the Least Sandpiper, the Long-toed is a dweller of marshy edges rather than open mud flats. It occurs by itself or in small, loose groups, feeding in water or in low vegetation. Like other marsh-dwelling shorebirds, it can hide effectively merely by crouching. When alert it usually stands tall, with neck stretched and body tilted up; this has been compared to the stance of a *Tringa* sandpiper and is rarely seen in Least. The Long-toed typically feeds in shallow water and among fairly dense low vegetation, picking prey items from mud or water surface. Where bare shorelines slant quickly downward, as at sewage ponds, it feeds in the open right at the water's edge with other stints. It is probably even more of a picker and less of a prober than the Least, as indicated by its slightly shorter, straighter bill, and its longer neck and legs fit it both for picking active prey from the substrate and for a slightly more aquatic foraging mode.

Structure (Figures 38, 41, 42)

Size very small. Bill short, fairly straight. Legs short, toes longer than in other stints. Wings just reach tail tip, primary projection none in adults and none to short in juveniles. Tail double-notched, pointed central rectrices projecting beyond others.

Plumage

Breeding. Bill black, often with pale base to lower mandible. Legs yellowish, rarely yellowish brown or greenish yellow. Fresh-plumaged adults brightly patterned above, with dark brown-centered feathers and extensive rich rufous edges on crown, mantle, scapulars, and tertials imparting variegated but predominantly striped pattern. Some scapulars tipped with white, most tertials entirely fringed with rufous; coverts vary from gray-brown to dark brown, fringed with pale buff (fringes vary to rufous at base and white toward tip). Prominent white supercilium, dark loral stripe usually expanded into smudge before eye, reddish ear coverts, and white throat. Breast whitish, finely streaked with black and washed with buff in some individuals. Rest of underparts white.

Lower back, central uppertail coverts, and central pair of rectrices black; outer uppertail coverts white, rest of tail gray. Wing brown above, with narrow white wing stripe formed by moderately broad tips of greater secondary coverts and quite narrow tips of greater primary coverts. Underwing and axillars white, gray-brown on flight-feather tips and greater primary coverts and darker brown on marginal coverts.

Early fall adults look much darker above, like Least Sandpipers, when large proportion of rufous fringes worn off.

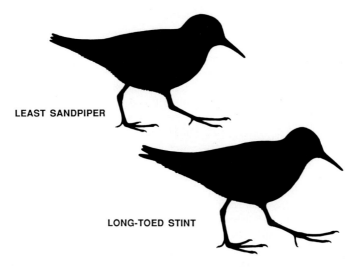

Fig. 41. Least Sandpiper and Long-toed Stint. Tarsus and middle toe both shorter than bill in Least, longer than bill in Long-toed. Silhouettes drawn in same posture to emphasize leg-length difference.

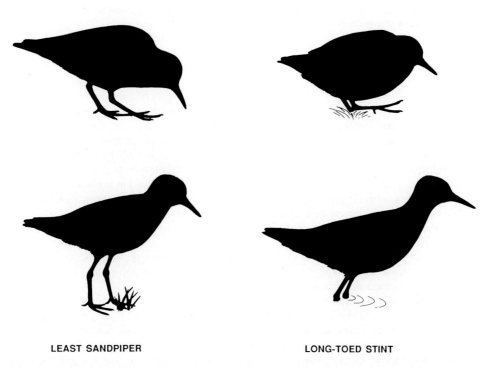

LEAST SANDPIPER LONG-TOED STINT

Fig. 42. Least Sandpiper and Long-toed Stint. Silhouettes traced to scale from photographs to show variation in poses. Foraging birds above, alert birds below. Note upright stance and longer neck of Long-toed, similarity of species when neck not extended.

Nonbreeding. Overall plain gray-brown; crown and hindneck streaked with blackish and feathers of mantle, scapulars, and some tertials conspicuously centered with dark brown, causing characteristic blotched appearance. Prominent white supercilium and dark loral stripe. Breast gray-brown, varying from uniformly colored to finely streaked. Most adults fully molted into this plumage by September.

Juvenal (Figure 43). Marked rather like breeding adult but much brighter above than any worn adult with which it might occur. Feather edgings vary from wide and bright rufous to narrower and duller, but less likely than adult to show buff breast or rufous cheeks. Usually show vivid white mantle lines, as conspicuous as in brightest Little Stints, and less well-defined scapular lines. All feathers of upperparts fringed with rufous, scapulars also with buff; coverts dark-centered and pale buff-fringed. White supercilium very distinct, dark loral stripe less so than in breeding adult and not usually expanded into smudge in front of eye. Breast light gray-brown with brown streaks.

Immature. Probably matures in first year, but no direct evidence, as few records of oversummering exist.

Identification

All Plumages. LONG-TOED STINTS, because of their basic similarity to LEAST SANDPIPERS (brownish, yellow-legged), should be compared with that species in detail. Differences independent of plumage appear to be:

• LONG-TOED often *forages with head up*, while LEAST typically does so with head down, giving the latter its "mousy" look. LONG-TOED looks long-necked and alert while foraging, probably more of a visual forager than is LEAST, but many birds act just like LEASTS.

• LONG-TOED when alert *stands up high*, with neck stretched and body tilted up; this posture is rarely seen in LEAST. Any bird of this type should be closely scrutinized. I have occasionally seen a LEAST SANDPIPER stand upright but, at its extreme, not as high as a LONG-TOED. The many photographs of LONG-TOED like this are, of course, of birds disturbed by the photographer. LONG-TOED looks small-headed and thin-necked

relative to LEAST, but this appearance stems from the postural difference between the two species rather than from any real discrepancy in head or neck size.

• LONG-TOED has *longer toes and legs* than LEAST. The difference can be striking, although difficult to determine in wading birds. Both the *tarsus and middle toe are distinctly longer than the bill* in LONG-TOED, while all three body parts are about the same length in LEAST. However, the toes are similarly proportioned in both species, the central toe no longer relative to the other toes in LONG-TOED than in LEAST. Bear in mind that LEAST also has long toes relative to the length of its tarsus, when compared with WESTERN and SEMIPALMATED.

• LONG-TOED has a *pale base to the lower mandible*. This character is not consistent; some LONG-TOED do not show it and at least one LEAST photograph I examined has the same pale area.

• LONG-TOED has a *slightly straighter bill*, but this is only an average difference with considerable overlap.

• With the same length bill on a slightly larger bird, the LONG-TOED STINT also looks *smaller-billed* than the LEAST.

• LONG-TOED is *slightly larger* than LEAST, but only in direct comparison would this be evident, when its longer legs would make it look larger yet.

It must be borne in mind that none of these differences is diagnostic by itself, but all could serve to call attention to a LONG-TOED. Other differences mentioned in published accounts are even more equivocal—for example, head shape, with an average difference (LONG-TOED forehead more sloped, LEAST more abrupt) but much overlap.

Breeding Plumage. Differences in this plumage are subtle but add up to different-looking birds.

• LONG-TOED look *striped above*, LEAST scalloped or mottled. In LONG-TOED the *mantle is striped*, all feathers with dark central stripes and parallel reddish edges, while in LEAST the same feathers look *scalloped*, as the pale edges are narrower and curve toward one another at the rear. This is a definitive field mark for the two species in both breeding and juvenal plumage, showing up clearly in live birds and photographs. In addition, the scapulars and tertials are similarly colored, brown with broad, complete rufous fringes, enhancing the striped effect. In LEAST the scapulars are much darker-centered, the outer and inner fringes differently colored, which gives them a black-blotched appearance not seen in LONG-TOED.

• In fresh-plumaged LONG-TOED the *tertials are broadly fringed with bright rufous*, in LEAST narrowly fringed with the same or slightly duller rufous. In worn autumn migrants, LONG-TOED are still brightly fringed, while the narrower fringes of LEAST have worn off, leaving them very dull and dark.

• LONG-TOED consistently shows a *reddish cap* in fresh plumage, equalled in brightness by very few LEASTS.

• The *supercilium is better defined* in LONG-TOED, typically a conspicuous white stripe. The *forehead is dark*, separating the two supercilia and producing a distinctly "capped" effect. The supercilium is *not* more conspicuous behind the eye, as indicated in some guides. In LEAST the forehead is usually pale and the supercilium relatively inconspicuous. There is enough variation to preclude this as a definitive field mark, although it is so in juveniles.

• A *dark smudge before the eye*, partially interrupting the supercilium, is characteristic of LONG-TOED although not invariably present; in LEAST the dark loral stripe is of about equal width throughout.

• The throat is immaculate white in LONG-TOED, clearly set off from the breast color, while it usually has small spots in LEAST. This is difficult to determine.

• LONG-TOED has a *pale breast, whitish or washed with buff*, while LEAST shows a brownish wash on the breast.

• The *breast streaks are fine* in LONG-TOED, whereas those in LEAST are coarse enough almost to be called spots; there is overlap.

Nonbreeding Plumage. Nonbreeding adults are the least likely plumage type to be observed in the Northwest. They are very similar to nonbreeding LEAST SANDPIPERS, differing in only one plumage character. In LONG-TOED the *back is heavily blotched*, the extensive blackish cen-

ters of the back feathers, scapulars, and tertials producing this effect. The LEAST usually has a dark central stripe at the most on each of these feathers, many individuals with no contrast at all. The two species just overlap; some individuals closely approximate the typical pattern of the other species. As when using some of the other field marks for this species, any black-blotched bird should be checked as a possible LONG-TOED and all other characteristics determined. Note that some nonbreeding-plumaged individuals of other species may have dark feather centers, typically LITTLE and much less often RUFOUS-NECKED.

Juvenal Plumage. Some of the differences in juvenal plumage parallel those in breeding plumage, while others are confined to juveniles.

• In LONG-TOED the *mantle is striped*, in LEAST scalloped, just as in adults. However, juvenile LONG-TOED in a few photographs that I have seen looked intermediate between these two conditions.

• LONG-TOED has *brighter, wider rufous edges and fringes on scapulars and tertials,* but the two species overlap in this characteristic. Worn LONG-TOED will certainly converge on the LEAST mode, and some LEAST are particularly bright.

• In LONG-TOED the *scapular line is usually buffy*, while in LEAST it is white. The mantle line is white in both.

• The *coverts are usually whitish-edged* in LONG-TOED, contrasting strongly with the rufous-fringed scapulars and tertials. In LEAST the contrast is less, as the coverts are usually buff-edged. The two species overlap slightly, but this difference is easily checked by someone looking over foraging LEASTS trying to find that one LONG-TOED if it is there.

• The *breast is pale gray with finer, more distinct streaks* in LONG-TOED, often with a buffy wash and indistinct streaks in LEAST.

• The *supercilium is more conspicuous* in LONG-TOED, with the dark forehead clearly defining its rounded and somewhat expanded front end. This bulb-shaped supercilium is diagnostic of LONG-TOED, as in LEAST the supercilium typically narrows to where it extends across the forehead. There is no difference in the shape

LEAST SANDPIPER

LONG-TOED STINT

Fig. 43. Juvenile Long-toed Stint and Least Sandpiper. Least with relatively narrow white supercilium extending forward onto forehead, scaly mantle pattern, blurry streaks on breast. Long-toed with broad white supercilium ending at forehead, striped mantle, fine streaks on breast.

of the loral stripe in juveniles of the two species, but in the LONG-TOED it is more likely to be interrupted just in front of the eye.

Juvenile LONG-TOED might be mistaken for juveniles of other brightly marked species such as WESTERN and LITTLE. They should be easily distinguished from those two species by their well-defined *fine breast streaks, very short primary projections, and pale legs*.

In Flight

This species looks like a LEAST SANDPIPER in flight but is somewhat larger, almost more comparable to a WESTERN. Flushed individuals may ascend rapidly or not but look long-necked in comparison with LEASTS. With longer legs, the *toe tips extend distinctly beyond the tail tip* in flight, not very evident in a bird flying rapidly away but useful in some situations and clearly showing in photographs.

Voice

The flight call is a rolling *prrrp*, lower pitched than that of the LEAST; the difference is not striking and might go unnoticed. Bear in mind that LEASTS occasionally give a lower-pitched call, so it is not enough in itself for a definitive identification.

Further Questions

Much has been made of the distinctive alert posture in the Long-toed Stint, and it would seem to be an adaptation to seeing over vegetation; many grassland species, as well as the marsh-dwelling Pectoral Sandpiper, show the same posture. Anyone observing Long-toed Stints should note whether they regularly perform this behavior when in the open. More easily achieved for Northwesterners, we should watch Least Sandpipers in different environments, in particular in dense vegetation, to see how often that species shows the same behavior. Anyone quantifying the time spent picking versus probing for both of these species would undoubtedly document differences.

Notes

None of a series of unpublished sight records of fall adults and juveniles from southwestern British Columbia and Oregon provides photographic documentation. Difficulty of distinguishing it from Least Sandpiper in any plumage will continue to make careful documentation necessary to establish any patterns of occurrence in the Long-toed Stint.

Although Hayman et al. (1986) considered this species to have a longer white wing stripe than the Least, wings of the two species look the same to me in the field and in flight photographs.

Breeding Least Sandpiper. Small, brown, yellow-legged stint; rich reddish-brown upperparts with prominent scapular blotches in fresh spring plumage; virtual lack of primary projection precludes confusion with larger, similarly colored calidridines. Victoria, BC, Tim Zurowski

Nonbreeding Least Sandpiper. Dull brownish-gray upperparts and breast, along with yellow legs, diagnostic of plumage and species; with juvenile Western Sandpiper. Ocean Shores, WA, 18 September 1983, Dennis Paulson

Photos

The juvenile from California in the photographs in *Birding* 20: 384 (1988), looks relatively dull, short-necked, and short-legged and would certainly be called a Least Sandpiper unless carefully examined.

References

Alström and Olsson 1989 (identification), Gilligan et al. 1987 (Oregon record), Kitson 1978 (identification), Myers et al. 1982 (general biology), Patten and Daniels 1991 (California record, identification).

LEAST SANDPIPER *Calidris minutilla*

After a session of mud-flat trudging, relax and enjoy a stint with this little bird, always tame and always of interest if watched for a few moments. Watch where a mousy brown form disappeared among the greenery; it will reappear. Listen for their high trills as they flush from the marshy margins. See how they make nearby Western Sandpipers look large. Appreciate the world's smallest shorebird, available at roadside puddles as well as picturesque estuaries.

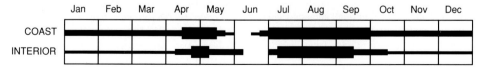

Distribution

The Least Sandpiper breeds in the great expanse of taiga (boreal forest) from Alaska and northern British Columbia across to the Atlantic in Nova Scotia. Migrating all across the continent, it winters on fresh and salt water from the southern United States to northern South America.

Northwest Status

For the most part it is difficult to record numbers of this small species, both on the mud flats when they are mixed with Western Sandpipers and in the dense salt-marsh vegetation in which they commonly forage. Nevertheless, if not the most common sandpiper, the Least is one of the most ubiquitous. Migrants appear early in Oregon, large numbers as early as mid March. They are abundant from mid April to early May on the Washington and British Columbia coasts, with smaller numbers to mid May and a few almost to the beginning of June. Concentrations present in Grays Harbor totalled 5,000 to 10,000 during the peak period in late April in 1981. A few birds have summered on the Fraser River delta.

In fall Leasts are common from early July through early October. Fall migration draws slowly to a close, with moderate numbers staying well into December, when they are seen on Christmas Bird Counts from Grays Harbor south through all the Oregon coastal bays. Smaller numbers, usually no more than a few birds, winter at coastal localities in northern Washington and southern British Columbia. Least Sandpipers that winter at Grays Harbor include both first-year and adult individuals of both sexes.

In the Willamette Valley, Least Sandpipers are common in migration and also winter in some numbers toward its southern end. They are also fairly common throughout the interior of the region during spring and fall. Adults peak in early May and July, juveniles from August through mid September in the Okanagan Valley. There are generally fewer records in spring than in fall in the interior, but at some localities they are equally common in both seasons. In eastern Washington they are more common than Westerns in spring (yet still in small numbers) and less common in fall, perhaps because the receding shorelines of summer provide better habitat for Westerns.

Recent spring censuses at Summer Lake have revealed astronomical numbers of this species, a concentration perhaps unequalled anywhere else. The highest counts elsewhere in the interior of the region are in the low hundreds. The high counts of thousands of juveniles at Salmon Arm are also extraordinary, with the next largest British Columbia counts involving flocks of 100 or more in the Okanagan Valley in early August and counts elsewhere still smaller. Leasts occasionally winter at interior localities, usually only a few birds.

COAST SPRING. *High counts* 1,200 at Yaquina Bay, OR, March 17, 1968; 3,000 at south jetty of Columbia River, OR, April 24, 1968; 3,000 at New River, OR, April 27, 1984; 500 at Medford, OR, April 30, 1984; 4,000 at Steveston, BC, May 2, 1923; 1,500 at Metchosin, BC, May 2-4, 1974; 2,500 at Long Beach, BC, April 30, 1983. *Late date* Ocean Shores, WA, June 6, 1989#.

COAST FALL. *Early date* Samish Flats, WA, June 20, 1974. *Adult high counts* 1,000 at Courtenay, BC, July 5, 1941; 3,000 at Iona Island, BC, July 7, 1977; 1,500 at Tillamook, OR, July 2, 1988. *Juvenile early date* Iona Island, BC, July 30, 1981. *Juvenile high counts* 2,000 at Iona Island, BC, September 2, 1972; 1,000 at Iona Island, BC, August 9, 1981; 3,000 at Leadbetter Point, WA, September 9, 1967; 1,200 at Grays Harbor, WA, August 9, 1979*; 9,000 at Tillamook and Netarts bays, OR, August 25-26, 1979; 4,000 at Tillamook, OR, August 18,1985.

COAST WINTER. *High counts* 500 at Fern Ridge Reservoir, November 27, 1965 (possibly late migrants); 1,000 at Port Orford, OR, December 27, 1981.

INTERIOR SPRING. *Early date* Malheur Refuge, OR, April 11, 1977. *High counts* 23,150 at Summer Lake, OR, May 1, 1987; 6,000 at Abert Lake, OR, April 27, 1989; 100 at Kamloops, BC, May 3, 1986. *Late date* White Lake, BC, June 7, 1975.

INTERIOR FALL. *Early date* Reardan, WA, July 1, 1973; Silver Lake, OR, July 1. *Adult high counts* 60 at Quilchena, BC, July 9, 1973; 50 at Prineville, OR, July 11, 1975; 30 at Bend, OR, July 3, 1983. *Juvenile early date* Okanagan, BC, July 25, 1916#. *Juvenile high counts* 5,000 at Salmon Arm, BC, September 6, 1970; 1,000 at Salmon Arm, BC, August 25, 1977.

INTERIOR WINTER. *High counts* 31 at Yakima

Breeding Least Sandpiper. Extreme of worn plumage, looking very black and white; dark upperparts and breast, yellow legs, and short primary projection still distinctive. Seattle, WA, 26 July 1983, Jim Erckmann

Juvenile Long-toed Stint. Yellow-legged stint with upright stance and long toes; dark forehead, pale base of lower mandible, striped mantle, and broadly and evenly fringed upperparts distinctions from juvenile Least Sandpiper. Japan, September 1984, Urban Olsson

Juvenile Least Sandpiper. White mantle stripe and broken white scapular stripe typical; note scalloped mantle feathers and white supercilium reaching bill. South jetty of Columbia River, OR, 17 August 1987, Jeff Gilligan

Juvenile Least Sandpiper. Much brighter juvenile, but same characteristics as August bird. Reifel Refuge, BC, 2 September 1982, Jim Erckmann

CHRISTMAS BIRD COUNTS	Five-Year Averages		
	74-78	79-83	84-88
Grays Harbor, WA	272	66	47
Leadbetter Point, WA	57	71	40
Coos Bay, OR	823	217	560
Medford, OR	39	20	15
Tillamook Bay, OR	102	95	98
Yaquina Bay, OR	109	75	143

River delta, WA, January 11, 1976; 21 at Scootenay Reservoir, WA, February 21, 1976; 40 at Malheur Refuge, OR, December 11, 1986 (21 on December 19).

Habitat and Behavior

Least Sandpipers forage higher in coastal environments than do Westerns and many other species, typically at the upper edge of the mud flats and in *Salicornia*, arrow-grass, and other low marsh vegetation. Frequently they are seen with look-alike Pectorals in both fresh and salt marshes, nothing visible but striped backs as they feed. Leasts and Westerns often separate out in freshwater ponds, with the Leasts on the dry mud, but they also forage in the Western domain of wet mud and water. Only very rarely do Leasts turn up on sandy ocean beaches.

Leasts feed by both picking and probing, probably depending in part on whether they are foraging on dry or wet mud. They may move fairly rapidly, but more typically they shuffle along in a partial crouch, with head down almost continually. They often forage in shallow water, both fresh and salt, picking prey too small to be readily identified by the onlooker and moving a bit faster and more alertly than when probing in mud.

They typically roost in or near their feeding areas among marsh vegetation, the individuals scattered and cryptic, and even when roosting in upland sites away from vegetation, they typically roost by themselves or in small groups. On one occasion I saw a hundred or more birds in a relatively dense aggregation, recalling Western Sandpipers, and they may roost like this regularly where abundant.

Structure (Figures 38-42, 44)

Size very small. Bill short, moderately thick at base, tapered to fine point, with distinct curvature

apparent in many but not all individuals. Legs short. Wings reach tail tip on average, primary projection short or lacking in adults and juveniles. Tail double-notched, pointed central rectrices projecting slightly beyond others.

Plumage

Breeding. Bill black, legs dull to bright yellowish. Mantle feathers black, edged with light brown. Scapulars black, with or without broad pale to bright buff bars; edged or fringed with light brown to pale buff, even rufous, and tipped with whitish in some. Tertials similarly variable, often narrowly fringed with light rufous; many retained from nonbreeding plumage. Most coverts plain gray-brown, those few replaced patterned dark brown and buff. Whitish supercilium and dark loral stripe fairly conspicuous. Breast light brown with conspicuous dark streaks, demarcation between it and white belly marked. Much variation in overall brightness, but all would be characterized as basically brown at a distance.

Lower back, central uppertail coverts, and central pair of rectrices black; outer uppertail coverts white, rest of tail gray. Wing brown above, with narrow white wing stripe formed by moderately broad tips of greater secondary coverts and quite narrow tips of greater primary coverts. Underwing and axillars white, gray-brown on flight-feather tips and greater primary coverts and darker brown on marginal coverts.

By early July, returning fall migrants have light edges variably worn off, majority of birds looking black-backed because of this. At this time, pale buff edges, when still present, have faded to whitish, and a few fall adults fairly conspicuously marked because of this.

Nonbreeding. Light brown on back and breast, slightly darker feather centers lending a faintly mottled appearance. Some birds look just about plain, others more conspicuously streaked. White supercilium only moderately well defined. This plumage assumed as early as mid August, wing molt extending from July to September.

Juvenal (Figure 43). Marked above with same colors as breeding adult but generally brighter. Crown and mantle feathers fringed with reddish-brown. Scapulars black, variably marked with reddish on bases, edged with buff and tipped with white. White mantle lines conspicuous, those on scapulars slightly less so. Tertials blackish, narrowly fringed with rufous, internal patterning on some. Coverts blackish, edged with buff to pale rufous. White supercilium and dark loral stripe conspicuous. Breast varies from light brown to pale buff; streaks finer and less distinct than in adults, in some individuals lacking from center of breast.

Immature. Molt into first-winter plumage under way in September. By mid October most juveniles with plain backs and worn, indistinctly edged tertials, indistinguishable from adults. Outer primaries molted in southern part of wintering range, probably not northern. Most if not all mature in first year, as few records of oversummering.

Identification

All Plumages.

• The LEAST SANDPIPER in all plumages can be distinguished from the other common stints by its *overall browner (darker) appearance* in any plumage. Breeding adults, juveniles, and non-breeding adults all are largely brownish above, even though marked with other colors, and in particular the *brown breast* allows distinction from WESTERN and SEMIPALMATED, which have a whitish breast in any plumage, either streaked or not. At a distance LEASTS look darker than any WESTERNS with which they might be associated, and the same holds true, although the differences may be less in breeding-plumaged adults, for SEMIPALMATED.

• LEASTS are *smaller and shorter-legged* than the other common stints, and they often feed with a characteristic hunched posture, like a mouse creeping through the marsh vegetation. When feeding or roosting with WESTERNS, the size difference is obvious.

• The *bill is shorter* than that of WESTERN and *finer* than that of SEMIPALMATED.

• The *legs are yellow*, but this must be tempered with the knowledge that mud-covered legs of any species can look different (light or dark

gray if covered by dried mud, orange-brown if covered by wet mud).

For differences from the very similar LONG-TOED STINT and the more dissimilar TEMMINCK'S STINT, two Eurasian members of this group with pale legs, see under those species. See also LITTLE STINT, the juvenile of which looks somewhat similar to that of the LEAST from above; it can be distinguished by black legs and a pale breast.

Breeding Plumage. In this plumage the *conspicuous black-blotched scapulars* stand out in comparison with the more muted tones of the rest of the upperparts. This contrast seems more typical of this species than any other stint.

In Flight (Figure 12)

Although tending to feed in different areas, LEASTS and WESTERNS cohabit often enough so that they at times fly in mixed flocks. When they do, LEASTS stand out by being *smaller and darker*. They look compact in flight, flying with wings less extended than WESTERNS; this is apparently the mode for the shorter flights they typically take when flushed from a marsh. They may ascend with weak, fluttery wingbeats at those times or may fly up rapidly with "towering" flight reminiscent of a SNIPE or SOLITARY SANDPIPER. When in flocks of WESTERNS, they seem to fly just as the WESTERNS do, but they often leave such flocks, as if their flight styles were not compatible. LEASTS probably move shorter distances between foraging and roosting sites than do WESTERNS, but in lengthy flight they fly rapidly and directly like other shorebirds.

Contrary to statements in some guides, WESTERN and LEAST appear to have similar wing stripes, the LEAST's slightly less contrasty than the WESTERN's. From below, the LEAST's wing linings are slightly more extensively dark, which, along with its darker breast, makes it look more patterned. This should be evident only in mixed flocks when comparison is possible, although the browner breast is always a good field mark for LEAST.

Voice

The LEAST's flight call is a high-pitched, clearly two-syllabled or slightly rolled *kree-eeet*, quite different from the WESTERN's squeak and the SEMIPALMATED's short single note.

Further Questions

While Least Sandpipers generally occur by the dozens rather than by the hundreds, thousands of them at times are reported at coastal localities in both spring and fall. The details of these occurrences would be of special interest. Are they spread out on mud flats with Westerns (I have observed this once), or do they occur in such numbers in their typical marsh-edge habitats? Are there species differences in feeding habits or aggressive behavior when they are mixed on mud flats? How often do they mix with Westerns in flight, and are they more likely to do so when both species are feeding together in numbers? What are typical flock sizes for flying Leasts?

On two occasions I have heard foraging Least Sandpipers making frequent soft vocalizations that sounded like tiny sneezes (an explosive *chuff*); the bill was opened at each sound. My impression was of a bird warning another one to keep out of its way! I was within a few yards of these birds on still days, and at greater distances or with any background noise I could not have heard them. Observers should watch (and listen) for this behavior, these short-distance calls perhaps in other shorebirds as well.

On another occasion I watched several Leasts bathing in a salt marsh while nearby Westerns were not doing so. Do shorebirds feeding away from the water come to it to bathe, while those feeding in the water are bathed already? This should be observable at any time where species of different foraging habits are feeding together.

Notes

The painting of the juvenile in Peterson (1990: 149) is too pale-breasted, thus losing the best distant field mark for distinguishing juvenile Least from Semipalmated and Western. It seems misleading to call the bill of this species "rather long" (Farrand 1983: 382). Also, the sexual dimorphism in color described by Miller (1984) is not apparent in examination of series of specimens.

Photos

Good comparisons between nonbreeding-plumaged Least and Western sandpipers are in Hosking and Hale (1983: 14 and 119), although each illustration names only one of the species. Juveniles unlabeled as such are in Bull and Farrand (1977: Pl. 223) and Udvardy (1977: Pl. 195). The first-winter bird in Chandler (1989: 107) is more advanced in molt than are juveniles seen in the Northwest at that date, probably indicating an individual already on its wintering grounds.

References

Baker 1977 (summer diet), Baker and Baker 1973 (summer and winter foraging behavior), Baldassarre and Fischer 1984 (diet and foraging behavior in migration), Brooks 1967 (prey choice in migration), Miller 1979a (display flights), Miller 1979b (functions of display flights), Miller 1983a (habitat and breeding cycle), Miller 1983b (aerial displays and vocalizations), Miller 1985 (parental behavior), Miller 1986 (breeding vocalizations), Moore 1912 (breeding behavior), Page 1974b (molt), Page and Bradstreet 1968 (fall migration), Page and Salvadori 1969 (fall weights), Spaans 1976 (molt).

WHITE-RUMPED SANDPIPER *Calidris fuscicollis*

Check out pond margins late in spring when shorebird migration seems over, especially after a spell of dry winds off the prairies. The best of luck might bring a surprise encounter with a rufous-and-gray bird like a Western Sandpiper but larger and more attenuated. Flushed with a squeak, as if equally surprised, it will expose its distinctive field mark.

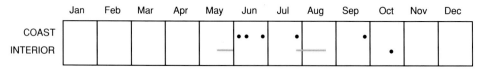

	Jan	Feb	Mar	Apr	May	Jun	Jul	Aug	Sep	Oct	Nov	Dec
COAST						• • •	•		•			
INTERIOR									•			

Distribution

The White-rumped Sandpiper nests on tundra in the North American Arctic from northern Alaska to Baffin Island. A long-distance migrant, it winters in southern South America. Most birds move through the Great Plains in spring and off the Atlantic coast in fall, later migrants than most other shorebirds in both seasons.

Northwest Status

Rare anywhere west of the Rockies, this species is known from the Northwest from only 15 records. Coastal records are all from British Columbia, interior records widely spread. The three July records presumably refer to southbound adults, but the later records could represent either age stage. All records involve single birds except the three seen at Stump Lake. Records are more numerous east of the Cascades than west, as would be expected in this eastern species, and indicate that White-rumped Sandpipers are most likely to appear in our area during the late-spring period when northbound migrants are common on the plains. However, the 1964 Washington record occurred after three days of southwesterly winds.

To the south, six of eight California records fall in the period May 17 to June 16, apparently also displaced northbound migrants, and the others include single adults in August and September. Fall records are especially unlikely in the Northwest, with the primary migration route off the Atlantic coast at that season. Fall records on the Great Plains are few but include adults. Both adults and juveniles are surprisingly late migrants in the Northeast, commonly lingering through October and even into early November.

COAST SPRING. Sooke River, BC, June 1, 1958; Iona Island, BC, June 15-16, 1981.

COAST FALL. Iona Island, BC, July 30, 1974, June 29 to July 1, 1989; Blackie Spit, BC, September 23, 1978.

INTERIOR SPRING. Ravalli Refuge, MT, May 1967; Hauser, ID, May 26, 1950; Reardan, WA, May 20, 1962, May 23, 1964; Quilchena, BC, June 1, 1973.

INTERIOR FALL. Grand Forks, BC, October 18, 1884; Stump Lake, BC, August 8, 1976 (3); Harvard, ID, July 25, 1981; Fortine, MT, August 19, 1985, August 2, 1989*.

Habitat and Behavior

An individual of this vagrant species could turn up anywhere, but the edges of freshwater lakes or sewage ponds would be the most appropriate habitats, with salt-water mud flats less likely. It usually feeds by probing in soft mud, often venturing into shallow water to do so. Surprisingly, its foraging habits are more like those of a Dunlin than those of the similarly shaped (and closely related?) Baird's Sandpiper. Where common, it usually collects in small flocks in migration and huge flocks on its wintering grounds.

Structure

Size small. Bill short, fairly slender and slightly decurved; legs short. Wings long, projecting well beyond tail tip; primary projection long in juveniles and only slightly shorter in breeding adults. Tail double-notched, central rectrices pointed and projecting.

Plumage

Breeding. Bill black, in most individuals with dull reddish at extreme base of lower mandible; legs black. Dark brown above with pale brown edges or fringes on most feathers. Edges on crown, mantle feathers, and upper scapulars often tinged with rufous, subtle rather than bright and in some birds virtually lacking. Ear coverts in some birds tinged with rufous. Often faintly indicated light buff mantle and scapular lines. Coverts brown with gray-brown fringes. White supercilium fairly conspicuous. Breast whitish with fine black stripes that continue prominently down sides, where mixed with or largely replaced by chevrons.

Lower back dark brown scalloped with light brown. Uppertail coverts white, some spotted with brown or black. Central pair of rectrices black, remainder gray-brown. Wing brown above with narrow white wing stripe formed by moderately broad tips of greater secondary coverts and quite narrow tips of greater primary coverts. Underwing and axillars white, gray-brown on flight-feather tips and greater primary coverts and darker brown on marginal coverts.

Most likely plumage to be seen in Northwest. Returning adults in early fall look darker and more blotchy above, as in other sandpipers with dark-centered, light-bordered feathers.

Nonbreeding. Upperparts gray-brown with dark feather centers varying from fine shaft streaks to prominent oval markings. White supercilium conspicuous against evenly colored head, breast, and upperparts. Breast washed with light gray-brown, breast streaks less conspicuous than in breeding plumage; sides sparsely streaked if at all. For a long-distance migrant, body plumage molted surprisingly rapidly, perhaps because of quick move to mud-flat habitats but lingering southbound migration. From early August on, some individuals in almost full nonbreeding plumage, usually with scattered traces of breeding plumage. Conversely, breeding plumage apparent on others well into October. Least likely plumage type in Northwest.

Juvenal. Upperparts a bit brighter and browner than in breeding adult, often with rufous edges to crown, ear coverts, mantle, scapulars (especially upper ones), and tertials. Coverts dark-centered, buff- to whitish-fringed, usually more conspicuously marked than in adults. Upperparts more scalloped overall than in breeding adults, white mantle and scapular lines more prominent. Prominent white supercilium and dark loral stripe. Cheeks, throat, and breast more faintly streaked than in breeding adults, sides rarely so; overall impression a bird more heavily marked above and much more lightly marked below. This plumage also molted while birds in migration in North America, with October and November individuals showing much first-winter plumage (eventually indistinguishable from adult).

Immature. Most immatures molt flight feathers during first winter and migrate north in breeding plumage in first summer, but some do not molt flight feathers and retain worn nonbreeding plumage through that period.

Identification

All Plumages. Only the BAIRD'S SANDPIPER shares the *attenuated look* caused by the *long wings projecting well beyond the tail tip*, and that quality distinguishes both of them from the smaller stints and most of the larger *Calidris* sandpipers. The bill in WHITE-RUMPED is slightly heavier and more curved than in BAIRD'S, but most birds will not need to be identified in this way. Both species of knots also have relatively

long wings but are considerably larger and stockier. The CURLEW SANDPIPER is also relatively long-winged but is longer-legged and somewhat curve-billed. Interestingly, the WHITE-RUMPED is colored somewhat like a WESTERN SANDPIPER in each of its plumages. Note that the *white rump* is not easy to see on a resting or feeding bird, visible only when the wings are lifted.

Breeding Plumage. From BAIRD'S SANDPIPER, this species can be readily distinguished by *prominently streaked sides, reddish markings on head and back*, and, at close range, usually *pale bill base*. It is colored more like a WESTERN SANDPIPER in this plumage but can be distinguished by its *larger size and relatively shorter bill with reddish base*, as well as its more attenuated look. Only rarely will one be as bright as a typical WESTERN, which usually has extensively rufous-based lower scapulars, while WHITE-RUMPED has these same feathers narrowly rufous-margined.

Nonbreeding Plumage. In this plumage WHITE-RUMPED is very similar to BAIRD'S, differing primarily in its *grayer, more uniform* appearance, as well as a slightly more prominent white supercilium and slightly heavier bill. The grayness is distinctive of WHITE-RUMPED, compared with any other small *Calidris*. BAIRD'S, although similarly dull in nonbreeding plumage, shows distinct brownish rather than gray tones. Neither species is likely to be seen in this plumage in the region, although WHITE-RUMPED is more likely in it than BAIRD'S in North America. The presence or absence of a white rump would have to be noted for definitive identification.

WHITE-RUMPED could be seen in nonbreeding plumage in late fall with DUNLINS or WESTERN SANDPIPERS. It is superficially like a DUNLIN because its breast is colored like its upperparts, but it is easily distinguished by its shorter bill and characteristic shape. It looks even less like a WESTERN because of these characteristics, as well as its distinctly larger size and darker breast.

Juvenal Plumage. Overall, WHITE-RUMPED is *contrasty above*, BAIRD'S paler, washed out, and buffy. More specifically, WHITE-RUMPED is marked with *reddish on crown, mantle, and upper scapulars*, while BAIRD'S is uniformly scalloped with pale buff to whitish. WHITE-RUMPED usually shows *white mantle and scapular lines*, while BAIRD'S does not.

WHITE-RUMPED is colored like both WESTERN and the more reddish SEMIPALMATED SANDPIPERS, from which it differs, of course, by size and shape. It differs further from WESTERN by its overall browner appearance above, usually without the clean reddish-and-gray contrast of the smaller species.

In Flight (Figure 12)

It is when it takes to the air that the WHITE-RUMPED is most easily distinguished from other small sandpipers by its *white rump*. The scattered dark spots on the rumps of some breeding-plumaged adults are insufficient to promote confusion with dark-rumped species. The only other small shorebird that is white-rumped rather than white-tailed is the CURLEW SANDPIPER, from which the WHITE-RUMPED differs by *smaller size, shorter legs, a less conspicuous wing stripe, and a narrower rump patch*. The wing stripe in WHITE-RUMPED is similar to those of the STINTS, in CURLEW almost as conspicuous as that of a DUNLIN. In WHITE-RUMPED the white rump patch looks about as broad from front to back as the darker tail behind it, in CURLEW considerably broader. In CURLEW SANDPIPER, but not in WHITE-RUMPED, the toe tips project conspicuously beyond the tail in flight.

Voice

The flight call of this species is distinctive, a mouselike squeak that is even thinner and higher-pitched than the similar call of the WESTERN. Flushed birds are likely to call.

Further Questions

The peculiarly late autumn migration of this species needs further understanding. It is one of the later-lingering shorebirds in the Northeast, surprising for birds that have such long flights ahead of them.

Notes

The extremely early date relative to the normal migration pattern of this species casts doubt on a record from Victoria, April 20, 1977.

References

Alström 1987 (identification), Cartar and Montgomerie 1987 (incubation behavior), Drury 1961 (breeding biology), Harrington et al. 1991 (migration), Holmes and Pitelka 1962 (behavior and taxonomic position), Parmelee et al. 1968 (breeding biology).

BAIRD'S SANDPIPER *Calidris bairdii*

Scaly-backed and buffy, long and low, juvenile Baird's Sandpipers are easy to pick out as they trip lightly along the upper beaches of autumn. Sand-colored sand coursers, they visit us briefly on their way to Patagonian pampas and are as likely to be seen on an alpine snow bank as a mud flat. Spring adults, always unexpected, are more likely in the interior.

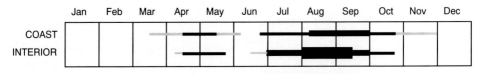

Distribution

The Baird's Sandpiper breeds on high-arctic tundra from northeastern Siberia to Baffin Island and winters in Andean and southern South America. Adults use the Great Plains migratory corridor spring and fall, but juveniles spread out to both coasts.

Northwest Status

Only a small percentage of the great numbers of individuals of this species that move up and down the Great Plains in migration reach the Northwest, and they tend to be more common in the interior because of their source. Numbers decrease from British Columbia and Montana

Breeding White-rumped Sandpiper. Medium black-legged calidridine with long primary projection; rufous on upperparts and strongly speckled breast and sides diagnostic of species and plumage; pale base of mandible more obvious on some birds. Scituate, MA, 1 June 1985, Wayne Petersen

Juvenile Baird's Sandpiper. Long wings, overall brownish look, and buff scallops diagnostic of species and plumage. Sandy Point State Park, MD, 13 September 1979, Richard Rowlett

Breeding Baird's Sandpiper. As White-rumped, but coarsely mottled upperparts and streaked, brownish-washed breast in this plumage; long primary projection not obvious because of wing position. Wales, AK, July 1978, Jim Erckmann

southwestward to Washington and Oregon. Adults move through from early April to mid May and are uncommon anywhere, more frequently recorded toward the east and seldom seen on the coast. The mid-April peak of spring migration is surprisingly early for an arctic breeder. They return during July in quite small numbers, females beginning the migration slightly before males.

From the end of July through September, fair numbers of juveniles pass through the region, with single birds and small groups seen regularly in appropriate habitats. High counts in coastal areas usually involve no more than a dozen or two birds, but hundreds of juveniles have been seen on a few occasions. These counts were well above normal peaks and indicative of year-to-year variation. Only a few have been seen after September, and the November records are unusually late for this species, so far from its South American winter range.

COAST SPRING. *Early date* Comox, BC, March 22, 1938. *High count* 12 at Saanich, BC, April 27, 1978. *Late date* Yaquina Bay, OR, June 7, 1983.

COAST FALL. *Early dates* Iona Island, BC, June 25, 1981; Ocean Shores, WA, June 28, 1970. *Adult high count* 8 at Iona Island, BC, July 16, 1982. *Juvenile early date* Ocean Shores, WA, August 8, 1981. *Juvenile high counts* 100 at Comox, BC, September 19, 1949; 268 at Iona Island, BC, August 6, 1977; 120 at Nehalem, OR, September 1, 1974. *Late dates* Comox, BC, October 27, 1928; Reifel Island, BC, November 18, 1978; Yaquina Bay, OR, November 28, 1976.

INTERIOR SPRING. *Early dates* Malheur Refuge, OR, April 8, 1982; Neppel, WA, April 8, 1936#. *High counts* 8 at Nyssa, OR, April 25, 1982; 8 at Bend, OR, April 15, 1984; 10 at Malheur Refuge, OR, April 22, 1989; 150 at Banks Lake, WA, spring 1963; 30+ at Kamloops, BC, April 14, 1985. *Late date* Vaseux Lake, BC, May 18.

INTERIOR FALL. *Early date* Rock Lake, WA, June 17. *Adult high counts* 25 at Douglas Plateau, BC, July 10, 1983; 14 at La Grande, OR, July 17, 1982; 18 at Lewiston, ID, July 31, 1958 (possibly including juveniles). *Juvenile early date*

Okanagan Landing, BC, July 19, 1915#. *Juvenile high counts* 300 at Creston, BC, August 11, 1948; 600 at Salmon Arm, BC, August 22, 1973; 75 at Kamloops, BC, September 12, 1982; 60 at Walla Walla River delta, WA, fall 1986; 35 at Bend, OR, August 14, 1985; 220 at Deer Flat Refuge, ID, August 17, 1990; 600 at Ninepipe and Pablo refuges, MT, August 25-26, 1981; 165 at Missoula, MT, September 12, 1986. *Late dates* Okanagan Landing, BC, October 18; Malheur Refuge, OR, October 14, 1983 & 1985.

Habitat and Behavior

Baird's Sandpipers prefer drier environments than many of their close relatives, often foraging on the upper beach or even in grassy areas, where they associate with golden-plovers. They are the only *Calidris* that regularly forages on dry, sandy areas such as sand dunes, at times with Buff-breasted Sandpipers. They also occur on mud flats with Western Sandpipers but rarely forage in the dense marsh vegetation chosen by Pectorals and Leasts. Baird's typically forage by moving rapidly along and picking up prey with quick bill jabs, an obvious correlate to feeding on relatively dry substrates. Because of their preferred feeding habitat, insects are the most common prey.

They look more alert than many other small sandpipers, with head up much of the time as they move, and they doubtless forage visually more than is typical of calidridines. The body is held horizontal even when the neck is extended, unlike many other open-country shorebirds that tilt the body up at an angle as the head rises.

Throughout its range Baird's is known as a montane autumn migrant, and the Northwest is no exception. Individuals and small flocks are seen at high alpine lakes and even feeding on invertebrates on snowbanks. Very few other shorebirds are so regular as migrants in the mountains. Flock size is usually small, often only single birds, but where the species is common, flocks can include dozens of birds. Groups of about ten have been the maximum at Ocean Shores.

Structure (Figure 44)

Size small. Bill short, fairly straight, and slender, small for bird of its size. Legs short. Wings long,

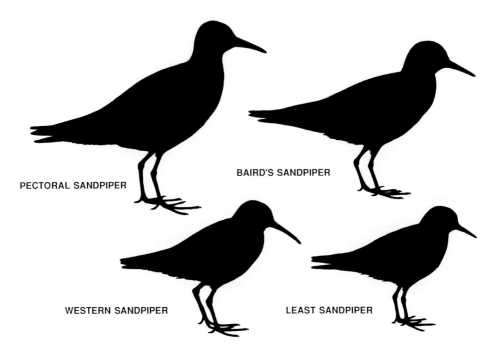

PECTORAL SANDPIPER

BAIRD'S SANDPIPER

WESTERN SANDPIPER

LEAST SANDPIPER

Fig. 44. Common *Calidris* shapes. Differences in size, bill size and shape, and wing projection.

extending well beyond tail tip; primary projection short to moderate in breeding adults and long in juveniles. Tail double-notched, central rectrices pointed and projecting slightly.

Plumage

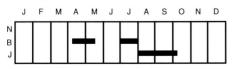

Breeding. Bill and legs black. Upperparts mottled black, brown, and buff. Crown and mantle feathers and scapulars blackish-centered and fringed with sandy brown, coverts and tertials entirely sandy brown. Faintly indicated paler supercilium and darker loral stripe. In fresh-plumaged spring migrants, feathers evenly and widely margined with warm, almost buffy-brown; paleness prevails. In worn individuals in early fall, pale fringes faded and narrower, allowing dark centers to dominate and present a more irregularly blotched appearance. Breast light brown, streaked with dark brown; throat and remainder of underparts white.

Lower back dark brown, scalloped with buff. Central uppertail coverts dark brown, outer ones with alternating brown and white chevrons. Central pair of rectrices dark brown, rest of tail brownish gray. Wing brown above with narrow whitish wing stripe formed by narrow tips of greater secondary coverts. Underwing gray-brown, with marginal coverts brown, median secondary coverts and bases of secondaries whitish, and axillars white.

Nonbreeding. Much plainer than breeding-plumaged adult, with overall brownish cast and little evidence of dark feather centers; supercilium and loral stripe slightly more conspicuous accordingly. Breast streaks finer and less conspicuous than in breeding birds. This plumage rarely attained in North America, but a specimen from Okanagan Landing, BC, July 16, 1915, well along in molt, and single August specimens from Colorado and Kansas in full nonbreeding plumage. Most adults in this plumage by mid October in South America, long after adult Baird's pass through Northwest.

Juvenal. A smoothly and evenly marked plumage, quite distinct from blotchy-looking fall adults. All feathers of upperparts medium to fairly dark brown, fringed with light brown to buffy-white. Even-width fringes on scapulars, coverts, and tertials produce more distinctly scaled appearance than in most other *Calidris* juveniles. No trace of mantle or scapular lines. Breast light buffy-brown, with fine and indistinct dark streaks less conspicuous than stripes of breeding-plumaged adults. Most commonly seen plumage in Northwest.

Buff fringes become narrower in later-migrating birds, virtually disappearing in October; few Baird's still in Northwest by that time. Juveniles apparently do not begin molt into first-winter plumage until on South American wintering grounds, then become indistinguishable from adults.

Immature. Most juveniles apparently with complete molt on wintering grounds, including flight feathers, but minority with very worn primaries in spring. Perhaps virtually all fly north in first summer.

Identification

All Plumages. BAIRD'S is one of a pair of species, along with WHITE-RUMPED, that share the characteristics of *relatively short legs and long wings*, producing an *attenuated look*. The *short, all-black bill and black legs* allow separation from several other mid-sized *Calidris* such as PECTORAL, SHARP-TAILED, ROCK, CURLEW, and STILT SANDPIPERS and DUNLIN.

Breeding Plumage. BAIRD'S in this plumage may be virtually identical in coloration to SEMIPALMATED, from which it is distinguished by size and shape as well as relatively smaller, *more slender bill*. With its brown, evenly streaked breast, it is somewhat similar to PECTORAL but differs in *overall lighter color with no hint of reddish above and paler, more finely striped breast*, as well as *smaller size and black legs*. BAIRD'S is superficially similar to SANDERLING in breeding plumage (similar size, with evenly marked back and breast) but is always duller, without the orange tones of the latter. See WHITE-RUMPED SANDPIPER for differences from that species.

Nonbreeding Plumage. As unlikely as it is to occur in our region, the observer should be cognizant of this plumage. Because of size and shape, only WHITE-RUMPED is extremely similar, and some individuals might not be distinguishable without checking rump color. Other possible sources of confusion would be the very common DUNLIN (BAIRD'S with much smaller bill and longer wings) and PECTORAL and SHARP-TAILED SANDPIPERS, also unlikely to occur in the Northwest in nonbreeding plumage (BAIRD'S smaller, with shorter, *black legs*).

Juvenal Plumage. Juvenile BAIRD'S are very distinctive, with their *buffy, scaly-backed* look. Only the much smaller, shorter-winged SEMIPALMATED SANDPIPER ever approaches this look closely.

In Flight (Figure 12)

BAIRD'S has *long-looking wings* with an *inconspicuous white wing stripe* but is otherwise similar to other small-to-medium *stripe-tailed Calidris* in flight. The pronounced *brownish breast* allows easy distinction from WESTERN SANDPIPER overhead, the *larger size and longer wings* from LEAST SANDPIPER. The wingbeats are perceptibly slower than in the smaller STINTS.

PECTORAL is the closest species in flight, and the females are about as close to BAIRD'S in size as they are to males of their own species. BAIRD'S look *lighter* than PECTORAL, with a slightly more conspicuous wing stripe, but at a distance they are difficult to distinguish. Both often call in flight, but even their flight calls are similar enough to be confusing.

Voice

BAIRD'S has a rolling flight call, similar to PECTORAL but a bit higher-pitched and more musical sounding. The roll is slightly slower, the individual notes better separated from one another, actually intermediate between LEAST and PECTORAL. BAIRD'S calls frequently in flight, perhaps because single individuals are typical, and lone shorebirds appear to call more than individuals in flocks. Nevertheless, they call much more frequently even when in flocks than do WESTERNS.

Further Questions

Why do Baird's commonly forage in mountain habitats in migration, while so few other shorebirds do so?

Notes

An unusually late record from Reardan, November 17, 1962, has not been considered sufficiently documented to accept. A December Oregon record (seven at Coos Bay, December 21, 1975) is even more dubious.

Most field guides have ignored the drab nonbreeding plumage of this species; only Hayman et al. (1986) treat it.

Photos

A juvenile unlabeled as such is in Bull and Farrand (1977: Pl. 225). The breeding-plumaged "Baird's Sandpiper" in Farrand (1988a: 187) is a very unusual individual of that species, if correctly identified, and it may instead be a White-rumped, from bill shape, brightness of markings above, and streaked sides. The "unidentified *Calidris*" photographed in Ontario on November 14, 1980 (Goodwin 1981), was probably a cold, fluffy, nonbreeding-plumaged Baird's Sandpiper, a long distance from where it should have been.

References

Alström 1987 (identification), Baldassarre and Fischer 1984 (diet and foraging behavior in migration), Dixon 1917 (breeding biology), Drury 1961 (breeding biology), Holmes and Pitelka 1968 (summer diet), Jehl 1979 (fall migration of age classes and molt), Miller et al. 1988 (breeding vocalizations), Norton 1972 (incubation schedules), Paulson 1983 (juvenile fledging and migration).

PECTORAL SANDPIPER *Calidris melanotos*

A salt-marsh stroll on a late-fall afternoon should provide Pectoral encounters. As they are approached, heads pop up from the pickleweed, necks grow longer and longer, and individuals and then small groups take flight, rolling their r's as if practicing Spanish for their winter sojourn. Double-sized versions of Least Sandpipers, they feature rich browns, striped backs, and yellow legs. The breeding male's inflatable breast gives it its name, and the sharp pattern of the same area is the distinctive field mark for the species.

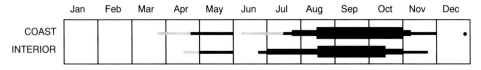

Distribution

The Pectoral Sandpiper breeds on wet tundra from central Siberia across the North American Arctic to Hudson Bay and winters in southern South America and rarely in Australia and New Zealand. Spring migration is primarily through the Great Plains but in some volume to the Atlantic coast, while fall migration occurs throughout the East, including over the Atlantic Ocean; juveniles wander in numbers to both coasts.

Northwest Status

This species is known in the Northwest only as a migrant, much more common in fall than in spring. Few are seen anywhere in the region in spring, from late April to the end of May. The few mid and late June records could represent summering birds. Adult males, which defend territories and attempt to mate promiscuously but contribute no parental care, average earlier than females in both northbound and southbound migration.

Southbound adults normally arrive in mid July and are present in small numbers to mid September, rather late for a long-distance migrant. Far more adults are seen in southern British Columbia than farther south, paralleling the situation in some other plains migrants. Adults seem no more common in the interior than on the coast, with a somewhat later spring migration period. Juveniles move through the Northwest in large numbers, locally common from late August through October, with counts of a hundred or more at some coastal marshes. They regularly linger into early November, and there are a few records in southern British Columbia right to the end of December.

Autumn counts in the interior usually range from a few birds to a few dozen, with high counts occasionally in the hundreds. Numbers seen in Idaho and western Montana seem consistently lower than in British Columbia, Washington, and Oregon, which seems peculiar for a species with its main migration route through the plains. However, this could be explained by a substantial contingent of juveniles from Siberia migrating down the American Pacific coast, leaving a hiatus between plains and Pacific birds. Like Pacific Golden-Plovers and Sharp-tailed Sandpipers, Pectorals diminish in numbers to the south along the Pacific coast, and they may cross the Pacific Ocean to some wintering ground other than the Americas. The juveniles migrate later than the juveniles of most other shorebirds of the region, surprising for a species that winters so far south, but the lateness of their migration would also be explained by their coming primarily from Siberia. This hypothesis is supported by the low numbers of Pectorals in 1989 and 1990, low years for Sharp-tailed Sandpipers.

COAST SPRING. *Early date* Iona Island, BC, March 25, 1979. *High counts* 6 at Banks, OR, April 24, 1986; 3 near Port Angeles, WA, May 15, 1983; 9 at Surrey, BC, May 19, 1987. *Late date* Leadbetter Point, WA, June 8, 1974.

COAST SUMMER. South jetty of Columbia River, OR, June 20, 1970; Cleland Island, BC, June 18, 1975; Iona Island, BC, June 23, 1975 (crippled).

COAST FALL. *Adult high counts* 30 at Cattle Point, BC, July 15, 1964; 54 at Iona Island, BC, July 23, 1973; 4 at Neah Bay, WA, July 28, 1970; 5 at Nehalem, OR, July 10, 1987. *Adult late date* Grays Harbor, WA, September 22, 1979 (3). *Juvenile early date* Aberdeen, WA, August 18, 1982*. *Juvenile high counts* 300 at Iona Island, BC, September 24, 1972; 400 at Iona Island, BC, October 2, 1984; 200+ at Ocean Shores, WA, September 14, 1985; 500 at Tillamook and Nehalem, OR, September 19, 1982. *Late dates* Vancouver, BC, December 26, 1962; Campbell River, BC, December 2, 1975; Nanaimo, BC, December 31, 1978 (3).

INTERIOR SPRING. *Early date* southeastern

WA, April 15. *High counts* 9 at Deadman's Lake, BC, May 12, 1985; 20 at Kamloops, BC, May 13, 1989. *Late date* White Lake, BC, May 31, 1961.

INTERIOR FALL. *Early date* Potholes, WA, June 29, 1953. *Adult high count* 30 at Homedale, ID, August 6, 1950. *Juvenile early dates* Okanagan, BC, September 6, 1916#; Pullman, WA, September 6, 1949#. *Juvenile high counts* 200 at Creston, BC, September 10, 1949; 100 at Kamloops, BC, September 14, 1980; 40 at Reardan, WA, September 11, 1964; 50 at Reardan, WA, fall 1966; 350 at Stinking Lake, OR, September 1975; 94 at Midland, OR, September 26, 1982; 40 at Lewiston, ID, September 14, 1955. *Late dates* Vernon, BC, December 7, 1987; Missoula, MT, November 23, 1962.

Habitat and Behavior

Pectoral Sandpipers inhabit salt marshes and forage among dense *Salicornia* and other vegetation, often moving out onto mud flats as well. They are found throughout the region in smaller numbers in freshwater marshes, where they forage on drying-up pond and lake margins and in grass and sedge wetlands. Pastures and agricultural fields moistened or flooded by rain or irrigation are also preferred habitats. Their most common associate is the Least Sandpiper, with almost identical habitat preference.

Pectorals feed by moving steadily over the substrate with heads down, picking surface prey very much as do Leasts or probing shallowly. Often feeding in dense vegetation, they spread out and disappear from view after landing. Alert for potential predators, when disturbed they stand upright with neck well extended. Flushed birds quickly collect into small flocks, flying relatively low rather than "towering" as some other sandpipers of the same habitats do. Roosts consist of small groups of scattered birds.

Structure (Figure 44)

Size small (female) to medium-small (male). Sexes differ more in size than in any other regional shorebird but Ruff, with females about two-thirds the size of males by weight. Wing lengths of sexes do not overlap, averages differing by 13 mm, but bill and tarsus lengths differ by only about 2 mm and overlap considerably. Single birds probably impossible to sex, but size difference evident when birds of both sexes together. Note that male Pectoral is Killdeer-sized, female Dunlin-sized.

Bill short, slightly decurved, and fairly bluntly pointed; legs short. Wings usually extend to or a bit beyond tail tip, primary projection short in breeding adults and long in juveniles. Tail wedge-shaped, central rectrices projecting well beyond others.

Plumage

Breeding. Bill entirely black (at least some adults) or marked with dull greenish, yellowish, or brown on basal half; legs dull yellow to greenish yellow. Overall coloration fairly bright, buff to reddish above. Crown, mantle, scapulars, and new tertials blackish-centered and fringed with buff or light brown to light rufous; coverts and old tertials medium brown with light brown fringes. Whitish supercilium and darker loral stripe inconspicuous. Foreneck and breast light brown with blackish stripes, augmented on lower breast by transverse markings in females. In males entire base of each feather dark, with dark streak and paired whitish spots at tip that produce dark and mottled look to lower breast. This sexual difference apparent on breeding grounds, where male breasts usually expanded in display, but may be less apparent in migration. Belly white, sides often sparsely streaked.

Lower back, central uppertail coverts, and central pair of rectrices black; outer uppertail coverts brown with white fringes; outer rectrices gray-brown, narrowly white-fringed. Wing brown above with indistinct whitish wing stripe formed by narrow tips of greater secondary coverts. Underwing gray-brown with brown marginal coverts and whitish median secondary and primary coverts; axillars white.

Bright fringes on upperparts faded to dull light brown in most returning adults in fall, although considerable variation in overall brightness. Some adults look particularly washed-out in September, others almost as bright as juveniles.

Nonbreeding. Dull version of breeding plumage, with most pale areas light brown. Little hint of bright buff or reddish except, in some individuals, on crown and on few scapulars and/or tertials. These birds not very different from worn breeding adults but of course would be fresh-feathered. No indication that Pectorals achieve this plumage in Northwest.

Juvenal. Brightest plumage of species, patterned basically like adult but with bright rufous edges to crown feathers, scapulars, and tertials in fresh plumage. Coverts darker-brown centered than those of adults, with buff to reddish fringes. White lines on both mantle and scapulars distinct; row of lower scapulars just outside lined row also have white outer edges, producing ghost of third line in some individuals. Moderately distinct whitish supercilium and dark loral stripe. Breast buffier, stripes finer than in adults. Some juveniles duller, with rufous edges darker and reduced in width, and a very few are dull, with almost no rufous. Contrarily, some autumn adults brighter and more similar to juveniles than others, with indications of mantle and scapular lines (probably never as conspicuous as those of juveniles). At close range, worn tertials of autumn adults usually obvious and allow easy distinction from juveniles.

Immature. Juveniles molt into first-winter plumage, becoming indistinguishable from adults, on wintering grounds. All probably migrate north in first year to breed, although some molt flight feathers and others do not.

Identification

All Plumages. The *boldly striped brown breast sharply demarcated from the white belly* is the best instant field mark for this species. Most other streaked brown sandpipers of its size range have more lightly striped (or streaked) or unstriped breasts that blend more gradually with the belly color. The most similar species are BAIRD'S and WHITE-RUMPED, from which PECTORAL can be further distinguished by *darker brown color, yellow legs, and less attenuated look*, and SHARP-TAILED (see that species).

Juvenal Plumage. The PECTORAL, SHARP-TAILED, and STILT SANDPIPERS are fairly large, striped, pale-legged *Calidris*, usually with *white mantle and scapular lines*. PECTORAL differs from STILT in its *shorter legs and shorter bill and reddish tones above* as well as heavily striped breast. The LEAST SANDPIPER is similarly colored but much smaller with virtually no primary projections and a diffusely streaked breast. See also juvenile RUFF, which has an unstriped buff breast, and juvenile SHARP-TAILED SANDPIPER, which is more similar.

In Flight (Figure 12)

PECTORAL is *stripe-tailed* with an *inconspicuous wing stripe*. SHARP-TAILED SANDPIPER is virtually identical from above, without the contrasting brown breast, and BAIRD'S SANDPIPER is similar from both above and below but smaller, longer-winged, and more buffy brown in appearance. More different are two other species also seen with PECTORALS at times: RUFF, with extensive white patches at the tail base, and BUFF-BREASTED SANDPIPER, with no white at all around the tail; both of these species have plain buffy breasts.

Voice

The PECTORAL's flight call is a low, rolling trill, not particularly musical. It is similar to but typically lower and faster than the comparable call of BAIRD'S SANDPIPER.

Further Questions

Adults are quite uncommon in the region both spring and fall, and it would be of interest to determine the sex, if possible, of any adults seen. Sex ratios in juvenile flocks would be of equal interest and perhaps easier to determine with lengthy observation, as both sexes would often be present. Also, observers in fall should always attempt to record age of birds seen (especially during August and September, when the two age classes overlap).

Notes

A record of 12 at Victoria, March 26, 1960, is considered inadequately documented.

Few field guides have noted how reddish this species appears in spring or how bright juveniles can be. A juvenile seen at Kent in September 1987 appeared from other field marks (bright rufous cap, vividly fringed tertials) to be a Sharp-tailed, but its breast was entirely streaked. If not a hybrid, it represented an extreme variant of the Pectoral.

Photos

Farrand (1983: 387-89) shows a worn adult as well as two juveniles, but the juveniles are examples of variation rather than fading. Armstrong (1983: 152) and Farrand (1988b: 164) show fresher-plumaged adults, while most other field guides show only juveniles (including birds in Udvardy [1977: Pl. 198], Hammond and Everett [1980: 123], and Keith and Gooders [1980: Pl. 100] not labeled as such). The "first-winter" bird in Chandler (1989: 113) is more likely an adult, as its greater coverts, tertials, and primaries look too unworn to be retained from a previous plumage.

References

Brooks 1967 (prey choice in migration), Hamilton 1959 (territoriality in migration), Holmes and Pitelka 1968 (summer diet), Myers 1982 (breeding biology), Norton 1972 (incubation schedules), Pitelka 1959 (breeding biology).

SHARP-TAILED SANDPIPER *Calidris acuminata*

A bright spot in an autumn salt-marsh flock of Pectoral Sandpipers may be this close relative with rufous cap and unstreaked buff breast. We have no clues why so many of these Siberia-to-Australia migrants appear on our coast every fall, but lack of understanding does not preclude appreciation.

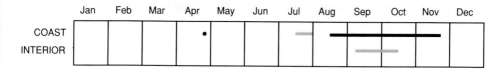

	Jan	Feb	Mar	Apr	May	Jun	Jul	Aug	Sep	Oct	Nov	Dec
COAST				•								
INTERIOR												

Breeding Pectoral Sandpiper. Medium, plain brown, long-necked, yellow-legged calidridine with heavily striped breast contrasting with white belly; mottled upperparts characteristic of plumage. Cambridge Bay, NT, 24 June 1975, Jim Erckmann

Juvenile Pectoral Sandpiper. As adult, but upperparts contrastily marked with white mantle and scapular lines and rufous fringes; pale bill base and yellow legs typical, crown more rufous than usual. Victoria, BC, Tim Zurowski

Juvenile Sharp-tailed Sandpiper. Above much like juvenile Pectoral, but breast bright buff and unstriped (also diagnostic of plumage); rufous crown and prominent supercilium typical. Attu, AK, 13 September 1983, Edward Harper

Distribution

The Sharp-tailed Sandpiper breeds in eastern Siberia and winters in Australasia, migrating by both continental and oceanic routes.

Northwest Status

Considering that this Siberian species is not known to winter in the New World, it is surprisingly numerous along the west coast. Migrants are locally fairly common in autumn in Alaska, with much smaller numbers trickling down to our region. Perhaps those individuals that move to the wrong hemisphere head out over the Pacific eventually, small numbers of them using our region to fatten up for a long flight to some tropical Pacific island or even Australia. Normally they do not return, as adults are only rarely seen anywhere in North America. Some juveniles may get caught up in flocks of Pectoral Sandpipers, which breed all across Siberia but mostly move eastward into western North America on their way to their wintering grounds. This eastward movement must be innate in some Sharp-taileds, however, as pure flocks of them have been seen in Alaska as well as mixed flocks of Pectorals and Sharp-taileds.

There is only a single spring record, and two July birds were probably adults although not cited as such. None of these records is documented by specimen or photograph, and the only other adult records south of Alaska come from southern California, where one or two were observed in late July 1988. Otherwise, all observations are of juveniles. In most years the first individuals are seen in late August or early September, numbers are usually greatest in October, and, being late migrants like the Pectoral Sandpipers with which they often occur, a few linger well into November and even December. More are seen in British Columbia and Washington—where they are dependably present at a number of localities—than in Oregon. Numbers vary from year to year, with especially few in 1989 and 1990, also correlated with Pectoral Sandpiper numbers.

Most records are coastal, well-distributed on both protected and outer coasts, and small numbers have been recorded in the Willamette Valley south to Eugene. The nine interior records span the period from September 15 to October 13.

COAST SPRING. Leadbetter Point, WA, April 26, 1979.

COAST FALL. *Adults* Iona Island, BC, July 22, 1967, July 23 and 27, 1976. *Juvenile high counts* 20 at Iona Island, BC, August 22, 1977; 14 at Reifel Island, BC, October 4, 1986; 6 at Leadbetter Point, WA, October 8-13, 1978; 3 at Sauvie Island, OR, October 11, 1980; 3 at Nehalem, OR, September 27, 1987. *Late date* Iona Island, BC, December 21, 1976.

INTERIOR FALL. Tranquille, BC, September 20, 1980; Swan Lake, BC, September 12, 1982; Kamloops, BC, September 18, 1988; Soap Lake, WA, September 15, 1972; Sunnyside, WA, October 13, 1973; Richland, WA, September 28 to October 5, 1975; Othello, WA, September 26, 1982; Yakima River delta, WA, October 12, 1986, September 26, 1990.

Habitat and Behavior

Sharp-tailed Sandpipers typically forage in coastal salt marshes, occasionally moving out onto adjacent mud flats. Less often they occur in freshwater marshes near the coast. They are typically but not always found with Pectoral Sandpipers and seem to occur on mud flats slightly more readily. The two species forage similarly when they are together and seem to exhibit about the same array of foraging techniques. Sharp-taileds occur in flocks of up to dozens of individuals where common but are found only as scattered individuals, or occasionally a few together, in the Northwest.

Structure

Size small. Bill short and fairly straight, legs short. Wings reach tail tip or fall slightly short of it, primary projection short in breeding adults and moderate to long in juveniles. Tail wedge-shaped, central feathers projecting well beyond others. Tail feathers indeed more sharply pointed than those of related species, including Pectoral Sandpiper; pointed rectrices may be adaptations for marsh-dwelling (as in Sharp-tailed Sparrow).

Plumage

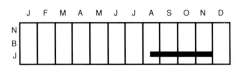

Breeding. Bill black or dark brown, in some individuals grayish or greenish at base of lower or both mandibles. Legs greenish yellow to olive-green. Crown feathers widely edged with bright rufous, fairly conspicuous white supercilium. Mantle rich medium brown vividly striped with black; most scapulars and tertials brown, fringed with bright rufous; some lower scapulars with whitish outer edges. Coverts dark brown, fringed with light brown. Narrow white supercilium, dark loral stripe expanded into smudge just before eye. Face and upper breast whitish streaked with dark brown, lower breast washed with buff and spotted with black. Sides and part of belly heavily marked with black chevrons, rest of underparts white.

Lower back dark brown, central uppertail coverts and central pair of rectrices dark brown with rufous fringes. Outer uppertail coverts brown with broad white fringes, outer rectrices brown with narrow white fringes. Wing brown above with narrow whitish stripe formed by narrow tips of greater secondary coverts. Underwing gray-brown with brown marginal coverts. Greater and median secondary coverts and tips of primary coverts on underwing whitish, axillars white.

Breeding plumage held from April through September, typically persisting late in autumn in this long-distance migrant; however, some adults in heavy body molt in September.

Nonbreeding. Much duller plumage, feathers of upperparts dark brown with gray-brown to buff fringes. Some scapulars and/or tertials narrowly edged with rufous in some individuals, as are crown feathers. Breast pale buff with scattered spots, belly white with scattered streaks on sides. No birds in this plumage reported from Northwest.

Juvenal. Legs average paler, more yellowish, than in adults. Color of upperparts much like breeding adult, with finely black-streaked, rich rufous cap and wide bright rufous fringes on scapulars and tertials; differs by well-defined white mantle and scapular lines. White supercilium conspicuous, dark loral stripe conspicuous or not. Underparts more lightly marked than adult. Breast and foreneck buff, rest of underparts including throat white. Band of fine dark streaks across foreneck, coarser streaks on sides of breast, and well-defined streaks on sides; typically fine streaks on undertail coverts. Body molt into first-winter plumage noted in migration in England as early as late October, but most juveniles pass through Northwest with little if any molt, and birds in this plumage seen as late as December.

Immature. Juveniles become indistinguishable from adults during first winter except for some retained, worn tertials in some individuals. All apparently migrate north in first summer, as indicated by absence from Australia during that period, although some molt flight feathers and others do not.

Identification

All Plumages. This species is a medium-sized sandpiper with *fairly short bill and short, yellow legs*. Only the PECTORAL and BUFF-BREASTED SANDPIPERS and juvenile RUFF are otherwise included in this group, and both BUFF-BREASTED and RUFF are scaly-backed rather than striped above. The juvenile STILT SANDPIPER, somewhat similar in pattern, has considerably longer bill and legs. Very similar to the PECTORAL, the SHARP-TAILED differs in its *slightly shorter and straighter bill* (male SHARP-TAILED and female PECTORAL overlap greatly, however) that is less often extensively pale at the base. The *legs are often greenish*, duller on the average than the PECTORAL's yellow legs, but the two species overlap.

Breeding Plumage. The *rufous cap, heavily spotted buffy breast, and large chevrons* ("boomerangs" in Australia) *on the sides* are a combination found in no other shorebird, and a look at almost any part of one of these birds should suffice for identification. Even worn individuals show all the characteristics clearly. This plumage also features a *conspicuous white eyering*, more evident than in any other *Calidris* species.

Nonbreeding Plumage. In this plumage only the markings of the underparts—*pale buff breast with scattered spots and streaks*—will allow easy separation from PECTORAL, which always has its sharply demarcated breast/belly line. The two species look identical from above.

Juvenal Plumage. This is the plumage to

know in the Northwest, where almost all individuals seen are juveniles. Again, the trick is to distinguish them from the very similar PECTORAL, from which they differ by *conspicuous rufous cap and unstriped bright buff breast*. Many PECTORALS have rufous-edged crown feathers, but the edges are not so wide and thus the cap not so red in that species; nevertheless, the brightest PECTORALS are as rufous-capped as some SHARP-TAILED. There are distinct streaks at the side of the SHARP-TAILED's breast, but they do not extend across the center as in the PECTORAL, and, because of this, the line between buff breast and white belly is not a sharp one. Because of considerable overlap, this may be the only diagnostic field mark (see under Notes). Note the fine streaks across the *foreneck* in the SHARP-TAILED.

Other more minor points, visible at close range for further confirmation or just the joy of seeing the bird well, include:

• an entire cap in SHARP-TAILED, usually a distinctly split supercilium in PECTORAL;

• a more conspicuous white supercilium in SHARP-TAILED, broad and well developed behind the eye;

• a more conspicuous white eye-ring in SHARP-TAILED; and

• broader and slightly paler rufous tertial fringes in SHARP-TAILED, producing a very brightly marked back (bright PECTORALS overlap with less bright SHARP-TAILED), often the best way to pick out a SHARP-TAILED in a flock of PECTORALS when all are foraging with heads down.

In Flight

SHARP-TAILED looks essentially like PECTORAL in flight and often cannot be distinguished from it. Juveniles differs on the average by their slightly brighter tertial edgings and more contrasty white wing stripes; one bird in a flock of PECTORALS was picked out by these characteristics.

Voice

The flight call is a liquid, single *kvik* or *wit* or double *kuvick* or *tuwit*, not dissimilar to the call of a BARN SWALLOW. My impression is that this species is less vocal than the PECTORAL.

Further Questions

On several occasions, Sharp-tailed have been reported as being aggressive to nearby Pectorals, but I once watched a Sharp-tailed run right through a flock of Pectorals to chase another Sharp-tailed 30 feet away. It would be of interest to determine if this species is indeed more aggressive than the Pectoral or if this has something to do with being a "loner" flocking with a different species. Australian observers could contribute to this by watching single Pectorals in flocks of Sharp-taileds.

Notes

Some observers have stated that Sharp-tailed is less likely than Pectoral to stand up high with neck extended. This may be a case of familiarity bias, with some of the many Pectorals seen looking long-necked. If the observer had seen Sharp-tailed foraging only on mud flats, there would have been no reason for these birds to stand up as high as Pectorals do at times when in dense vegetation. Sharp-tailed has also been said to be chunkier, shorter-necked, and flatter-crowned, but I have been unable to see these differences in birds foraging together, and a series of photographs of two such individuals shows no such differences. Also, some observers have called Sharp-tailed slightly larger than Pectoral, al-though in fact each sex of the Sharp-tailed is smaller than the comparable sex in Pectoral, males more than females because there is less dimorphism in Sharp-tailed. Presumably these observers were comparing male Sharp-tailed and female Pectoral.

The painting of a breeding Sharp-tailed Sandpiper in Peterson (1990: 147) is much too dull for this striking plumage. Contrary to Farrand (1983: 388), breeding-plumaged Sharp-taileds look quite unlike juvenile Pectorals. That book also calls the Sharp-tailed a geographic replacement for the Pectoral in Siberia; in fact, their breeding ranges overlap entirely. Presumably the two species diverged in isolation on the two continents, then the Pectoral reinvaded Siberia as a breeding species and now breeds even west of the westernmost Sharp-tailed.

Photos

A juvenile unlabeled as such is in Udvardy (1977: Pl. 199). The "Sharp-tailed Sandpiper" in Hosking and Hale (1983: 123) is a juvenile Long-toed Stint.

References

Britton 1980 (identification), Hindwood and Hoskin 1954 (winter ecology), Webb and Conry 1979 (Colorado record and identification).

PURPLE SANDPIPER *Calidris maritima*

The PURPLE SANDPIPER might occur in the Northwest, as it breeds in the eastern Canadian Arctic west to Melville Island (as well as east to northwestern Siberia) and winters around the North Atlantic as far south as Virginia and rarely Florida on the American coast. Stragglers reach the Great Lakes with regularity, with westernmost records from south-central Saskatchewan, Minnesota, Iowa, Oklahoma, and Texas.

This species is closely related to the ROCK SANDPIPER, is identically proportioned, has a similar molt and migration schedule, and has been considered the same species by some authors. In fact, it is more similar to some subspecies of the ROCK SANDPIPER than they are to each other (see that species). So far no individual of this species group has been observed between the Pacific Coast and the eastern edge of the Great Plains, but any bird of the group in the Northwest interior should be carefully checked and captured if possible, as PURPLE might be as likely as or more likely than ROCK in such a situation. The converse may be true, of course, and one or more of the PURPLE SANDPIPERS recorded from central North America could have been ROCK SANDPIPERS!

In breeding plumage, PURPLE SANDPIPERS are colored most like duller individuals of the Aleutian subspecies (*couesi*) of ROCK SANDPIPER; however, most birds have the breast evenly streaked, either heavily or lightly, without the conspicuously darker lower breast of the typical ROCK SANDPIPER. Duller birds are not much brighter than those in nonbreeding plumage. In the most brightly colored birds, the scapulars are broadly fringed with white, buff, or rufous. Most parts of the bird are colored like the same parts of a ROCK SANDPIPER. It appears, however, that the legs and bill base may become brighter in breeding plumage, even orange, while those of the ROCK become duller, toward gray. This difference would be useful only for summer birds, especially unlikely to occur in the Northwest.

In nonbreeding plumage, most likely to be seen in our region, PURPLE SANDPIPERS on the average have plainer and slightly paler breasts than ROCK SANDPIPERS, with less distinct spotting at the lower edge of the breast. I have yet to see a ROCK SANDPIPER (except the big, paler-gray *ptilocnemis* from the Bering Sea) with as pale and unspotted a breast as is the case in many PURPLE SANDPIPERS, but some PURPLES are colored just like typical ROCKS. The legs and bill base of even nonbreeding PURPLES tend more toward orange than do those of ROCKS, and orange legs on an interior bird would be at least weak confirmation of its being a PURPLE.

Juveniles, highly unlikely to be seen here, are colored much like juvenile ROCK SANDPIPERS but are typically more heavily streaked on the breast and not washed with buff in the same area.

Although some identification guides stress the PURPLE'S "weak wing stripe" in flight, the wing stripe is in fact moderately conspicuous, about as much so as that of the ROCK. Finally, if flight calls of the two species are different, no one has yet so determined. A series of photographs would be the minimum evidence necessary to document the occurrence of this species in the region, should a likely bird be observed in the interior, and it would probably be almost impossible to pick one out of a flock of ROCK SANDPIPERS on the coast.

Note. The "winter" individual in Hammond and Everett (1980: 120) is a juvenile, still showing nestling down on its head.

Reference
McKee 1982 (winter feeding).

ROCK SANDPIPER *Calidris ptilocnemis*

Dunlin-shaped and Surfbird-colored, this least common of the "rockpipers" can be missed in flocks of larger, showier Black Turnstones and Surfbirds on wave-washed jetties and rocky shores. Its drab, dark winter plumage is relieved briefly in spring by bright rufous backs and dark breast patches; then the birds disappear, not to return until late autumn.

	Jan	Feb	Mar	Apr	May	Jun	Jul	Aug	Sep	Oct	Nov	Dec
COAST	■	■	■	■						■	■	■

Distribution
The Rock Sandpiper breeds primarily on upland tundra on Bering Sea islands and the adjacent mainland of the Chukotsk Peninsula and western Alaska. Most island birds are residents, the mainland birds wintering south along the Pacific coast to northern California.

Northwest Status
These birds winter along the coast from early October (mid September on northern Vancouver Island, late October in Oregon) to early May. Only rarely have individuals been seen in other months, but a few, probably unusually early migrants, have been reported in August and early September on British Columbia and Washington coasts. In Victoria it appears to be more common in the latter part of the winter, from February to May, but elsewhere numbers seem uniform throughout the period, for example at the Ocean Shores jetty, where 10 to 25 birds formerly were present throughout the winter (fewer now).

This species is much less common and widespread than the ubiquitous Black Turnstone and the locally common Surfbird. It is virtually lacking from anywhere on protected waters east of Victoria, no more than one or two birds turning up here and there all the way south to the Seattle area. Generally Rock Sandpipers decrease from north to south in the region, which is south of the primary wintering range of the species. Like the Surfbird, this species is unusual in being fairly common on the coast but unrecorded from the interior.

Christmas Bird Count data indicate a decrease in regional wintering populations in recent years, perhaps the clearest trend of any common Northwest species (and this is before the major oil spills in its range). Four of 5 possible highest counts were obtained during the 1974-78 period and 5 lowest counts during the 1984-88 period. The average number of Rock Sandpipers recorded on the 5 counts analyzed was 15.6 for 1974-78, 10.4 for 1979-83, and 3.6 for 1984-88.

COAST SPRING. *High count* 82 at Victoria, BC, May 6, 1974. *Late date* Ocean Shores, WA, May 24, 1983.

COAST FALL. *Early dates* Trevor Channel, BC, August 9, 1968; Mitlenatch Island, BC, August 4, 1973; Ocean Shores, WA, August 12, 1972; Westport, WA, August 11, 1974. *High count* 60 at Port Hardy, BC, September 18, 1940.

COAST WINTER. *High counts* 70 at Victoria, BC, January 27, 1953; 90 at Chain Islets, BC, December 27, 1980; 50 at Ocean Shores, WA, March 26, 1977; 18 at Tillamook Bay, OR, December 20, 1980.

CHRISTMAS BIRD COUNTS	Five-Year Averages		
	74-78	79-83	84-88
Victoria, BC	13	15	7
Grays Harbor, WA	33	24	6
Coos Bay, OR	7	2	
Tillamook Bay, OR	10	9	4
Yaquina Bay, OR	11	2	1

Habitat and Behavior

Perhaps because it is less common, this sandpiper is even less likely to be seen off rock substrates than the Black Turnstone or Surfbird—the epitome of an ecologically stereotyped species. It is one of the most cryptic shorebirds, very difficult to see on the rocks it matches so well. Although commonly flying with the other rock species, Rock Sandpipers may feed in separate little groups away from them. It forages in typical sandpiper fashion, moving slowly about on mussel and barnacle beds and especially on large mats of foliose algae. Fairly short-billed for a sandpiper, it finds its prey—a wide variety of motile invertebrates—by sight and captures it by picking. Roosts are on rocks near feeding grounds, just above the high-tide spray with Surfbirds and Black Turnstones.

Structure

Size at high end of small. Bill medium and slightly decurved, legs short. Wings relatively short, falling well short of tip of tail; primary projection moderate to long in juveniles and adults. Tail rounded, varying to wedge-shaped, with slightly projecting pointed central rectrices.

Plumage

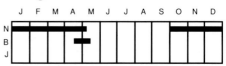

Breeding. Bill black, in some individuals yellowish at base. Legs olive to gray (gray color seen only late in spring in our region). Crown, mantle, and scapulars black-centered and rufous-edged; mantle and scapulars also whitish-tipped. Both brightness and extent of rufous vary considerably (see below under Subspecies). Coverts and tertials dark gray-brown with whitish fringes, probably all feathers from nonbreeding plumage. Sides of head and throat vary from dark, with poorly defined supercilium, to pale with prominent supercilium and dark lores and ear coverts. Breast mottled dark gray to blackish on white, black more concentrated into solid patch on lower breast. Belly white, with gray streaks and spots all across it.

Lower back, central uppertail coverts, and central pair of rectrices black; outer uppertail coverts white; outer rectrices brownish gray, narrowly white-fringed. Wing dark brownish gray above with white wing stripe formed from broad white tips on greater secondary coverts, narrower white tips on inner greater primary coverts, white bases of secondaries, and white linear markings on outer webs of inner primaries. Paler inner webs of outer primaries enhance stripe, which varies in width. From below, tips of secondaries, most of primaries, and all but tips of greater primary coverts gray-brown; marginal coverts dark gray-brown; median primary coverts, secondary coverts, and axillars white.

Breeding plumage assumed during April, with birds in latter part of that month and early May usually fully molted. Some birds leave area while apparently still in nonbreeding plumage (immatures?). At least one August bird in breeding plumage.

Nonbreeding (Figure 36). Bill with distinct yellow base (mostly yellow at extreme), legs dull to fairly bright yellow. Head and breast gray, mantle feathers and scapulars black with more or less extensive gray fringes. More narrowly fringed individuals look black-blotched above, more broadly fringed ones entirely gray. Coverts and tertials gray with whitish fringes. Lower breast, sides, and, in some individuals, entire belly heavily spotted with dark gray; spots often arranged in lines. Molt into nonbreeding plumage during August and September on or near breeding grounds, after which adults move south into our region. As in Dunlin, flight feathers molted before migration.

Juvenal. Bill and legs as in nonbreeding adults. This plumage, perhaps never seen in Northwest, much like same plumage of other *Calidris*, with overall effect of conspicuously scalloped upperparts. Crown striped black on brown; mantle feathers black, fringed with buff to rufous. Scapulars and tertials black and coverts dark gray, all fringed with buff. Face and hindneck gray with poorly defined white supercilium. Upper breast gray-brown, lower breast buff, and belly white; breast and sides heavily streaked with brown.

Juveniles, like adults, undergo complete body molt on or near breeding grounds and arrive in our region in first-winter plumage. This plumage indistinguishable from that of adult, although some earliest-migrating individuals retain some rufous-fringed coverts or tertials and other remnants of juvenal plumage such as a speckled whitish breast. Such birds seen as late as early November, contrasting with otherwise drab gray individuals.

Immature. Virtually no individuals seen in summer south of breeding range, so maturity probably occurs in first summer, and first-year birds may be indistinguishable from adults.

Subspecies

The issue of what subspecies of Rock Sandpiper winters in the Northwest is not fully settled. The three North American subspecies breed on the Pribilof, St. Matthew, and Hall islands (*C. p. ptilocnemis*); on St. Lawrence and Nunivak islands and the western Alaska mainland from the Seward Peninsula to the Yukon delta (*C. p. tschuktschorum*); and on the Aleutian Islands and the tip of the Alaska Peninsula (*C. p. couesi*). The present thought is that *ptilocnemis* and *couesi* are resident, the latter possibly moving as far as Kodiak Island in winter; and that *tschuktschorum* is migratory, wintering in the range of the other two and farther south. In the literature, both *couesi* and *tschuktschorum* have been cited from the Northwest, the former only before *tschuktschorum* had been designated as the migratory subspecies. The picture is even more complicated because some populations, for example birds on St. Lawrence Island, are not typical of the subspecies to which they have been assigned. In addition, the subspecies *C. p. quarta* of the Commander and Kurile islands off Siberia has been reported twice as a vagrant to western Alaska.

The American subspecies can be distinguished in breeding plumage (Table 31), but there appears to be considerable variation, particularly in

Table 31. Characteristics of Rock Sandpiper Subspecies

Subspecies	Size	Nonbreeding Plumage	Breeding Plumage			
			Fringes on Upperparts	Head/ Upper Breast	Spot on Ear Coverts	Patch on Lower Breast
ptilocnemis	larger	paler	broad, buff	pale	well defined	well defined
tschuktschorum	smaller	darker	broad, rufous	pale	well defined	well defined
couesi	smaller	darker	narrow, rufous	dark	obscure	obscure

Breeding Rock Sandpiper. Medium rock-inhabiting calidridine with medium-length bill and relatively short wings; reddish upperparts, dark ear spot, and black breast patch diagnostic of species and plumage; legs turning gray on this date; typical brightly marked *tschuktschorum*. Ocean Shores, WA, 26 April 1980, Dennis Paulson

Breeding Rock Sandpiper. Typical *couesi*, duller above than *tschuktschorum* with heavily marked neck and little evidence of ear spot. Ocean Shores, WA, 3 May 1986, Jim Erckmann

couesi (part of the apparent variation may be because of late-migrant specimens of *tschuktschorum* taken within the range of *couesi*). The variation is so great in British Columbia and Washington specimens and photographs of birds in breeding plumage that it seems almost certain that both *tschuktschorum* and *couesi* visit our region. Our birds vary from dull and dark, with narrow rufous fringes above and heavily marked head and upper breast, as is typical of *couesi*, to much brighter and paler, with wide rufous fringes above and essentially white face and upper breast relieved only by dark ear patches and breast blotch, as is typical of *tschuktschorum*. Measurements of specimens indicate they are too small to be the Pribilof subspecies *ptilocnemis*. Further work with specimens will be necessary to resolve this question. The very brightly colored *quarta*, unlikely to occur in the Northwest, might not be distinguishable from *tschuktschorum* in the field.

Hayman et al. (1986) stated that *ptilocnemis* was distinctive in its large size, pale coloration, and well-defined ear patch and breast blotch but that the other subspecies were more like Purple

Sandpipers. This is only partly the case, as many individuals of *tschuktschorum* are as brightly marked as *ptilocnemis*, although darker; they look superficially more like *ptilocnemis* than like *couesi*. The underwing of *ptilocnemis* is paler than in the other subspecies, with both marginal and greater primary coverts paler, looking exactly like a Dunlin underwing. The difference is not as great in specimens I have examined, however, as was illustrated in Hayman et al. (1986). A distinctive feature of *ptilocnemis* may be its tendency to be a prober as well as a picker. On the Pribilof Islands, Rock Sandpipers commonly occur on sandy and muddy substrates, where they probe like Dunlins; they also feed on rocks there. Note that Dunlins are uncommon in the Pribilofs.

Although *tschuktschorum* is considered the migratory subspecies, the others also move between breeding and wintering grounds, *ptilocnemis* to the shore of Bristol Bay and as far south on the Alaska coast as Juneau and Wrangell. In any plumage the latter subspecies should be distinguishable from the others by its larger size and paler coloration and should be watched for in Rock Sandpiper flocks in this region. The breast/belly contrast in nonbreeding

Identification

All Plumages. The *slightly decurved medium bill, short legs, and tail extending beyond wings* are all structural field marks that would indicate this species even in silhouette. On the rocks it frequents, nothing else shows this combination: BLACK TURNSTONE has a shorter bill, SURFBIRD is larger with a shorter bill, and WANDERING TATTLER is larger with a slightly longer, straight bill; all of these species have wing tips and tail tips about even. Bear in mind that this species occasionally wanders away from rocks and other species wander to them. See PURPLE SANDPIPER for a very similar species not recorded from the region.

Breeding Plumage. Although quite variable, the plumage characteristics allow distinction from any other species. They include *much rufous on mantle and scapulars and either an entirely dark, streaked and mottled head and breast or dark ear patch and breast blotch on pale background*. The only other similar-sized and -plumaged bird is the DUNLIN, which is much brighter rufous above, lacks the ear patch and dark breast markings, and instead shows a conspicuous black belly patch.

Nonbreeding Plumage. With *dark gray back and breast, heavily spotted sides, and yellowish legs*, it is *smaller and longer-billed* than the similarly colored SURFBIRD. It can be distinguished from BLACK TURNSTONE by being *paler, longer-billed*, yellow-legged, and spotted beneath. It is also rather similar to a DUNLIN in this plumage but is *darker, with spotted sides and yellow legs*.

Juvenal Plumage. In the unlikely event of the occurrence of a bird in this plumage in the Northwest, it could be distinguished by *dark, buff- and rufous-scalloped back, heavily streaked breast, and yellow legs and bill base*. It is as similar to the PECTORAL SANDPIPER as any other species in this plumage, but it is much darker and scalloped instead of streaked above, and the breast streaks are not sharply demarcated from the belly.

In Flight (Figure 12, 37)

ROCK SANDPIPERS always look *dark with conspicuous white wing stripes and striped tail*, without the white tail bases of the similarly dark SURFBIRDS and BLACK TURNSTONES with which they usually fly. They are much darker than DUN-

Nonbreeding Rock Sandpiper. Smallest, droopiest-billed, and shortest-legged of rock shorebirds; streaked breast and rufous-fringed scapulars retained from juvenal plumage indicate first-winter bird. Ocean Shores, WA, 4 October 1980, Dennis Paulson

plumage is more like that in a Dunlin, and the sides are similarly less spotted than in *couesi* or *tschuktschorum*. If my observations in the Pribilofs are indicative of the habitat preference of this subspecies, it might be as likely to occur with Dunlins as with the other rock shorebirds.

It would be impossible to distinguish *couesi* in nonbreeding plumage from *tschuktschorum*, and literature records of *couesi* in nonbreeding plumage from the Pribilofs are more likely to represent *tschuktschorum*. The only listed difference between *couesi* and *tschuktschorum*—the former darker above because of narrower gray fringes (Gabrielson and Lincoln 1959)—does not hold in the specimens I have examined.

Curiously, the width of the wing stripe varies substantially in Washington winter specimens, and, as *tschuktschorum* and *couesi* cannot be distinguished in nonbreeding plumage, we have no way of knowing if this variation is geographic or individual. The wing stripe is wider in *ptilocnemis* than in the other forms, but one Washington specimen, clearly not *ptilocnemis* because of its color and size, has a stripe almost as wide as in that subspecies.

LINS in flight, with wing stripes typically wider (the possibly occurring Pribilof subspecies is as pale as a DUNLIN but grayer, with a much more conspicuous wing stripe). The rump and center of the long tail are quite dark, contrasting substantially with the pale gray of the outer tail feathers.

Voice

This species is rarely heard in the Northwest, perhaps not needing a flocking call when in the middle of flocks of noisy BLACK TURNSTONES. I described one call heard in Washington as "a soft *peep* like the second part of a SEMIPALMATED PLOVER call," another as "a *creek* note somewhat like a subdued COMMON SNIPE call." Birds in monospecific flocks were very vocal in late summer in the Pribilofs. The call there is *chu-ree*, often rolled in the middle, like a hybrid between a SEMIPALMATED PLOVER and a BAIRD'S SANDPIPER.

Further Questions

The subspecies question remains an important one. Birds in breeding plumage should be assessed in terms of variation, described by fairly detailed notes or good photographs. As rock shorebirds are relatively tame, photography is easy for anyone willing to inch her or his way across enough jetty rocks. Late April or early May would be the appropriate time. Individuals of the Pribilof subspecies *ptilocnemis* should be sought in wintering flocks, and much more information is needed about possible differences in behavior or vocalizations between that subspecies and the others.

Incoming fall flocks (late September and October) should be scrutinized for brightly marked, late-molting juveniles. And, of considerable interest, any Rock Sandpiper seen between mid May and mid September should be carefully described and photographed if possible to determine age and plumage status.

Are Rock Sandpiper populations pulling back from the Northwest (evidence for global warming?), or does the decrease in our peripheral populations indicate a decrease in Rock Sandpipers overall? This species needs to be watched.

Notes

Most identification books have stressed the yellow legs and bill base of this species, but in breeding birds the bill may be entirely black, the legs fairly dark gray. The illustrations in Hayman et al. (1986) are misleading in this instance, as they show one subspecies in breeding plumage with dark legs and bill and the others with yellow legs and bill base. The white wing stripe is formed by the white secondary bases as well as the tips of the greater coverts; no published illustration of this species in flight has been accurate in this regard.

Photos

Published photographs of breeding-plumaged Rock Sandpipers from Alaska include *C. p. ptilocnemis* from St. Paul, Pribilof Islands, in Farrand (1983: 393); *tschuktschorum* from Wales, Seward Peninsula, in Armstrong (1983: 154); *tschuktschorum* from Old Chevak, Yukon Delta, in Terres (1980: 799); and *couesi* from Nelson Lagoon, Alaska Peninsula, in Udvardy (1977: Pl. 211). The "Redshanks" in flight in Hosking and Hale (1983: 131) are Rock or Purple sandpipers.

References

Conover 1944a (subspecies), Gabrielson and Lincoln 1959 (subspecies distribution), Hanna 1921 (breeding biology), Myers et al. 1982 (general biology).

DUNLIN *Calidris alpina*

Dun-colored blankets of Dunlins warm the chilly shores at times when other shorebird species have flown to warmer climes. The frumpy browns of winter gradually metamorphose to the dazzling red backs and bold black belly patches of spring, as the flocks join those of Western Sandpipers in a stream flowing ever northward to tundra breeding territories. Enjoy the colors and the buzzy Dunlin songs of spring, as there will be no trace of them in the legions returning in October.

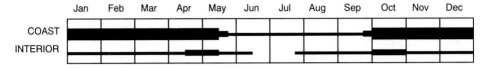

Distribution

The Dunlin breeds around the world at high latitudes, North American populations confined to arctic tundra from western Alaska to Hudson Bay. The range swings farther south in western Europe than elsewhere, with many birds breeding in non-tundra habitats. The species winters on northern-hemisphere coasts, in the New World from British Columbia and New England south to northern Mexico on the Pacific coast and Texas and Florida on the Atlantic and Gulf coasts.

Northwest Status

Although *the* winter shorebird of the Northwest coast, this species is actually quite locally distributed considering its abundance. Few birds are encountered in some estuaries that look just like others that support flocks of thousands or tens of thousands. Perhaps the advantage of occurring in sizable flocks as a strategy to detect and avoid raptors is greater than the advantage of extra food that might be available to small groups that pioneered new mud flats. Thus the Fraser River delta, Bellingham and Skagit bays, Grays Harbor, Willapa Bay, and the Columbia River estuary all consistently support flocks of 5,000 or more Dunlins during the winter. Wintering birds include both adults and immatures, with a substantially higher proportion of the former, and both sexes, with a slightly higher proportion of males. A higher proportion of immatures in November than later in winter on the outer coast of Washington was thought to indicate continued southbound migration of that age class.

In addition to the marine populations, flocks of dozens to hundreds of birds winter in freshwater environments in the Puget Sound lowlands and lower parts of the Willamette Valley. Up to 2,000 Dunlins feed at Burnaby Lake, near Vancouver, and high counts have reached thousands in the Willamette Valley. Dunlins are also common during spring migration in that area.

Unlike most shorebirds of our region, there is no early fall influx of adult Dunlins, although a flock of 15 at Ocean Shores on August 4, 1979*, and not seen the week before, may have been unusually early migrants. Normally, small flocks move into the region in late September, and not until mid October do the really big flocks, containing adults and juveniles already molted into nonbreeding plumage, begin to arrive from the north. British Columbia's Boundary Bay receives the largest influx, but southbound migrants also accumulate on the outer coast; most birds move into protected estuaries as winter ensues.

Wintering Dunlins from the California coast apparently move north to the more favorable foraging grounds of our estuaries as early as late January. For example, numbers on Leadbetter Point increased from a midwinter low of about 5,000 to 12,500 on January 31 and 19,000 on February 26, 1979. In the Fraser River estuary, numbers decrease from December to January and increase again in February to peak in March. Outer-coast birds are further augmented by even larger numbers in April and early May, with late April peaks. By mid May most of them are gone, and only scattered individuals (up to a half-dozen birds in one place) are seen throughout the summer on the coast, often in odd plumages and presumably nonbreeding subadults and/or injured birds. Slightly larger numbers have summered at Leadbetter Point and up to hundreds in the Fraser River estuary.

Breeding Dunlin. Medium calidridine with long, droopy bill and short, black legs; mostly rufous upperparts and large black belly patch diagnostic of species and plumage. Churchill, MB, June 1984, Linda M. Feltner

Nonbreeding Dunlin. Brownish upperparts and breast diagnostic of nonbreeding plumage. Seattle, WA, 22 March 1992, Dennis Paulson

Juvenile Dunlin. Streaked breast and black-patched belly diagnostic of plumage (not usually seen in Northwest), as are rufous-fringed coverts and tertials; gray-brown feathers of first winter appearing on mantle and upper scapulars. Juneau, AK, September 1978, Robert Armstrong

Dunlins are uncommon in eastern Washington and eastern Oregon, where high spring counts do not exceed a dozen or two birds except at large lakes in south-central Oregon (where they are at times common). They are rare in eastern British Columbia, Idaho, and western Montana. Most spring records occur from late April to mid May. Dunlins are more common in spring than fall in Idaho and interior Washington and Oregon, about equally rare in both seasons in interior British Columbia and Montana. Their rarity in autumn may relate to their late migration period, when freshwater environments in the interior are shutting down; alternatively, fewer observers may be searching shorebird habitats.

Fall migration dates for Malheur Refuge (July 24 to September 1, peak August 20-25) and Fortine (August 11 to September 1) are surprising and entirely at variance with the timing of migration anywhere on the coast. Records in Idaho in August and September are similarly anomalous. Most records in the interior of British Columbia and Washington fall in October and later, as on the coast. Perhaps the early interior birds are from a different breeding population than those contributing to the Pacific coast wintering populations, forced into a different schedule because of their interior migration route.

Only occasionally are Dunlins found in the interior of the region in winter, with scattered records from the Kamloops area and Okanagan Valley in British Columbia and the Columbia River Basin in Washington.

Christmas Bird Count data indicate a decrease in regional wintering populations in recent years. Ten of 23 possible highest counts were obtained during the 1974-78 period and 11 lowest counts during the 1984-88 period. At Grays Harbor, count totals varied from 40,000 to 95,500 from 1974 to 1978, then from 12,000 to 28,000 from 1979 to 1986; only 2,100 were reported in 1987, probably an incomplete census. Counts from elsewhere along the Pacific coast during the same period showed a similar drop in numbers from southern Oregon south through California.

COAST SPRING. *High counts* 12,000 at south jetty of Columbia River, OR, May 3, 1969; 30,000 at Grays Harbor, WA, May 1, 1976; 31,000 at Grays Harbor, WA, April 27, 1981*.

COAST SUMMER. *High counts* 415 at Tsawwassen, BC, July 11, 1965; 300 at Crescent Beach, BC, August 28, 1965.

COAST FALL. *High counts* 75,000 at Crescent Beach, BC, October 30, 1976; 14,200 at Leadbetter Point, WA, November 4, 1978; 22,000 at Ocean Shores, WA, October 27, 1979*.

COAST WINTER. *High counts* 60,000 at Mud Bay, BC, December 1, 1978; 100,000 at Grays Harbor, WA, December 17-18, 1960; 19,000 at Leadbetter Point, WA, February 26, 1979; 20,000 at the mouth of the Columbia River,

WA/OR, December 21, 1980; 7,600 at Tillamook Bay, OR, December 18, 1976; 11,000 at Coos Bay, OR, December 17, 1977; 2,500 at Corvallis, OR, December 23, 1980.

INTERIOR SPRING. *Early dates* Malheur Refuge, OR, March 28, 1985; Okanagan Landing, BC, April 5, 1941. *High count* 1,680 at Summer Lake, OR, April 21-22, 1987. *Late date* Sirdar, BC, June 14, 1981.

INTERIOR FALL. *Early dates* Malheur Refuge, OR, July 24, 1973; Nampa, ID, September 9, 1979; Fortine, MT, August 10 and 21, 1975.

INTERIOR WINTER. *High counts* 7 at Vernon, BC, December 29, 1962; 7 at Kamloops, BC, December 25, 1979; 10 at Kamloops, BC, 1987-88; 70 at Yakima River delta, WA, January 1, 1982.

CHRISTMAS BIRD COUNTS	Five-Year Averages		
	74-78	79-83	84-88
Comox, BC	1,037	1,124	467
Deep Bay, BC	440	356	331
Duncan, BC	259	307	118
Ladner, BC	25,407	19,376	27,980
Nanaimo, BC	170	12	1
Pitt Meadows, BC	264	350	159
Vancouver, BC	10,272	2,341	1,943
Victoria, BC	134	279	200
White Rock, BC	8,778	8,578	9,474
Bellingham, WA	3,170	5,832	5,112
Everett, WA	522	1,204	2,082
Grays Harbor, WA	63,164	19,892	20,130
Kitsap County, WA	286	261	261
Leadbetter Point, WA	6,410	10,370	10,255
Seattle, WA	18	60	76
Sequim-Dungeness, WA	2,553	814	3,150
Tacoma, WA	167	234	140
Coos Bay, OR	4,318	2,318	1,338
Corvallis, OR	200	739	1,085
Eugene, OR	1,422	165	1,249
Sauvie Island, OR	1,887	684	105
Tillamook Bay, OR	4,848	4,189	3,409
Yaquina Bay, OR	439	86	239

Habitat and Behavior

Small groups are scattered everywhere in coastal habitats, primarily on mud flats but in small numbers on gravelly and sandy beaches, rocky points, and flooded fields near the shore. Perhaps their very abundance spreads Dunlins out into many peripheral habitats, although the difficulty for a shorebird in making a living during a Northwest winter may be in part responsible. In any case, Dunlins at times are found with Sanderlings, Killdeers, and Surfbirds as well as, more typically, Black-bellied Plovers and Western Sandpipers. Mud flats are optimal foraging habitats, but by no means are Dunlins found on all of them. With low tides occurring at night during the winter, Dunlins are seen roosting during much of the winter daylight period. Roost sites include ocean beaches, spits, islands, log booms, and upland fields, at times the birds gathering in huge flocks, packed densely enough to recall a brown carpet.

When a roosting flock flushes, of regular occurrence even in the absence of predators, they put on an aerial ballet that must be seen to be believed, much less appreciated. At any time during the winter these flocks can be observed in Northwest bays, and, during that season, it is safe to assume that huge flocks in the distance are Dunlins. They twist and turn in the air, spreading and coalescing, bulging at any point and flowing into a new shape, the brown upperparts and white underparts visible alternately. At a distance only the white flashes can be seen, disappearing and reappearing even as their shape changes. Cinematographic studies have shown that both flashing and changes of flight direction are initiated at one edge of the flock and are not simultaneous, as it appears to us. It is apparently normal behavior for flocks to remain in the air for many minutes on end, even when not harassed, although the function of this behavior remains obscure. The best shows are put on when a Merlin or Peregrine Falcon is in pursuit.

Dunlins forage like typical long-billed sandpipers, moving slowly over the substrate with head down, probing or picking for invertebrates. They feed in mud both exposed and covered by shallow water, and forage similarly in moist soil in plowed fields. Some feeding motions are dowitcherlike, with rapid stitching in one spot. Amphipod crustaceans are important food in many areas.

Structure (Figure 45)

Size small. Bill medium, distinctly drooped toward tip. Bill length of females averages about 4 mm longer than that of males but substantial

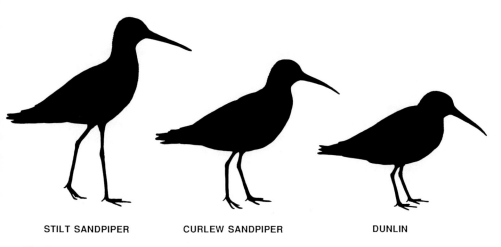

STILT SANDPIPER   CURLEW SANDPIPER   DUNLIN

Fig. 45. Long-billed *Calidris* shapes. Differences in bill shape, neck length, and leg length.

overlap; very long-billed or short-billed individuals should be assignable to sex (but subspecies vary, see below). Legs short. Wings reach tail tip, primary projection moderate to long in adults and juveniles. Tail double-notched, with projecting, pointed central rectrices.

Plumage

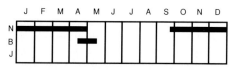

Breeding. Bill and legs black. Bright in breeding plumage, with rich reddish back. Crown and mantle feathers rufous (crown slightly duller) with black streaks, hindneck grayer. Fresh scapulars and tertials rufous, variably marked with black central and subterminal markings; lower scapulars in some individuals unmarked rufous. Many scapulars and fresh tertials with buff tips that produce slight spangling on upperparts. Coverts and some unmolted tertials retain brown of winter. Face, neck, and breast white with dark streaks; faintly darker loral stripe (but face looks unmarked at a distance). Large, usually solid black, blotch occupies entire center of belly; sides and undertail coverts white.

Lower back and central uppertail coverts brown, fringed with rufous; outer uppertail coverts white. Central pair of rectrices brown, outer ones brownish gray. Wing brown above; white

wing stripe formed from bases of inner secondaries, moderately broad tips of greater secondary coverts, narrower tips of inner greater primary coverts, and white linear markings on outer webs of inner four or five primaries; pale inner webs of outer primaries add to conspicuousness of stripe. Underwing coverts and axillars white; tips of secondaries and most of primaries light gray-brown, marginal coverts slightly darker brown.

Breeding plumage acquired during April, black belly patch before rufous back. Autumn molt on or near breeding grounds, so fall migrants arrive in Northwest in nonbreeding plumage. Even flight feathers molted prior to migration, all primaries replaced by end of August. Wing molt before migration unusual among arctic/subarctic-breeding shorebirds, shared only by Rock and Purple sandpipers.

Nonbreeding (Figure 15). Very dull plumage, with virtually unmarked gray-brown upperparts, neck, and breast and white belly and undertail coverts. Coverts fringed with whitish and breast indistinctly streaked with darker brown in fresh plumage. By late winter, covert fringes worn off and breast streaks more distinct. Faint whitish supercilium and dark loral stripe furnish only pattern on head.

Juvenal. Much brighter than nonbreeding adult. Crown reddish brown, streaked with black; mantle feathers and scapulars black with rufous or whitish fringes, not dissimilar to pattern of juvenile White-rumped Sandpiper. White mantle lines apparent in some individuals, less often white scapular lines. Coverts and tertials gray-brown with buff fringes. Whitish supercilium and dark loral stripe. Neck and upper breast buffy brown with dark brown streaks. Belly blotched with dark brown and black corresponding to breeding adult's black belly patch; posterior belly and undertail coverts white. Like adult breeding plumage, replaced by drab nonbreeding plumage in Alaska and not normally seen in our region.

A few juveniles have not yet completed body molt upon arrival here in September and early October, at beginning of migration period. May be recognizable by retention of few dark, buff-fringed scapulars and/or black belly spots, these feathers already replaced by adults.

Immature. First-winter birds indistinguishable from adults in field. Western Alaskan juveniles grow wing and tail feathers in July and August, adults one to two months later. Juvenile feathers often less sturdy, thus always a bit more worn in first-year birds, conspicuously so by following spring. In hand, particularly if comparisons possible, age classes easily distinguishable by degree of wear of primaries and rectrices. Differences in lesser wing coverts, discussed in literature for other populations of species, useless for distinguishing age classes in the field in this region.

Most Pacific coast Dunlins must breed in first year, as relatively few summering individuals seen anywhere in winter range. Similarly small proportion of eastern Dunlin populations spends summer on Gulf of Mexico coast, apparently first-year individuals. Individuals seen during summer in Northwest usually exhibit features of both breeding and nonbreeding plumages, often combining plain brown back with black belly patch. May be sick or injured, first-year birds that failed to migrate, and/or early fall migrants.

Subspecies

Only one subspecies is known to occur in the Northwest, the population that breeds in western Alaska (*C. a. pacifica*). This is the largest, longest-billed, and most richly colored subspecies of the species. *C. a. pacifica* has been characterized by the presence of a narrow, unmarked white band between the streaked breast and the black belly patch, but many Northwest specimens that appear to be this subspecies because of their long

bills and unstreaked undertail coverts have breast streaks contiguous with their belly patches, so this characteristic is not diagnostic.

The northern Alaskan birds that winter in Asia (*C. a. sakhalina*) are colored like *pacifica* but very slightly smaller, averaging about 3-4 mm shorter in bill length than those of *pacifica*. They could readily occur in our region as vagrants but could only be distinguished from individuals of *pacifica* in the hand by knowledge of both their sex and bill length; a male would look short-billed for a Dunlin, however. Unlike other North American Dunlins, *sakhalina* may migrate to the wintering grounds while still in breeding plumage; for example, I saw large numbers of Dunlins in breeding plumage at Osaka, Japan, in early August 1980. Thus a breeding-plumaged Dunlin in August, if an arriving migrant rather than a summering bird, would be suspect.

Finally, birds that breed in central Canada and winter in eastern North America (*C. a. hudsonia*) are the same size and color as *pacifica* but typically have finely streaked posterior sides and undertail coverts. Individuals of *pacifica* occasionally have a few very fine shaft streaks in that area, but *hudsonia* are distinctly streaked there. Furthermore, individuals of *hudsonia* in nonbreeding plumage have small but distinct brown dots arranged in lines along the sides that are rarely if ever present in *pacifica*. A Dunlin reminiscent of a Rock Sandpiper in this way is likely to be a *hudsonia*, again only present here as a vagrant (perhaps most likely in the interior). Juveniles of the three subspecies, unlikely to occur here in any case, have not been distinguished.

One additional note: European Dunlins are smaller than American ones and tend to be duller, even buff instead of rufous above, with more heavily marked breasts, the breast markings blending with the belly patch in extreme cases. The slightly smaller *alpina* from Scandinavia and the considerably smaller and shorter-billed *arctica* from northeastern Greenland have been recorded from the Atlantic coast of North America, although neither is likely to reach the Northwest. Birds from European populations migrate earlier in autumn than North American ones, molting after they migrate like most other shorebirds.

Identification

All Plumages. In all plumages a *medium-length, droopy-tipped bill and black legs* are a unique combination among common small Northwest shorebirds. Only the casual CURLEW SANDPIPER shares these characteristics.

Breeding Plumage. The *rich rufous upperparts, light face and breast, and black belly patch* are an unmistakable combination, even if the distinctive bill is not seen. Only a ROCK SANDPIPER in breeding plumage is anything like this species, and its back is less bright, its face and neck are dark (or, if pale, there is a distinct dark ear patch), and the dark patch on the underparts is on the breast rather than belly. RUDDY TURNSTONES are similarly bright rufous, black, and white but are short-billed and black on the head and breast, not the belly.

Nonbreeding Plumage. This is the plainest of our winter shorebirds, a smallish sandpiper with *drab gray-brown upperparts and breast, droopy-tipped bill, and short black legs.* The only other common species at all like it is the WESTERN SANDPIPER, and the DUNLIN's *larger size, longer bill, and brownish breast* easily distinguish it, especially in the mixed flocks that are typical during the winter. The LEAST SANDPIPER is colored somewhat more like a DUNLIN but is even smaller than the WESTERN, with shorter bill and yellow legs. See also CURLEW and STILT SANDPIPERS.

Juvenal Plumage. Although individuals in this plumage are unlikely to occur here, they would be recognizable by their *diffuse black belly patch* together with the *brightly marked upperparts* typical of juvenile *Calidris*. DUNLIN size and bill shape should provide clues as well.

In Flight (Figure 12)

This smallish sandpiper is *stripe-winged and stripe-tailed, brown above* (fall and winter) *or reddish above with black belly patch* (spring). In comparison with species with which it might be seen, the wing stripe is more conspicuous than in the smaller WESTERN and LEAST SANDPIPERS, less conspicuous than in the SANDERLING, and slightly less conspicuous than in the ROCK SANDPIPER, the last two about the same size. The DUNLIN's browner coloration distinguishes it from the paler SANDERLING and darker, grayer

ROCK SANDPIPER. The winter color is similar to that of SEMIPALMATED PLOVER, but the DUNLIN is larger than that species, with long bill and different tail pattern.

Voice

The flight call is a rasping *cheeezp*, fairly loud and far carrying. Of regional shorebirds, this call is most like the WESTERN SANDPIPER'S call but more raspy. Birds often sing a rhythmically pulsed buzzy song in spring migration.

Further Questions

Observers should look for Dunlins in August and September to see if juveniles ever appear in our region before they molt. To determine if breeding-plumaged adults ever do the same, an area would have to be checked regularly to make sure there were no summering individuals. A Dunlin in full breeding plumage in August might as likely be a vagrant of the northern Alaskan subspecies *sakhalina* as an early-migrating *pacifica*, but only capture would allow distinction.

Observers who enjoy watching Dunlin flocks can also be on the lookout for individuals of the eastern subspecies *hudsonia*. Check for well-defined undertail covert streaks in breeding-plumaged birds and finely spotted sides in nonbreeding-plumaged birds. A specimen or excellent photograph would be needed to confirm such a record, which is probably more likely in the eastern interior of the region.

A few Dunlins have been seen on the Gulf coast in late October and early November with extensive remnants of breeding plumage, including almost full black belly patches, but completely molted primaries. I know of no instances of this in Pacific coast birds, and observers should watch for and document such individuals just to determine their frequency in this population. Anyone handling large numbers of Dunlins in the fall should check age ratios, as there is some evidence that immatures arrive in numbers before adults do on the Pacific coast.

The decrease in numbers in wintering Dunlins at Grays Harbor and elsewhere along the Pacific coast in the last decade is cause for some concern. Populations should be closely monitored, at least by regular analysis of Christmas Bird Counts if not by more detailed censuses.

A significant ecological question to answer is why Dunlins winter successfully so far north compared with other mud-flat sandpipers. Or, to turn the question around, why can't the other species winter here? Does competitive interaction with the Dunlin limit Western Sandpipers and dowitchers, for example, or do Dunlins feed on particular organisms or in particular ways that fit them especially well to winter at these latitudes? Low tides in winter occur at night; is this when all or most Dunlin feeding takes place during that season?

Photos

Keith and Gooders (1980: Pl. 90) show a dull juvenile, Hammond and Everett (1980: 122) a bright one. The juvenile in Farrand (1983: 395) is well into its first-winter plumage; the immature in Armstrong (1983: 156) is entirely into it. The "immature" in Udvardy (1977: Pl. 231) is a nonbreeding adult, its streaked sides indicating the *hudsonia* subspecies. Many of the "Sanderlings" in Hosking and Hale (1983: 116-17) are Dunlins, a good comparison at rest and in flight.

References

Baker 1977 (summer diet), Baker and Baker 1973 (summer and winter foraging behavior), Bengtson and Svensson 1968 (prey choice in migration), Brennan et al. 1984 (sexing by measurements), Browning 1977 (geographic variation in North American populations), Buchanan 1988 (status in southern Puget Sound), Buchanan et al. 1986a (age and sex composition of wintering populations), Buchanan et al. 1986b (predation by Peregrines), Buchanan et al. 1988 (predation by Merlins), Burton 1974 (winter foraging behavior), Davis 1980 (flock coordination in flight), Dwight 1900 (molt and plumages), Ferns 1981 (identification of European subspecies), Fry 1980 (winter ecology), Goss-Custard 1970 (winter foraging and spacing behavior), Goss-Custard et al. 1977 (diet), Greenwood 1983 (geographic variation in primary molt), Holmes 1966a (breeding ecology and annual cycle), Holmes 1966b (molt cycle on Pacific coast), Holmes 1966c (summer feeding ecology), Holmes 1970 (biology of arctic and subarctic populations), Holmes 1971a (breeding and molt cycles), Holmes and Pitelka 1968 (summer diet), Kaiser and Gillingham 1981 (weight), Kus et al. 1984 (population biology and Merlin predation), Lifjeld 1984 (feeding ecology in migration), Littlefield and Cornely 1984 (migration dates at Malheur Refuge), MacLean and Holmes 1971 (distribution and characteristics of North American subspecies), Miller 1983c (aerial displays and vocalizations), Norton 1972 (incubation schedules), Page 1974a (age, sex, molt, and migration), Potts 1984 (flock coordination in flight), Rands and Barkham 1981 (winter feeding behavior), Ruiz et al. 1989 (winter population structure), Senner 1976 (diet in migration), Soikkeli 1967 (breeding and population biology), Soikkeli 1970a (breeding-ground dispersal), Soikkeli 1970b (population biology), Strauch 1967 (spring migration in western Oregon).

CURLEW SANDPIPER *Calidris ferruginea*

Described at times as an elegant Dunlin, this handsome sandpiper deserves even better than that. An encounter with a delicately scalloped juvenile would be thrilling in its own right, but to come upon a fully plumaged adult, chestnut from head to tail, on a Northwest beach would be to have stepped into fortune's path. The bill is as noteworthy for its slenderness as for its slight curvature.

	Jan	Feb	Mar	Apr	May	Jun	Jul	Aug	Sep	Oct	Nov	Dec
COAST					•			▨▨▨				
INTERIOR					•							

Distribution

The Curlew Sandpiper breeds on the arctic tundra of northern Siberia and winters widely in Africa, southern Asia, and Australia.

Northwest Status

This Siberian species visits our region rarely but just about annually in recent years, with a total of 21 records. Eight have been recorded from southern British Columbia (4 in the Vancouver area and 4 from southern Vancouver Island), 5 from Washington (3 from the outer coast, one from Dungeness, and one from the interior), and 9 from Oregon (8 coastal and one Willamette Valley). There are thus 13 records from the outer coast, 7 from protected waters, and one from the Willamette Valley in the coastal subregion. Records are surprisingly numerous, considering that this species is one of the rarer Siberian shorebirds in western Alaska. It has bred in that state, however, and it is possible that some of our migrants are from a North American breeding population.

There are only two spring records, both in mid May and including the region's only interior record. Fall records range from July 11 to October 5, but, surprisingly for a vagrant, about half of them concern adult birds, largely or partially in breeding plumage. A bird seen at Tillamook on July 17, 1983, was a juvenile from its description, an amazingly early date for this age class, with the next earliest one August 20. Some September birds were noted to be juveniles, but the October 5 bird was an adult.

COAST SPRING. Leadbetter Point, WA, May 17, 1983.

COAST FALL. Iona Island, BC, July 30-31, 1977; Comox, BC, July 11, 1981; Victoria, BC, July 14-24, 1981; Iona Island, BC, August 31 to September 1, 1981, September 17, 1983, September 4, 1984; Long Beach, BC, September 20, 1987; Pacific Rim National Park, BC, September 2, 1990; Ocean Shores, WA, October 5, 1979; Dungeness, WA, July 29, 1984; Ocean Shores, WA, September 19, 1990; Sauvie Island, OR, August 11-18, 1965; Yaquina Bay, OR, July 21, 1976; Seven Devils Wayside, OR, August 16, 1976; south jetty of Columbia River, OR, September 16, 1982; Tillamook, OR, July 17, 1983; Bandon, OR, July 27 to August 2, 1985;

Tillamook, OR, August 17-18, 1985, August 20-24, 1985; south jetty of Columbia River, OR, September 23, 1986.

INTERIOR SPRING. Ephrata, WA, May 10, 1972.

Habitat and Behavior

Birds have been seen in grassy freshwater marshes, *Salicornia* salt marshes, and sewage lagoons, and on mud flats and sand beaches. They occur in large flocks where common but would be expected singly in this region. Their associates have ranged from stints (Western Sandpipers on several occasions) to yellowlegs and dowitchers. Curlew Sandpipers feed like other long-billed calidridines by moving along steadily and probing or stitching in shallow water or wet mud for small invertebrates. Longer-legged, they feed in water more often than does the similar Dunlin, and among the calidridines they are exceeded in foraging depth only by the Stilt Sandpiper.

Structure (Figure 45)

Size small. Bill medium to long, slender, and fairly evenly curved downward from about basal third, averaging slightly (4 mm) longer in female. Legs medium, again slightly (2 mm) longer in female. Sexes similar in size otherwise. Wings extend beyond tail tip, primary projection moderate to long in breeding adults and juveniles. Tail squarish, central rectrices projecting only very slightly and not pointed as in many other *Calidris*.

Plumage

Breeding. Bill and legs black. Some sexual dimorphism. Males rich rufous all over with exception of gray-brown coverts and many retained tertials. Crown and mantle brown, broadly streaked with black; scapulars and molted tertials vary from brown to mostly black with paired gray-brown to white spots at tips. Scattered dark bars on sides and undertail coverts.

Lower back brown; uppertail coverts white barred with black, brown, and some rufous. Tail gray-brown, narrowly fringed with white. Wing brown above; white wing stripe formed from moderately broad tips of greater secondary co-

Juvenile Curlew Sandpiper. Long-winged calidridine with long, slightly decurved bill and dark legs; evenly pale-scalloped upperparts and buff-washed breast characteristic of plumage. England, late September 1985, Urban Olsson

verts, narrower tips of inner greater primary coverts, and white linear markings on outer webs of inner five primaries. Pale inner webs of outer primaries add to conspicuousness of stripe. Underwing mostly whitish, axillars white; tips of flight feathers light gray-brown, marginal coverts slightly darker brown.

Females colored similarly, but duller above with more brown and less rufous and slightly lighter rufous below with scattered black bars and largely white belly. Supercilium may be distinctly paler than rest of head.

Molt into breeding plumage occurs during March and April. Newly molted spring individuals in April have many underpart feathers tipped with white; these tips already worn off in many May birds. Upperparts brown to mostly blackish by late summer, underparts entirely dark. Some breeding plumage retained in many birds into September, but only a few show traces in October. Several adults observed in this region in fall were molting into nonbreeding plumage, and an October bird was largely in nonbreeding plumage with scattered rufous feathers below. Because feathers of underparts molted last, relatively easy to distinguish adults of dark-bellied species such

as this when only a few breeding-plumage feathers remain late in fall.

Nonbreeding. Light gray-brown above; scapulars, tertials, and coverts with narrow white fringes when fresh. Head and mantle streaked with slightly darker gray-brown. Moderately conspicuous white supercilium and brown loral stripe. Cheeks and breast streaked with gray-brown, breast washed with light gray-brown, and rest of underparts white. Uppertail coverts white, with no trace of bars present in breeding birds. Many birds fully in this plumage by September.

Juvenal. Basically like nonbreeding plumage but upperparts distinctly scaly. Crown slightly darker than in adult; mantle feathers brown with light gray-brown fringe, scapulars light brown with brown subterminal and whitish terminal fringe, and tertials brown with whitish fringe. Breast less obviously streaked than in adult but washed with light brown to buff, belly may also be buffy.

Immature. Molt into first-winter plumage, indistinguishable from nonbreeding adult, takes place on wintering grounds; most individuals molt at least outer primaries during first winter. Most remain in worn nonbreeding plumage on wintering grounds in southern hemisphere for first summer, a few showing traces of breeding plumage.

Identification

All Plumages. This *curve-billed Calidris* is *relatively long-necked, long-legged, and long-winged* compared with others in its size category such as DUNLIN and ROCK SANDPIPER. This greater length, presumably an adaptation for wading, gives it a certain tringine elegance. The bill is considerably longer than that of a ROCK SANDPIPER but the same length as that of a DUNLIN, although more slender and evenly curved instead of drooped at the tip. The bill is also more slender and evenly curved than that of a STILT SANDPIPER, and the legs are considerably shorter.

Breeding Plumage. Among small or medium sandpipers, only the RED KNOT, DOWITCHERS, and RED PHALAROPE are similarly *cinnamon below* like this species. The CURLEW SANDPIPER is distinctly smaller than the KNOT or DOWITCHERS, with *bill long and slightly curved* rather than straight and short or very long, respectively, as in those species. Only at a distance might it be mistaken for a RED PHALAROPE, but it lacks the short, yellow bill, white cheeks, brightly striped back, and distinctive behavior of the latter species.

Nonbreeding Plumage. In this plumage CURLEW SANDPIPERS are colored somewhat like DUNLINS, from which they must be carefully distinguished. The *paler coloration, especially the breast,* is one of the best marks, although some DUNLINS are paler-breasted than others. With this difference, as well as the rather substantial shape difference, there is little likelihood that one familiar with DUNLINS would confuse the two species. Several of the region's fall adults, as late as early October, have shown patches of breeding plumage on the underparts.

Juvenal Plumage. This *scaly-backed* plumage is distinct from all white-bellied calidridines except STILT and BAIRD'S SANDPIPERS and RUFF. From STILT, CURLEW differs by having *shorter, black legs, more slender, evenly decurved bill, and longer wings* (extending beyond tail tip in CURLEW, to it in STILT). It is between RUFF and BAIRD'S in size, differing from both by its *longer,*

curved bill and *gray-brown* rather than buffy coloration. It differs further from RUFF by its black legs.

In Flight (Figure 12)

The *white rump* (subdued in breeding plumage and difficult to see against the gray tail under some lighting conditions), together with the *conspicuous white wing stripes,* distinguish it from all but the even rarer WHITE-RUMPED SANDPIPER. From that species it differs by its *slightly more conspicuous wing stripe, larger white rump patch, longer bill, and longer legs.* See that species for more details. The RUFF can look white-rumped in flight, but the white is in the shape of a "U" or "V" and does not take up the entire rump area.

Voice

The usual flight call is a musical trill, often written *chirrip* and unlike any other regularly occurring Northwest species, although somewhat like the call of the MONGOLIAN PLOVER.

Further Questions

All breeding-plumaged Curlew Sandpipers in the Northwest should be identified to sex if possible, as there is much interest in determining the distribution and migration patterns of the two sexes in shorebird species. Males leave the breeding grounds before females and probably migrate south slightly earlier. Note the age class in autumn birds, because, from the present record, temporal separation between adults and juveniles seems less marked in this species than in most shorebirds of the region.

Photos

The "winter" birds in Hammond and Everett (1980: 122) are juveniles, feeding with a juvenile Dunlin (the size comparison is inappropriate for the Northwest, as Dunlins are smaller in Europe).

References

Elliott et al. 1976 (migration and molt), Hindwood and Hoskin 1954 (winter ecology), Holmes and Pitelka 1964 (breeding behavior and taxonomic relationships), Jackson 1918 (molt), Lifjeld 1984 (feeding ecology in migration), Melville 1981 (measurements, weight, and molt),

Paton et al. 1982 (molt), Pearson et al. 1970 (weight), Puttick 1978 (winter diet), Puttick 1979 (winter foraging behavior), Puttick 1981 (sexual differences in foraging behavior), Thomas and Dartnall 1971a (winter feeding behavior), Thomas and Dartnall 1971c (molt).

STILT SANDPIPER *Calidris himantopus*

Fancifully called a dowitcher-yellowlegs hybrid, this species looks like a Lesser Yellowlegs and often associates with that species. However, it feeds like a dowitcher, making it unmistakable when one turns up at a favored sewage lagoon or farm pond. Breeding-plumaged adults are distinctive but rarely seen in the Northwest.

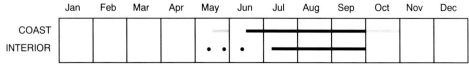

Distribution

The Stilt Sandpiper breeds on arctic tundra from northern Alaska to Hudson Bay and winters in southern South America, with outlying winter populations around the Caribbean.

Northwest Status

Juveniles of this species regularly defect from the main migration route across the Great Plains to visit the Northwest. They are not common anywhere in the region, although they are present annually in small numbers in thoroughly scrutinized areas such as Vancouver. Probably the concentration of observers near the coast is responsible for more birds being seen there than in the interior of the region, but there are many more records from the Vancouver/Puget Sound area than from the outer coast. The scarcity of records from Oregon indicates a diminution of numbers toward the south, following a pattern typical of interior migrants.

The few spring records include three from coastal British Columbia and five from the interior, centering around late May. Fall records that certainly or presumably refer to adults include at least 16 from coastal British Columbia, Washington, and Oregon (June 20 to July 28) and ten from interior Washington, Idaho, and Montana (July 7 to August 13). The coastal migration of juveniles normally occurs during the last half of August and first half of September, rarely extending into October. Stilt Sandpipers are sufficiently uncommon in the region so that single birds are usually seen, but small flocks of up to five to ten individuals are not infrequent.

COAST SPRING. Burnaby Lake, BC, May 20, 1968; Saanich, BC, May 27-29, 1978; Iona Island, BC, June 2-3, 1990.

COAST FALL. *Early date* Iona Island, BC, June 20, 1980. *Juvenile early dates* Boundary Bay, BC, August 16, 1987; Swantown, WA, August 17, 1985*. *High counts* 36 at Iona Island, BC, August 21, 1985; 22 at Boundary Bay, BC, August 16, 1987; 13 at Crockett Lake, WA, August 18, 1974; 12 at Crockett Lake, WA, September 17, 1987; 8 at Coos Bay, OR, September 1-2, 1982. *Late dates* Iona Island, BC, October 25, 1978; Kent, WA, October 17, 1987; Bandon, OR, October 22, 1989.

INTERIOR SPRING. Rupert, ID, May 13, 1919 (also two others from that state); Reardan, WA, May 24, 1973; Okanagan Landing, BC, June 9, 1927.

INTERIOR FALL. *Early date* Pablo Refuge, MT, July 7, 1979. *Juvenile early date* Reardan, WA, August 9, 1960#. *Juvenile high counts* 15 at Vernon, BC, August 22, 1932; 20 at Kamloops, BC, August 20, 1978; 10 at Moses Lake, WA, September 12, 1935; 45 at the Walla Walla River delta, WA, fall 1985; 10 at Lind Coulee, WA, August 26, 1987; 10 at the Walla Walla River delta, WA, August 30, 1987; 13 at Reardan, WA, August 26, 1989; 13 at Ninepipe Refuge, MT, August 25, 1981; 25 at Somers, MT, fall 1985. *Late date* Summer Lake, OR, September 28, 1980.

Habitat and Behavior

Stilt Sandpipers are usually seen in fresh or brackish marshes or sewage lagoons and only rarely on mud flats, even when in coastal areas. They commonly associate with Lesser Yellowlegs and dowitchers, species with which they are best compared. They feed primarily by probing, often with a rapid stitching action like other probing *Calidris*, and typically in shallow water, even up to their bellies; the head is very often submerged during the process. They move forward rather slowly, working one spot before progressing onward. Less commonly they pick insects from the water's surface like a yellowlegs.

Flocks can total up to hundreds of individuals where they are common, but group size in this region is always small. At freshwater sites, roosting takes place at the water's edge, often in mixed flocks.

Structure (Figure 45)

Size small, looks larger because of long legs. Bill medium to long, straight but slightly drooped and widened at tip. Legs long, longest of calidridines. Outer and middle toes with evident web between them, only web-footed calidridine besides Western and Semipalmated sandpipers. Wings extend beyond tail tip, primary projection moderate in breeding adults and long in juveniles. Tail double-notched, central feathers slightly projecting and pointed but not so much as in most *Calidris*.

Plumage

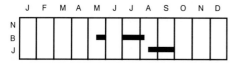

Breeding. Bill black, legs greenish (typical on breeding grounds) to dull yellow or yellowish brown (commonly in migration). Overall impression of mottled back and barred underparts, with bright rufous ear coverts. Males average brighter than females but with much overlap, probably distinguishable only in mated pairs. Feathers of upperparts dark brown with light brown to pale buff edges or fringes, crown often with much rufous. Hindneck dark-striped; mantle and scapulars dark with narrow fringes, scapulars with broader white tips (barely divided into paired spots). Pale fringes fade to whitish during summer and become narrower, so upperparts more contrasty and scaly in autumn adults than in

spring ones. Coverts and tertials (apparently not replaced in spring) gray-brown, latter darker-centered. Face and throat whitish with brown streaks, entire underparts whitish to light buff, heavily barred with brown. Ear coverts rufous, usually bright and distinct, although in some individuals crown almost as bright. Whitish supercilium contrasts strongly with darker parts of head, loral stripe brown and moderately well-defined.

Lower back brown, uppertail coverts white with dark brown bars and spots; tail gray-brown, inner webs of rectrices whitish or not. In some individuals, one or more rectrices, especially central ones, apparently molted in spring and very differently colored from others—whitish with contrasty dark brown bars. Wing brown above, narrow white tips of greater secondary coverts form very indistinct wing stripe. Underwing mostly brown; only median secondary coverts and margins of greater secondary coverts white, marginal coverts darker brown. Axillars white, some with brown central streak.

Birds in this plumage begin autumnal molt in August and may be in virtually complete non-breeding plumage by end of that month. Northwest adults all in full breeding plumage.

Nonbreeding. Legs vary from dull greenish or brownish yellow to bright yellow. A plain plumage, gray-brown above and white below. Entire upperparts light gray-brown, tertials darker. Scapulars, coverts, and tertials narrowly fringed with white, visible at close range. White supercilium distinct, dark loral stripe less so. Neck and breast lighter and grayer than back, fairly heavily streaked with black. Sides with brown streaks, outer undertail coverts with brown chevrons. Uppertail coverts white, in many individuals with scattered brown spots and bars. Tail gray-brown; inner webs of rectrices whitish, in some individuals sparsely barred with brown.

Juvenal. Legs colored as in nonbreeding adults. Head and underparts superficially similar to those of nonbreeding adults, but upperparts highly patterned, appearing either striped or scalloped depending on exact nature of feather edging. In most highly colored individuals, crown streaked with dark brown and mantle feathers, scapulars, and tertials similarly dark, often darkening from base toward tip. Mantle feathers, scapulars, and tertials fringed with buff or whitish, and in relatively few individuals crown feathers, upper scapulars, and tertials fringed with rufous. White mantle lines often conspicuous, scapular lines not usually evident. Coverts also darker than in nonbreeding adults, therefore with more contrasty whitish fringes.

Breast varies from buff to pale gray, with brown streaks varying from only barely evident to fairly conspicuous. Some individuals finely but heavily spotted and striped on sides of breast, streaks extending down sides. Rest of underparts white, but younger juveniles may be entirely washed with buff below. Uppertail coverts usually white, less often spotted than in adults; tail as in nonbreeding adults. Some birds less contrasty all over but all more heavily patterned than nonbreeding adults. Most juveniles pass through Northwest early enough so that little molting into first-winter plumage noted, but feathers of mantle and scapulars can be replaced by plain first-winter feathers by September. This molt apparently protracted, with many juvenile feathers retained into at least November.

Immature. By midwinter much like adults, but only outer primaries molted by most individuals. Maturity and breeding plumage attained in first spring.

Identification

All Plumages. The *slightly droopy bill and long, pale legs* (darker in breeding plumage) distinguish this species from all other regional sandpipers. The habit of feeding by *constant probing,* usually in water, is distinctive of this species. Of medium-sized sandpipers in the area at the same time as STILT SANDPIPERS, only DOWITCHERS feed similarly, and STILTS are *long-legged and short-billed* in comparison. LESSER YELLOW-LEGS, with approximately similar size and proportions, feed very differently as they move rapidly through the water picking prey from or below the surface.

Breeding Plumage. No other Northwest shorebird is *mottled above and heavily barred below, with rufous ear coverts.* The distinctive bill and long, greenish legs confirm the identification.

Nonbreeding Plumage. STILTS can be distinguished from DUNLINS, with similar bill, forag-

ing behavior, and overall appearance in this plumage, by their *long, yellowish legs*. In addition, the STILT is slightly paler and grayer than the DUNLIN, with a more obviously streaked breast. The two exhibit only slight seasonal overlap, but those STILT SANDPIPERS that overlapped with DUNLINS in late September and October would most likely be in a similar plumage.

Juvenal Plumage. STILT SANDPIPERS are patterned like a number of other juvenile calidridines that are *striped or scaled above*, including BAIRD'S, WHITE-RUMPED, and CURLEW SANDPIPERS and RUFF. WHITE-RUMPED and CURLEW may feed similarly, and if both bill and legs are under water, further study will be necessary. Of these species only the CURLEW has a long, slightly decurved bill, and the STILT's *longer, yellow legs* will again surely identify it. It is much *longer-billed* than the yellow-legged RUFF or the black-legged BAIRD'S and WHITE-RUMPED. RUFFS and BAIRD'S are usually dry-land feeders, moving rapidly forward and taking prey by picking, but either species may move into the STILT's watery domain at times. Note that all of these species are uncommon to rare except BAIRD'S.

In Flight (Figure 10)

Plain-winged and white-tailed, this species falls in a group that otherwise contains two species of YELLOWLEGS and WILSON'S PHALAROPE. Juveniles and nonbreeding adults look very much like LESSER YELLOWLEGS in flight, only the *slightly smaller size, slightly shorter legs, droopy bill, and hint of a wing stripe* (YELLOWLEGS shows no wing stripe at all) allowing differentiation. From WILSON'S PHALAROPE, STILT is distinguished by *heavier bill, considerably longer legs, and darker coloration*, especially the head and breast. Only the extreme toe tips project beyond the tail in the PHALAROPE, the entire toes just project beyond the tail in STILT, and a bit of the tarsus as well is usually visible in YELLOWLEGS.

Bear in mind that some STILT SANDPIPERS have somewhat darker rumps and may have some tail barring; nevertheless, the tail itself looks paler than the back.

Voice

This species is very quiet in migration, the occasionally heard flight call a single low *tew* reminis-

Juvenile Stilt Sandpiper. Medium calidridine with long, droopy bill, long neck, and long, pale legs; contrastily fringed upperparts diagnostic of plumage; leg color obscured by mud, drooped wings not typical. Reifel Refuge, BC, 2 September 1982, Jim Erckmann

cent of a single LESSER YELLOWLEGS note but not nearly so loud. This call or one like it may be rolled or trilled also, even reminiscent of the call of a CURLEW SANDPIPER.

Further Questions

This species not only looks like a Lesser Yellowlegs in flight but often occurs with it in migration, and it may be a yellowlegs mimic. With head down much of the time while feeding, the Stilt could certainly benefit from the alertness and alarm calls of the yellowlegs, and it may more readily flock with it by looking like it. Observers could note the associates of Stilt Sandpipers in the Northwest and how often Stilts actually fly with yellowlegs when feeding birds are flushed. Also, the frequency of alarm calls in the two species could be compared at these times.

Notes

A record of six at Salmon Arm, October 28, 1970, is so unusually late that I consider it dubious.

Juveniles usually have paler legs than at least breeding adults, contrary to the statement in Far-

rand (1983: 398). Several accounts have overemphasized differences in flight pattern between adults and juveniles or between breeding and nonbreeding adults. Both adults and juveniles have narrow wing stripes in fresh plumage, but that of the adult has worn off by the time the two are likely to be seen together in fall. Similarly, adults are more heavily spotted on the rump, but both age stages look white-tailed at a distance. Of the silhouettes in Hayman et al. (1986: 384), the right one is surely intended to be a Curlew Sandpiper.

Photos

The "winter plumage" individual in Bull and Farrand (1977: Pl. 200) is a juvenile.

References

Baker 1977 (summer diet), Baldassarre and Fischer 1984 (diet and foraging behavior in migration), Burton 1972 (foraging behavior in migration), Fix 1979 (occurrence in Oregon and identification), Jehl 1973 (breeding biology and taxonomic relationships), Miller 1983c (aerial displays and vocalizations).

SPOONBILL SANDPIPER *Eurynorhynchus pygmeus*

This casually occurring species is a stint in every way but one, its astonishing bill. Because of great plumage similarities, one assumes a bird like the Rufous-necked Stint as the Spoonbill's ancestor, and one can only wonder at the direction taken by the evolution of this amazing appendage. The rufous head and breast of adults simulates Rufous-necked, while a juvenile might be mistaken for a Western Sandpiper. A glimpse of the bill should remove any doubts.

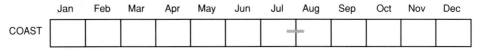

	Jan	Feb	Mar	Apr	May	Jun	Jul	Aug	Sep	Oct	Nov	Dec
COAST												

Distribution

Spoonbill Sandpipers breed on tundra in far eastern Siberia and winter locally on the coast of southern Asia; numbers have been found only in Bangladesh.

Northwest Status

The Spoonbill Sandpiper apparently does not occur anywhere in very large numbers. With only three records from Alaska (adults in May, June, and August), it would be considered unlikely to occur in the Northwest, but nevertheless an adult has been photographically documented from British Columbia. Other small Siberian sandpipers probably occur in the region more often, but, with identification so definitive, individuals of this species are more likely to be both detected and believable.

COAST FALL. Iona Island, BC, July 30 to August 3, 1978.

Habitat and Behavior

Spoonbill Sandpipers feed primarily on mud flats, where they forage by sweeping the bill back and forth through shallow water or wet mud while moving forward slowly—a feeding method much like that of their namesakes, the spoonbills. Only recently has this foraging behavior been noted; before that, we could only wonder about the bizarre bill shape. Notwithstanding this specialization, they also pick like other small sandpipers. Perhaps because they occur in such small numbers, individuals of this species regularly associate with stints, in particular the Rufous-necked, with which they share range and habitat. The Iona Island bird was with Western Sandpipers.

Structure (Figure 38)

Size very small. Bill short but massive for such a small sandpiper, from above widening rapidly in terminal third to' broad, diamond-shaped "spoon," sharply angled at tip and sides. This prominent expansion about three times width of bill at base but from side scarcely evident, visible only as slight projection on lower edge of bill at about two-thirds length. Legs short. Wings typically extend just beyond tail tip, primary projection moderate in juveniles (less in adults?). Tail double-notched, central rectrices pointed and projecting beyond others.

Plumage

Breeding. Bill and legs black. Similar to same plumage of Rufous-necked Stint, with bright reddish head, breast, and upperparts. Unlike that species, males average only slightly brighter than females, with much overlap. Crown and mantle rufous, striped with dark brown. Scapulars dark brown to blackish, fringed with rufous. Coverts gray-brown and tertials darker brown, both narrowly fringed with whitish when fresh. Supercilium may be prominent, whitish to buff, or may be largely obscured by rufous. Sides of head, throat, and breast rich rufous, in some birds intermixed with white. Usually dark spots at sides of breast that may extend across breast behind rufous. Belly white, in some individuals with scattered dark spots on sides.

Lower back dark brown, central uppertail coverts black, and outer uppertail coverts white. Central pair of rectrices black, rest of tail light gray-brown. Wing brown above, white wing stripe formed from narrow tips of greater secondary coverts. Underwing light gray-brown, secondary coverts and axillars white and marginal coverts slightly darker brown.

Nonbreeding. Upperparts, including ear coverts, brownish gray with dark streaks, much like coloration of black-legged stints. Forehead typically extensively white, with split supercilium and poorly defined loral stripe. Underparts white, dark streaks on sides of breast.

Juvenal. Head and upperparts boldly patterned, almost looking black and white, but often marked with rufous. Crown and mantle brown to gray-brown, striped with black; hindneck conspicuously paler. Scapulars black with pale buff to whitish fringes, some feathers in some individuals with rufous fringes. Coverts and tertials vary from gray-brown to dark brown, tertials averaging darker than coverts; all fringed with whitish. Forehead white; prominent supercilium typically split, and, in some birds, with dark diagonal line from eye interrupting it. Blackish loral stripe varies from poorly to well defined; ear coverts usually brown, forming distinct patch in most birds. Underparts white, with or without dark streaks at sides of breast; breast washed with pale buff in some birds. This plumage borne during August and September at least. Further plumage changes not documented.

Identification

All Plumages. The *spatulate bill* will be all the identification mark needed if it can be seen. The "spoon" is not highly obvious from the side, but there is still something peculiar about the bill in comparison with the spoonless stints. Sandpipers of other species with mud or sand caked on their bill might be called SPOONBILLS.

Breeding Plumage. In fresh or worn breeding plumage, this species looks very much like a RUFOUS-NECKED STINT, and a look at the bill would be necessary to identify it. The more brightly marked RUFOUS-NECKED are much more rufous above than typical SPOONBILLS.

Nonbreeding Plumage. Again, the SPOONBILL is so much like other black-legged stints in this plumage that the bill is the only definitive mark, although the forehead averages more extensively white than in most stints.

Juvenal Plumage. In this plumage, coloration is distinctive in addition to bill shape. Among the

stints, the SPOONBILL stands out in its *usual lack of rufous* anywhere on the plumage. A small number of individuals, however, do show scattered rufous-fringed scapulars. The only black-legged stint that regularly lacks rufous is the SEMIPALMATED, which tends to have more warm buff tones, but not always. From this species and any of the others, the SPOONBILL can usually be distinguished by its *extensive white forehead, split supercilium, and dark ear coverts.* The *dark crown* usually contrasts strongly with the pale hindneck. Other species such as RUFOUS-NECKED and WESTERN can vary in this direction, as can SEMIPALMATED, but the combination seems especially characteristic of SPOONBILL. SPOONBILLS also typically have *fewer streaks on the breast and longer primary projections* than WESTERN and SEMIPALMATED.

In Flight

Again, this species falls in with stints in its appearance in flight, probably not to be distinguished unless the *spoon bill* is seen. The wing stripe is reported to be slightly more conspicuous than in other stints, but this needs confirmation.

Voice

The flight call has been reported as "a shrill *wheet wheet*" and "a quiet rolled *preep*," both descriptions suggesting to me the call of a LEAST SANDPIPER.

Further Questions

Anyone who encounters an individual of this species should carefully observe its foraging behavior to see if the specialized spoon is typically used as described above. Some birds on their arctic breeding grounds were said to feed like other small sandpipers by picking insects from the water surface.

Notes

There is still controversy about a juvenile reported from Tillamook Bay, August 20-21, 1982. Reasonably good photographs have been studied and considered a Western Sandpiper with sand on its bill by some experts, a Spoonbill Sandpiper by others. The Oregon Bird Records Committee has not accepted the record.

Juvenile Spoonbill Sandpiper. Stintlike calidridine with spatulate bill; pale-fringed upperparts and white breast diagnostic of plumage; extensive white forehead, split supercilium, and dark ear coverts characteristic of species. Kyushu, Japan, mid September 1984, Urban Olsson

References

Dixon 1918 (breeding biology and early North American records), Sauppe et al. 1978 (British Columbia record).

BROAD-BILLED SANDPIPER *Limicola falcinellus*

The Broad-billed Sandpiper is another of the Siberian shorebirds that is of improbable occurrence in the region, based on its Alaska records: a total of seven birds in the western Aleutians between August 19 and September 6 of three years. This pattern of autumn-only occurrence is unusual compared with the more normal occurrence of Siberian species in spring in that area. It is normally a coastal species, characteristic of mud flats and foraging by moving fairly rapidly and picking as well as probing. It is usually uncommon, never forming large flocks.

The BROAD-BILLED shows a distinctive constellation of characteristics. It is a small calidridine, size about halfway between WESTERN SANDPIPER and DUNLIN. Claims that it is about DUNLIN size are based on comparisons with small European DUNLINS; in fact, it is about two-thirds the size of an American DUNLIN, closer in weight to a WESTERN and easily mistaken for a STINT.

The bill is as distinctive as the name implies, about as long relative to its size as that of a DUNLIN but broad at the base for a small sandpiper and slightly drooped at the tip. It averages 2-3 mm longer in females than in males and from any view looks a bit heavy for the size of the bird.

It is black, pale at the extreme base in at least some juveniles. The legs are short, varying from gray or greenish to yellowish.

In all plumages a diagnostic field mark is the split supercilium, with a distinct white stripe extending onto the crown from the primary supercilium. This mark is especially obvious in juveniles; usually but not always visible in breeding-plumaged adults, in which it may be lost in the very dark crown in heavily marked individuals; and often obscure in nonbreeding adults. Brightly marked individuals might briefly recall a JACK SNIPE, of similar size but even more vividly striped on both head and upperparts.

In nonbreeding plumage the BROAD-BILLED is often compared with a DUNLIN, but it is much smaller and paler than American individuals of that species, with a white breast. It is much more like a STINT in the same plumage but can be recognized by its long, heavy bill, finely streaked breast, and more conspicuous white supercilium.

Juveniles are colored much like juvenile STINTS. They are heavily marked above with buff to rufous edges on scapulars and tertials, with well-developed white mantle lines and usually white scapular lines. The breast is pale buff to white, streaked on the sides. The color pattern is most like that of the brightly marked LITTLE STINT, less like the rufous-and-gray WESTERN SANDPIPER, and distinctly different from the browner-backed and -breasted LEAST SANDPIPER.

Breeding-plumaged adults of Siberian BROAD-BILLED SANDPIPERS (*L. f. sibirica*) are quite differently colored than those of European populations of the species (*L. f. falcinellus*) illustrated in most field guides. They are strongly patterned with rufous, black, and white above, much more reddish than European birds. White mantle lines are more conspicuous than in dark STINTS. The breast is cinnamon, marked with black specks and white fringes, unlike the heavily blackish-streaked breast of European BROAD-BILLEDS.

In flight BROAD-BILLED looks like the small marsh-inhabiting *Calidris*, with a narrow, not very conspicuous, white wing stripe and a stripe-tailed look. A possibly useful field mark is furnished by the dark lesser wing coverts, producing a darker area on the wing base corresponding to the "epaulets" of a male Red-winged Blackbird; this is not always conspicuous. The flight call is a light trill—soft, short, and high-pitched.

BUFF-BREASTED SANDPIPER *Tryngites subruficollis*

Ploverlike with its short bill, this bird nevertheless walks about in its upland habitats like a typical sandpiper. With breast colored like dry wheat, back speckled like the stubble, and yellow legs like grass stalks, it blends well with its environment. Check dry coastal meadows every autumn during the brief Pacific-coast movement of juveniles.

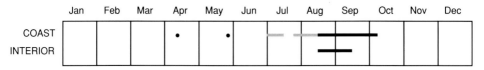

Distribution
The Buff-breasted Sandpiper breeds on dry tundra in the northern Alaskan and Canadian Arctic and winters on the pampas of southern South America. Adults migrate both north and south across the Great Plains, but juveniles are more widespread, with a well-developed flight south across the western Atlantic to South America. Far fewer deviate to the west of the normal route.

Northwest Status
As in other shorebirds that migrate primarily up and down the center of the continent, adults of this species stay well away from our region. Unlike the other species, however, there is scarcely a record of an adult from the Northwest, only two in spring and two presumed adults in July. The April record from Tillamook is quite anomalous, both out of range and early. Juveniles move west all the way to the coast, where they occur annually in small numbers from mid August to mid September.

Juvenile Buff-breasted Sandpiper. Short-billed upland calidridine with buff underparts and bright yellow legs; contrastily fringed upperparts diagnostic of plumage. Tillamook, OR, September 1985, Jeff Gilligan

There are numerous records from the well-studied Vancouver and Victoria areas. Birds are seen essentially every autumn at Grays Harbor and Willapa Bay, usually only one or two at a time, but in 1979 there were 20 in one flock at Ocean Shores. On the Oregon coast there are fewer records, but the species is seen just about annually. Farther south and away from the outer coast there are very few records, for example one each from the Seattle and Portland areas. Two at Medford, September 3, 1979, furnished the only Willamette Valley record.

Interior records range from August 14 to October 5 but are surprisingly few, considering that most Buff-breasted breed and migrate to the east of the region. There are seven records from southern interior British Columbia, four from eastern Washington, one from eastern Oregon, one from Idaho, and two from western Montana. Probably the relative paucity of both observers and appropriate habitat in the interior contribute to this; these birds should shun that part of the region in which upland habitats are excessively dry in autumn.

COAST SPRING. Tillamook, OR, April 12, 1981; Leadbetter Point, WA, May 31, 1984.

COAST FALL. *Adults* Iona Island, BC, July 11, 1976 (2), July 7, 1979. *Juvenile early date* Ocean Shores, WA, July 30, 1984. *High counts* 7 at Vancouver, BC, September 7-13, 1985; 11 at Ocean Shores, WA, August 27, 1978; 20 at Ocean Shores, WA, September 5, 1979; 14 at Florence, OR, September 11, 1985. *Late dates* Iona Island, BC, October 8, 1974; Clatsop Beach, OR, October 17, 1988.

INTERIOR FALL. *Early date* Tranquille, BC, August 14, 1977 (probably juvenile). *Earliest definite juvenile* Okanagan, BC, August 22, 1932. *Late dates* Reardan, WA, September 18, 1988; Prineville, OR, September 24 to October 5, 1990; Ninepipe Refuge, MT, September 18, 1977.

Habitat and Behavior

Buff-breasted Sandpipers in the coastal subregion are primarily found on lawns, edges of roads, and open, sparsely vegetated areas such as sand dunes and weedy fields, but they often forage along the upper edge of sandy beaches and occasionally wander out onto mud flats with their calidridine relatives. In the interior they should be looked for at watered golf courses and cemeteries. They forage at a walk, moving steadily over the ground, through and around grass clumps, often with head elevated and bobbing like a pigeon; "dainty" has been applied to their movements. The foraging path may be somewhat erratic, with quick directional changes as in phalaropes. They may also run and stop much like a plover, perhaps a response to disturbance, but other sandpipers do not usually do this in any circumstance. Foraging is by picking, and insects are the typical prey, although surface invertebrates such as crustaceans are taken at the shore.

Structure

Female small, male at low end of medium-small. Males about a third larger than females, but size difference insufficient to allow determination of sex of single individuals and needing close scrutiny to do so with individuals in flocks. Bill short and slender, tapering to fine point; legs short. Wings extend to tail tip, primary projection short in adults and moderate in juveniles. Tail wedge-shaped.

Plumage

Adult. Bill black, legs fairly bright yellow to orange-yellow. Overall appearance entirely buff, marked with black above. Evenly and coarsely patterned above; all feathers blackish with broad, pale buff fringes. No seasonal variation, although heavily worn fall adults with narrower and more faded fringes. Underparts, including face, extensively buff; breast on average darker than belly and slightly mottled, with small dark spots at bend of wing.

Lower back and uppertail coverts black with broad buff fringes. Rectrices gray-brown, central pair darker; narrow subterminal black band and pale buff tip. Most rectrices finely barred with black, especially on inner webs, but not in all individuals. Wing brown above, secondaries narrowly black-tipped with inner webs mottled. Primaries extensively black-tipped, forming dark line along rear edge of wing; greater primary coverts similarly dark-tipped. Most of primaries, terminal half of secondaries, terminal half of greater primary coverts, and marginal coverts heavily freckled with brown, forming most complexly patterned underwing of any of our shorebirds. Center of underwing and axillars white.

Juvenal. Feathers of upperparts dark brown (slightly paler at bases), narrowly fringed with pale buff; darker, more scalloped, and more finely marked than adults, every feather clearly outlined. Underparts about like adult but averaging slightly paler, especially on belly. Tail less patterned than that of adult, with poorly defined subterminal band and lacking fine black bars.

Immature. Matures in first summer, when plumage identical to adult.

Identification

All Plumages. The *entirely buff underparts*, together with the *short bill, unmarked face, and yellow legs* allow easy distinction of this species, usually seen in a characteristic habitat in any case. The few species at all similar include UP-LAND and perhaps PECTORAL SANDPIPERS and RUFF. The BUFF-BREASTED is distinguished from the UPLAND by its much smaller size and shorter neck and from UPLAND and PECTORAL by its buff underparts, with no breast streaking. Juvenile RUFFS have scalloped upperparts and most are buffy-breasted. However, the scalloped markings are reddish buff or rufous rather than the pale sandy buff of BUFF-BREASTED; there is a faintly indicated facial pattern, with a fine dark streak behind the eye; the belly is invariably white; and the legs are usually dull yellowish to greenish. RUFFS are also larger, longer-necked, and longer-legged than BUFF-BREASTED.

In Flight (Figure 9)

BUFF-BREASTED looks *plain brown above*, virtually unpatterned at a distance, and shows *white underwings that contrast with buff underparts*. It is much smaller than other plain-backed brown species such as UPLAND SANDPIPER and GOLDEN-PLOVERS, and it has a shorter tail than the sandpiper and an even less well-developed wing stripe than the plovers. The most similar species in flight is the EURASIAN DOTTEREL, also seen in the same habitat at the same time of year but only casual in occurrence; it has a conspicuous black-and-white tail tip and prominent pale supercilium. Note that the DOTTEREL, like the BUFF-BREASTED, has a buffy breast and whitish wing linings.

Voice

This is generally a silent species, the calls soft enough to be scarcely audible at moderate distances. The usual flight call is a low, single *tu* or a short trill something like that of a BAIRD'S SANDPIPER.

Further Questions

Observers in the interior might conduct special searches for this species on grass lawns in late August and early September, as it may occur more regularly in the eastern part of the region than is indicated by present knowledge.

Notes

The record of 200 reported at Tofino, May 2, 1974, is so unlikely that I have not accepted it. Similarly, a record from Westport, December 19, 1920, involves no more than a casual mention of a sighting and, as such, cannot be given full credibility.

Contrary to the information presented in some guides, adults and juveniles are not always distinguishable by the color of their underparts.

Photos

Juveniles in Bull and Farrand (1977: Pl. 224) and Keith and Gooders (1980: Pl. 116) are not labeled as such.

References

Campbell and Gregory 1976 (British Columbia records), Cartar and Lyon 1988 (mating system), Myers 1979 (courtship behavior), Myers 1980b (territoriality and flocking), Oring 1964 (courtship behavior), Paulson and Erckmann 1985 (nesting associations), Prevett and Barr 1976 (courtship behavior), Pruett-Jones 1988 (display behavior and spacing), Sutton 1967 (nesting behavior), Vinicombe 1983 (identification).

RUFF *Philomachus pugnax*

Ruffs are rare but regular runabouts, moving through the lesser calidridines at a rapid pace. They are distinctive in shape, with head seeming too small and bill too short for a bird of their size. Sex distinction is easy: females are of dowitcher and males of Greater Yellowlegs size. Most of our birds are juveniles, buff-scalloped and buff-breasted.

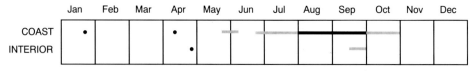

Distribution

The Ruff is an Old-World species, breeding from northern Europe to Siberia in arctic and subarctic tundra and meadows. It winters primarily in southern Europe and Africa, with smaller numbers spread through southern Asia to Australia.

Northwest Status

After the Sharp-tailed Sandpiper, this species is the most numerous of Northwest shorebirds of Siberian origin. This is surprising, as eastern Siberian birds, which should be the source of our records, typically move directly westward toward Europe in autumn before turning southward toward Africa. Considered entirely a Eurasian species, it is nevertheless reported widely across North America every year, and there may be a small North American breeding population. A single nest was reported from northern Alaska in 1976, with no males seen in the area (the egg examined was infertile).

The Ruff is a rare migrant in Alaska, with records concentrated in fall, and a regular migrant farther south on the Pacific coast, also primarily in autumn. There are only three spring records from the coast, from the Vancouver area, and a smattering of records of adults in fall (June 26 to August 30), almost all in British Columbia and including at least three males still in partial breeding plumage in August. Males take no part in parental care and leave the breeding grounds before females; they should therefore furnish more of the early fall records in this region.

Juveniles are seen every fall on the coasts of British Columbia, Washington, and Oregon, through most of August and September and less often in October. There were from 5 to 21+ records per autumn in the region during the period 1981-90, an amazing number if the birds are indeed coming from Siberia. As many as a dozen or more were present at Iona Island in autumn 1986 and four at Ocean Shores in autumn 1979, these two areas the Ruff spots of the region. Over half the Ruff records from Washington come from Ocean Shores, while Oregon Ruff records are spread throughout its coastal estuaries from year to year. Year-to-year variation in autumn numbers (5 records in 1985, 21+in 1986, 2 in 1987, 4 in 1988, 11 in 1989, and 8 in 1990) may be correlated with wind directions at higher latitudes.

As Ruffs winter with some regularity in California, it is not surprising that there is a winter record from southern Oregon. Only three have been encountered to date in the interior, a male in breeding plumage in spring and two juveniles in fall.

COAST SPRING. Iona Island, BC, May 27-29, 1976; Comox, BC, May 23 to June 7, 1980; Reifel Island, BC, April 8-14, 1989.

COAST FALL. *Early date* Reifel Island, BC, June 26, 1974. *Juvenile early date* Aberdeen, WA, August 18, 1982*. *Late dates* Iona Island, BC, October 17, 1987; Ocean Shores, WA, October 27, 1976; Yaquina Bay, OR, October 31, 1984.

COAST WINTER. Coquille, OR, January 18, 1980.

INTERIOR SPRING. St. Andrews, WA, April 27, 1986.

INTERIOR FALL. Reardan, WA, September 22, 1972, September 27, 1987.

Habitat and Behavior

Most Ruff records come from coastal ponds, lagoons, and estuaries, where the birds feed on mud flats and in flooded fields and salt marshes. They often feed up to their belly in fresh water but typically stay above the shoreline when at salt water. They apparently prefer Pectoral Sandpiper habitats and often associate with them, even joining Pectoral flocks when flushed. Most records involve single birds, but on several occasions two have been seen together. Ruffs often forage at a run with head up when on open substrates, easily picked out among walking sandpipers and run-and-stop plovers, but they also move methodically through marsh vegetation with head down, much like a typical *Calidris* sandpiper.

Structure (Figure 46)

Size medium-small in female, medium in male. Males much larger than females, greatest sexual dimorphism in size of any Northwest shorebird. Average female weight about 60 percent of male, sexes easily distinguished by size. In comparison with birds with which they are likely to be seen, males about size of Greater Yellowlegs and one-third larger than golden-plovers. Females about size of dowitchers and one-fourth to one-fifth larger than Killdeers, male Pectoral Sandpipers,

and Lesser Yellowlegs; males about twice as large as these last three species. Remember that these comparisons refer to body bulk, not length or height.

Bill short to medium, legs medium. Wings project to tail tip, primary projection virtually lacking in adults and short in juveniles. Tail rounded, central rectrices rounded and projecting only slightly beyond others.

Plumage

Breeding Male. Bill pale, usually pinkish or orange but in some individuals gray or yellow; some individuals show black tip. Legs usually orange or pink. Individual variation in plumage sufficient virtually to defy description. Males molt in late winter into "breeding" plumage they will hold for summer, then with supplemental spring molt add neck ruff and head tufts that prepare them for complex display behavior. Head

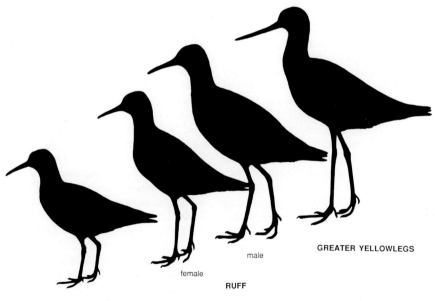

Fig. 46. Ruff size dimorphism. Female next to female Pectoral Sandpiper and male next to Greater Yellowlegs.

(with tufts), neck ruff, breast, and back all vary in color, with ground color black, rufous, buff, white, or some lighter or darker variation on these themes. Ground color may be overlain with barring or spotting of almost infinite variability in shape and size; markings vary from coarse, producing contrasty pattern, to very fine, producing some apparently intermediate color.

Plumage areas vary independently, so back may be rufous-and-brown barred, ruff dark chestnut, head tufts white, and breast black, for example. Nevertheless, males usually "dark" or "light." Scapulars and tertials typically rufous or gray-brown, extensively tipped or more finely barred with dark brown or black; often finely vermiculated along much of their length, varying more rarely to black or black and white. Lower belly always white, but breast color and pattern may extend down sides to base of legs.

Lower back and central uppertail coverts typically marked like scapulars and tertials of same individual. Outer uppertail coverts long and white. Rectrices gray-brown, usually central pair and often others conspicuously barred and/or mottled. Wing brown above, narrow wing stripe formed from whitish tips of greater secondary coverts. Underside of wing light gray-brown, paler toward bases of flight feathers. Greater and median secondary coverts, terminal half of greater primary coverts, and axillars white; marginal coverts mixed brown and white.

In addition to developing complex plumage, breeding males lose feathers on forehead and lores and develop "warts" there, usually yellowish or reddish.

Breeding Female. Bill black, in some individuals with reddish base. Also more variable than other shorebird species, although much less so than breeding males. Typically upperparts vary from cinnamon or rufous to gray; mantle feathers, scapulars, and tertials variably marked with black bars or wider subterminal markings (usually more extensive in gray birds). Neck and breast similarly vary from rufous to buff to gray, relatively plain or fairly heavily spotted or barred with brown or black. Remainder of underparts white. A few individuals approach males in intensity of coloration, with largely rufous or blackish foreparts or vividly striped upperparts, but none shows head tufts or neck ruffs.

Nonbreeding. Bill black, in some individuals brown at base. Sexes look similar in this plumage. Entire upperparts plain gray-brown with darker feather centers. Some individuals with fairly wide, prominent black bars on some or all mantle feathers, scapulars, coverts, and tertials; tertial bars may be particularly conspicuous. Neck and upper breast light gray-brown, faintly to moderately barred with darker gray-brown. Throat white, contrasting with brownish crown; in some individuals cheeks and forehead also white. Lower breast and belly white. Some males may be extensively white on neck and breast, rarely with entirely white head and breast. This is seen in winter but may actually represent early acquisition of part of breeding plumage.

Juvenal. Legs generally duller than in adults—yellowish to greenish—but situation very complex, with substantial individual variation. Legs apparently change color with age in some individuals but not in others. Crown reddish brown, streaked with blackish. Mantle feathers, scapulars, coverts, and tertials either entirely blackish or gray-brown with blackish subterminal markings, in either case fringed by buff or rufous. Bright markings of upperparts allow ready distinction from nonbreeding-plumaged adult. Poorly defined pale buff or whitish supercilium, narrow dark postocular stripe. Neck and breast vary from gray to buffy brown, neck streaked and breast in some individuals obscurely barred with slightly darker gray-brown; belly white.

Immature. Molt into first-winter plumage, similar to nonbreeding adult plumage, takes place from September to November, although tertials not molted until winter or even spring. Those molted in December and January replaced by nonbreeding plumage, those from February onward by breeding plumage. About 15 percent of juveniles molt outermost primaries during first winter. Most males leave Africa in first summer, at least to migrate north if not to attempt breeding, while at least some females oversummer. First-year males arrive on breeding grounds in partial breeding plumage, without full ruffs and head tufts of adults.

Identification

All Plumages. RUFFS in any plumage are dis-

Juvenile Ruff (right). Large, long-legged calidridine with fairly short, straight bill and small-headed look; fringed upperparts and buffy breast diagnostic of plumage; size in comparison with juvenile Ruddy Turnstone indicates female. South Bay, WA, 16 September 1989, Robert Sundstrom

tinctively shaped birds, their *plump shape* a consequence of a fairly short bill for their size and an apparently small head, long neck, and longish legs for a calidridine. They often *forage at a run*, a characteristic behavior that may allow them to be picked out at great distances among similar-sized shore-foraging species.

Breeding Plumage. Because of its *rufous, black, and/or white ruffs*, a male in breeding plumage could be mistaken for no other shorebird. Females are usually *distinctively patterned above* and, less often, below, in this plumage, often with *contrasty breast markings*. At their dullest they are somewhat like birds in nonbreeding plumage (see below). Be aware that birds in molt could look strangely variegated.

Nonbreeding Plumage. Birds in this plumage are also distinctively patterned above, usually with *boldly barred tertials*. The breast may be marked as in breeding plumage, especially with black bars, or may be entirely plain gray-brown, recalling a DUNLIN, but the bird is larger, with relatively *shorter bill and longer, pale legs*. Birds with orange legs would be distinctive by this characteristic alone.

Juvenal Plumage. This plumage is the one most often seen in the Northwest. At a distance juveniles look superficially similar to PECTORAL and SHARP-TAILED SANDPIPERS but differ from both of them in their uniformly *buff-scalloped upperparts*, while the two smaller species are

prominently striped above. Some RUFFS have mantle feathers edged in such a way as to look striped, but they never show the whitish stripes contrasting with buff to rufous feather edges shown by the other two species. They differ further from PECTORALS in their *unmarked, buffy to gray breast*, although similar to SHARP-TAILEDS in this character. The crown is never so rufous as it is in SHARP-TAILED, and the *supercilium is less distinct* than in most calidridines. Furthermore, the *postocular stripe is more distinct than the loral*, while in juvenile *Calidris* the loral stripe is typically more distinct. And, of course, shape and size differences will be obvious when PECTORALS are around for comparison, as they usually are when the RUFF is in a salt marsh.

The only other similarly patterned and colored species is the BUFF-BREASTED SANDPIPER, from which the RUFF differs in having a *gray-buff breast and extensively white belly*, rather than the rich, extensively buff underparts of the former species. The pale markings of the upperparts are slightly darker and more linear (less obviously scalloped) than in BUFF-BREASTED, between the pattern of that species and that of PECTORAL. Female RUFFS are about one and one-half times as large as male BUFF-BREASTEDS, with much longer neck and legs.

In Flight (Figure 12)

Typical of marsh calidridines, the RUFF shows

narrow, white wing stripes, but its most distinctive flight character is furnished by the unusually long, white outer uppertail coverts that either form *large, white patches on either side of the rump* or overlap to form a *U- or V-shaped white patch* in the same area. This variation exists between individuals, but it might occur in the same individual depending on how the feathers are arranged. The toes project substantially beyond the tail tip, while they do not do so in other stripe-tailed calidridines such as PECTORAL and DUNLIN. Because of their ruffs, breeding-plumaged males look particularly bulky in front.

A "lazy" flight is distinctive at times, the bird moving with slow or irregular wingbeats and depressed wings, interspersed with glides.

Voice

Ruffs are very quiet shorebirds, but both feeding and flying birds give low, whistled or grunting calls occasionally.

Further Questions

With this being one of the shorebirds in which the sex can usually be determined, it would be of great value if all Ruff observations were accompanied by sex designation, in particular if a nearby shorebird could be used for size comparison. It would also be of interest to know if individual Ruffs can show both separated and joined white rump patches.

Notes

I consider a published record of a Ruff from Tillamook, June 2, 1984, inadequately documented; photos and description also match a Pectoral Sandpiper.

Photos

For some of the variation in breeding males, see Hammond and Everett (1980: 124) and Keith and Gooders (1980: Pls. 161-63). For variation in breeding females, see Chandler (1989: 130-31). The "winter" bird in Hammond and Everett (1980: 124) is a juvenile; it illustrates the ease with which a shorebird can stand in one direction and face in another!

References

Andersen 1944, 1948, 1951 (breeding biology, population biology, and plumages), Chandler 1987b (female variation), Gibson 1977 (Alaska nesting), Hogan-Warburg 1966 (social behavior), Höglund and Lundberg 1989 (polymorphism and size variation), Koopman 1986 (molt and weight), Lank and Smith 1987 (male mating strategies), Lifjeld 1984 (feeding ecology in migration), Pearson 1981 (molt and population dynamics), Pearson et al. 1970 (weight), Prater 1982 (identification), Schmitt & Whitehouse 1976 (size, molt, and leg color), Shepard 1976 (display behavior), van Rhijn 1973 (male mating strategies), van Rhijn 1983 (male mating strategies), van Rhijn 1985 (evolution of mating system), van Rhijn 1991 (monograph).

DOWITCHERS Limnodromini

Three species are included in this group, two of which occur in North America and are common in the Northwest (Table 32).

Dowitchers are medium-small to medium, very long-billed sandpipers. They are like calidridines in their substantial seasonal plumage changes as well as their courtship displays and breeding habits but like snipes in their feeding adaptations. Breeding plumage is speckled brown above and bright rufous below, nonbreed-

Table 32. Dowitcher Distribution and Habitat Preference

Species	Hemisphere	Breeding Range	Breeding Habitat	Winter Range	Winter Habitat
Short-billed	W	subarctic	taiga	N/equatorial	bay
Long-billed	W	arctic	tundra	N	bay/marsh

ing plumage gray-brown above and paler below. Juveniles are intermediate between the two adult plumages in the complexity and brightness of their plumage pattern. Foraging is by deep probing with rather slow forward progression, typically in tight flocks. American dowitchers breed on arctic tundra and subarctic muskeg and migrate to both coastal beaches and estuaries and freshwater marshes and lakes.

References

Conover 1941 (species and subspecies distinctions), Jehl 1963 (migration in New Jersey), Lenna 1969 (identification), Miller et al. 1984 (song comparisons), Nehls 1989 (status in Oregon), Newlon and Kent 1980 (identification), Nisbet 1961 (identification), Pitelka 1950 (variation, taxonomy, and identification), Rowan 1932 (species and subspecies differences), Wallace 1968 (identification), Wilds and Newlon 1983 (species and subspecies identification).

SHORT-BILLED DOWITCHER *Limnodromus griseus*

Medium-small shorebirds with ultralong bills, dowitchers are readily recognizable. Feeding birds move mostly their heads, like animated sewing machines, and progress across the flats slowly as they sample deep in the mud. Rufous beneath in spring, both species stand out among their white-bellied relatives, and their white backs mark them clearly in the air. This locally abundant coastal migrant can usually be distinguished from the very similar Long-billed Dowitcher by close-range field marks if it fails to utter its distinctive "tututu."

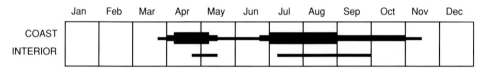

Distribution

The Short-billed Dowitcher breeds in wet clearings in the boreal forest in two distinct populations: southern Alaska and Yukon, and central Canada. The Caribbean, Mexico, and the north coast of South America comprise its primary wintering grounds.

Northwest Status

In both spring and fall migration this species is locally common in the Northwest, the adults moving through rapidly. Spring migrants appear early in April and are common from mid April to early May, peaking in the tens of thousands in Grays Harbor during the last week of April. The only concentrations in British Columbia are at Tofino, on the west coast of Vancouver Island, where large estimates of dowitchers almost surely pertain to Short-billed. Only Grays Harbor, Willapa Bay, and probably Tofino harbor really large numbers of this species, with high counts of adults elsewhere on the region's coasts in the lower hundreds. Although other mud-flat shorebirds are common in the Fraser River estu-

ary in spring, dowitchers are not; the published high count is 300. Similarly, few are present on the coast of Oregon, with the peak count 200. Even at Grays Harbor, numbers vary greatly, and migration peaks of fewer than a thousand birds have been recorded in some years. Thus this species equals the Red Knot in the extreme localization of its primary staging grounds in the region.

Adults are locally common from the end of June to mid July and occur sparingly into late August; again, few are recorded from anywhere but the outer coast. Their numbers tail off and merge with much smaller concentrations of juveniles, which peak in August and early September in the low hundreds and persist in small numbers into October, rarely later. There is no evidence that they regularly winter in the Northwest, although they have been listed on occasional Christmas Bird Counts from Vancouver south. Some of these records may be valid, as recent observations of dowitchers with Short-billed calls in southwestern Washington indicate at least occasional wintering. A few summer regularly in

Grays Harbor and Willapa Bay, and recently studied flocks included numbers of both first-year birds in nonbreeding plumage and adults in breeding plumage, both sexes in each category.

The Short-billed is primarily a bird of salt water, with only small numbers in coastal fresh water. Even in the Willamette Valley and Puget Sound areas relatively few are seen, and east of the Cascades it is rare. The smattering of records from eastern British Columbia, eastern Washington, eastern Oregon, and western Montana are mostly sight records, of which older ones especially are dubious. With better understanding of field marks, and occasional vocalizations for confirmation, the species is now reported regularly in the interior both spring and fall.

Interior specimen records include nine from Osoyoos, British Columbia, May 11, 1922; one from near Lamont, Washington, September 5, 1980; two from Potholes Reservoir, Washington, August 23, 1990, and August 7, 1991; one from Summer Lake, Oregon, July 24, 1947; and at least seven from northern Idaho, one on May 6 and the others from July 13 to September 26. The 40 Long-billed Dowitchers collected over the same period in Idaho indicates a ratio of 5.7:1, certainly a lower ratio of Long-billed to Short-billed than in eastern Washington, where I have seen only 10 definite Short-billed among many hundreds of Long-billed. Other than the series collected at Osoyoos, the most observed at once were six (four photographed) near Potholes Reservoir, Washington, May 12, 1990*, and the surprising total of 25 at Malheur Refuge, Oregon, September 9, 1989.

COAST SPRING. *Early dates* Westport, WA, April 7, 1934#; Willapa Bay, WA, March 20, 1966 (sight record). *High counts* 200 at Yaquina Bay, OR, April 19, 1977; 200 at New River, OR, April 27, 1984; 200 at Tillamook, OR, May 9, 1984; 34,000 at Grays Harbor, WA, April 27, 1981; 26,000 at Willapa Bay, WA, April 23, 1983; 300 at Iona Island, BC, April 24, 1982; 10,000 to 15,000 at Tofino, BC, May 2, 1974.

COAST SUMMER. *High count* 70 at Leadbetter Point, WA, June 8, 1974.

COAST FALL. *High count* 12,000 at Grays Harbor, WA, July 10, 1979. *Juvenile early date* Ocean Shores, WA, August 4, 1979. *Late date*

Ocean Shores, WA, November 17, 1979.

COAST WINTER. Willapa Bay, WA, December 19, 1987 (50); Long Beach, WA, January 30, 1988 (3), January 8, 1989 (2).

INTERIOR SPRING. *Early date* Klamath Falls, OR, April 28, 1980 (6, sight record). *Late dates* Reardan, WA, May 26, 1980 (sight record); White Lake, BC, May 18, 1986 (sight record).

INTERIOR FALL. *Early dates* Bend, OR, July 7, 1983 (sight record); Lewiston, ID, July 13, 1956#. *Juvenile early date* Potholes Reservoir, WA, August 7, 1991#. *Late dates* Potlatch, ID, September 26, 1956#; west of Missoula, MT, October 4, 1986 (photo).

Habitat and Behavior

Short-billed is the dowitcher of salt water, most common on mud flats in our region. Rarely do flocks show up in other situations, but I have seen a few on the ocean beach on occasion and in great numbers on a flooded golf course (both roosting and feeding) after a rain storm.

Habitat, distribution, and seasonal occurrence can be guides to identification of the two species of dowitchers, but they are not definitive. The Short-billed is a saltwater species, while dowitchers on fresh water are likely to be Long-billed. The Long-billed commonly visits the marine environment, but even there it is more frequent at pools and in salt-marsh vegetation than is Short-billed; both feed on mud flats. Huge flocks of dowitchers are usually Short-billed (they could contain Long-billed), while small groups could be either species.

Dowitchers typically forage by probing deeply and continually in mud or soft soil (inundated or not) while advancing relatively slowly. They feed as closely together as any of our shorebirds, at times almost touching. Their probing is more vigorous than that of similarly probing calidridines, and they are appropriately described as "animated sewing machines." In other situations they jab the bill shallowly into mud while moving more rapidly, but they do not show stitching behavior in the manner of Dunlins and Stilt Sandpipers. At their most active they run about between brief bouts of probing; presumably this is in areas where prey is less common or more scattered or both. At times shorebirds in Grays

Harbor feed as if choreographed: Short-billed Dowitchers in the water, Dunlins at the waterline, Western Sandpipers on the wet mud, Semipalmated and Black-bellied plovers on the drier mud, and Least Sandpipers adjacent to the salt marsh.

Structure (Figure 47)

Size medium-small. Bill long to very long, legs short. Wings extend about to tail tip, primary projection short in adults and juveniles. Tail approximately square in adults, wedge-shaped with projecting central pair in juveniles. Dowitchers only regional shorebirds with distinct difference in tail shape in two age classes.

Plumage

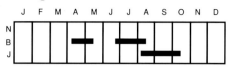

Breeding (Figure 48). Bill entirely black or tinged gray or greenish at base, legs light green. Overall appearance heavily mottled buff and black above and largely reddish to orange below. Crown black, feathers edged with rufous; rest of head rufous with scattered brown streaks. Loral and short postocular stripe dark, fairly prominent supercilium rufous or partially white. Mantle feathers, scapulars, scattered fresh coverts, and

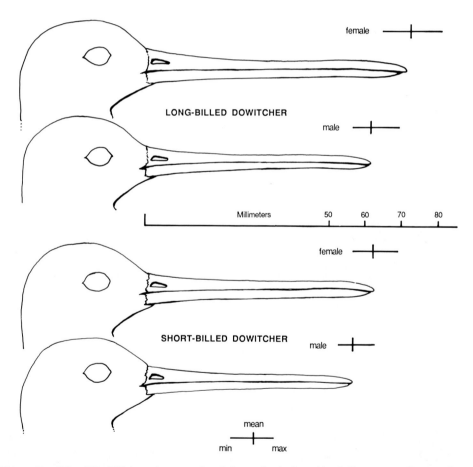

Fig. 47. Dowitcher bills. Bill shown is average length for species; horizontal bar indicates range. Great overlap shown, also extreme length of female Long-billed.

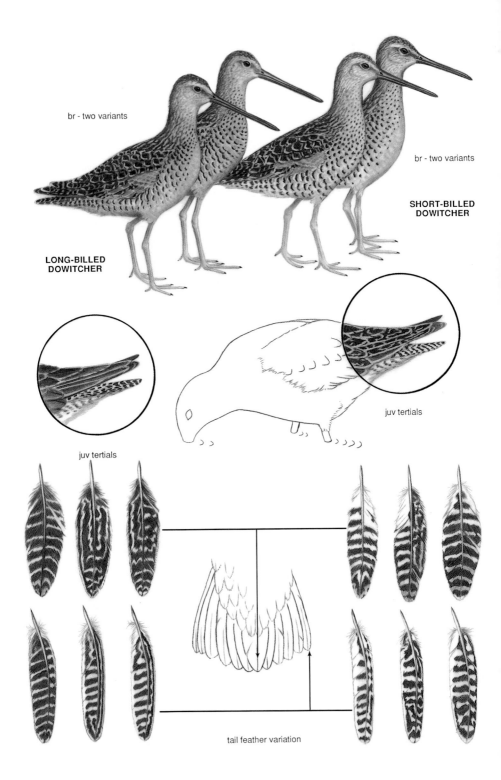

br - two variants

br - two variants

**SHORT-BILLED
DOWITCHER**

**LONG-BILLED
DOWITCHER**

juv tertials

juv tertials

tail feather variation

fresh tertials black with buffy to whitish fringes and reddish submarginal marks. Fringes form conspicuous mantle lines, indistinct scapular lines. Remainder of coverts and tertials brownish-gray feathers from nonbreeding plumage. Sides extensively spotted and barred; breast in most individuals spotted, bars or spots may extend over entire underparts in more heavily marked individuals. Undertail coverts colored like belly or with more extensive white.

Lower back white with black spots, uppertail coverts black and white barred; rectrices barred or mottled with dark brown and white. In some birds central pair of rectrices cleanly barred black and pale rufous, obviously new feathers. Wing brown above and paler gray-brown below, most markings not visible at a distance. From above greater secondary coverts with narrow white tips, forming scarcely visible wing stripe. Secondaries tipped with white and, especially inner ones, narrowly barred with white near tips, forming pale stripe at rear edge of wing. From below fine whitish fringes on coverts and blackish, white-centered and -fringed marginal coverts may be visible, but overall impression is of dark underwings. Axillars white, variably barred with brown.

Earliest spring migrants may be largely in nonbreeding plumage (specimen from Westport, April 13, 1933), but most have attained breeding plumage by that time (for example, almost all birds in flock of 19 at Tokeland, April 8, 1989*). Returning adults in fall darker above, buff edges worn off many feathers of upperparts. Adults begin body molt soon after arrival, underparts showing more and more white and then upperparts more and more gray. In northern part of region body molt essentially complete in some individuals by mid August.

Fig. 48. Dowitchers. Typical breeding Long-billed with barring on sides of breast, unmarked center of belly. Typical Short-billed with spots on sides of breast, barred belly. Variants (behind) show Long billed with more heavily barred breast, Short-billed with more lightly barred breast. Juvenile tertials plain in Long-billed and bright and contrasty in Short-billed. Central and outermost tail feathers drawn from three individuals of each species; Long-billed with generally greater extent of dark pigment but much variation and overlap.

Nonbreeding. Bill greenish gray at base. Upperparts and breast brownish gray, breast with scattered dark spots; belly white, sides with coarse gray bars. Distinct white supercilium, dark loral stripe, and short, dark postocular stripe. Posterior sides and undertail coverts spotted and barred with black.

Juvenal (Figures 48, 49). Bill more extensively pale greenish gray at base than in nonbreeding adult, quite distinct from that of breeding adult. Plumage brighter than in nonbreeding adult, with darker crown and dark, buff-striped mantle. Blackish scapulars and tertials with reddish fringes and submarginal markings. Coverts dark brown with buff fringes. Breast gray to rich buff, buff in some individuals extending onto belly, which is otherwise white. While still on breeding grounds, juvenile Short-billed may be rich rufous below, this color faded by the time they reach our region. Early-migrating juvenile might be mistaken for breeding adult because of this.

Usual contrast between buff breast and white belly and striped appearance of upperparts, as well as fresher plumage, allows easy distinction of juveniles from breeding adults during overlap in August and September. More than in most shorebirds, juvenile Short-billed Dowitchers quite different from both breeding and nonbreeding adults.

Juveniles of both species begin molt into first-winter plumage while in Northwest, and Short-billed may be well along in this molt by early September. By October many individuals of both species indistinguishable from nonbreeding adults except for tertials. Tertials retained into October or November, even into midwinter in some individuals, providing distinction (easy in Short-billed but more subtle in Long-billed) from nonbreeding-plumaged adults at close range.

Immature. Most Short-billed Dowitchers mature in first year, but at least some spend first summer on wintering grounds in usually fresh (acquired by a spring molt) nonbreeding plumage; often with few feathers of breeding plumage, especially scapulars and tertials. A few individuals attain most of breeding plumage but still fail to migrate. Primaries of these individuals very worn in early summer and in molt in midsummer. In hand, birds in breeding plumage at least two years old typically show fresh central rectrices with pale rufous ground color, one-year-olds worn rectrices with white ground color.

Subspecies

Two subspecies of Short-billed Dowitcher have been recorded from our region. The Alaska-breeding *L. g. caurinus* is the common coastal migrant, and it is known sparingly from the interior, where it seems more common than the central-Canadian-breeding *L. g. hendersoni*. Most of ten spring adult specimens (nine taken from the same flock) from Osoyoos were considered *caurinus* by Pitelka (1950); two individuals were more like *hendersoni*. Specimens examined by me from northern Idaho include two adult (July 13-28) and three juvenile (August 19 to September 26) *caurinus* and one adult (May 6) and one juvenile (September 10) *hendersoni*. I have examined the juvenile specimen from Lamont and confirmed that it is *hendersoni*, as determined by Weber (1985). The two juvenile specimens from the Potholes Reservoir and the adult specimen from Summer Lake appear to be *caurinus*. The May 1990 birds photographed near Potholes Reservoir looked like *caurinus*, but a breeding-plumaged dowitcher I saw by itself east of Ellensburg, May 6, 1984*, showed plumage characteristics of *hendersoni*, as did one in a flock of Long-billed at Reardan, July 16, 1990*. With its central and relatively low-latitude (for a sandpiper) breeding range, *hendersoni* is an unlikely but clearly not impossible migrant this far west.

Typical individuals of both adults and juveniles can be distinguished in the hand and, under favorable conditions, in the field by the wider, paler fringes on the mantle feathers, scapulars, and tertials in *hendersoni*. Juveniles of *hendersoni* average paler on the breast than those of *caurinus*, although some of them retain entirely buff underparts well into August (*caurinus* does not seem to do this). Fresh breeding adults of *hendersoni* are usually entirely reddish below, and the ventral spots are small and scattered, usually absent from the belly; they look very different from the Long-billed with which they may flock. Individuals of *caurinus* are usually heavily spotted on the breast and lightly spotted on the belly, which is often extensively white. There is overlap between the two forms.

Identification

All Plumages. DOWITCHERS are *very long-billed* for their size, the bill about twice the head length, only the COMMON SNIPE equalling their bill-to-body-size ratio. DOWITCHERS can be distinguished from SNIPES, with which they sometimes associate, by their *unstriped crown*, finely striped or unstriped back (gray in winter, mottled in summer), and reddish or gray rather than brown breast. STILT SANDPIPERS feed much like DOWITCHERS and occur regularly with them but have shorter bills like DUNLINS and longer yellowish legs like YELLOWLEGS.

The identification problem here of course is between the two species of dowitchers (Table 33). Not until 1950 was it generally accepted that the two were separate species, and in any plumage they look and act about the same. Winter birds are almost impossible to distinguish by plumage characteristics, most breeding adults should be identifiable under good viewing conditions, and juveniles are relatively easy to differentiate compared with the other two plumages.

• A characteristic used at times in the hand and worth studying in the field is the *tail barring*—usually more extensively white in SHORT-BILLED. A typical SHORT-BILLED may have white or pale cinnamon bars as wide as or wider than the black bars between them, and many individuals have irregular patterns. LONG-BILLED are more conservative, with black bars wider than the pale ones (as much as twice as wide in some) and less often any deviation from simple crossbarring. However, Pacific coast dowitchers overlap considerably in this characteristic, more than recent literature would imply.

• The names of the species of course indicate another field character, the *bill length*. The problem entailed here is shown clearly in Figure 47, in which it can be seen that only the longest-billed female LONG-BILLED will be recognizable as such in the field. Even within sexes there is overlap between the two species, and the great majority of individuals will fall in the zone of overlap. The variation in bill length in males and the overlap between the two species is great enough so that even birds with seemingly short bills could be LONG-BILLED. As the two species sometimes flock together, this character cannot be used simply to identify a flock as the shorter-

Table 33. Dowitcher Identification

	Short-billed	Long-billed
All Plumages		
Bill length	51-69 mm	54-81 mm, females especially long
Primary projection	short	none
Dark tail bars	relatively narrow	relatively broad
Flight call*	staccato *tu tu tu*	high *peep*
Distribution	primarily coast	coast/interior
Season	migrant, few after September	migrant, common through October, winters locally
Breeding		
Anterior sides*	spotted	barred
Underparts	often spotted, with whitish belly (*hendersoni* uniform rufous)	uniform rufous
Juvenal		
Overall color	brighter	duller
Tertials*	heavily marked	nearly plain
Nonbreeding		
Breast	lightly spotted	unmarked

*most definitive characters

billed or longer-billed species.

• The length of the *primary projection* appears greater on the average in Pacific coast SHORT-BILLED (*L. g. caurinus*) than in LONG-BILLED. Some photographs of birds in fresh juvenal plumage show the primaries exposed by as much as a half-inch in SHORT-BILLED, while similar photographs of LONG-BILLED show a quarter-inch or less of primary projection. This difference is clearly indicated in a series of photographs I have of two juveniles of each species together. This characteristic could be of value in identifying birds in fresh nonbreeding plumage, but it needs more field testing before it can be declared definitive. If it is found to work, then a Pacific coast dowitcher with a longer wing tip, which shows up at a distance as a blackish point at the posterior end of the bird, would be SHORT-BILLED. This field mark probably would not work to distinguish interior or Atlantic coast SHORT-BILLED DOWITCHERS (*L. g. hendersoni* and *L. g. griseus*), with their shorter wings, from LONG-BILLED.

Breeding Plumage.
• In fresh breeding plumage SHORT-BILLED

DOWITCHERS average *paler below* and may have some *white areas on the lower belly*, but molting complicates assessment of this. LONG-BILLED DOWITCHERS in spring seem invariably to have entirely rich rufous underparts (birds early in the season have pale-tipped feathers), as do many SHORT-BILLED. Nevertheless, when seen together on closer examination, the lighter, less completely colored individuals are found to be SHORT-BILLED, while darker, more intensely colored birds are LONG-BILLED. By August most adult SHORT-BILLED are in molt and look whitish on the belly, while LONG-BILLED remain in full breeding plumage, looking considerably darker.

• Some SHORT-BILLED are heavily spotted over the entire belly, but LONG-BILLED and most SHORT-BILLED have the center of the belly clear and unspotted. Although distinctive, this is difficult to see in the field.

• The best close-range characteristic is the nature of the *markings on the side of the breast*, spotted in SHORT-BILLED ("S in S" is a mnemonic device) and barred in LONG-BILLED. This characteristic should work in the vast majority of individuals, but rare exceptions occur. Remember that *both* species are barred on the flanks.

RED KNOTS occur with dowitchers and are roughly similar in size and rufous underparts but with much shorter bills (see that species for other distinctions).

Nonbreeding Plumage. As adults and juveniles of both species of dowitchers molt into nonbreeding plumage, they become less separable. On the average, in nonbreeding SHORT-BILLED the *gray on the breast is paler, less extensive, and with a sprinkling of small dark spots*, that of LONG-BILLED slightly darker, more extensive, and without spots. Many birds will nevertheless be confusing.

Again, RED KNOTS are of similar size but have short bills (comparison under that species). When compared with DUNLINS in winter feeding or roosting flocks, DOWITCHERS appear obviously *larger, longer-billed, and with gray-brown sides*, with none of the contrast between breast and belly shown by DUNLINS. This last characteristic is an excellent one for sleeping birds with bills hidden.

Juvenal Plumage. Juveniles of the two dowitcher species differ in general appearance, although they overlap.

• A typical SHORT-BILLED is bright buff to almost rufous on the breast, with similarly bright markings on mantle, scapulars, coverts, and tertials; it shows *high contrast* all over, at first glance almost as bright as a breeding adult. The cap is usually dark, strongly set off from the rest of the head, and the breast may be spotted. A typical LONG-BILLED is duller, with gray, buff-washed, and less often spotted breast and less contrasty cap and back. The buff markings above are usually confined to the mantle and scapulars and are somewhat less extensive on those feathers. In mixed flocks (which are not all that common) the two can be easily separated.

• At closer range a virtually certain field mark is the *color and pattern of the tertials*—very dark brown with bright buff markings in the SHORT-BILLED and grayer with paler buff edges in the LONG-BILLED. The lower scapulars are typically marked like the tertials in both species, and the greater coverts are also more brightly fringed in SHORT-BILLED and plainer in LONG-BILLED. This difference is a substantial contributor to the more contrasty look of the juvenile SHORT-

BILLED. The pale markings on SHORT-BILLED tertials occur as both edgings and crossbars or other markings nearer the center of the feathers, and commonly the buff edge is paralleled by another similar line within the feather. The paler markings on the tertials of LONG-BILLED juveniles are confined to the edges or are lacking entirely, making the major part of the visible wing very plain. However, occasional juvenile LONG-BILLED show additional markings, some perhaps as well-defined as those of more sparsely marked SHORT-BILLED. Usually the ground color of these feathers is still gray-brown rather than the dark brown of a SHORT-BILLED, but a few individual birds might prove difficult to identify.

The great majority of juvenile dowitchers seen in the Northwest in August will be SHORT-BILLED, in September a mixture of the two, and in October LONG-BILLED. Retained tertials will allow identification of young SHORT-BILLED well into the winter.

In Flight (Figure 10)

The *white back*, extending forward as a point between the wings, is a sure mark for a DOWITCHER in flight, along with (or preceded by) the *very long bill*. Only the very rare SPOTTED REDSHANK and a few other Old-World *Tringa* that might visit the region have this marking, also more obscurely present in the much larger and casually occurring Siberian WHIMBREL. DOWITCHERS differ from other white-backed species also by the fairly conspicuous *white rear wing borders*, a good field mark from below. See under LONG-BILLED DOWITCHER for differences from COMMON SNIPE.

At Grays Harbor, where huge flocks of SHORT-BILLED DOWITCHERS, DUNLINS, and WESTERN SANDPIPERS pass in review in spring, DOWITCHERS more often fly in lines, the smaller species in bunches.

Voice

The best "field mark" for all plumages is the call. SHORT-BILLED seem almost invariably to call *tututu*, a staccato series of low, musical notes a bit faster than but similar to those of a LESSER YELLOWLEGS. A series of three notes is typical, with some variation to two and four. Some birds give

longest tertial

SHORT-BILLED

LONG-BILLED

Fig. 49. Juvenile dowitcher tertials. Juvenile Long-billed tertials plain with paler fringe or, rarely, with internal markings. Short-billed slightly darker, with conspicuous fringes and internal markings.

the notes in a longer, quicker series. LONG-BILLED typically call *keek* or *peep*, a single much higher note that may be repeated several or even numerous times in rapid succession. Interestingly, juvenile BLACK-NECKED STILTS have an almost identical call. Anyone fortunate enough to hear dowitchers (not always the case) should be confident in assigning them to species, and statements in the literature that individuals occasionally give the call of the other species are probably erroneous. Dowitchers commonly give their display song in flight in spring, especially in flocks heading to roost. The song, similar in both species but with minor differences, is a broken trill ending in a buzzy *dowitcher*.

Further Questions

An early spring (February and March) movement of dowitchers through the Willamette Valley is thought to include some Short-billed, but this needs confirmation. Interior dowitchers should always be carefully studied and documented to

get a better handle on the actual frequency of Short-billed, with much more information on prevalence of the two subspecies; more specimens are needed. The interior of the Northwest is peripheral to the main migration range of either subspecies, and it would be fascinating to attempt to correlate particular records with prevailing wind directions at the time of their occurrence. Records of Short-billed from this region from December through March also need to be well documented (now birders will have to carry tape recorders as well as cameras!), as the species normally winters no farther north than southern California.

A peculiarity of the Short-billed is the low proportion of juveniles to adults in fall. Daily counts of juveniles rarely exceed *one-tenth* the number of the adults that are present a month earlier, at least in censuses at Grays Harbor and Leadbetter Point. Do Short-billed Dowitchers have a particularly low breeding success compared with other shorebirds (in most of which somewhere near as many juveniles as adults are seen on a daily basis in our region)? Do juveniles usually overfly our region? Is the ratio merely an artifact of an extremely concentrated peak—both spatial and temporal—of adults? What is the situation up and down the Pacific coast?

Plumage characteristics of summering birds should be recorded to compare with the situation reported for other subspecies in the East, and these flocks should be carefully checked for Long-billed Dowitchers.

Notes

Two of the nine Short-billed specimens listed from Idaho by Burleigh (1972) were found to be Long-billed on recent reexamination: Lewiston, September 24, 1953, and August 24, 1954. Also, subspecific determinations on a few Idaho specimens by both Burleigh (1972) and Weber (1985) were different from mine. I consider the "*hendersoni*" listed by Burleigh from Lewiston, August 19, 1953, and Potlatch, September 26, 1956, to be *caurinus*. I consider the "*caurinus*" listed by Burleigh from Lewiston, May 6, 1954, to be *hendersoni*. Of the specimens listed by Weber, I consider USNM 463208 *hendersoni* and USNM 464980, 465899, 463209, and 464396 *caurinus*. USNM 463207 is *L. scolopaceus*. USNM 464980, listed by Weber from October 26, 1956, was actually collected September 26.

The dowitchers censused by Herman and Bulger (1981) in Grays Harbor were not determined to species, and the popular view since then has been that both species are present in large numbers in the impressive spring flocks. In fact, the great majority are Short-billed (Wilson 1989).

As Pacific coast Short-billed Dowitchers (*caurinus*) have very slightly longer wings than Long-billed (adult female mean wing length 150 mm versus 146 mm) but are certainly no larger, it might be expected that the wings extend slightly farther relative to the tail tip in the Short-billed, as was suggested by Nisbet (1961). Probably birds with wings extending beyond the tail tip are more likely to be Short-billed, those with wings falling obviously short of it Long-billed, but this can only be indicative and not a conclusive field mark (as in too many other dowitcher characteristics). Some writers have noted differences in the markings on the undertail coverts—spotted in Short-billed and barred in Long-billed, but birds in the Northwest cannot be separated by this feature in any plumage.

Prater et al. (1977: 116) and Hayman et al. (1986: 361) illustrate the extreme of pale coloration in Short-billed Dowitcher tertials (I have seen none with pale markings as extensive as shown in the latter); the tertials usually look dark with light markings. Similarly, Prater et al. (1977: 116) show wider white bars in the tail of both species than is typical. The colored illustrations in Hayman et al. (1986: 181) are much more representative. Peterson (1990: 138) is the latest of many authors to claim that the two dowitcher species can be distinguished by the width of their tail bars; this is not always the case, especially on the Pacific coast. The inland subspecies *hendersoni*, with less black on the tail, is perhaps the source of many of the claims of the distinctiveness of this character.

Photos

"Summer" Short-billed Dowitchers in some guide books are actually juveniles (Udvardy 1977: Pl. 212; Armstrong 1980: 135; Terres 1980: 792). The juvenile in Farrand (1983: 405) is so bright and buffy above and below, it is doubtless a *hendersoni*. The breeding-plumaged bird in Bull and Farrand (1977: Pl. 212) is also of that race. The breeding adult "*L. g. griseus*" in Farrand (1983: 403) is more typical of *L. g. caurinus* and, as the photographer lives in southwestern British Columbia, probably is that subspecies.

References

Baker 1977 (summer diet), Baker and Baker 1973 (summer and winter foraging behavior), Burger et al. 1979 (aggressive behavior in migration), Burton 1972 (foraging behavior in migration), Loftin 1962 (summering), Mallory and Schneider 1979 (foraging and aggressive behavior in migration), Miller et al. 1983 (breeding vocalizations), Weber 1985 (interior Washington and Idaho records), Wilson 1989 (Grays Harbor migration).

LONG-BILLED DOWITCHER *Limnodromus scolopaceus*

Probing the depths of freshwater marshes as well as those of coastal estuaries, a dowitcher in the interior or in the winter ought to be this species if paying attention to the conventions of dowitcher occurrence. Observers who likewise pay attention can distinguish it from Short-billed by subtle plumage characters and especially its "peep" call.

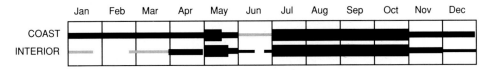

Distribution

The Long-billed Dowitcher breeds on arctic tundra in western and northern Alaska and winters on both interior and coastal habitats in southern United States and Mexico.

Northwest Status

Unlike its coastal relative, this species is common all across the region during migration. It rarely numbers in the thousands or tens of thousands as does the Short-billed, but daily peaks of up to a few hundred birds have been noted in many areas both spring and fall. The large numbers reported in October indicate a sizable late migration, easy to document after Short-billed have departed, but there have been no unequivocal records of large numbers of Long-billed among the hordes of Short-billed in spring.

Long-billed breed farther north and are apparently later in their migration schedule in spring, on the move from late April to mid May in British Columbia and Washington, somewhat earlier in Oregon. At freshwater ponds in Seattle that have been watched on a daily basis for several years, almost all records lie in the first two weeks of May (extremes April 25 and May 27). A few migrants linger into June, and fewer yet remain for part of the summer; there is no summering population comparable to that in the Short-billed. Adults appear in early July, almost as early as Short-billed, but peak later and are still present into mid September, while adult Short-billed are not usually seen after mid August. Juvenile Long-billed appear in mid August, again averaging slightly later than Short-billed, and they remain common through October, when Short-billed have become quite uncommon. Unlike the Short-billed, juveniles are seen in larger numbers than adults.

In some areas, flocks of dowitchers can be seen during the winter, and all that have been collected and virtually all that have been heard have been Long-billed. Most wintering birds are on the coast, although numbers have been seen at that season in the Puget Sound lowlands and Willamette Valley. Some large flocks seen in December diminished or disappeared during the winter and may have been on their way south. No midwinter specimens have been examined, but two from November from southern British Columbia are first-year birds.

The seasonal occurrence of this species in the interior is similar to that on the coast but probably better known, as virtually all interior dowitchers can safely be called Long-billed. Both spring and fall migration peaks are distinctly later than that of Short-billed on the coast. A bird in mid June cannot be allocated to either spring or fall.

Migrants are much more common in the Okanagan Valley of British Columbia in fall than in spring, although this discrepancy is not so evident in all parts of the region. Fall migration is protracted in this subregion, beginning in early July and lasting very late for a shorebird in the interior, regularly to about the end of October. Dowitchers peak in the tens of thousands at Malheur Refuge, adults perhaps undergoing wing molt as they stage there, but they are much less common elsewhere in the interior.

There are a few records of Long-billed Dowitchers at least into midwinter during mild winters in the interior (Washington, Oregon, and Idaho).

COAST SPRING. *High counts* 250 at Corvallis, OR, May 9, 1964; 274 at Medford, OR, May 11,

Breeding Short-billed Dowitchers. Smallish sandpiper (medium calidridine size) with very long bill and short legs; reddish underparts diagnostic of plumage and spotted sides of breast diagnostic of species; heavily marked underparts indicate *caurinus* subspecies; far left and right birds probably females, center rear bird probably male, from bill length. Near Potholes Reservoir, WA, 12 May 1990, Jo Anne Rosen

Nonbreeding Short-billed Dowitcher. Plain upperparts diagnostic of plumage; fine spots on breast and short primary projection indicative of species. Florida, mid January 1987, R. J. Chandler

Juvenile Short-billed Dowitcher. Fringed upperparts and buff breast diagnostic of plumage; brightly marked tertials and greater coverts diagnostic of species; relatively narrow fringes indicate *caurinus*. Victoria, BC, Tim Zurowski

Juvenile (left) and breeding Long-billed Dowitchers. Fringed upperparts and paler underparts diagnostic of juvenile; unmarked tertials and greater coverts distinguish from juvenile Short-billed. Seattle, WA, 3 September 1981, Dennis Paulson

Nonbreeding Long-billed Dowitchers. Plain upperparts diagnostic of plumage; plain medium gray breast and lack of primary projection indicative of species; reddish on scapulars of left-hand birds retained from juvenal plumage. Reifel Refuge, BC, 24 October 1970, Dennis Paulson

1984; 200 at Seattle, WA, May 13, 1981; 170 at Reifel Island, BC, May 8, 1972; 150 at Iona Island, BC, May 1, 1979; 150 at Iona Island, BC, May 15, 1983.

COAST FALL. *Adult high counts* 120 at Iona Island, BC, August 28, 1973; 120 at Iona Island, BC, August 2, 1976; 160 at Delta, BC, July 29, 1987; 220 at Sauvie Island, OR, July 21, 1984. *Juvenile early date* Ocean Shores, WA, August 8, 1983*. *Juvenile high counts* 1,200 at Reifel Island, BC, October 13-23, 1973; 1,000 at Delta, BC, September 21, 1985; 1,550 at Reifel Island, BC, October 11, 1987; 163 at Grays Harbor, WA, October 4, 1980*; 420 at Sauvie Island, OR, October 2, 1982; 400 at Sauvie Island, OR, September 9, 1986; 1,000+ at Sauvie Island, OR, October 15, 1987.

COAST WINTER. *High counts* 683 at Reifel Island, BC, in 1975-76; 400 at Woodland, WA, January 31, 1984; 300 at Eugene, OR, in 1968-69; 480 at Sauvie Island, OR, in 1982-83.

INTERIOR SPRING. *Early dates* Malheur Refuge, OR, February 22, 1975; Pullman, WA, March 15, 1973. *High counts* 290 at Lee Metcalf Refuge, MT, May 13, 1979*; 100 at Bruneau State Park, ID, May 19, 1982; 779 at Summer Lake, OR, May 1, 1987; 500 at Turnbull Refuge, WA, May 12, 1965; 200 at Kamloops, BC, May 12, 1982. *Late date* Moses Lake, WA, June 6, 1950.

INTERIOR SUMMER. Potholes, WA, June 14, 1951.

INTERIOR FALL. *Early dates* Columbia Refuge, WA, June 20, 1986; Malheur Refuge, OR, June 27, 1987. *Adult high counts* 32 at Kamloops, BC, July 27, 1950; 21 at Reardan, WA, July 16, 1966; 17,800 at Malheur Refuge, OR, August 31, 1973 (may include some juveniles); 5,262 at Summer Lake, OR, August 3,

1987. *Juvenile early date* Potholes Reservoir, WA, August 28, 1982*. *Juvenile high counts* 100 at Salmon Arm, BC, September 20, 1970; 70 at Kamloops, BC, October 3, 1984; 100+ at Reardan, WA, September 23, 1957; 200 at Yakima River delta, WA, October 19, 1974; 5,000 at Malheur Refuge, OR, October 5, 1979; 120 at Nyssa, OR, October 10, 1982; 100 at Ninepipe Refuge, MT, September 29, 1978.

INTERIOR WINTER. Klamath Falls, OR, until early January 1987 (13); Springfield, ID, to January 20, 1975 (8).

Habitat and Behavior

Long-billed Dowitchers are not abundant anywhere (with Malheur Refuge and Summer Lake notable exceptions) but are fairly common and widespread in many habitats, from freshwater marshes and drying lake shores to salt marshes and mud flats. I have not found them on ocean beaches where Short-billed occur sparingly, but they often feed with that species in any of its preferred habitats. A few Short-billed may turn up in flocks of Long-billed in marshes, while the two may be equally common on mud flats except briefly in spring and early fall when the huge flocks of Short-billed dominate this habitat.

As far as we know, the two species of dowitchers forage identically (see Short-billed Dowitcher) and often forage together.

Structure (Figure 47)

Size medium-small. Bill very long, distinctly longer in females than males (average difference 10 mm), greater dimorphism than Short-billed; legs short. Wings extend about to tail tip, virtually no primary projection in adults and juveniles. Tail approximately square in adults, wedge-shaped with projecting central rectrices in juveniles.

Plumage (Figures 15, 48, 49)

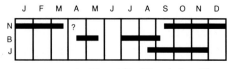

Bill and legs as in Short-billed Dowitcher. Plumage changes much as in Short-billed Dowitcher, with minor differences (see both Description and

CHRISTMAS BIRD COUNTS	Five-Year Averages 74-78	79-83	84-88
Ladner, BC	149	39	15
Vancouver, BC	8	20	18
Grays Harbor, WA	48	22	11
Coos Bay, OR	15	16	63
Eugene, OR	37	6	31
Sauvie Island, OR	91	206	48
Tillamook Bay, OR	11	13	27

Identification under that species). In addition to plumage differences enumerated above, feathers of underparts more likely to be tipped with white in breeding plumage in Long-billed, giving it frosted appearance. Spring molt lagging slightly behind that of Short-billed; for example, most of 17 Long-billed at Tokeland, April 8, 1989*, in nonbreeding plumage, while most of 19 Short-billed at same time and place in breeding plumage. Adult Long-billed may molt more rapidly than Short-billed in fall and typically look gray-headed (more so than Short-billed) while passing through Northwest. Molt into nonbreeding plumage begins in late August, and by late September many adult Long-billed in virtually complete nonbreeding plumage.

Many (all?) adult Long-billed on their way south pause in August and September in United States (so far documented only from southern Idaho in Northwest) to molt both body and wing feathers before continuing southward. Full wing molt in mid migration unusual in shorebirds, presumably something about ecology of this species allows lingering at moderate interior latitudes to accomplish this. Juveniles begin molting into first-winter plumage during late September and October but retain tertials into November, in some birds into December. Great majority of Long-billed apparently mature in first summer, molt into breeding plumage, and migrate north.

Identification

See under Short-billed Dowitcher.

In Flight (Figures 10, 50)

Distinctive dowitcher characters are the *very long bill and white back*. Any dowitcher that calls in flight can be identified (see under Short-billed Dowitcher), but dowitchers are often frustratingly silent. The LONG-BILLED is more likely than the SHORT-BILLED to be seen in the same habitat as the COMMON SNIPE. From above the white markings allow easy distinction from a SNIPE in flight, but from below the two birds are more similar. DOWITCHERS show relatively *little contrast between breast and belly*, while SNIPES have a dark breast and white belly.

Voice

The flight call is a single or repeated *peep* or *keek*, when flushed often a rapid, twittered series. See also under SHORT-BILLED DOWITCHER. I hear LONG-BILLED much more often than SHORT-BILLED relative to their numbers. Either they are always more vocal than the other species, or, being less common, they are more likely to be by themselves and thus calling for company.

Further Questions

The wing molt undergone by Long-billed Dowitchers in the southern part of the region is anomalous, the only case known to me of a shorebird molting primaries on migration in the Northwest (avocets and phalaropes do so farther south and perhaps in this region). More information is needed. Do individual birds remain for long periods while in molt? Are other shorebirds doing the same thing at these rich interior lakes?

Notes

The counts of thousands of dowitchers from Tofino in early May in 1974-77, were listed as "mostly Long-billed." They were much more likely to be Short-billed and are so considered here, but that area should be rechecked to confirm this.

Photos

The juvenile in Farrand (1983: 405) is especially bright, probably just off the breeding grounds; most of them seen in the Northwest are much duller. The "winter adults" in Keith and Gooders (1980: Pl. 112) and Farrand (1988b: 149) are juveniles. The "immature" in Udvardy (1977: Pl. 239) is a nonbreeding adult and may be a Short-billed from the obvious spots on its breast.

References

Baldassarre and Fischer 1984 (diet and foraging behavior in migration), Conover 1926 (breeding behavior).

SNIPES Gallinagonini

This group is a large one, with 18 species distributed throughout the world. It is unusual among sandpipers but reminiscent of lapwings and *Charadrius* plovers in having a variety of tropical species closely related to the temperate and subarctic ones. Only two species have been recorded from North America (Table 34), one of them an accidental; both are discussed here. Sight records of the Pin-tailed Snipe from the western Aleutians have not been accepted as valid.

Snipes are small to medium sandpipers that inhabit marshes, where they wade with short legs and probe deeply with long, straight bills. Only dowitchers are similar in shape. Staying in marshes as they do, snipes show no seasonal plumage change, and their fairly dark brown, conspicuously pale-striped upperparts camouflage them well in their reedy environment. Some tropical species are forest inhabitants, converging on the woodcocks in appearance.

Reference

Tuck 1972 (complete monograph).

JACK SNIPE *Lymnocryptes minimus*

This small snipe has been recorded once to the north (Alaska in spring) and once to the south (northern California on November 20) of our region. Secretive, almost raillike as it walks about or crouches in dense low marsh vegetation, it is unlikely to be seen but should be readily recognizable on the ground or in flight. It looks superficially like a COMMON SNIPE but is considerably smaller, about half the bulk. It differs further in having a much shorter bill, slightly longer than the head (about twice as long as head in COMMON SNIPE) and extensively pale at the base; no median crown stripe; conspicuous bright buff back stripes (whitish buff in COMMON and not quite so vivid); and a narrow, dark, pointed tail (broader, rounded, and orange-tipped in COMMON).

The JACK SNIPE is about the size of a YELLOW RAIL, with a back pattern somewhat similar to that species. The YELLOW RAIL has barred sides and, of course, a short, yellow bill, but a look at the head might be necessary to assure which of these rare and secretive species was in sight! The JACK SNIPE forages like a COMMON SNIPE but also bounces like a DIPPER at times as it feeds.

In flight it is similarly raillike, with more rounded wings and fluttery flight than the COMMON SNIPE and without the white wing patches of the YELLOW RAIL. When flushed, typically at close range, the JACK SNIPE rises silently, usually flies only a short distance, and settles again. The occasionally given call is low and weak, unlike the loud *scaip* of the COMMON.

COMMON SNIPE *Gallinago gallinago*

This cryptic year-round marsh dweller is usually flushed before it is seen, its harsh call attracting attention as it zigzags into the sky. The very long bill and other snipe characteristics may be seen during its erratic flight. The most satisfying encounters come in spring in the moist meadows of the interior, where the music of snipes winnowing from above alerts the observer to watch for others capping the fence posts and marsh hummocks below.

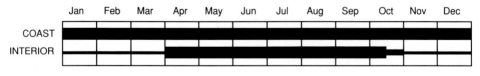

Distribution

The Common Snipe breeds widely in northern North America, from arctic tundra to midcontinent marshes. It winters from the southern half of the United States south to the northern edge of South America.

Northwest Status

The snipe breeds throughout the lowlands of the region, most commonly in the extensive interior marshes. It has been reported to over 5,000 feet elevation in the breeding season in both British Columbia and Montana. Migration occurs primarily from late March to early May and again from September to early November but is difficult to document because of large numbers of wintering birds. The few records of concentrations in spring are higher in the interior than on the coast, but the only substantial autumn concentrations are reported near the coast. Presumed fall migrants have been seen as early as the end of July in areas in which they are not known to breed.

During migration and winter many snipes appear in areas that were not adequate as nesting habitat, and the mass of the population swings west of the Cascades. Even as far north as Vancouver, many dozens of snipes may be seen in a single flooded field during winter. Early April to mid October represents the period of abundance in the interior, although a few birds winter anywhere there is a bit of open water, all the way north to British Columbia's Okanagan Valley.

The Willamette Valley and coast of Oregon play host to amazing concentrations of this species in winter, with tallies on Oregon Christmas

Table 34. Snipe and Woodcock Distribution and Habitat Preference

Species	Hemisphere	Breeding Range	Breeding Habitat	Winter Range	Winter Habitat
Jack Snipe	E	subarctic	marsh/bog	N/equatorial	marsh
Common Snipe	both	sub./temperate	marsh/tundra	N/equatorial	marsh
American Woodcock	W	temperate	woodland	N	woodland

Adult Common Snipe. Long-billed, short-legged sandpiper of marshes with vividly striped head and upperparts. Thorp, WA, April 1985, Richard Droker

Bird Counts the highest for the continent in 1980 and 1983. Apparently this part of North America (including northern California) features just the habitats and climate that appeal to snipes. These populations are variable from year to year, however, as Tillamook had from 3 to 226 and Corvallis from 8 to 435 snipes in the other years from 1979 to 1987.

COAST SPRING. *High count* 72 at Comox, BC, April 16, 1954.

COAST FALL. *High counts* 350 at Reifel Island, BC, September 17, 1965; 180 at Iona Island, BC, October 23, 1971; 200 at Pitt Lake, BC, August 26, 1976.

COAST WINTER. *High counts* 546 at Tillamook Bay, OR, December 20, 1980; 595 at Corvallis, OR, December 20, 1983.

INTERIOR SPRING. *High count* 200 at Pablo Refuge, MT, May 15, 1970.

	Five-Year Averages		
CHRISTMAS BIRD COUNTS	74-78	79-83	84-88
Duncan, BC	41	10	17
Ladner, BC	14	13	13
Nanaimo, BC	9	10	1
Vancouver, BC	24	113	60
Victoria, BC	44	29	21
Bellingham, WA	67	8	28
Grays Harbor, WA	59	50	21
Leadbetter Point, WA	4	4	11
Seattle, WA	18	26	10
Sequim-Dungeness, WA	2	27	15
Tacoma, WA	61	27	6
Coos Bay, OR	58	19	22
Corvallis, OR	71	187	124
Dallas, OR	30	39	41
Eugene, OR	75	145	85
Medford, OR	67	77	46
Portland, OR	80	56	38
Roseburg-Sutherlin, OR	26	99	94
Salem, OR	82	107	39
Sauvie Island, OR	115	87	38
Tillamook Bay, OR	72	119	96
Yaquina Bay, OR	79	7	83

Habitat and Behavior

Snipes are birds of open marshlands, preferring sedge meadows to dense cattails, although they move into dense marshes where the vegetation has broken down and the substrate is exposed. Any grassy field at least partially flooded is appropriate habitat too. Salt marshes are inhabited sparingly during migration and winter. They may forage either in fairly dense, low marsh vegetation or in the open in plowed fields, but they do not use mud flats. They move slowly while probing deeply and fairly rapidly in the substrate, in soft soil or mud and in water up to their belly. When not feeding they crouch cryptically for long periods. For the most part they appear to remain in small areas, flying primarily when flushed. Snipes are relatively solitary birds, but large numbers of them may occur loosely associated in favorable feeding areas, in particular flooded farmlands. Occasionally a group is seen roosting together on high ground, in the manner of a high-tide roost of coastal species, and at times they coalesce into small flocks in flight.

On their breeding grounds male Common Snipes display aerially (winnowing) in a "roller-coaster" flight. The musical sound (*whi whi whi whi WHI WHI WHI WHI* etc.) they make, seeming to rise and fall, is produced by the widespread outermost tail feathers vibrating as the bird gains speed on each dive. Sounding hollow and ethereal, the winnowing is made even more mysterious by coming from high in the air, the snipe a speck in the sky or entirely invisible. The male may give a vocal "song" during its final descent, dropping from the air with wings held high, or it may sing similarly during a low, rapid flight over the marsh or while perched on a fence post or even utility pole. These conspicuous birds convey a very different impression from the skulkers of winter.

Nests are well concealed, usually among dense grass or sedge hummocks at the edges of marshes, flooded fields, and bogs. The usual clutch size is four, the eggs incubated only by the female. Oddly, the male remains in the area, and each parent cares for one or more of the young until they fledge. This is one of the few shorebirds in which a parent that does not incubate is involved in further parental care.

Structure

Size medium-small. Bill long to very long, legs short. Wings do not reach tail tip, primary projection lacking. Tail rounded.

Plumage

Adult. Bill blackish at tip, gradually becoming grayish toward base; legs gray to greenish. Head pattern contrasty with white median crown stripe, wide, dark brown lateral crown stripe, white supercilium, dark eye stripe, and slightly narrower, paler brown stripe on each side of face. Neck heavily striped with brown; mantle mostly blackish with wide, cream-colored line on either side. Scapulars blackish with scattered transverse reddish markings and cream-colored to buff outer edges, producing on either side (typically) continuous line bordering upper scapulars and interrupted one bordering lower scapulars. Thus as many as six conspicuous pale stripes on upperparts. Tertials dark brown, irregularly and profusely barred with light brown to buff and fringed with whitish. Coverts gray-brown, barred with darker brown and tipped with whitish. Breast suffused with light brown, it and sides barred with dark brown; belly white. Undertail coverts light brown, usually barred with black. Some individuals more heavily barred underneath, including almost entire belly.

Lower back dark brown, uppertail coverts mottled reddish brown and black. Rectrices complexly patterned; central ones largely black and outer ones barred black and white, with extensive rufous at tips of most. Wing dark brown above and below, marked only with whitish tips of most coverts and narrow white tips of secondaries; looks unmarked at a distance. Underwing dark gray-brown; greater coverts with white tips, innermost ones and axillars evenly barred with white. Median, lesser, and marginal coverts darker brown, all evenly barred with white.

Wing molt primarily during August on breeding grounds.

Juvenal. Snipes scarcely change plumage during their lives, juveniles looking exactly like adults in the field. Even in hand, average differences in wing-covert patterns are difficult to see. Juvenile coverts average more richly colored than those of adults, bases darker brown and tips brighter buff to rufous. Adults usually with dark brown shaft streak extending to tip of each median and lesser secondary covert, juveniles without this streak but instead with faint black margin to same feathers.

Subspecies

The North American *G. g. delicata* is the regularly occurring Northwest subspecies. The Old World *G. g. gallinago* is a regular migrant through the western Aleutian Islands and has probably bred there; it is also recorded from some of the central Bering Sea islands. It is as likely a vagrant in the Northwest as many other Siberian shorebirds but will be impossible to document without a specimen. The only difference apparent in the field is the slightly broader white trailing edge of the wing in flight in *gallinago* (this was evident to me the first time I saw individuals of this race), but this would be difficult if not impossible to document. In the hand it could be seen that *gallinago* has 14 tail feathers, the outer ones usually more than 9 mm in width, while *delicata* has 16 tail feathers, the outer ones usually less than 9 mm in width. The winnowing song of *gallinago* is lower-pitched and of a different quality than that of *delicata*, but the flight calls sound about the same.

Identification

All Plumages. The snipe is a *short-legged, long-billed* sandpiper, adapted for moving slowly over moist ground while probing for deep-dwelling invertebrates. No other similar-sized shorebirds approach its bill length except for the two species of dowitchers (the long-billed is about equal), and it can be easily distinguished from them by its *white-striped head and back.* Juvenile dowitchers, brown above with white belly, have light supercilia and obscure buffy stripes on the back, but with nothing like the conspicuousness of the snipe's stripes. dowitchers have obscurely barred sides in adult plumages, which otherwise look quite unsnipelike. snipes tend to forage in dense vegetation, dowitchers out in the open, but they can be seen together in both fresh and salt marshes, where they feed similarly.

If the bill were not seen, a pectoral sandpiper, with its brown, striped back and brown breast, might be mistaken for a snipe, but the snipe is a stockier, shorter-legged bird, with more vivid stripes and heavily *barred sides.*

In Flight (Figures 9, 50)

When snipes flush they typically "tower" high

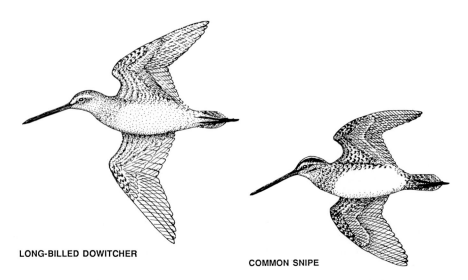

LONG-BILLED DOWITCHER

COMMON SNIPE

Fig. 50. Dowitcher and snipe in flight. Nonbreeding dowitcher is relatively long-winged, snipe more contrasty, with darker breast, underwings, and undertail.

in the air, often to disappear in the distance. In flight overhead they share their *very long-billed* silhouette with DOWITCHERS but have a *dark breast, wing linings, and undertail coverts contrasting with a white belly.* DOWITCHERS in any plumage are rather uniformly colored beneath. SNIPE wings are slightly shorter than those of DOWITCHERS, giving the birds a less racy look. From above they look all brown, and the back stripes and orange on the tail may be seen, very different from the DOWITCHER'S white back and gray tail.

Voice

The flight call is a loud *scaip* (could be construed as *snipe*), typically given only once as they fly away. Less often a bird flying past will call. The breeding-ground vocalization is a loud, continued series of notes (*chip-er-chip-er-chip* etc.).

See under Habitat and Behavior for a description of the characteristic winnowing.

Further Questions

Although of low probability in the region, the Siberian subspecies *gallinago* might be detected with regular examination of snipe hunters' bags by wildlife personnel. Sex (by dissection) and age ratios could also be determined in this way.

References

Brooks 1967 (prey choice in migration), Burton 1974 (foraging behavior), Dwyer and Dobell 1979 (age determination), Fritzell et al. 1979 (diet in migration), Mason and Macdonald 1976 (breeding biology), Senner and Mickelson 1979 (diet in migration), Sutton 1981 (displays), Williamson 1950 (distraction display).

WOODCOCKS Scolopacini

There are six species of woodcocks, of which two have been reported from North America. One species, widespread in the eastern half of the continent, has occurred in the Northwest and is covered herein (Table 34). Woodcocks are birds of temperate-zone deciduous forests, although a few inhabit tropical forests of southeast Asia. The tropical woodcocks are among the most poorly known shorebirds.

These sandpipers are of medium size and very similar to snipes and dowitchers in their feeding adaptations, with long bills and short legs, but they are adapted to woodland environments rather than marshes or mud flats. Because of this they are grouse-colored, heavily patterned with darker and lighter brown above and rich cinnamon, buff, or barred black and white below. Migrations occur between similar habitats, and there is no seasonal plumage change.

Because they stay within the woodlands, their wings are rounded and grouselike, adapted for bursts of quick flight to escape predators and quite different from the usual pointed, long-distance shorebird wing. Birds in such dense environments have a difficult time spotting predators, and the eyes of woodcocks are set relatively far back on the head, to enhance their vision when their bill is buried in the ground.

AMERICAN WOODCOCK *Scolopax minor*

As its name implies, this is no ordinary shorebird, and shorebird enthusiasts scrutinizing open wetlands should be the last to find it. With its plumpness, rounded wings, dead-leaf coloration, and woodland habitat, a woodcock might register as a long-billed grouse.

	Jan	Feb	Mar	Apr	May	Jun	Jul	Aug	Sep	Oct	Nov	Dec
COAST			•									
INTERIOR										•		

Distribution

The American Woodcock breeds in eastern North America, from southern Canada south almost to the Gulf of Mexico and west to the eastern edge of the Great Plains. It winters on the Atlantic and Gulf coastal plains.

Northwest Status

This species is rare west of its breeding range and is no more than a vagrant as far west as the Rocky Mountains. There are two sight records from the Northwest, acceptable because the bird is unmistakable: one in winter from the British Columbia coast and one in fall from Montana. Another October record from eastern Montana indicates fall as the season in which to watch for such vagrants. Unfortunately, the British Columbia record could stem from the release of woodcocks into California some time during that period in a misguided attempt to establish them on the Pacific coast.

COAST WINTER. North Surrey, BC, March 5, 1960.

INTERIOR FALL. Ninemile Creek, MT, October 1, 1983.

Habitat and Behavior

The American Woodcock is solitary, secretive, and active at night, thus considerably less likely to be seen than other shorebirds of the region. It is very much a woodland species but also occurs in wet thickets adjacent to woodlands. It probes in soft soil for earthworms and other invertebrates like a snipe or turns over leaves and detritus like a turnstone.

Structure

Size medium. Bill long, tapered from fairly thick base. Legs short. Wings rounded, fall short of tail tip; no primary projection. Three outermost primaries shorter and much narrower than others, function in sound production. Tail rounded.

Plumage

Adult. Bill dull pinkish, extreme tip black. Legs dull pinkish. Forehead gray-brown, crown and nape crossed by three narrow buff bars, each in front of broad black bar. Narrow black lines through lores and across lower cheek, also, in some individuals, from bill through center of forehead. Hindneck gray-brown. Exposed parts of mantle feathers and scapulars mostly black, barred and tipped with cinnamon. Both upper and lower scapulars broadly edged with pale gray, forming wide mantle and scapular lines. Tertials and coverts gray-brown, finely barred with buffy brown; tertials also with small black blotches near tips. Chin white, otherwise entirely cinnamon below from throat to undertail coverts.

Lower back finely barred brown and cinnamon. Central uppertail coverts mostly black barred with cinnamon; outer uppertail coverts mostly cinnamon, longer ones finely barred with brown. Rectrices black, dotted with cinnamon on outer webs and broadly tipped with gray. Wing brown, flight feathers darker than coverts; lesser and median coverts of underwing and axillars same cinnamon color as underparts.

Juvenal. Just-fledged juveniles with throat and foreneck same gray-brown as hindneck, contrasting more with white chin than in adult (may not allow distinction later in season). Seem more likely to have black forehead stripe developed.

Identification

Even if a vagrant WOODCOCK were to turn up in a more typical shorebird habitat, its *black head bars, mottled upperparts, and cinnamon underparts* would distinguish it from DOWITCHERS and SNIPES, the other long-billed species in its size range. If the bill weren't seen, one walking in the woods would more likely be mistaken for a miniature GROUSE.

In Flight

The *long bill, rounded wings, and uniform brown coloration*, with tawny underparts and wing linings, are distinctive in flight, quite different from a SNIPE's pointed wings and contrasting brown-and-white underparts. The habitat and grouselike explosive flushing are also distinctive, as is the sound made by the wings.

Voice

No vocalization when flushed but male has somewhat nighthawklike *peent* call during courtship. Narrowed outer primaries presumably responsible for whistling or twittering sound made when flushed and during courtship flight (male's outer primaries slightly narrower than female's).

References

Davis 1970 (territoriality), Diefenbach et al. 1990 (winter distribution patterns and site fidelity), Krohn 1971 (summer activity patterns), Martin 1964 (sexing and ageing), Owen and Krohn 1973 (molt and weight), Pettingill 1936 (general biology), Pitelka 1943 (territoriality and display), Rabe et al. 1983 (foraging strategies), Sheldon 1967 (general biology), Smith and Barclay 1978 (distribution).

PHALAROPES Phalaropodini

This distinctive tribe is made up of three small swimming sandpipers, two worldwide and one American (Table 35). The bills are straight and very fine in two species and thicker, more like other sandpipers, in the third. For such a small group, bill and leg proportions vary substantially (Appendix 3). The Red-necked and Wilson's are long-billed species, the Red much shorter-billed, certainly correlated with feeding differences. The more terrestrial Wilson's has relatively longer legs than the other two species. The toes are lobed for swimming, and the birds spend much of their life on the water. They swim rapidly and jerkily, heads bobbing back and forth like seagoing pigeons.

The Wilson's feeds regularly on the shore, but the other two species spend their nonbreeding season mostly at sea (and can be called "sea phalaropes" to distinguish them from Wilson's); migration is long-distance in all species. Seasonal plumage change is substantial, associated in part with seasonal habitat change and in part with the evolution of colorful plumage for sexual competition. All species feed by picking tiny invertebrates from or just under the surface. They also feed by spinning around on the water's surface (Figure 51), causing minivortices that draw their prey up in the water column.

Phalaropes are the only sandpipers other than Spotted in which the females perform courtship behavior, and the only sandpipers in which only the males incubate the eggs and care for the young. This is an excellent group with which to get out of the habit of calling all birds on nests "her." Also distinctive is the reversed sexual dimorphism of phalaropes, the females larger and more brightly colored than the males. Polyandry (females mating with more than one male) has been reported in all three species, as might be expected from the size and color dimorphism, but monogamy is nevertheless the mode.

The Wilson's Phalarope in nonbreeding plumage looks superficially like a tringine, the other two species somewhat like calidridines, but the

Fig. 51. Spinning phalarope. Spinning behavior typical of all three phalarope species.

relationship of this group to the other sandpiper tribes is obscure. The Wilson's is certainly sufficiently different from the other two species to deserve its own genus, and it is surprising that all three were lumped by the American Ornithologists' Union (1983).

References

Blomqvist 1983b (bibliography), Dittmann and Zink 1991 (phylogeny), Höhn 1969 (natural history and hormone effects), Meinertzhagen 1925 (distribution), Wetmore 1925 (food).

Table 35. Phalarope Distribution and Habitat Preference

Species	Hemisphere	Breeding Range	Breeding Habitat	Winter Range	Winter Habitat
Wilson's	W	temperate	marsh	equatorial/S	marsh
Red-necked	both	arctic/subarctic	tundra	equatorial	open ocean
Red	both	arctic	tundra	N/equatorial/S	open ocean

WILSON'S PHALAROPE *Phalaropus tricolor*

Sailing over the open water among striding stilts and avocets or sneaking among the sedges, this species is a feature of alkaline interior marshes. Flashy females pursue drab males until they convince one to be a single parent, then abandon them to migrate south. The males follow soon after, then the juveniles, assembling in huge flocks to molt and prepare for their long journey to Argentina.

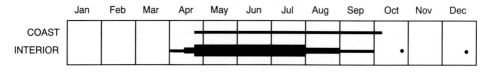

Distribution

Wilson's Phalarope breeds across much of western North America and winters in southern South America in similar marshy habitats. Many birds, especially juveniles, reach the Atlantic coast in migration, but the bulk of the population moves

south through the West, gathering in huge flocks at some inland lakes used as staging areas.

Northwest Status

This species breeds widely in the interior of the region, occurring all over eastern Washington, southeastern Oregon, southern Idaho, and western Montana and locally in southern British Columbia and northern Idaho. In Oregon it breeds in some Cascade lakes, for example Diamond Lake, to over 5,000 feet in elevation. It also breeds at least occasionally west of the Cascades, most recently and regularly at Iona Island and near Victoria. It has expanded its range greatly in the past half-century, moving from the Great Plains across the Rockies to the coast, but it also has pulled back recently from some former breeding sites, including Nisqually Refuge (to about 1975) and Hoover's Lakes (until 1969). It is typical for Wilson's Phalaropes to breed in ephemeral marshes, nesting populations moving about the countryside from year to year.

These birds are fairly late migrants, many arriving as much as a month after the avocets with which they cohabit. Average arrival dates on breeding grounds vary from the last week of April in southern Oregon and southern Idaho to the second week of May farther north. The average arrival date over 12 years at Fortine was May 11, the range May 3-19. The adults' stay is relatively short, and females both arrive and leave slightly before males. Flocks in mid July usually contain only males and juveniles, and even they disappear shortly thereafter from drying-up nesting marshes. After breeding they congregate at large lakes in the southern part of the region, staging grounds for southward migration; numbers have reached the thousands at these localities. Juveniles show a more scattered distribution.

Migrants turn up almost anywhere in the area, much more often in spring than in fall west of the Cascades and somewhat more so in the northern interior. Birds arrive west of the Cascades at about the end of April, but records are concentrated surprisingly late, in late May and early June; thus birds in June should not be interpreted as breeding without further evidence. Also, at least females could be in southbound migration by mid June.

Other than at the staging grounds, few are seen in fall migration, with a smattering of records in July, August, and early September away from breeding sites. The latest record from the interior, December 19, was during an exceptionally mild winter. Adults are quite uncommon west of the

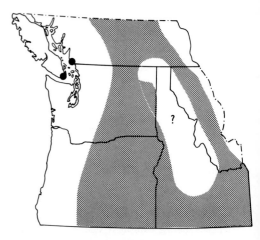

Wilson's Phalarope breeding distribution

Cascades in fall, the juveniles that first appear in mid July constituting most of the western records.

COAST SPRING. *Early dates* Jackson County, OR, April 20; Surrey, BC, April 4, 1960; Serpentine River, BC, April 17, 1965. *High counts* 51 at Ankeny Refuge, OR, May 13, 1986; 30 at Iona Island, BC, May 14, 1973.

COAST FALL. *High counts* 11 at Iona Island, BC, September 7, 1975; 12 at Blaine, WA, July 14, 1975. *Late date* Comox, BC, October 2, 1928.

INTERIOR SPRING. *Early date* Malheur Refuge, OR, April 2, 1964. *High count* 112 at Fortine, MT, May 22, 1977.

INTERIOR FALL. *Adult high counts* 13,000 at Malheur Refuge, OR, July 3, 1977; thousands at Abert Lake, OR, July 25, 1982; thousands at American Falls Reservoir, ID, June 28, 1979. *Juvenile high count* 46 at Vernon, BC, August 30, 1950. *Late dates* southeastern WA, October 26; Summer Lake, OR, December 19, 1989.

Habitat and Behavior

For nesting habitats Wilson's Phalaropes choose small freshwater marshes, sedge meadows, or marshy areas at lake edges. They regularly occur with Black-necked Stilts and American Avocets but move into denser vegetation to forage. In migration they may be seen anywhere in fresh water, much more rarely in coastal habitats.

Wilson's and Red-necked phalaropes are often seen together in migration in the interior. Wilson's swim like other phalaropes but whirl about less than the other species. They forage more commonly in shallow water or on shore than the other phalaropes, and they run about more actively and erratically than other shoreline sandpipers.

Wilson's Phalaropes feed by pecking at the surface, by bill-pushing (thrusting the open bill into the water while advancing rapidly, perhaps virtual plankton feeding), or by tipping up like little dabbling ducks. Individuals in shallow water may feed by scything like a Lesser Yellowlegs, but they move the bill rapidly back and forth through the water two or three times, in contrast with the yellowlegs' single scything movement on each pass. Even on land they may feed with sideways swipes of the bill.

Courtship is vigorous, with frequent pursuit flights and aggression against intruding females. Males are guarded as moving resources until the eggs are laid. Nests are shallow scrapes, usually well hidden in dense low grasses and sedges near water. Four eggs are typically laid, the male incubating them and taking care of the young until they fledge.

Structure

Size small, males about 25 percent smaller than females, difference evident in field when sexes together (thus affording sexual identification of birds not in breeding plumage). Bill medium and quite slender. Legs short, toes relatively longer than in other phalaropes; toes slightly webbed at base and with lateral flanges, but not lobed as in other phalaropes. Wings extend to tail tip or just beyond, primary projection absent in breeding adults and moderate in juveniles. Tail doublenotched; central rectrices projecting slightly beyond others as in many *Calidris*, differently shaped than in other phalaropes.

Plumage

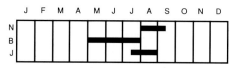

Breeding. Bill black, legs brownish gray to black. Sexual dimorphism in color pronounced. Females chestnut, rufous, gray, and white. Gray above, palest on crown and mantle, darker on scapulars and tertials, and darkest on coverts. Hindneck, short supercilium (no more than spot

Breeding male (left) and female Wilson's Phalarope. Large, often terrestrial phalarope with long neck and slender bill; dark stripe on neck and rusty-washed breast diagnostic of species and plumage; male conspicuously duller than female. Near Potholes Reservoir, WA, 17 May 1980, Dennis Paulson

in some birds), throat, and posterior underparts white. Foreneck and upper breast rich rufous, in some birds shading to gray-brown on sides of breast. Wide black stripe runs from each eye down side of neck, changing to dark chestnut on either side of mantle and extending as narrow chestnut stripe along lower scapulars.

Lower back gray-brown; uppertail coverts white, rearmost ones with gray-brown tips. Rectrices gray with white centers and tips, latter disappear with wear. Wing brownish gray above, white marginal coverts and dark, white-tipped greater primary coverts create obscure pattern. Underside of flight feathers and bases of greater primary coverts pale brownish gray, rest of underwing and axillars mostly white; brown bases of many coverts create mottled effect.

Pattern of female fairly complex, but little individual variation; males quite variable, however. Brightest males look superficially like females, differing by dark gray crown (best characteristic for sexual distinction), longer white supercilium (extending behind eye), and more mottled upperparts, feathers with darker centers than those of females. Many males duller than this, grayish brown with little indication of reddish, as if sepia tone versions of the brighter ones. Dullest males, distinctly in minority, differ from nonbreeding-plumaged individuals primarily by leg color and darker, mottled back.

Nonbreeding (Figure 52). Legs olive to yellow or orange-yellow, beginning to change color as soon as breeding over in July and beginning to change again as early as January in adults, all arriving in Northwest with dark legs. Plain gray above and white below, with white supercilium and darker gray postocular stripe. Scapulars and tertials narrowly fringed with white, which wears off by late winter. Molt very early in autumn, some adults in full nonbreeding plumage by August, while others still in very worn breeding plumage. Females typically molt slightly before males. Wing molt occurs in at least some birds in southern part of region, perhaps only in those using big lakes for staging areas.

Juvenal. Legs yellowish, as in nonbreeding adult. Crown and upperparts brown, hindneck gray. All feathers of upperparts—mantle, scapulars, tertials, and coverts—fringed with buff, creating strongly scalloped or, in some individuals,

lined (especially on mantle) pattern. Fairly distinct dark eye stripe and white supercilium. Underparts entirely white, fresh-plumaged individuals with buff-washed breast.

Immature. As in adult, autumn molt of juvenile occurs quite early, with complete replacement of body plumage of many birds by end of August. Still recognized as immature by scalloped coverts and buff-edged tertials, but by midwinter two age stages similar. Primary and tail molt in midwinter in first-year birds, molt into full breeding plumage in first spring.

Identification

All Plumages. Phalaropes are typically seen *swimming*, although WILSON'S *also feeds on shore* more than do the other two species. Even when on shore, its *erratic foraging actions* furnish distant recognition of a phalarope. It is a *large, long-necked, thin-billed* phalarope, considerably larger than the RED-NECKED and thinner-billed than the RED PHALAROPE.

Breeding Plumage. Breeding females are unmistakable with *pale gray crown, black-and-rufous neck stripe, and cinnamon breast*. Males are duller, darker versions of the same, although the dullest is a dull bird indeed, with its strongest field marks its *phalarope habits, needle bill, white supercilium, and dark eye stripe*. The *plain back* is the best distinguishing mark from a male RED-NECKED PHALAROPE, which has a similar head and neck pattern but vividly striped back.

Nonbreeding Plumage. The *plain, light gray upperparts* of this species, together with its *phalarope habits, fine, straight bill, and pure white breast* in this plumage (and in first-winter plumage), allow it to be distinguished from other straight- and slender-billed sandpipers in its size range such as LESSER YELLOWLEGS and SOLITARY SANDPIPER. From the other PHALAROPES it can be distinguished by its *narrow postocular stripe*, from RED-NECKED further by its *plain upperparts*, and from RED by its *thin bill*.

Juvenal Plumage. This is the palest of the juvenile phalaropes, with *pale buff markings above* rather than the vivid dark buff of the other two species. It is never so vividly lined as the RED-NECKED nor so dark-breasted as the RED. Compared with the other phalaropes, the *postocular stripe is dull and narrow* in WILSON'S,

Nonbreeding Wilson's Phalarope. Plain gray up-perparts, entirely white underparts, and inconspicuous postocular stripe diagnostic of species and plumage; one worn, brownish scapular from breeding plumage just visible; gray leg color retained from breeding plumage, may also be yellow by this time. Mono Lake, CA, July 1980, J. R. Jehl, Jr.

First-winter Wilson's Phalarope. As nonbreeding adult, but coverts and tertials from pale-fringed juvenal plumage. Jamaica Bay Wildlife Refuge, NY, late August 1984, Urban Olsson

Juvenile Wilson's Phalarope. Superficially like dull breeding male, especially head and neck pattern, but scapulars, coverts, and tertials conspicuously fringed with pale buff. Mono Lake, CA, 2 August 1982, J. R. Jehl, Jr.

although it can be quite distinct. In juvenile sandpipers other than PHALAROPES and RUFF, the loral stripe is more conspicuous than the postocular.

In Flight (Figure 10)

A member of the *plain-winged, white-tailed* group of shorebirds, it can be distinguished from LESSER YELLOWLEGS and STILT SANDPIPER in its size range by *short legs* (the tips of the toes at most projecting beyond the tail) and, in nonbreeding plumage, *pale gray back and white breast*. The rich colors of breeding-plumaged birds can often be seen in flight.

Voice

Both sexes give a nasal grunting call (*ernt*) on the breeding grounds that may be used as a contact call, but it is given especially frequently by courting females. The grunting call can also be heard in migration, but it does not carry very far.

Further Questions

There may be autumn staging areas for this species in southern Oregon and Idaho, comparable to the well-known one at Mono Lake, California, where they undergo molt before leaving for South America; this needs to be checked. It would be of interest to document differences in feeding behavior between Wilson's and Red-necked phalaropes where they occur together on the big interior lakes. And the significance of the substantial variation in males defies understanding.

Photos

The "winter plumage" bird in Bull and Farrand (1977: Pl. 204) is a breeding-plumaged male, a good example of the erroneous conclusion "if it isn't bright it must not be in breeding plumage." The "immature" in Udvardy (1977: Pl. 200) is an adult in fresh nonbreeding plumage. The "bird molting into winter plumage" in Farrand (1983: 411) is a juvenile with just a few new scapulars of its first-winter plumage, while the "juvenile" in Farrand (1988a: 175) has completed that molt. The "winter" bird in Hammond and Everett (1980: 127) is probably in first-winter plumage, from the contrast between fresh, pale scapulars and more worn, darker tertials.

References

Baldassarre and Fischer 1984 (diet and foraging behavior in migration), Bomberger 1984 (nest-site placement), Burger and Howe 1975 (molt

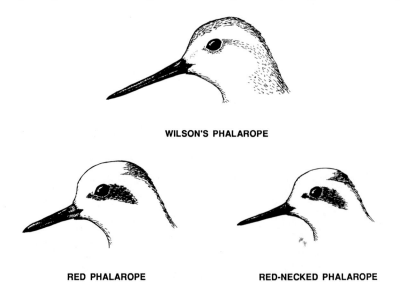

WILSON'S PHALAROPE

RED PHALAROPE   RED-NECKED PHALAROPE

Fig. 52. Nonbreeding phalaropes. Wilson's with long, slender bill, gray crown, conspicuous supercilium; other two species with shorter bills, white crown (dark in juvenile), dark ear patch. Red-necked with slender bill, Red with thick bill.

and winter feeding behavior), Colwell and Oring 1988a (sex ratio and mate competition), Colwell and Oring 1988b (breeding biology), Höhn 1967 (breeding biology), Höhn and Barron 1963 (summer diet), Holmes 1939 (winter distribution), Howe 1975a (courtship behavior), Howe 1975b (breeding behavior), Jehl 1981 (autumn staging), Johns 1969 (breeding behavior), Kagarise 1979 (breeding biology), Mahoney and Jehl 1985 (adaptations to salinity), Murray 1983 (breeding biology), Reynolds et al. 1986 (spring arrival), Siegfried and Batt 1972 (foraging behavior).

RED-NECKED PHALAROPE *Phalaropus lobatus*

Bobbing and fluttering, riding the fiercest waves, this shorebird is actually the smallest of the seabirds. The widespread seagoing rafts are not duplicated inland, but enough of these dainty plankton-pickers are seen away from the ocean to fulfill the fantasies of phalarope-fanciers.

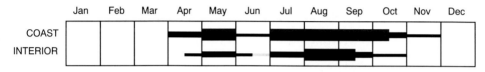

Distribution

The Red-necked Phalarope breeds worldwide in low arctic latitudes and winters very locally on tropical seas. Most migration is offshore, many Pacific birds within sight of land. The species is also locally numerous in the western interior, and smaller numbers move continentwide.

Northwest Status

This species is a migrant through the Northwest, common offshore and at times along the coast, less common but regularly reported elsewhere in the region. The main migration period falls in May in spring. Females arrive in numbers before males on the breeding grounds and may migrate through our region slightly earlier; however, they are mixed together in most flocks I have examined. Some June records are difficult to categorize and may indicate occasional summering. Fall migration extends from July to mid September, less commonly through October. Females leave the breeding grounds earlier than males, but there is no information about the migration schedules of the two sexes farther south.

The largest flocks in the region have been seen off the west side of Vancouver Island during most of May and from late July to early September. The July and late August/September dates probably represent migratory peaks of adults and juveniles, respectively. Note that adult peaks were recorded in Queen Charlotte Strait and juvenile peaks off Victoria, at the other end of Vancouver

Island. Numbers recorded have varied greatly from year to year on Washington offshore trips.

Coastal daily counts of 1,000 or more have been made from April 24 to May 14 on the outer coasts of Oregon and Washington, primarily during strong westerly winds. Although a regular migrant in protected waters south and east of Victoria, large numbers are not usually seen in that area, but 1,000 appeared off Seattle on one occasion. Individuals have been recorded at times in "wrecks" of Red Phalaropes, and such a wreck in late October 1930 on the Oregon coast involved primarily Red-necked Phalaropes.

Migration in the interior occurs primarily in May and from July through September, peaking in August and early September (probably juveniles). Records usually involve a few to a few dozen birds, but flocks of several hundred have been seen. The spring record at Lake Lenore involved an unusually large number, perhaps storm-blown birds. Small numbers regularly summer on large lakes at Malheur Refuge, with a maximum of 25 in 1982.

COAST SPRING. *Early dates* south jetty of Columbia River, OR, March 27, 1971; Yaquina Bay, OR, April 14, 1982 (200). *High counts* thousands off Westport, WA, May 3, 1975; 1,140 off Westport, WA, May 9, 1982; 20,000 at Cleland Island, BC, May 15, 1969; 17,000 near Port San Juan, BC, May 5, 1985. *Late dates* Yaquina Bay, OR, June 7, 1983 (500); Destruction Island, WA, June 10, 1916 (flew into light).

Breeding female Red-necked Phalarope. Small, slender-billed sea phalarope; dark head and back, white throat, and rufous neck diagnostic of species and plumage; plain head with white upper eyelid distinctive of female. Wales, AK, 10 June 1978, Dennis Paulson

COAST SUMMER. South jetty of Columbia River, OR, June 20-24, 1965; Tillamook, OR, June 11, 1983; Leadbetter Point, WA, June 12, 1971, June 17, 1975; Saanich, BC, June 26, 1976.

COAST FALL. *Adult high counts* 8,000 in Queen Charlotte Strait, BC, July 29, 1970; 4,500 in Queen Charlotte Strait, BC, July 14, 1975; 4,500 in Queen Charlotte Strait, BC, July 4, 1982; 3,129 off Westport, WA, July 26, 1984. *Juvenile early dates* Mason County, WA, August 11, 1917#; Beaver Creek, OR, August 14, 1947#. *Juvenile high counts* 4,000 at Victoria, BC, August 31, 1974; 2,500 at Victoria, BC, September 2, 1975; 3,000 at Victoria, BC, September 13, 1978; 1,957 off Westport, WA, September 7, 1980 (with over 1,000 unidentified phalaropes); 1,000 at Seattle, WA, September 4, 1985. *Late dates* Victoria, BC, November 25, 1979; Tillamook Bay, OR, November 26, 1960.

INTERIOR SPRING. *Early dates* Malheur Refuge, OR, April 17, 1985; Klamath Falls, OR, April 17. *High counts* 300 at Lake Lenore, WA, spring 1963; 30 at Reardan, WA, May 20, 1975; 20 at Osoyoos, BC, May 27, 1979. *Late dates* Missoula, MT, June 14, 1959; Redmond, OR, June 12, 1924.

INTERIOR FALL. *Early date* Lewiston, ID, June 18, 1955. *Juvenile early date* Okanagan Landing, BC, July 28, 1915#. *High counts* 75 at Creston, BC, September 10, 1947; 200 at Salmon Arm, BC, August 28 to September 2, 1970; 200 at Alvord Lake, OR, late August 1979; 60 at National Bison Range, MT, July 24, 1963. *Late dates* Vernon, BC, October 27, 1963; Walla Walla, WA, October 28, 1955; Hatfield Lake, OR, November 17, 1984.

Habitat and Behavior

This is a bird of open water bodies, from ponds to large lakes to storm-tossed ocean swells. Upwellings and convergence lines (tide rips), where floating micro-organisms are concentrated, are highly favored, and birds often feed at the edges of kelp beds. Birds blown to the coast by storms may be seen anywhere, even in small ditches. Sewage ponds are preferred habitats where they are available, full of little invertebrates to delight a phalarope's palate. Flocks are typically small, but at times, especially on convergence lines, there are dozens to hundreds of birds spread out across the ocean's surface. They drift with tidal currents and then fly upcurrent to drift again.

Feeding consists of short jabs to pick minute invertebrates from or just below the surface. Spinning is common in shallow water but not on the ocean, where it is probably nonfunctional. The Red-necked is much less likely than Wilson's to feed on shore, but occasionally one is seen running about erratically.

Breeding male Red-necked Phalarope. As breeding female, but typically duller, with white and rufous supercilium. Wales, AK, 10 June 1978, Dennis Paulson

Structure

Size small. Females average larger than males, but size difference less than in other phalaropes and not evident in field. Bill short to medium, straight, and very fine. Legs short, toes extensively webbed at base and with scalloped lateral flanges. Wings reach tail or fall slightly short of it, primary projection short in breeding adults and moderate in juveniles. Tail rounded.

Plumage

Breeding. Bill black, feet dark gray to black. Females slaty gray above, gray extending onto sides of breast and sides striped with same color. Broad rufous mantle and scapular lines; some scapulars and coverts narrowly white-tipped, tertials white-edged. Spot above each eye and throat and upper foreneck white, contrasting strongly with dark head and hindneck. Prominent chestnut patch on either side of neck extending across foreneck in some individuals to border throat irregularly from behind. In others dark gray of breast extends forward to throat, splitting chestnut into two patches. Chestnut patch also varies in forward extent, in some birds extending onto

head toward but not contacting supercilium. Variation sufficient that dullest females look much like males, some birds perhaps not sexable in field. In worn-plumaged summer birds, scapular and mantle stripes have faded slightly, and white tips and edges have disappeared from scapulars, tertials, and coverts.

Males usually obviously duller and more heavily patterned than females, with upperparts slightly darker gray, emphasizing pale back stripes. Rufous neck patch varies from about as large and bright as in typical females to virtually lacking, replaced by gray. Supercilium better developed in males, generally white before and over eye and rufous behind it; best sexual distinction joining of rufous neck patch and white supercilium in males. Gray feathers of upperparts and breast narrowly white-tipped, further breaking up pattern. Tertials and, in some individuals, scapulars gray-brown to reddish brown with blackish centers and narrow white or buff tips and/or fringes, again more complexly patterned than in females. Pattern becomes simpler in males by midsummer, when many pale fringes worn off.

Lower back, central uppertail coverts, and central pair of rectrices blackish. Outer uppertail coverts black and white barred, outer rectrices gray. Wing dark gray above, wing stripe formed from broad white greater secondary covert tips and narrow white greater primary covert tips. Wing fairly dark from beneath, white patch formed from white on inner greater secondary coverts diminishing to narrow tips on inner primary coverts and another patch formed from inner median secondary coverts. Marginal coverts dark gray, axillars white.

All birds in spring in Northwest in breeding plumage, but molt out of this plumage occurs very quickly in both sexes, beginning on arctic breeding grounds. Large numbers seen off Grays Harbor on July 28 in both 1982 and 1983 in nonbreeding plumage (thus in "winter" plumage in midsummer!). Other individuals retain much breeding plumage well into August. Wing molt usually in midwinter, but begins in August during migration in birds visiting Mono Lake in California. Tail molt also occurs during migration.

Nonbreeding (Figure 52). Gray and white striped above and entirely white below. Head mostly white, with narrow, dark gray patch from eye to ear (combination postocular stripe and auricular spot). Nape, narrow stripe down hindneck, and upperparts generally medium gray, scapulars and tertials darker and edged with white to produce striped back pattern. Outer uppertail coverts entirely white. White edges on scapulars and tertials wear off sufficiently so stripes less evident by midwinter, birds looking like this unlikely to be seen here. Tertials molt fairly late, dull brownish and contrasting with fresh gray scapulars until late in fall.

Juvenal. Brightly marked on and near breeding grounds, blackish above with bright buff mantle and scapular stripes and white-edged tertials; scapular and mantle stripes fade to pale buff in older individuals. Crown and postocular stripe/auricular spot blackish, forehead white. Breast washed with brown. Uppertail coverts as in nonbreeding adults. Brown-breasted birds seldom seen in our region, as brown fades quickly. Just as in adults, molt in juvenile phalaropes earlier and more rapid than in other shorebirds, many juveniles begin to replace upperpart feathers with gray first-winter equivalents by mid August. By September majority of juveniles in molt but not yet in full first-winter plumage, achieved south of our region. Even when gray-backed,

juveniles distinguished from nonbreeding adults through most of fall by dark crown and dark, white-edged tertials.

Immature. Most individuals apparently molt flight feathers during first winter, mature in first year, and attain full breeding plumage. Minority delays maturation for another year, with worn flight feathers and nonbreeding plumage in first summer.

Identification

All Plumages. Almost always *on the water*, its aquatic behavior characterizes it as a phalarope. Only the RED will be seen with it on the ocean, and it differs from that species (and the WILSON'S) in *smaller size*. From the RED it differs further by its much *more slender bill* (Table 36).

Breeding Plumage. Birds of both sexes in this plumage can be distinguished from the only similar species, WILSON'S PHALAROPE, by their *brightly buff-striped, dark gray upperparts*, with contrasty *white throat and reddish neck patch*. In some birds the head and neck pattern look superficially like that of WILSON'S, but the much darker color of the head and upperparts are diagnostic. RED-NECKED never shows the whitish

Table 36. Sea Phalarope Identification

	Red-necked	Red
All Plumages		
Size	smaller	larger
Bill shape*	very slender	relatively thick
Bill color	black	yellow in breeding adults
Flight call	hard tic, tic	similar, higher pitched
Distribution	offshore, regularly inland	offshore, near coast after storms
Season	migrant, common into November	migrant, common into October
	absent in winter	present in winter
	common in spring	rare in spring
Breeding		
Cheeks	dark	white
Throat	white	dark
Back	few wide stripes	many narrow stripes
Underparts	breast dark, belly white	mostly chestnut
Nonbreeding		
Back*	striped (feathers white-edged)	unstriped

*most definitive characters

nape patch of WILSON'S, and its back is always more vividly striped.

Nonbreeding Plumage. The vivid *dark ear patch* on the otherwise pale head of this and the RED PHALAROPE are unique to this pair of oceanic species (think of the ear patches worn for seasickness), and the RED-NECKED can be distinguished by its more *slender bill and striped back*. Nonbreeding adult RED-NECKED are paler above than juveniles, and illustrations and texts of some field guides have inappropriately contrasted dark juvenile RED-NECKED with pale adult RED. Adult RED-NECKED are considerably closer to the color of adult RED, in fact, necessitating a look at bill shape or back stripes.

Juvenal Plumage. Juveniles are also recognized as phalaropes by their *aquatic habits and dark ear patches*. RED-NECKED are *smaller and much more vividly striped above* than juvenile WILSON'S or RED.

In Flight (Figure 12)

In breeding plumage the overall *dark gray head and upperparts* are diagnostic of this species. In nonbreeding and juvenal plumages these birds are superficially similar to some calidridine sandpipers, for example DUNLINS and ROCK SANDPIPERS, because of their *conspicuous wing stripes and striped tails*. If the *dark ear patches* can be seen, sandpipers can be eliminated. The *wing stripes are both narrow and conspicuous* in comparison to similar-sized sandpipers. See RED PHALAROPE for distinction from that species.

Voice

This species is moderately noisy in migration, its intermittent but distinctive calls simple and hard (*tic, tic*). Those of the RED PHALAROPE are similar but a bit higher-pitched.

Further Questions

Much remains to be learned about the itinerary of this species in the Northwest. Are there differences in timing of fall migration or molt between ocean-going birds and those in the interior, the latter perhaps on their way to staging grounds at Mono Lake, California? Although earlier arrival of females than males on their arctic breeding grounds has been established, determination of the sex ratio in spring flocks would tell us whether this discrepancy was already established at moderate latitudes. Sex-ratio counts in early fall would be of even greater interest, to determine whether the males catch up with the females as they progress southward.

Notes

I have considered two published winter records inadequately documented: Tokeland, January 25, 1964, and Yaquina Bay, December 29, 1979.

Contrary to the statement in Farrand (1983: 412), the white wing stripes on the two seagoing phalaropes are approximately similar. The juvenile illustrated in National Geographic Society (1987: 121) is browner than birds that are usually seen in our region, and the "winter" bird darker than most nonbreeding adults. In the breeding female the white on the lower eyelid is usually invisible when the eye is open, so there is a single white spot above the eye rather than an eye-ring. The illustrations in Hayman et al. (1986: 163) are misleading in emphasizing a black crown in nonbreeding plumage, when in fact adults in that plumage usually have a white crown.

Photos

Most photos in guides are of females, but a breeding-plumaged male is shown in Hosking and Hale (1983: 63); the "male" in Udvardy (1977: Pl. 202) looks like a female, a duller bird than the bright female opposite it. The "winter adults" in Udvardy (1977: Pl. 237), Bull and Farrand (1977: Pl. 203), Keith and Gooders (1980: Pl. 106), and Farrand (1988b: 153) are juveniles that have lost their early buffiness by fading. The late August bird in Chandler (1989: 190 [upper photo]) labeled "juvenile" is actually an adult; it has largely molted into nonbreeding plumage but still shows the dark forehead and lores and entirely dark tertials of the breeding adult. The juvenile illustrated in Armstrong (1983: 135) has just replaced a couple of its scapulars. The "nonbreeding" Red Phalarope in Armstrong (1983: 137) is not typical of that species and may be a nonbreeding adult Red-necked Phalarope. It was with juvenile Red-necked and was obviously larger (more so than could be explained by sexual dimorphism), but its back pattern and bill are more typical of Red-necked than Red. Hybridization is a possibility.

References

Baker 1977 (summer diet), Brown and Gaskin 1988 (pelagic ecology), Hildén and Vuolanto 1973 (breeding biology), Höhn 1968 (breeding biology), Höhn 1971 (breeding biology), Jehl 1986 (ecology during migration), Reynolds 1987 (plumage variation and mating system), Reynolds and Cooke 1988 (philopatry and nest-site fidelity), Reynolds et al. 1986 (spring arrival), Schamel and Tracy 1988 (age distinction), Schamel and Tracy 1991 (nest-site fidelity and philopatry), Tinbergen 1935 (breeding behavior), Wahl 1975 (Washington pelagic records).

Nonbreeding Red-necked Phalarope. Black face patch and gray-and-white striped upperparts diagnostic of species and plumage; note worn, brownish primaries and other remnants of breeding plumage. Mono Lake, CA, August 1980, J. R. Jehl, Jr.

Juvenile Red-necked Phalarope. As nonbreeding adult, but upperparts much darker, with well-defined mantle stripes; gray feathers of first-winter plumage appearing first in scapular tract; buff on breast and upperparts typical of younger juveniles mostly faded away. Victoria, BC, August 1985, Tim Zurowski

Nonbreeding Red Phalarope. Black face patch and plain gray upperparts diagnostic of species and plumage; bill with extreme base pale; wing molt completed. San Diego, CA, December 1982, J. R. Jehl, Jr.

First-winter Red Phalarope. As nonbreeding adult, but black on crown, streaks at front of mantle, and reddish tinge on neck retained from juvenal plumage (all will disappear later in winter); bill entirely black: Ocean Shores, WA, 27 October 1979, Dennis Paulson

RED PHALAROPE *Phalaropus fulicaria*

This most seagoing of all shorebirds is to be sought well away from solid ground. The persistent boat tripper will eventually see it on the pelagic realm, and the seasick-prone can cover the coast during autumn storms. A good look at strikingly colored breeding adults or crisp, gull-gray winter birds is well worth the effort.

	Jan	Feb	Mar	Apr	May	Jun	Jul	Aug	Sep	Oct	Nov	Dec
COAST												
INTERIOR				•		•	•					

Distribution

The Red Phalarope breeds on high-arctic tundra around the world, wintering primarily off western South America and Africa but also in moderate numbers off western North America.

Northwest Status

Primarily an offshore migrant in the Northwest, this phalarope is regularly seen on pelagic trips anywhere along the coast. Large numbers are

seen only occasionally, perhaps whenever conditions (water and/or weather) are such that the birds move coastward from their main area of migration over warmer waters far out in the Pacific. The Red has been seen only about half as often as the Red-necked on trips off Westport and almost always in much smaller numbers. It is seldom seen on spring trips and tends to occur later than the Red-necked, the largest numbers in mid May. Perhaps most migration takes place late in May, when there have been relatively few boat trips. My largest count at Ocean Shores was on June 1, indicative of quite late migration. This is the latest spring date for large numbers in the region, but there are several other coastal records through the month of June, with no clear indication whether the birds were in northward or southward migration or summering.

Highest numbers on the Washington coast in fall have been noted off Westport in late August and early September, probably mostly adults, and at Leadbetter Point in late October and early November, probably juveniles. Much higher numbers have been recorded off the Oregon coast, on several occasions in the thousands, and the species must normally be more common off that state. The largest concentration of all, however, was off Victoria, an amazing occurrence as the species is usually recorded only in small numbers in protected waters.

Additional information comes from "wrecks" of this species, large numbers of phalaropes that appear at or even behind the coast after storms. Wrecks were reported from the coast of Oregon in late September 1920, in late October 1921, 1930, 1934, and 1982, and in mid November 1959 and 1969. The October 1982 wreck extended to southern Washington, and the October 1934 wreck left birds all along the coast, north to both sides of Vancouver Island. During some of these wrecks, hundreds to thousands of phalaropes were driven inland, some to the Willamette Valley. They are often encountered in surprising places for days after such an event. With the exception of the unusual occurrence at Victoria, the species is normally much rarer away from the outer coast, only occasionally seen in protected waters and even less often in the interior. Most near-coastal records have been made very late in spring (late May and early June) and late in fall (October and November).

Apparently the great numbers of birds late in fall are immatures. All of 42 specimens from the October 1934 wreck on the Oregon coast are first-year individuals, the sexes equally represented.

Red Phalaropes winter fairly far north in the Pacific Ocean, and migrants remain in our waters quite late, regularly in small numbers into mid December at least. Only a few records are at hand from January, February, and March in most winters, but in 1985-86 there were obviously thousands offshore in midwinter, from the numbers seen after a January 18 storm on the southern Washington and Oregon coasts.

There are only a few records of undoubted adults east of the Cascades, including a spring migrant in May and two individuals in June that could have been either spring or fall migrants. Another two dozen records from western Montana, Idaho, and the interior of British Columbia, Washington, and Oregon include primarily if not entirely juveniles.

COAST SPRING. *Early dates* Yaquina Bay, OR, April 14, 1982 (50); Ocean Shores, WA, April 18, 1976; Barkley Sound, BC, May 8, 1969. *High counts* 200 off Westport, WA, May 16, 1976; 372 at Ocean Shores, WA, June 1, 1980; 1,000 off Triangle Island, BC, May 17, 1974.

COAST FALL. *Juvenile early date* Tokeland, WA, August 24, 1960#. *High counts* 5,000 off Victoria, BC, November 11, 1982; 319 off Westport, WA, September 7, 1980*; 220 off Westport, WA, August 19, 1981*; hundreds at Leadbetter Point, WA, October 28 to November 5, 1982; thousands at Yaquina Bay, OR, mid November 1969; thousands off Newport, OR, July 21, 1974 ; 500 off Brookings, OR, September 20, 1980; 4,000 at Boiler Bay, OR, December 10, 1987; 15,000 at Cape Arago, OR, November 25, 1990.

INTERIOR SPRING. Long Creek, OR, May 3, 1987.

INTERIOR SUMMER. Malheur Refuge, OR, June 25, 1961; Anatone, WA, June 14-15, 1981.

INTERIOR FALL. *Early date* Creston, BC, August 1, 1951. *High count* 3 at Walla Walla, WA, September 23-27, 1983. *Late dates* Vernon, BC, October 29, 1934; Fort Klamath, OR, October 31, 1882.

Habitat and Behavior

This is a bird of the open ocean, although it seems perfectly at home on coastal ponds when storm-blown. These ponds, after all, are little different from the tundra pools on which it spends several months each year. In the interior it is most likely to occur on larger bodies of water. On the average, it occurs farther offshore than the Red-necked, although there is total overlap. The short, relatively wide bill can be used as a plankton scoop, although feeding is usually by picking individual organisms from the surface. Foraging behavior and general habits at sea are very much like those of the Red-necked.

The Red occurs typically in smaller flocks than the Red-necked, at times a few in flocks of the latter species. Individuals are occasionally seen running along the beach, an environment to which they are clearly not adapted. Phalaropes are not Sanderlings, and they look about as awkward as a duck in the same circumstance! Such birds run slowly, with short legs at midbody and head cocked to one side, often landing near detritus as they would on the water.

Structure

Size small. Males slightly smaller than females, about three-fourths of their weight, but difference not obvious in field. Bill short, straight, thick (for phalarope), and blunt-tipped. Legs short, toes extensively webbed at base and with scalloped lateral flanges. Wings extend to tail tip, primary projection short to nonexistent in breeding adults and moderate in juveniles. Tail rounded, almost graduated.

Plumage

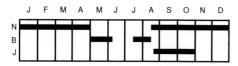

Breeding. Bill yellow with black tip. Legs (rarely visible) vary from grayish olive, with yellower lobes, to yellowish. Females with vivid head pattern from black crown, white cheeks, and black chin. Only source of variation is chin, where some individuals show little black. Upperparts basically black with vivid buff stripes from conspicuous edges on scapulars and tertials. Evenly striped pattern in some individuals, in others widest stripe occurs on either side of mantle. Coverts medium gray, finely fringed by white; mostly covered by long scapulars. Prominent white tips to greater coverts often exposed, forming conspicuous white stripe along edge of wing. Most females entirely rich chestnut below, often with scattered white feathers on belly, but some individuals with more extensively white

bellies (not visible on swimming birds).

Lower back gray. Central uppertail coverts blackish; outer ones rufous, with or without black stripes. Central pair of rectrices blackish, outer ones gray. Wing gray above, conspicuous wing stripe formed from broad white tips to greater secondary coverts and narrow white tips to inner greater primary coverts. Underwing, including axillars, most coverts, and primary bases, white; remainder of flight feathers gray; marginal primary coverts darker, forming dark bar at fore edge of wing.

Males duller versions of females, varying from almost as bright to very much duller. Bill slightly duller with more black on tip. Best sexual distinction furnished by crown, brown- and black-streaked in males. This characteristic alone distinguishes even the most femalelike of males. White cheek patch varies from as conspicuous as in female (although usually narrower) to virtually absent; in some birds represented only by white supercilium. Back looks more finely striped than that of female, as all mantle feathers dark-centered and buff-edged. Pale stripes reddish buff, in contrast with yellow-buff stripes of female. In some males stripe on either side of mantle better defined, wider, and paler than others.

Chin varies from white to chestnut, more rarely blackish. Males typically more extensively white below than females, darkest individuals about like lightest-bellied females and lightest individuals largely white below. Many look largely chestnut because white belly not visible in water. Some males on breeding grounds with many feathers retained from nonbreeding plumage, may be first-year birds.

Birds heading north in spring migration usually in full breeding plumage. Oddly, one in this plumage in Willapa Bay, January 19-21, 1986, among hundreds in normal plumage for season.

Nonbreeding (Figure 52). Bill black with extreme base usually pale. Plain medium gray above and white below. In fresh plumage tertials and coverts narrowly fringed with white. Head white with black postocular stripe and ear coverts that combine to produce "ear patch." Dark gray bar across back of head connected to gray mantle by narrow line down hindneck. Outer uppertail coverts white.

Molt into nonbreeding plumage begins as soon

Breeding female Red Phalarope. Larger, thick-billed sea phalarope; dark crown, white cheeks, yellow bill, and rufous underparts diagnostic of species and plumage; black crown diagnostic of female. Cambridge Bay, NT, 14 June 1975, Dennis Paulson

as birds leave arctic breeding grounds, earlier in females than males; difficult to find one in full breeding color when they arrive in Northwest in late July. All 18 birds off Westport on August 19, 1979, in nonbreeding plumage except for few with some reddish beneath. Hundreds off Westport on August 27, 1988, nonbreeding adults, with surprising lack of juveniles. Traces of breeding coloration persists on underparts of some adults even into November, after molt of rest of body feathers. Specimens of adults from October and November with retained red on underparts have not molted flight feathers, while those entirely white beneath have done so, perhaps indicating two classes of birds, ongoing migrants and winterers. Adults with wing molt not yet completed in Northwest as late as early January, but adults in other areas with finished wing molt as well as body molt by October.

Juvenal. Bill entirely black. Richly colored, more like dull breeding-plumaged male than like nonbreeding adult. Upperparts colored much like those of breeding male: crown and mantle feathers, scapulars, and tertials black with buffy-brown edges. Coverts black with white fringes. Blackish ear patch like that of nonbreeding adult. Supercilium and throat white; cheeks, foreneck, and breast buffy to light brown or reddish brown; rest of underparts white. Uppertail coverts as in nonbreeding adult.

Molt into first-winter plumage begins soon after fledging, many already partially gray-backed on arrival in August. Scapulars molt first, producing gray patch on either side of upperparts and then largely gray back, still with dark crown and streaked mantle. Birds in this condition common through September and October. Brown retained on foreneck well into September, even October in some birds, but lost quite early in others. By late October many and by November most juveniles with mantle and crown feathers replaced respectively by gray and white of first-winter plumage.

Breeding male Red Phalarope. As breeding female, but duller overall, with streaked crown and reduced white cheek patch. Wales, AK, 6 June 1978, Dennis Paulson

These birds now like nonbreeding adults but distinguished by dark, narrowly white-fringed coverts and tertials.

Immature. Molt in winter quarters poorly known, but some individuals remain in nonbreeding plumage during first summer and do not molt flight feathers. Perhaps majority mature and molt into breeding plumage in first year.

Identification

All Plumages. A persistent *swimmer*, this bird should be easily recognized as a phalarope. When on land it *runs awkwardly*. It is the only phalarope with a *thick bill*, diagnostic in any plumage.

Breeding Plumage. From a long distance females are unmistakable in this plumage with their *black crown, white cheeks, yellow bill, brightly striped back, and chestnut underparts*. Similarly colored males are as easily distinguished, but duller ones can still be recognized by their *evenly black-and-buff striped upperparts and partially*

chestnut underparts. Rich reddish shorebirds such as CURLEW SANDPIPERS and RED KNOTS have bills differently shaped and colored, mottled upperparts, no white on the face, and are not usually swimming.

Nonbreeding Plumage. Pale gray with black ear patch, this species is like no other shorebird except the RED-NECKED PHALAROPE. From that species it is distinguished by *larger size, thick bill, and slightly paler upperparts with no stripes*. The RED's pale legs would distinguish it from the dark-legged RED-NECKED in the unlikely event that leg color could be seen. The black ear patch and *crazy locomotion* easily distinguish RED PHALAROPES blown onto a beach with SANDERLINGS after a storm.

Juvenal Plumage. Juveniles, rarely if ever seen in our region in full plumage, would be distinguished by aquatic habits, thick bill, brown breast, and *evenly striped upperparts*. Juvenile RED-NECKED always show obvious wider mantle and scapular stripes. Molting RED PHALAROPES, largely gray but with some stripes on the upperparts, could be mistaken for RED-NECKED. The contrast between the new, pale feathers and the older, dark feathers is greater in the RED.

In Flight (Figure 12)
From above juveniles and nonbreeding adults are *stripe-winged and stripe-tailed* like calidridine sandpipers. The wing stripes are especially conspicuous in this species, and light gray first-winter and adult birds in flight look much like SANDERLINGS, from which they could be distinguished by their *dark ear patches, gray wings, and habit of lighting on the water*. SANDERLINGS over the water are usually heading somewhere in rapid, direct flight, while phalaropes may fly slowly and erratically while moving only a short distance. Both sea phalaropes are often seen in flight over the ocean, and RED is easily distinguished by being *paler and larger* in nonbreeding plumage. This difference is apparent in young birds, as most juvenile REDS in migration have largely gray backs while juvenile RED-NECKEDS are still mostly dark-backed.

Breeding-plumaged RED PHALAROPES show no white at the sides of the rump in flight and are thus *stripe-winged and dark-tailed*, a rare combination. RED KNOTS, vaguely similar in flight, have much less conspicuous wing stripes. RED-NECKED PHALAROPES, common at sea with REDS, differ by showing obvious white at the sides of the rump as in nonbreeding plumage. In breeding-plumaged REDS seen from below, the contrast between *dark belly and white wing linings* is distinctive. In any plumage RED shows more white under the wing than RED-NECKED.

RED PHALAROPES fly in loose flocks, the individuals scattered or in lines, while RED-NECKED often fly in bunches, in some cases as densely packed as any small sandpiper. The two species may flock together, although they tend to remain separated.

Voice
Not often audible at sea, and seemingly less vocal than the RED-NECKED PHALAROPE, this species has a *pit* call similar to the call of that species but slightly higher-pitched and more musical. Somewhere between RED-NECKED PHALAROPE and

SANDERLING would describe it best.

Further Questions
It would be of interest to document differences in feeding behavior between Red and Red-necked phalaropes when they are together. Anyone finding dead phalaropes after a late fall or winter wreck should save them for a museum, so we can learn more about the occurrence of sex and age classes in the region.

Notes
The illustrations in Hayman et al. (1986: 163) are misleading in emphasizing a black crown in nonbreeding plumage; adults in that plumage usually have a white crown.

Photos
Most photographic guides show breeding-plumaged females, which could lead naïve observers to think that half the birds in spring had not yet attained breeding plumage. The winter-plumaged bird in Bull and Farrand (1977: Pl. 205) shows remnants of breeding plumage on its head, neck, and tertials. The "winter adults" in Udvardy (1977: Pl. 236) and Farrand (1988b: 152) are juveniles well along in their molt into first-winter plumage, that in Hammond and Everett (1980: 126) has the same molt completed, and the juvenile in Keith and Gooders (1980: Pl. 105) is just starting it. See under Red-necked Phalarope for comments on the "nonbreeding Red Phalarope" photo in Armstrong (1983: 137).

References
Bengtson 1968 (breeding behavior), Briggs et al. 1984 (pelagic ecology), Brown and Gaskin 1988 (pelagic ecology), Dodson and Egger 1980 (summer feeding ecology), Haney and Stone 1988 (foraging behavior), Höhn 1971 (breeding behavior), Kistchinski 1975 (breeding biology), Mayfield 1978a (breeding biology), Mayfield 1978b (variation in breeding conditions), Myers 1981 (winter distribution of age and sex classes), Ridley 1980 (feeding ecology and breeding behavior), Schamel and Tracy 1977 (polyandry and site tenacity), Schamel and Tracy 1987 (latitudinal variation in breeding biology), Schamel and Tracy 1991 (nest-site fidelity and philopatry), Wahl 1975 (Washington pelagic records).

Appendix 1

Gazetteer of Northwest Localities in Text

Localities are given with reference to the closest map locality (**boldface** in list) in Fig. 1. Distances are in air rather than road miles. All localities on salt water are asterisked.

A

Aberdeen, WA* - Grays Harbor County, E end of Grays Harbor

Abert Lake, OR - Lake County, 30 mi. N Lakeview

Acequia, ID - Minidoka County, 6 mi. NE Rupert

Admiralty Inlet, WA* - inlet between Port Townsend and Whidbey Island

Alvord Lake, OR - Harney County, 92 mi. E Lakeview

American Falls Reservoir, ID - Power County, 15 mi. W Pocatello

Anatone, WA - Asotin County, 20 mi. S Lewiston, ID

Ankeny Refuge, OR - Linn County, 13 mi. NE Corvallis

Arco, ID - Butte County, 66 mi. W Idaho Falls

Ashland, OR - Jackson County, 13 mi. SE Medford

Asotin, WA - Asotin County, 6 mi. S Lewiston, ID

Auburn, WA - King County, 10 mi. E Tacoma

B

Bainbridge Island, WA* - Kitsap County, in Puget Sound opposite Seattle

Bandon, OR* - Coos County

Banks, OR - Washington County, 23 mi. W Portland

Banks Lake, WA - Grant County, 44 mi. N Moses Lake

Barkley Sound, BC* - E of Ucluelet

Beach Grove, BC* - on Boundary Bay

Bear Valley, OR - Grant County, 39 mi. NW Burns

Beaver Creek, OR - Lincoln County, 13 mi. ESE Newport

Bellingham, WA* - Whatcom County

Bellingham Bay, WA* - Whatcom County, at Bellingham

Bend, OR - Deschutes County

Blackie Spit, BC* - on Boundary Bay

Blaine, WA* - Whatcom County, 22 mi. NNW Bellingham

Boardman, OR - Morrow County, 70 mi. WSW Walla Walla, WA

Boiler Bay, OR* - Lincoln County, 14 mi. N Newport

Boise, ID - Ada County

Bottle Beach, WA* - Grays Harbor County, S side of Grays Harbor

Boundary Bay, BC*

Bowerman Basin, WA* - Grays Harbor County, N side of Grays Harbor

Brookings, OR* - Curry County, 93 mi. S Coos Bay

Browning Inlet, BC* - 29 mi. SW Port Hardy

Brownsville, OR - Linn County, 19 mi. SE Corvallis

Bruneau State Park, ID - Owyhee County, 55 mi. SE Boise

Bumping Lake, WA - Yakima County, 42 mi. WNW Yakima

Burnaby Lake, BC - 8 mi. E Vancouver

Burns, OR - Harney County

C

Camas Refuge, ID - Jefferson County, 33 mi. N Idaho Falls

Campbell River, BC* - 24 mi. NW Courtenay

Cattle Point, BC* - at Victoria

Central Ferry, WA - Garfield County, 31 mi. W Pullman

Chain Islets, BC* - off Victoria

Chatham Island, BC* - off Victoria

Chesterman Beach, BC* - 18 mi. NW Ucluelet

Chewelah, WA - Stevens County, 46 mi. NW Spokane

Clatsop Beach, OR* - Clatsop County, 44 mi. N Tillamook

Cleland Island., BC* - 27 mi. NW Ucluelet

Coburg, OR - Lane County, 6 mi. N Eugene

Cohasset, WA - Grays Harbor County, 2 mi. S Westport

Colebrook, BC - 14 mi. SE Vancouver

Columbia Lake, BC - 84 mi. NNE Creston

Columbia River estuary, WA* - Pacific County, mouth of Columbia River

Columbia Refuge, WA - Grant/Adams counties, 16 mi. S Moses Lake

Colville, WA - Stevens County, 67 mi. NW Spokane

Comox, BC* - 4 mi. E Courtenay

Coos Bay, OR* - Coos County

Coquille, OR - Coos County, 14 mi. S Coos Bay

Corvallis, OR - Benton County

Coupeville, WA* - Island County, 7 mi. NNE Port Townsend

Courtenay, BC*

Cow Lake, WA - Adams County, 51 mi. SW Spokane

Crater Lake, OR - Klamath County, 51 mi. NNW Klamath Falls

Crescent Beach, BC* - on Boundary Bay
Creston, BC
Crockett Lake, WA* - Island County, 6 mi. NE Port Townsend

D

Damon Point, WA* - Grays Harbor County, at Ocean Shores
Deadman's Lake, BC - Okanagan Valley
Dee Lake, BC - 11 mi. SE Vernon
Deer Flat Refuge, ID - Canyon County, 24 mi. SW Boise
Delta, BC* - 12 mi. SSE Vancouver
Destruction Island, WA* - Jefferson County, off coast 18 mi. S La Push
Diamond Lake, OR - Douglas County, 61 mi. E Roseburg
Dodson Road, WA - Grant County, 12 mi. SW Moses Lake
Douglas Plateau, BC - 29 mi. SSE Kamloops
Downy Lake, OR - Wallowa County, 60 mi. E La Grande
Drayton Harbor, WA* - Whatcom County, 22 mi. NNW Bellingham
Duncan, BC* - 30 mi. NW Victoria
Dungeness, WA* - Clallam County
Dungeness Spit, WA* - Clallam County, at Dungeness

E

Edgewood, BC - 33 mi. SW Nakusp
Ellensburg, WA - Kittitas County, 28 mi. N Yakima
Enterprise, OR - Wallowa County, 40 mi. E La Grande
Ephrata, WA - Grant County, 18 mi. NW Moses Lake
Eugene, OR - Lane County
Eureka, MT - Lincoln County, 42 mi. NE Libby

F

Federal Way, WA - King County, 6 mi. NE Tacoma
Fern Ridge Reservoir, OR - Lane County, 9 mi. W Eugene
Finley Refuge, OR - Benton County, 12 mi. S Corvallis
Flattery Rocks, WA* - Jefferson County, off coast 18 mi. N La Push
Fleming Island, BC* - 16 mi. E Ucluelet
Florence, OR* - Lane County, 43 mi. N Coos Bay
Forest Grove, OR - Washington County, 21 mi. W Portland
Fort Flagler State Park, WA* - Jefferson County, 2 mi. SE Port Townsend
Fort Klamath, OR - Klamath County, 33 mi. NNW Klamath Falls
Fortine, MT - Lincoln County, 41 mi. NE Libby

Four Lakes, WA - Spokane County, 11 mi. SW Spokane
Fraser River, BC* - at Vancouver
Frenchman Hills, WA - Grant County, 23 mi. SW Moses Lake
Fruitland, ID - Payette County, 45 mi. NW Boise

G

Gaston, OR - Washington County, 24 mi. W Portland
Georgia Strait, BC* - between mainland and southern Vancouver Island
Gold Lake Bog, OR - Lane County, 62 mi. SE Eugene
Grand Forks, BC - 46 mi. E Osoyoos
Grangeville, ID - Idaho County, 55 mi. SE Lewiston
Grant Bay, BC* - 30 mi. SW Port Hardy
Grays Harbor, WA* - Grays Harbor County
Grays Lake, ID - Caribou County, 52 mi. E Pocatello
Great Chain Island, BC* - off Victoria

H

Hamma Hamma River, WA - Jefferson County, 39 mi. W Seattle
Hansen Lagoon, BC* - 39 mi. W Port Hardy
Harney Basin, OR - Harney County, 24 mi. S Burns
Harney Lake, OR - Harney County, 24 mi. S Burns
Hart Lake, OR - Lake County, 30 mi. NE Lakeview
Harvard, ID - Latah County, 39 mi. NE Lewiston
Hatfield Lake, OR - Deschutes County, at Bend
Hauser, ID - Kootenai County, 21 mi. ENE Spokane, WA
Hauser Lake, ID - Kootenai County, 21 mi. ENE Spokane, WA
Haynes Point, BC - at Osoyoos
Hermiston, OR - Umatilla County, 49 mi. WSW Walla Walla, WA
Homedale, ID - Owyhee County, 37 mi. W Boise
Hood River, OR - Hood River County, 59 mi. E Portland
Hoover's Lakes, OR - Jackson County, 7 mi. N Medford
Hoquiam, WA* - Grays Harbor County, NE shore of Grays Harbor

I

Idaho Falls, ID - Bonneville County
Iona Island, BC* - at Vancouver
Irrigon, OR - Morrow County, 49 mi. WSW Walla Walla, WA

J

Jordan River, BC* - 32 mi. W Victoria
Joseph, OR - Wallowa County, 42 mi. E La Grande
Juan de Fuca Strait, WA* - strait between Washington and Vancouver Island

K

Kahlotus, WA - Franklin County, 48 mi. SE Moses Lake

Kalaloch, WA* - Jefferson County, 25 mi. SSE La Push

Kamloops, BC

Kelowna, BC - 27 mi. S Vernon

Kent, WA - King County, 17 mi. S Seattle

Klamath Falls, OR - Klamath County

Kootenai Refuge, ID - Boundary County, 34 mi. NNE Sandpoint

L

La Grande, OR - Union County

La Push, WA* - Jefferson County

Ladner, BC* - 13 mi. S Vancouver

Lake Chatcolet, ID - Benewah County, 39 mi. SE Spokane, WA

Lake Lenore, WA - Grant County, 27 mi. NNW Moses Lake

Lake McDonald, MT - Flathead County, 65 mi. NNE Polson

Lakeview, OR - Lake County

Lamont, WA - Whitman County, 39 mi. SW Spokane

Leadbetter Point, WA* - Pacific County

Lee Metcalf Refuge, MT - Ravalli County, 24 mi. S Missoula

Lewiston, ID - Nez Perce County

Libby, MT - Lincoln County

Lind Coulee, WA - Grant County, 11 mi. SE Moses Lake

Logan Valley, OR - Grant County, 39 mi. N Burns

Long Beach, BC* - 12 mi. NW Ucluelet

Long Beach Peninsula, WA* - Pacific County

Long Creek, OR - Grant County, 66 mi. SW La Grande

Long Valley, ID - Valley Co., 63-90 mi. N Boise

Lopez Island, WA* - San Juan County, San Juan Islands

Lowden, WA - Walla Walla County, 12 mi. W Walla Walla

Lower Klamath Lake, OR - Klamath County, 17 mi. S Klamath Falls

Lower Klamath Refuge, OR - Klamath County, 18 mi. SSE Klamath Falls

M

Malheur Refuge, OR - Harney County, 19-53 mi. S Burns

Manning Park, BC - 50-70 mi. W Osoyoos

Mann's Lake, ID - Nez Perce County

Manzanita, OR* - Tillamook County, 19 m. N Tillamook

Manzanita Beach, OR* - Tillamook County, 19 m. N Tillamook

Medford, OR - Jackson County

Medicine Lake, ID - Kootenai County, 39 mi. ESE Spokane, WA

Merrill, OR - Klamath County, 17 mi. SE Klamath Falls

Metchosin, BC* - 8 mi. SW Victoria

Midland, OR - Klamath County, 8 mi. S Klamath Falls

Minidoka Refuge, ID - Blaine/Cassia counties, 50 mi. WSW Pocatello

Missoula, MT - Missoula County

Mitlenatch Island, BC* - 22 mi. N Courtenay

Moscow, ID - Latah County, 23 mi. N Lewiston

Moses Lake, WA - Grant County

Mount Vernon, WA - Skagit County, 24 mi. S Bellingham

Mud Bay, BC* - on Boundary Bay

Mukilteo, WA* - Snohomish County, 24 mi. N Seattle

N

Nakusp, BC

Nampa, ID - Canyon County, 18 mi. W Boise

Nanaimo, BC*

National Bison Range, MT - Lake County, 26 mi. S Polson

Neah Bay, WA* - Clallam County, W end of Juan de Fuca Strait

Nehalem, OR* - Tillamook County, 19 mi. N Tillamook

Neppel, WA - Grant County, near Moses Lake

Netarts Bay, OR* - Tillamook County, 6 mi. SW Tillamook

New River, OR* - Curry County, 12 mi. S Bandon

Newport, OR* - Lincoln County

Ninemile Creek, MT - Missoula County, NW of Missoula

Ninepipe Refuge, MT - Lake County, 18 mi. S Polson

Nisqually Refuge, WA* - Thurston County, 9 mi. E Olympia

North Bay, WA* - Grays Harbor County, N side of Grays Harbor

North Cove, WA* - Pacific County, N side of Willapa Bay

North Surrey, BC - near Vancouver

North jetty of Columbia River, WA* - Pacific County, mouth of Columbia River

Nyssa, OR - Malheur County

O

Oak Bay, BC* - at Victoria

Ocean Park, WA* - Pacific County, 28 mi. S Westport

Ocean Shores, WA* - Grays Harbor County

Ochoco Reservoir, OR - Crook County, 36 mi. NE Bend

Okanagan, BC - 5 mi. WSW Vernon

Okanagan Landing, BC - 5 mi. WSW Vernon

Okanagan Valley, BC - from Vernon to Osoyoos
Oliver, BC - 11 mi. NNW Osoyoos
Olympia, WA* - Thurston County
Osgood, ID - Bonneville County, 10 mi. N Idaho
Falls
Osoyoos, BC
Osoyoos Lake, BC - at Osoyoos
Osprey Lake, BC - 30 mi. WNW Penticton
Othello, WA - Adams County, 21 mi. S Moses Lake
Ovando, MT - Powell County, 42 mi. E Missoula

P

Pablo Refuge, MT - Lake County, 6 mi. S Polson
Pacific Rim National Park, BC* - W side of Vancouver Island, NW & SE Ucluelet
Padilla Bay, WA* - Skagit County, 17 mi. S Bellingham
Panama Flats, BC* - near Victoria
Pahsimeroi Valley, ID - Custer County, 34-55 mi. S
Salmon
Penn Cove, WA* - Island County, 7 mi. NNE Port
Townsend
Penticton, BC
Pitt Lake, BC - 27 mi. NE Vancouver
Pocatello, ID - Bannock County
Point Grenville, WA* - Grays Harbor County, 24 mi.
N Ocean Shores
Point New, WA* - Grays Harbor County, N side of
Grays Harbor
Polson, MT - Lake County
Port Angeles, WA* - Clallam County, 14 mi. W
Dungeness
Port Hardy, BC*
Port Orford, OR* - Curry County, 30 mi. S Bandon
Port San Juan, BC* - 50 mi. W Victoria
Port Townsend, WA* - Jefferson County
Portland, OR - Multnomah County
Potholes, WA - Grant/Adams counties, area south of
Moses Lake
Potholes Reservoir, WA - Grant County, at Moses
Lake
Potlatch, ID - Latah County, 36 mi. N Lewiston
Prescott, WA - Walla Walla County, 16 mi. N Walla
Walla
Prineville, OR - Crook County, 28 mi. NE Bend
Protection Island, WA* - Jefferson County, 7 mi. W
Port Townsend
Pullman, WA - Whitman County
Pulteney Point, BC* - 48 mi. S Port Hardy

Q

Queen Charlotte Strait, BC* - between mainland and
northern Vancouver Island
Quilchena, BC - 36 mi. S Kamloops

R

Ravalli Refuge, MT - Ravalli County, 37 mi. S Missoula
Reardan, WA - Lincoln County, 22 mi. W Spokane
Redmond, OR - Deschutes County, 16 mi. NNE Bend
Reifel Island, BC* - 12 mi. S Vancouver
Restoration Point, WA* - Kitsap County, 7 mi. W Seattle
Richland, WA - Benton County, 48 mi. WNW Walla
Walla
Richmond, BC* - at Vancouver
Ridgefield Refuge, WA - Clark County, 21 mi. N
Portland, OR
Rigby, ID - Jefferson County, 13 mi. NE Idaho Falls
Roberts Bank, BC* - at Vancouver
Roberts Creek, BC* - 24 mi. NW Vancouver
Robinson Lake, ID - Latah County, 26 mi. N Lewiston
Rock Lake, WA - Whitman County, 35 mi. SSW Spokane
Roseburg, OR - Douglas County
Rupert, ID - Minidoka County

S

Saanich, BC* - 7 mi. N Victoria
Salmon, ID - Lemhi County
Salmon Arm, BC - 27 mi. N Vernon
Samish Flats, WA* - Skagit County, 13 mi. S Bellingham
San Juan Island, WA* - San Juan County, San Juan Islands
San Juan Islands, WA* - San Juan County
Sandpoint, ID - Bonner County
Sauvie Island, OR - Multnomah County, 14 mi.
NNW Portland
Scootenay Reservoir, WA - Franklin County, 32 mi.
SSE Moses Lake
Sea Island, BC* - at Vancouver
Seal Rocks, OR* - Lincoln County, 10 mi. S Newport
Seaside, OR* - Clatsop County, 38 mi. N Tillamook
Seattle, WA* - King County
Sequim Bay, WA* - Clallam County, 8 mi. SE
Dungeness
Serpentine Fen, BC - 43 mi. SE Vancouver
Serpentine River, BC - 43 mi. SE Vancouver
Seven Devils Wayside, OR* - Coos County, 12 mi. N
Bandon
Shady Cove, OR - Jackson County, 19 mi. N Medford
Shuswap Falls, BC - 20 mi. E Vernon
Sidney, BC* - 15 mi. N Victoria
Sidney Island, BC* - 12 mi. N Victoria
Sidney Spit, BC* - 15 mi. N Victoria
Siletz Bay, OR* - Lincoln County, 20 mi. N Newport
Silver Lake, OR - Lake County, 74 mi. NE Klamath
Falls

Sirdar, BC - 12 mi. N Creston
Skagit Bay, WA* - Skagit County, 33 mi. S Bellingham
Skagit Flats, WA* - Skagit County, 31 mi. S Bellingham
Smith Island, WA* - Island County, 14 mi. N Port Townsend
Soap Lake, WA - Grant County, 22 mi. NW Moses Lake
Somers, MT - Flathead County, 28 mi. N Polson
Sooke, BC* - 15 mi. W Victoria
Sooke River, BC* - 15 mi. W Victoria
South jetty of Columbia River, OR* - Clatsop County, at mouth of Columbia River
Spokane, WA - Spokane County
Spokane Valley, WA - Spokane County, E Spokane
Springfield, ID - Bingham County, 18 mi. NW Pocatello
St. Andrews, WA - Douglas County, 40 mi. N Moses Lake
Stevensville, MT - Ravalli County, 25 mi. S Missoula
Steveston, BC* - 8 mi. S Vancouver
Stinking Lake, OR - Harney County, 22 mi. S Burns
Stratford, WA - Grant County, 20 mi. N Moses Lake
Stump Lake, BC - 22 mi. S Kamloops
Summer Lake, OR - Lake County, 49 mi. NW Lakeview
Summerland, BC - 9 mi. N Penticton
Summit Creek, BC - at Creston
Sunnyside, WA - Yakima County, 31 mi. SE Yakima
Sunset Beach, OR* - Clatsop County, 42 mi. N Tillamook
Surrey, BC - 15 mi. SE Vancouver
Swan Lake, BC - 5 mi. N Vernon
Swantown, WA* - Island County, 13 mi. N Port Townsend
Sycan Marsh, OR - Lake County, 53 mi. NE Klamath Falls

T

Tacoma, WA * - Pierce County
Tacoma watershed, WA - King County, 30 mi. E Tacoma
Tahkenitch Creek, OR* - Douglas County, 31 mi. N Coos Bay
Thorn Creek Reservoir, ID - Gooding County, 66 mi. WNW Rupert
Tillamook, OR * - Tillamook County
Tillamook Bay, OR* - Tillamook County, at Tillamook
Tofino, BC* - 20 mi. NW Ucluelet
Tokeland, WA* - Pacific County, N shore of Willapa Bay
Tranquille, BC - 11 mi. WNW Kamloops
Trevor Channel, BC* - 22 mi. E Ucluelet

Triangle Island, BC* - 52 mi. W Port Hardy
Tsawwassen, BC* - 16 mi. S Vancouver
Turnbull Refuge, WA - Spokane County, 18 mi. SW Spokane
Turtle Island, BC* - 12 mi. E Ucluelet
Twelve Mile Slough, WA - Adams County, 59 mi. E Moses Lake
Tzartus Island, BC* - 24 mi. E Ucluelet

U

Ucluelet, BC *
Ukiah, OR - Umatilla County, 43 mi. WSW La Grande

V

Vancouver, BC *
Vantage, WA - Kittitas County, 39 mi. WSW Moses Lake
Vaseux Lake, BC - 18 mi. N Osoyoos
Vernon, BC
Victoria, BC *

W

Waldport, OR* - Lincoln County, 15 mi. S Newport
Walla Walla, WA - Walla Walla County
Walla Walla River delta, WA - Walla Walla County, 28 mi. W Walla Walla
Wallula, WA - Walla Walla County, 27 mi. W Walla Walla
Wapato Lake, OR - Washington County, 23 mi. W Portland
Wenas Lake, WA - Yakima County, 16 mi. NNW Yakima
Westport, WA * - Grays Harbor County
White City, OR - Jackson County, near Medford
White Lake, BC - 20 mi. NNW Osoyoos
Willapa Bay, WA * - Pacific County
Willow Creek, WA - Columbia County, 34 mi. NE Walla Walla
Willow Lake, WA - Columbia County, 34 mi. NE Walla Walla
Witty's Lagoon, BC* - 7 mi. WSW Victoria
Woodburn, OR - Marion County, 28 mi. S Portland
Woodland, WA - Clark County, 27 mi. N Portland, OR
Wye Lake, BC - 4 mi. NW Vernon

Y

Yakima, WA - Yakima County
Yakima River delta, WA - Benton County, 42 mi. WNW Walla Walla
Yaquina Bay, OR* - Lincoln County, at Newport

Appendix 2

Status of Northwest Shorebirds

This table summarizes the status of all species recorded from the region, divided into coastal (including lowlands west of Cascades) and interior (east side of Cascades east) subregions. The following terminology is used, modified from a system originated at the Royal British Columbia Museum.

Regular species (recorded every year in region)

Very abundant (VA) - over 1,000 individuals per day in best localities; typically in large flocks

Abundant (A) - 200-1,000 individuals per day in best localities; typically in large flocks

Very common (VC) - 50-200 individuals per day in best localities; typically in moderate-sized flocks

Common (C) - 20-50 individuals per day in best localities; typically in small flocks

Fairly common (FC) - 7-20 individuals per day in best localities; typically in quite small flocks or solitary

Uncommon (U) - 1-6 individuals per day in best localities; any species at least this common can be seen yearly by active observer

Very uncommon (VU) - seen widely, more than 6 individuals per season in the region, but not every day, even in best localities; even active observers could miss this species

Rare (R) - 1-6 individuals per season in the entire region, likely to be missed by most observers without special effort

Irregular species (not recorded every year in region)

Very rare (VR) - over 6 records, but not recorded every year; very unlikely to be seen

Casual (cas) - 2-6 records

Accidental (acc) - only 1 record

The table reflects the *typical* status from year to year, not including unusually large concentrations recorded no more than a few times. The seasons are shorebird seasons, not calendar months, as migration periods of different species may occur at quite different times. An asterisk (*) indicates a regular species that is quite locally distributed, either overall or in that region at that season; irregular species would obviously be local.

Bold face indicates a species that breeds in one or both subregions of the region.

	Coast, west				Interior			
	W	Sp	Su	F	W	Sp	Su	F
Black-bellied Plover	A	A	FC*	VA	-	VU	-	U
American Golden-Plover	-	R	-	C	-	VR	-	U
Pacific Golden-Plover	VR	R	-	C	-	-	-	-
Mongolian Plover	-	-	-	cas	-	-	-	-
Snowy Plover	U*	FC*	FC*	FC*	-	VR	FC*	-
Semipalmated Plover	U	A	U*	A	-	VU	cas	U
Piping Plover	-	-	-	-	-	-	-	acc
Killdeer	C	C	C	C	U	C	C	C
locally abundant in winter, especially in Willamette Valley, Oregon								
Mountain Plover	cas	-	-	cas	-	cas	cas	acc
Eurasian Dotterel	-	-	-	cas	-	-	-	-

	Coast, west				Interior			
	W	Sp	Su	F	W	Sp	Su	F
American Oystercatcher	-	-	-	-	-	acc	-	-
Black Oystercatcher	FC	FC	**FC**	FC	acc	-	-	-
Black-necked Stilt	-	VU	acc	cas	-	U	**FC***	-
American Avocet	-	R	**cas**	R	acc	FC	**C**	A*
locally very abundant in fall in southeastern Oregon								
Greater Yellowlegs	U*	FC	R	FC	VU*	FC	**VR**	FC
Lesser Yellowlegs	VR	VU	cas	FC	cas	U	-	FC
Spotted Redshank	-	cas	-	cas	-	-	-	-
Solitary Sandpiper	-	U	**VR**	VU	-	VU	**R**	U
Willet	U*	VU	cas	VU	-	VU	**U***	VU
locally fairly common in winter on southern Washington and Oregon coasts								
Wandering Tattler	VR	FC	cas	FC	-	acc	-	cas
Gray-tailed Tattler	-	-	-	acc	-	-	-	-
Spotted Sandpiper	VU	U	**U**	U	VR	U	**U**	VU
Terek Sandpiper	-	-	-	acc	-	-	-	-
Upland Sandpiper	acc	cas	-	VR	-	cas	**VU***	VR
Whimbrel	VU	VC	**FC***	VC	-	VR	-	VR
Bristle-thighed Curlew	-	acc	-	-	-	-	-	-
Far Eastern Curlew	-	-	-	acc	-	-	-	-
Long-billed Curlew	VU	VU	R	VU	-	FC	**FC**	FC*
a substantial wintering flock at Tokeland, Washington								
Hudsonian Godwit	-	cas	-	VR	-	acc	-	VR
Bar-tailed Godwit	-	VR	-	R	-	-	-	-
Marbled Godwit	FC*	FC*	R*	FC*	-	FC*	-	FC*
locally common in winter on Willapa Bay, Washington, and Coos Bay, Oregon								
Ruddy Turnstone	VU	C	-	FC	-	cas	-	VR
uncommon in winter in Oregon								
Black Turnstone	C	C	VR	C	-	-	-	acc
Surfbird	C	C	-	C	-	-	-	-
Great Knot	-	-	-	cas	-	-	-	-
Red Knot	VU	FC*	R*	U	-	VR	-	R
abundant at Grays Harbor in spring								
Sanderling	A*	A*	VU	A*	acc	R	-	U
Semipalmated Sandpiper	-	VU	-	U	-	U	-	U
common in fall at Iona Island, British Columbia; rare in interior Oregon								
Western Sandpiper	C	VA	U*	VA	cas	U	-	C
locally abundant in winter on outer coast; uncommon in winter in northern part of range; locally very abundant in fall in southeastern Oregon								
Rufous-necked Stint	-	-	-	VR	-	-	-	-
Little Stint	-	-	-	cas	-	-	-	-
Temminck's Stint	-	-	-	acc	-	-	-	-
Long-toed Stint	-	-	-	acc	-	-	-	-
Least Sandpiper	FC	VC	-	VC	cas	C	-	C
locally common to very common in winter on outer coast; uncommon in winter in northern part of range								
White-rumped Sandpiper	-	cas	-	cas	-	cas	-	cas
Baird's Sandpiper	-	VU	-	FC	-	U	-	FC
locally common in fall in eastern interior								
Pectoral Sandpiper	-	VU	cas	C	-	VR	-	FC
locally very common in fall on coast								
Sharp-tailed Sandpiper	-	acc	-	U	-	-	-	VR
Rock Sandpiper	FC	FC	-	FC	-	-	-	-

	Coast, west				Interior			
	W	Sp	Su	F	W	Sp	Su	F
Dunlin	VA	VA	U*	VA	VR	U	-	VU
flock often winters at Yakima River delta, Washington								
Curlew Sandpiper	-	acc	-	R	-	acc	-	-
Stilt Sandpiper	-	cas	-	VU	-	cas	-	VU
Spoonbill Sandpiper	-	-	-	acc	-	-	-	-
Buff-breasted Sandpiper	-	cas	-	VU	-	-	-	VR
Ruff	acc	cas	-	VU	-	acc	-	cas
Short-billed Dowitcher	cas	A*	FC*	A	-	R	-	VU
very abundant in spring at Grays Harbor and Willapa Bay, Washington, and Tofino, British Columbia								
Long-billed Dowitcher	FC*	C	cas	VC	VR	VC	-	VC
very abundant in some falls at Malheur Refuge, Oregon; locally very common in winter in some years, especially in Willamette Valley, Oregon								
Common Snipe	FC	U	U	FC	U	FC	FC	FC
American Woodcock	-	acc	-	-	-	-	-	acc
Wilson's Phalarope	-	U	VU	VU	-	FC	FC	FC
very abundant in fall at southern Oregon lakes								
Red-necked Phalarope	-	A	cas	A	-	FC	-	FC
locally very abundant in spring and fall offshore								
Red Phalarope	U	FC	cas	VC	-	cas	-	VR
very abundant in some falls off coast								

Appendix 3

Shorebird Weights and Measurements[1]

	Weight	Wing	Tail	Tarsus	Bill	Tail/ Wing	Tarsus/ Wing	Bill/ Wing	Tarsus/ Tail
Oriental Pratincole	—	191	78	34	14	0.41	0.18	0.07	0.44
Black-bellied Plover	210	189	75	44	30	0.40	0.23	0.16	0.59
American Golden-Plover	145	182	65	43	23	0.36	0.24	0.13	0.66
Pacific Golden-Plover	130	165	61	43	23	0.37	0.26	0.14	0.70
Mongolian Plover	60	135	51	31	16	0.38	0.23	0.12	0.61
Snowy Plover	41	105	44	23	14	0.42	0.22	0.13	0.52
Wilson's Plover	57	122	47	30	21	0.39	0.25	0.17	0.64
Common Ringed Plover	53	131	56	24	14	0.43	0.18	0.11	0.43
Semipalmated Plover	48	125	56	24	13	0.45	0.19	0.10	0.43
Piping Plover	55	123	50	23	13	0.41	0.19	0.11	0.46
Little Ringed Plover	39	116	57	25	13	0.49	0.22	0.11	0.44
Killdeer	90	167	93	36	20	0.56	0.22	0.12	0.39
Mountain Plover	107	153	65	40	21	0.42	0.26	0.14	0.62
Eurasian Dotterel	108	153	67	36	16	0.44	0.24	0.10	0.54
American Oystercatcher	632	259	102	57	84	0.39	0.22	0.32	0.56
Black Oystercatcher	648	246	100	51	72	0.41	0.21	0.29	0.51
Black-winged Stilt	183	240	79	121	62	0.33	0.50	0.26	1.53
Black-necked Stilt	166	220	71	108	64	0.32	0.49	0.29	1.52
American Avocet	316	225	84	95	91	0.37	0.42	0.40	1.13
Common Greenshank	180	192	77	61	56	0.40	0.32	0.29	0.79
Greater Yellowlegs	170	198	74	63	56	0.37	0.32	0.28	0.85
Lesser Yellowlegs	76	161	60	51	36	0.37	0.32	0.22	0.85
Marsh Sandpiper	74	141	56	52	40	0.40	0.37	0.28	0.93
Spotted Redshank	158	169	64	57	59	0.38	0.34	0.35	0.89
Wood Sandpiper	61	128	49	38	29	0.38	0.30	0.23	0.78
Green Sandpiper	80	145	57	35	35	0.39	0.24	0.24	0.61
Solitary Sandpiper	51	136	54	31	30	0.40	0.23	0.22	0.57
Willet	257	209	79	65	62	0.38	0.31	0.30	0.82
Wandering Tattler	109	174	74	33	39	0.43	0.19	0.22	0.44
Gray-tailed Tattler	108	166	68	32	38	0.41	0.19	0.23	0.47
Common Sandpiper	48	112	53	24	25	0.47	0.21	0.22	0.45
Spotted Sandpiper	41	107	47	24	24	0.44	0.22	0.22	0.51
Terek Sandpiper	73	135	52	29	47	0.39	0.21	0.35	0.56
Upland Sandpiper	170	173	82	50	29	0.47	0.29	0.17	0.61
Little Curlew	180	186	72	50	43	0.39	0.27	0.23	0.69
Eskimo Curlew	—	218	80	44	53	0.37	0.20	0.24	0.55
Whimbrel	380	236	92	58	87	0.39	0.25	0.37	0.63
Bristle-thighed Curlew	433	234	94	59	89	0.40	0.25	0.38	0.63
Far Eastern Curlew	800±	317	110	89	175	0.35	0.28	0.55	0.81
Long-billed Curlew	587	276	103	81	154	0.37	0.29	0.56	0.79

	Weight	Wing	Tail	Tarsus	Bill	Tail/Wing	Tarsus/Wing	Bill/Wing	Tarsus/Tail
Black-tailed Godwit	291	190	72	66	78	0.38	0.35	0.41	0.92
Hudsonian Godwit	255	208	75	60	82	0.36	0.29	0.39	0.80
Bar-tailed Godwit	320	232	78	57	97	0.34	0.25	0.42	0.73
Marbled Godwit	371	234	82	73	106	0.35	0.31	0.45	0.89
Ruddy Turnstone	108	156	61	26	23	0.39	0.17	0.15	0.43
Black Turnstone	119	155	61	26	23	0.39	0.17	0.15	0.43
Surfbird	160	177	65	30	24	0.37	0.17	0.14	0.46
Great Knot	155	189	65	35	43	0.34	0.19	0.23	0.54
Red Knot	120	169	59	32	36	0.35	0.19	0.21	0.54
Sanderling	54	127	50	26	26	0.39	0.20	0.20	0.52
Semipalmated Sandpiper	26	95	40	21	18	0.42	0.22	0.19	0.53
Western Sandpiper	27	99	42	23	25	0.42	0.23	0.25	0.55
Rufous-necked Stint	26	105	43	20	18	0.41	0.19	0.17	0.47
Little Stint	25	98	40	21	18	0.41	0.21	0.18	0.53
Temminck's Stint	24	99	47	18	17	0.47	0.18	0.17	0.38
Long-toed Stint	24±	94	38	22	18	0.40	0.23	0.19	0.58
Least Sandpiper	21	91	38	19	19	0.42	0.21	0.21	0.50
White-rumped Sandpiper	44	124	49	24	24	0.40	0.19	0.19	0.49
Baird's Sandpiper	42	127	50	23	23	0.39	0.18	0.18	0.46
Pectoral Sandpiper* ♂	86	144	58	29	30	0.40	0.20	0.21	0.50
Pectoral Sandpiper* ♀	58	131	52	27	28	0.40	0.21	0.21	0.52
Sharp-tailed Sandpiper	67	135	52	30	26	0.39	0.22	0.19	0.58
Purple Sandpiper	70	130	57	23	30	0.44	0.18	0.23	0.40
Rock Sandpiper	70	121	54	22	29	0.45	0.18	0.24	0.41
Dunlin	58	123	55	26	38	0.45	0.21	0.31	0.47
Curlew Sandpiper	57	133	46	30	38	0.35	0.23	0.29	0.65
Stilt Sandpiper	58	133	50	41	40	0.38	0.31	0.30	0.82
Spoonbill Sandpiper	28±	103	39	21	22	0.38	0.20	0.21	0.54
Broad-billed Sandpiper	37	109	37	22	31	0.34	0.20	0.28	0.59
Buff-breasted Sandpiper* ♂	71	136	58	32	20	0.43	0.24	0.15	0.55
Buff-breasted Sandpiper* ♀	53	129	53	30	19	0.41	0.23	0.15	0.57
Ruff* ♂	180	191	66	50	35	0.35	0.26	0.18	0.76
Ruff* ♀	109	158	54	43	31	0.34	0.27	0.20	0.80
Short-billed Dowitcher	105	148	54	37	60	0.36	0.25	0.41	0.69
Long-billed Dowitcher	105	144	57	39	67	0.40	0.27	0.47	0.68
Jack Snipe	50	112	51	24	40	0.46	0.21	0.36	0.47
Common Snipe	100	134	58	33	68	0.43	0.25	0.51	0.57
American Woodcock	167	137	56	32	69	0.41	0.23	0.50	0.57
Wilson's Phalarope* ♂	50	125	50	32	31	0.40	0.26	0.25	0.64
Wilson's Phalarope* ♀	68	136	54	34	34	0.40	0.25	0.25	0.63
Red-necked Phalarope	35	109	49	20	22	0.45	0.18	0.20	0.41
Red Phalarope	56	134	64	22	22	0.48	0.16	0.16	0.34

[1]Measurements are means, from many sources; some weights are estimated (±).

*Sexes differ sufficiently in weight to justify listing separately.

Appendix 4

Earliest and Latest Migration Dates for Regular Northwest Shorebirds

Species	Coast				Interior			
	Spring		Fall		Spring		Fall	
	Earliest	Latest	Earliest	Latest	Earliest	Latest	Earliest	Latest
Black-bellied Plover	-	-	-	-	Apr. 9	May 30	Aug. 3	Nov. 7
golden-plovers (2 species)	-	-	-	-	Apr. 22	June 10	July 18	Nov. 12+
Snowy Plover	-	-	-	-	Feb. 27	-	-	Sep. 9
Semipalmated Plover	-	-	-	-	Apr. 28	June 10+	July 4	Oct. 6+
Black-necked Stilt	Apr. 6	May 17+	Aug. 7	Sep. 6	Apr. 5	-	-	Oct. 26
American Avocet	Mar. 31	June 18+	July 2	Oct. 26+	Feb. 15	-	-	Dec. 16
Greater Yellowlegs	-	-	-	-	-	May 21	June 11	-
Lesser Yellowlegs	-	-	-	-	Apr. 10	June 4	June 11	Nov. 4+
Solitary Sandpiper	Apr. 8	June 10	June 27	Oct. 26	Apr. 9	May 30	June 25	Oct. 26
Willet	-	-	-	-	Mar. 21	-	-	Oct. 20
Wandering Tattler	-	-	-	-	0	June 8	Aug. 25	Sep. 5
Upland Sandpiper	May 5	June 3	July 16	Oct. 6+	Apr. 30	May 31	July 30	Sep. 13
Whimbrel	-	-	-	-	Apr. 9	June 11	June 21	Aug. 30
Long-billed Curlew	-	-	-	-	Mar. 3	-	-	Nov. 19
Hudsonian Godwit	Apr. 30	June 8	June 22	Oct. 27	May 10	June 8	July 7	Oct. 2
Bar-tailed Godwit	Apr. 25	June 10	Aug. 1	Oct. 31	0	0	0	0
Marbled Godwit	-	-	-	-	Apr. 23*	June 3	June 18	Oct. 6+
Ruddy Turnstone	-	June 7+	July 1	-	May 11	June 4	June 21	Sep. 29
Black Turnstone	-	June 9+	June 26	-	0	0	Aug. 28	Sep. 8
Surfbird	-	June 4	June 28	-	0	0	0	0

Species								
Red Knot	-	-	-	-	May 4	June 6	July 11	Oct. 10
Sanderling	-	-	-	-	May 5	June 7	July 6	Oct. 11+
Semipalmated Sandpiper	Apr. 14	June 7	June 20	Sep. 28	May 3	May 29	July 8	Sep. 28
Western Sandpiper	-	-	-	-	Apr. 2*	May 24	June 28	Nov. 10+
Least Sandpiper	-	June 6+	June 20	-	-	June 7	July 1	-
White-rumped Sandpiper	June 1	June 16	July 30	Sep. 23	May 20	June 1	July 25	Oct. 18
Baird's Sandpiper	Mar. 22	June 7	June 25	Oct. 27+	Apr. 8	May 18	June 17	Oct. 18
Pectoral Sandpiper	Mar. 25	June 8	-	Dec. 31	Apr. 15	May 31	June 29	Dec. 7
Sharp-tailed Sandpiper	Apr. 26	0	July 22	Dec. 21	0	0	Sep. 12	Oct. 13
Rock Sandpiper	-	May 24	Aug. 4	-	0	0	0	0
Dunlin	-	-	-	-	Apr. 5*	June 4	July 24	-
Curlew Sandpiper	May 17	0	July 11	Oct. 5	May 10	0	0	0
Stilt Sandpiper	May 20	June 3	June 20	Oct. 25	May 13	June 9	July 7	Sep. 28
Buff-breasted Sandpiper	Apr. 12	May 31	July 7	Oct. 17	0	0	Aug. 14	Oct. 5
Ruff	May 23	June 7	June 26	Oct. 31+	Apr. 27	0	Sep. 22	Sep. 27
Short-billed Dowitcher	Mar. 20	-	-	Nov. 17	Apr. 28	May 26	July 7	Sep. 26
Long-billed Dowitcher	-	-	-	-	Mar. 15	June 6+	June 20	-
Wilson's Phalarope	Apr. 4	-	-	Oct. 2	Apr. 2	-	-	Oct. 26+
Red-necked Phalarope	Mar. 27	June 10+	-	Nov. 26	Apr. 17	June 14+	June 18	Nov. 17
Red Phalarope	Apr. 14*	-	-	-	May 3	-	Aug. 1*	Oct. 31

- unable to determine
0 no records in category
* isolated record(s) before this date
+ isolated record(s) after this date

Appendix 5

Northwest Shorebird Size Groups (by Weight)

	grams		
Very small		Pectoral Sandpiper (male)	86
Least Sandpiper	21	Killdeer	90
Long-toed Stint	24	Common Snipe	100
Temminck's Stint	24	Short-billed Dowitcher	105
Little Stint	25	Long-billed Dowitcher	105
Semipalmated Sandpiper	26	Mountain Plover	107
Rufous-necked Stint	26	Eurasian Dotterel	108
Western Sandpiper	27	Gray-tailed Tattler	108
Spoonbill Sandpiper	28	Ruddy Turnstone	108
		Wandering Tattler	109
Small		Ruff (female)	109
Red-necked Phalarope	34	Black Turnstone	119
Broad-billed Sandpiper	37	Red Knot	120
Little Ringed Plover	39		
Snowy Plover	41	*Medium*	
Spotted Sandpiper	41	Pacific Golden-Plover	130
Baird's Sandpiper	42	American Golden-Plover	145
White-rumped Sandpiper	44	Great Knot	155
Semipalmated Plover	48	Spotted Redshank	158
Common Sandpiper	48	Surfbird	160
Jack Snipe	50	Black-necked Stilt	166
Wilson's Phalarope (male)	50	American Woodcock	167
Solitary Sandpiper	51	Greater Yellowlegs	170
Common Ringed Plover	53	Upland Sandpiper	170
Buff-breasted Sandpiper (female)	53	Common Greenshank	180
Sanderling	54	Little Curlew	180
Piping Plover	55	Ruff (male)	180
Red Phalarope	56	Black-winged Stilt	183
Wilson's Plover	57	Black-bellied Plover	210
Curlew Sandpiper	57	Eskimo Curlew	?
Pectoral Sandpiper (female)	58		
Dunlin	58	*Large*	
Stilt Sandpiper	58	Hudsonian Godwit	255
Mongolian Plover	60	Willet	257
Wood Sandpiper	61	Black-tailed Godwit	291
Sharp-tailed Sandpiper	67	American Avocet	316
Wilson's Phalarope (female)	68	Bar-tailed Godwit	320
Purple Sandpiper	70	Marbled Godwit	371
Rock Sandpiper	70	Whimbrel	380
		Bristle-thighed Curlew	433
Medium-small			
Buff-breasted Sandpiper (male)	71	*Very large*	
Terek Sandpiper	73	Long-billed Curlew	587
Marsh Sandpiper	74	American Oystercatcher	632
Lesser Yellowlegs	76	Black Oystercatcher	648
Green Sandpiper	80	Far Eastern Curlew	800
Oriental Pratincole	?		

Literature Cited

Common and scientific names of birds are standardized with capitalized first letters and in italics, respectively, no matter how they are presented in the reference title.

Allen, A. A., and H. Kyllingstad. 1949. The eggs and young of the Bristle-thighed Curlew. *Auk* 66: 343-50.

Allen, J. A. 1980. The ecology and behavior of the Long-billed Curlew in southeastern Washington. *Wildlife Monogr.* 73: 1-67.

Alström, P. 1987. The identification of Baird's and White-rumped Sandpipers in juvenile plumage. *Birding* 19: 10-13.

Alström, P., and U. Olsson. 1989. The identification of juvenile Red-necked and Long-toed Stints. *Brit. Birds* 82: 360-72.

American Ornithologists' Union. 1957. *Check-list of North American Birds.* Am. Ornithol. Union, Baltimore.

—. 1983. *Check-list of North American Birds.* Am. Ornithol.Union, Lawrence.

Andersen, F. S. 1944. Contributions to the breeding biology of the Ruff (*Philomachus pugnax*). *Dansk Ornithol. Foren. Tids.* 38: 26-30.

—. 1948. Contributions to the biology of the Ruff (*Philomachus pugnax* [L.]) II. *Dansk Ornithol. Foren. Tids.* 42: 125-48.

—. 1951. Contributions to the biology of the Ruff (*Philomachus pugnax* [L.]) III. *Dansk Ornithol. Foren. Tids.* 45: 145-73.

Armstrong, R. H. 1980. *A Guide to the Birds of Alaska.* Alaska Northwest Publ. Co., Anchorage.

—. 1983. *A New, Expanded Guide to the Birds of Alaska.* Alaska Northwest Publ. Co., Anchorage.

Ashkenazie, S., and U. N. Safriel. 1979a. Breeding cycle and behavior of the Semipalmated Sandpiper at Barrow, Alaska. *Auk* 96: 56-67.

—. 1979b. Time-energy budget of the Semipalmated Sandpiper *Calidris pusilla* at Barrow, Alaska. *Ecology* 60: 783-99.

Ashmole, M. J. 1970. Feeding of Western and Semipalmated Sandpipers in Peruvian winter quarters. *Auk* 87: 131-35.

Baker, M. C. 1974. Foraging behavior of Black-bellied Plovers (*Pluvialis squatarola*). *Ecology* 55: 162-67.

—. 1977. Shorebird food habits in the eastern Canadian Arctic. *Condor* 79: 56-62.

Baker, M. C., and A. E. M. Baker. 1973. Niche relationships among six species of shorebirds on their wintering and breeding ranges. *Ecol. Monogr.* 43: 193-212.

Baldassarre, G. A., and D. H. Fischer. 1984. Food habits of fall migrant shorebirds on the Texas high plains. *J. Field Ornithol.* 55: 220-29.

Banks, R. C. 1977. The decline and fall of the Eskimo Curlew, or why did the curlew go extaille? *Am. Birds* 31: 127-34.

Barter, M., A. Jessop, and C. Minton. 1988. Red Knot *Calidris canutus rogersi* in Australia. *Wader Study Group Bull.* 54: 17-20.

—. 1989. Red Knot *Calidris canutus rogersi* in Australia. Part 2. Biometrics and moult in Victoria and North-western Australia. *Wader Study Group Bull.* 56: 28-35.

Bateson, P. P., and E. K. Barth. 1957. Notes on the geographical variation of the Ringed Plover *Charadrius hiaticula* L. *Nytt. Mag. Zool.* 5: 20-25.

Bayer, R. D. 1984. Oversummering of Whimbrels, Bonaparte's Gulls, and Caspian Terns at Yaquina Estuary, Oregon. *Murrelet* 65: 87-90.

Bengtson, S.-A. 1968. Breeding behaviour of the Grey Phalarope in West Spitzbergen. *Vår Fågelvärld* 27: 1-13.

Bengtson, S.-A., and B. Svensson. 1968. Feeding habits of *Calidris alpina* L. and *C. minuta* Leisl. (Aves) in relation to the distribution of marine shore invertebrates. *Oikos* 19: 152-57.

Bent, A. C. 1927. Life histories of North American shore birds. Order Limicolae (Part 1). *U. S. Natl. Mus. Bull.* 142.

—. 1929. Life histories of North American shore birds. Order Limicolae (Part 2.) *U. S. Natl. Mus. Bull.* 146.

Bergman, G. 1946. The Turnstone, *Arenaria i. interpres* (L.), in its relationship to the environment. *Acta Zool. Fenn.* 47: 1-136.

Bergstrom, P. W. 1988a. Breeding biology of Wilson's Plovers. *Wilson Bull.* 100: 25-35.

—. 1988b. Breeding displays and vocalizations of Wilson's Plovers. *Wilson Bull.* 100: 36-49.

Beven, G., and M. D. England. 1977. Studies of less familiar birds 181. Turnstone. *Brit. Birds* 70: 23-32.

Blomqvist, S. 1983a. Bibliography of the genera *Calidris* and *Limicola*. *Ottenby Bird Obs. Spec. Rep.* No. 3, Degerhamn, Sweden.

—. 1983b. Bibliography of the genus *Phalaropus*. *Ottenby Bird Obs. Spec. Rep.* No. 4, Degerhamn, Sweden.

Bock, W. J. 1958. A generic review of the plovers (Charadriinae, Aves). *Bull. Mus. Comp. Zool.* 118: 27-97.

—. 1959. The status of the Semipalmated Plover. *Auk* 76: 98-100.

Bomberger, M. L. 1984. Quantitative assessment of the nesting habitat of Wilson's Phalarope. *Wilson Bull.* 96: 126-28.

Bonham, D. M., and B. A. Cooper. 1979. *Birds of West-central Montana*. Privately published.

Boswall, J., and B. N. Veprintsev. 1985. Keys to identifying Little Curlew. *Am. Birds* 39: 251-54.

Branson, N. J. B. A., and C. D. T. Minton. 1976. Moult, measurements and migrations of the Grey Plover. *Bird Study* 23: 257-66.

Branson, N. J. B. A., E. D. Ponting, and C. D. T. Minton. 1979. Turnstone populations on the Wash. *Bird Study* 26: 47-54.

Brearey, D., and O. Hildén. 1985. Nesting and egg-predation by Turnstones *Arenaria interpres* in larid colonies. *Ornis Scand.* 16: 283-92.

Breiehagen, T. 1989. Nesting biology and mating system in an alpine population of Temminck's Stint *Calidris temminckii*. *Ibis* 131: 389-402.

Brennan, L. A., J. B. Buchanan, C. T. Schick, S. G. Herman, and T. M. Johnson. 1984. Sex determination of Dunlins in winter plumage. *J. Field Ornithol.* 55: 343-48.

Briggs, K. T., K. F. Dettman, D. B. Lewis, and W. B. Tyler. 1984. Phalarope feeding in relation to autumn upwelling off California. In *Marine Birds: Their Feeding Ecology and Commercial Fisheries Relationships,* ed. D. N. Nettleship, G. A. Sanger, and P. F. Springer. Canad. Wildlife Serv. Special Publ.

Britton, D. 1980. Identification of Sharp-tailed Sandpipers. *Brit. Birds* 73: 333-45.

Brooks, W. S. 1967. Food and feeding habits of autumn migrant shorebirds at a small midwestern pond. *Wilson Bull.* 79: 307-15.

Brown, R. G. B. 1962. The aggressive and distraction behaviour of the Western Sandpiper *Ereunetes mauri*. *Ibis* 104: 1-12.

Brown, R. G. B., and D. E. Gaskin. 1988. The pelagic ecology of the Grey and Red-necked Phalaropes *Phalaropus fulicarius* and *P. lobatus* in the Bay of Fundy, eastern Canada. *Ibis* 130: 234-50.

Browning, M. R. 1974. Notes on the hypothetical list of Oregon birds. *Northwest Sci.* 48: 166-71.

—. 1975. The distribution and occurrence of the birds of Jackson County, Oregon, and surrounding areas. *N. Am. Fauna,* No. 70. U. S. Fish and Wildlife Service, Washington.

—. 1977. Geographic variation in Dunlins, *Calidris alpina,* of North America. *Canad. Field-Nat.* 91: 391-93.

Brunton, D. H. 1988a. Sexual differences in reproductive effort: Time-activity budgets of monogamous Killdeer (*Charadrius vociferus*). *Anim. Behav.* 36: 705-17.

—. 1988b. Energy expenditure in reproductive effort of male and female Killdeer (*Charadrius vociferus*). *Auk* 105: 553-64.

—. 1988c. Sequential polyandry by a female Killdeer. *Wilson Bull.* 100: 670-72.

Buchanan, J. B. 1988a. The abundance and migration of shorebirds at two Puget Sound estuaries. *Western Birds* 19: 69-78.

—. 1988b. Migration and winter populations of Greater Yellowlegs, *Tringa melanoleuca,* in western Washington. *Canad. Field-Nat.* 102: 611-16.

Buchanan, J. B., L. A. Brennan, C. T. Schick, S. G. Herman, and T. M. Johnson. 1986a. Age and sex composition of wintering Dunlin populations in western Washington. *Wader Study Group Bull.* 46: 37-41.

Buchanan, J. B., S. G. Herman, and T. M. Johnson. 1986b. Success rates of the Peregrine Falcon (*Falco peregrinus*) hunting Dunlin (*Calidris alpina*) during winter. *Raptor Res.* 20: 130-31.

Buchanan, J. B., C. T. Schick, L. A. Brennan, and S. G. Herman. 1988. Merlin predation on wintering Dunlins: Hunting success and Dunlin escape tactics. *Wilson Bull.* 100: 108-18.

Bucher, J. E. 1978. On sexing American Avocets and Long-billed Curlews. *Inland Bird Banding News* 50: 15-18.

Buck, W. F. A., A. J. Greenland, J. G. Harrison, and R. E. Scott. 1966. Semipalmated Sandpiper in Kent and the problem of identification. *Brit. Birds* 59: 543-47.

Bull, J., and J. Farrand, Jr. 1977. *The Audubon Society Field Guide to North American Birds.* Eastern Region. Alfred A. Knopf, New York.

Burger, J. 1980. Age differences in foraging Black-necked Stilts in Texas. *Auk* 97: 633-36.

Burger, J., and M. Gochfeld. 1991. Human activity influence and diurnal and nocturnal foraging of Sanderlings (*Calidris alba*). *Condor* 93: 259-65.

Burger, J., D. C. Hahn, and J. Chase. 1979. Aggressive interactions in mixed-species flocks of migrating shorebirds. *Anim. Behav.* 27: 459-69.

Burger, J., and M. Howe. 1975. Notes on winter feeding behavior and molt in Wilson's Phalaropes. *Auk* 92: 442-51.

Burger, J., and B. L. Olla. 1984a. *Shorebirds: Migration and Foraging Behavior.* Behavior of Marine Animals, Vol. 5. Plenum Press, New York.

—. 1984b. *Shorebirds: Breeding Behavior and Populations.* Behavior of Marine Animals, Vol. 6. Plenum Press, New York.

Burger, J., and J. Shisler. 1978. Nest site selection of Willets in a New Jersey salt marsh. *Wilson Bull.* 90: 599-607.

Burleigh, T. D. 1972. *Birds of Idaho.* Caxton Printers, Caldwell, Idaho.

Burton, P. J. K. 1972. The feeding techniques of Stilt Sandpipers and dowitchers. *Trans. San Diego Soc. Nat. Hist.* 17: 63-68.

—. 1974. *Feeding and the Feeding Apparatus of Waders: A Study of Anatomy and Adaptations in the Charadrii.* Brit. Mus. (Nat. Hist.), London.

Buss, I. O. 1951. The Upland Plover in southwestern Yukon Territory. *Arctic* 4: 204-13.

Buss, I. O., and A. S. Hawkins. 1939. The Upland Plover at Faville Grove, Wisconsin. *Wilson Bull.* 51: 202-20.

Butler, R. W., and R. W. Campbell. 1987. The birds of the Fraser River delta: populations, ecology and international significance. *Canad. Wildlife Serv.,* Occ. Pap. No. 65.

Butler, R. W., and G. W. Kaiser. 1988. Western Sandpiper migration studies along the west coast of Canada. *Wader Study Group Bull.* 52: 16-17.

Butler, R. W., G. W. Kaiser, and G. E. J. Smith. 1987. Migration chronology, length of stay, sex ratio, and weight of Western Sandpipers (*Calidris mauri*) on the south coast of British Columbia. *J. Field Ornithol.* 58: 103-11.

Butler, R. W., and J. W. Kirbyson. 1979. Oyster predation by the Black Oystercatcher in British Columbia. *Condor* 81: 433-35.

Butler, R. W., B. G. Stushnoff, and E. McMackin. 1986. The birds of the Creston Valley and southeastern British Columbia. *Canad. Wildlife Serv., Occ. Pap.* No. 58.

Byrkjedal, I. 1987. Antipredator behavior and breeding success in Greater Golden-Plover and Eurasian Dotterel. *Condor* 89: 40-47.

—. 1989a. Nest habitat and nesting success of Lesser Golden-Plovers. *Wilson Bull.* 101: 93-96.

—. 1989b. Nest defense behavior of Lesser Golden-Plovers. *Wilson Bull.* 101: 579-90.

Cadman, M. 1979. Territorial behaviour in American Oystercatchers *Haematopus palliatus*. *Wader Study Group Bull.* 27: 40-41.

Cairns, W. E. 1982. Biology and behavior of breeding Piping Plovers. 1982. *Wilson Bull.* 94: 531-45.

Cairns, W. E., and I. A. McLaren. 1980. Status of the Piping Plover on the east coast of North America. *Am. Birds* 34: 206-8.

Campbell, R. W., N. K. Dawe, J. M. Cooper, G. W. Kaiser, M. C. E. McNall, and I. McT. Cowan. 1990. *Birds of British Columbia: Nonpasserines.* Royal British Columbia Museum, Victoria.

Campbell, R. W., and P. T. Gregory. 1976. The Buff-breasted Sandpiper in British Columbia, with notes on its migration in North America. *Syesis* 9: 123-30.

Campbell, R. W., and R. E. Luscher. 1972. Semipalmated Plover breeding at Vancouver, British Columbia. *Murrelet* 53: 11-12.

Campbell, R. W., M. G. Shepard, and R. H. Drent. 1972. Status of birds in the Vancouver area in 1970. *Syesis* 5: 137-67.

Campbell, R. W., M. G. Shepard, and W. C. Weber. 1973. *Vancouver Birds in 1971.* Vancouver Nat. Hist. Soc., Vancouver, B. C.

Cannings, R. A., R. J. Cannings, and S. G. Cannings. 1987. *Birds of the Okanagan Valley, British Columbia.* Royal Brit. Columbia Mus., Victoria.

Cartar, R. V. 1984. A morphometric comparison of Western and Semipalmated Sandpipers. *Wilson Bull.* 96: 277-86.

Cartar, R. V., and B. E. Lyon. 1988. The mating system of the Buff-breasted Sandpiper: Lekking and resource defense polygyny. *Ornis Scand.* 19: 74-76.

Cartar, R. V., and R. D. Montgomerie. 1987. Day-to-day variation in nest attentiveness of White-rumped Sandpipers. *Condor* 89: 252-60.

Chandler, R. J. 1987a. Yellow orbital ring of Semipalmated and Ringed Plovers. *Brit. Birds* 80: 241-42.

—. 1987b. Plumages of breeding female Ruffs. *Brit. Birds* 80: 246-48.

—. 1989. *The Facts on File Field Guide to North Atlantic Shorebirds.* Facts on File, New York.

Chapman, B.-A., J. P. Goossen, and I. Ohanjanian. 1985. Occurrences of Black-necked Stilts, *Himantopus mexicanus,* in western Canada. *Canad. Field-Nat.* 99: 254-57.

Choate, E. A. 1985. *The Dictionary of Bird Names.* Harvard Common, Cambridge.

Clark, N. A. 1977. The weights, moult and morphometrics of Spotted Redshanks in Britain. *Wader Study Group Bull.* 21: 22-26.

Colston, P., and P. Burton. 1988. *A Field Guide to the Waders of Britain and Europe, with North Africa and the Middle East.* Hodder & Stoughton, London.

Colwell, M. A., and L. W. Oring. 1988a. Sex ratios and intrasexual competition for mates in a sex-role reversed shorebird, Wilson's Phalarope (*Phalaropus tricolor*). *Behav. Ecol. Sociobiol.* 22: 165-73.

—. 1988b. Breeding biology of Wilson's Phalarope in southcentral Saskatchewan. *Wilson Bull.* 100: 567-82.

Connor, J. 1988. *The Complete Birder.* Houghton Mifflin Co., Boston.

Connors, P. G. 1983. Taxonomy, distribution, and evolution of golden plovers (*Pluvialis dominica* and *Pluvialis fulva*). *Auk* 100: 607-20.

Connors, P. G., J. P. Myers, C. S. W. Connors, and F. A. Pitelka. 1981. Interhabitat movements by

Sanderlings in relation to foraging profitability and the tidal cycle. *Auk* 98: 49-64.

Conover, B. 1943. The races of the Knot (*Calidris canutus*). *Condor* 45: 226-28.

——. 1944a. The North Pacific allies of the Purple Sandpiper. *Field Mus. Nat. Hist., Zool. Ser.* 29: 169-79.

——. 1945. The breeding Golden Plover of Alaska. *Auk* 62: 568-74.

Conover, H. B. 1926. Game birds of the Hooper Bay region, Alaska (concluded). *Auk* 43: 303-18.

——. 1941. A study of the dowitchers. *Auk* 58: 376-80.

——. 1944b. The races of the Solitary Sandpiper. *Auk* 61: 537-44.

Cramp, S., and K. E. L. Simmons, eds. 1983. *The Birds of the Western Palearctic,* Vol. III. *Waders to Gulls.* Oxford Univ. Press, Oxford.

Crawford, D. N. 1978. Notes on Little Curlew on the subcoastal plains, Northern Territory. *Aust. Bird Watcher* 7: 270-72.

Davis, F. W. 1970. Territorial conflict in the American Woodcock. *Wilson Bull.* 82: 327-28.

Davis, J. M. 1980. The coordinated aerobatics of Dunlin flocks. *Anim. Behav.* 28: 668-73.

Dean, W. R. J. 1977. Moult of Little Stints in South Africa. *Ardea* 65: 73-79.

Deane, C. D. 1944. The broken-wing behavior of the Killdeer. *Auk* 61: 243-47.

Demaree, S. R. 1975. Observations on roof-nesting Killdeers. *Condor* 7: 487-88.

Dementiev, G. P., N. A. Gladkov, and E. P. Spangenberg. 1969. *Birds of the Soviet Union,* Vol. 3. Israel Program for Scientific Translations, Jerusalem.

Diefenbach, D. R., E. L. Derleth, W. M. Vander Haegen, J. D. Nichols, and J. E. Hines. 1990. American Woodcock winter distribution and fidelity to wintering areas. *Auk* 107: 745-49.

Dittmann, D. L., and R. M. Zink. 1991. Mitochondrial DNA variation among phalaropes and allies. *Auk* 108: 771-79.

Dixon, J. 1917. The home life of the Baird Sandpiper. *Condor* 19: 77-84.

Dixon, J. S. 1918. The nesting grounds and nesting habits of the Spoon-billed Sandpiper. *Auk* 35: 387-404.

——. 1927. The Surfbird's secret. *Condor* 29: 3-16.

——. 1933. Nesting of the Wandering Tattler. *Condor* 35: 173-79.

Dodson, S. I., and D. L. Egger. 1980. Selective feeding of Red Phalaropes on zooplankton of arctic ponds. *Ecology* 61: 755-63.

Dorio, J. C., and A. E. Grewe. 1979. Nesting and brood rearing habitat of the Upland Sandpiper. *J. Minn. Acad. Sci.* 45: 8-11.

Drent, R., G. F. Van Tets, F. Tompa, and K. Vermeer. 1964. Breeding birds of Mandarte Island, British Columbia. *Canad. Field-Nat.* 78: 208-63.

Drury, W. H., Jr. 1961. The breeding biology of shorebirds on Bylot Island, Northwest Territories, Canada. *Auk* 78: 176-219.

Dukes, P. A. 1980. Semipalmated Plover: New to Britain and Ireland. *Brit. Birds* 73: 458-64.

Dunn, P. O., T. A. May, M. A. McCollough, and M. A. Howe. 1988. Length of stay and fat content of migrant Semipalmated Sandpipers in eastern Maine. *Condor* 90: 824-35.

Dwight, J., Jr. 1900. The moult of the North American shore birds (Limicolae). *Auk* 17: 368-85.

Dwyer, T. J., and J. V. Dobell. 1979. External determination of age of Common Snipe. *J. Wildlife Manage.* 43: 754-56.

Elliott, C. C. H., M. Waltner, L. G. Underhill, J. S. Pringle, and W. J. A. Dick. 1976. The migration system of the Curlew Sandpiper *Calidris ferruginea* in Africa. *Ostrich* 47: 191-213.

Espin, P. M. J., R. M. Mather, and J. Adams. 1983. Age and foraging success in Black-winged Stilts *Himantopus himantopus*. *Ardea* 71: 225-28.

Evanich, J. 1989. A review of the Semipalmated Sandpiper in Oregon. *Oregon Birds* 15: 109-11.

Evans, P. R. 1975. Moult of Red-necked Stints at Westernport Bay, Victoria. *Emu* 75: 227-29.

Evans, P. R., D. M. Brearey, and L. R. Goodyer. 1980. Studies on Sanderling at Teesmouth, NE England. *Wader Study Group Bull.* 30: 18-20.

Farrand, J., Jr. 1977. What to look for: Eskimo and Little Curlews compared. *Am. Birds* 31: 137-38.
—. 1983. *The Audubon Society Master Guide to Birding.* Vol. 1, *Loons to Sandpipers.* Alfred A. Knopf, New York.
—. 1988a. *Eastern Birds (An Audubon Handbook).* McGraw-Hill Book Co., New York.
—. 1988b. *Western Birds (An Audubon Handbook).* McGraw-Hill Book Co., New York.
Ferguson-Lees, I. J. 1959. Photographic studies of some less familiar birds 95. Terek Sandpiper. *Brit. Birds* 52: 85-90.
Ferns, P. N. 1978. Individual differences in the head and neck plumage of Ruddy Turnstones (*Arenaria interpres*) during the breeding season. *Auk* 95: 753-55.
—. 1981. Identification, subspecific variation, ageing and sexing in European Dunlins. *Dutch Birding* 3: 85-99.
Fix, D. M. 1979. Occurrence and identification of the Stilt Sandpiper in Oregon. *Oregon Birds* 5: 6-13.
Fleischer, R. C. 1983. Relationships between tidal oscillations and Ruddy Turnstone flocking, foraging and vigilance behavior. *Condor* 85: 22-29.
Flint, V. E. 1972. The breeding of the Knot on Vrangelya (Wrangel) Island, Siberia; comparative remarks. *Proc. Western Found. Vert. Zool.* 2: 27-29.
Forsythe, D. M. 1970. Vocalizations of the Long-billed Curlew. *Condor* 72: 213-24.
—. 1973. Growth and development of Long-billed Curlew chicks. *Auk* 90: 435-38.
Frank, P. W. 1982. Effects of winter feeding on limpets by Black Oystercatchers, *Haematopus bachmani. Ecology* 63: 1352-62.
Fritzell, E. K., G. A. Swanson, and M. I. Meyer. 1979. Fall foods of migrant Common Snipe in North Dakota. *J. Wildlife Manage.* 43: 253-57.
Fry, K. 1980. *Aspects of the Winter Ecology of the Dunlin (Calidris alpina) on the Fraser River Delta.* Canad. Wildlife Serv., Pacific & Yukon Region.
Furniss, O. C. 1933. Observations on the nesting of the Killdeer Plover in the Prince Albert District in central Saskatchewan. *Canad. Field-Nat.* 47: 135-38.
Gabrielson, I. N., and S. G. Jewett. 1940. *Birds of Oregon.* Oregon State College, Corvallis.
Gabrielson, I. N., and F. C. Lincoln. 1959. *The Birds of Alaska.* The Stackpole Co., Harrisburg, Pennsylvania.
Gerritsen, A. F. C., and A. Meiboom. 1986. The role of touch in prey density estimation by *Calidris alba. Netherlands J. Zool.* 36: 530-62.
Gibson, D. D. 1977. First North American nest and eggs of the Ruff. *Western Birds* 8: 25-26.
—. 1981. Migrant birds at Shemya Island, Aleutian Islands, Alaska. *Condor* 83: 65-77.
Gibson, D. D., and B. Kessel. 1989. Geographic variation in the Marbled Godwit and description of an Alaska subspecies. *Condor* 91: 436-43.
Gibson, F. 1971. The breeding biology of the American Avocet (*Recurvirostra americana*) in central Oregon. *Condor* 73: 444-54.
—. 1978. Ecological aspects of the time budget of the American Avocet. *Am. Midland Nat.* 99: 65-82.
Gill, R. E., Jr., C. M. Handel, and L. A. Shelton. 1983. Memorial to a Black Turnstone: An exemplar of breeding and wintering site fidelity. *N. Am. Bird Bander* 8: 98-101.
Gill, R. E., Jr., B. J. McCaffery, and T. G. Tobish. 1988. Bristle-thighed Curlews, biologists and bird tours—a place for all. *Birding* 20: 148-55.
Gilligan, J., O. Schmidt, H. Nehls, and D. Irons. 1987. First record of Long-toed Stint in Oregon. *Western Birds* 18: 126-28.
Gochfeld, M. 1971. Notes on a nocturnal roost of Spotted Sandpipers in Trinidad, West Indies. *Auk* 88: 167-68.
Gollop, J. B., T. W. Barry, and E. H. Iversen. 1986. Eskimo Curlew. A Vanishing Species? *Saskatchewan Nat. Hist. Soc., Spec. Publ.* No. 17.
Goodwin, C. E. 1981. Ontario region. *Am. Birds* 35: 176-79.

Goriup, P. D. 1982. Behaviour of Black-winged Stilts. *Brit. Birds* 75: 12-24.

Goss-Custard, J. D. 1970. Feeding dispersion in some overwintering wading birds. In *Social Behavior in Birds and Mammals,* ed. J. H. Crook. Academic Press, London.

Goss-Custard, J. D., R. E. Jones, and P. E. Newberry. 1977. The ecology of The Wash I. Distribution and diet of wading birds (Charadrii). *J. Appl. Ecol.* 14: 681-700.

Grant, P. J. 1981. Identification of Semipalmated Sandpiper. *Brit. Birds* 74: 505-9.

—. 1982. *Gulls: A Guide to Identification.* T. and A. D. Poyser, Calton.

—. 1984. Identification of stints and peeps. *Brit. Birds* 77: 293-315.

—. 1986. Four problem stints. *Brit. Birds* 79: 609-21.

Gratto, C. L. 1988. Natal philopatry, site tenacity, and age of first breeding of the Semipalmated Sandpiper. *Wilson Bull.* 100: 660-63.

Gratto, C. L., F. Cooke, and R. I. G. Morrison. 1981. Hatching success of yearling and older breeders in the Semipalmated Sandpiper *Calidris pusilla. Wader Study Group Bull.* 33: 37-38.

Gratto, C. L., and R. I. G. Morrison. 1981. Partial postjuvenile wing moult of the Semipalmated Sandpiper *Calidris pusilla. Wader Study Group Bull.* 33: 33-37.

Gratto, C. L., R. I. G. Morrison, and F. Cooke. 1985. Philopatry, site tenacity, and mate fidelity in the Semipalmated Sandpiper. *Auk* 102: 16-24.

Graul, W. D. 1971. Observations at a Long-billed Curlew nest. *Auk* 88: 182-84.

—. 1973a. Possible functions of head and breast markings in Charadriinae. *Wilson Bull.* 85: 60-70.

—. 1973b. Adaptive aspects of the Mountain Plover social system. *Living Bird* 12: 69-94.

—. 1974. Vocalizations of the Mountain Plover. *Wilson Bull.* 86: 221-29.

—. 1975. Breeding biology of the Mountain Plover. *Wilson Bull.* 87: 6-31.

Graul, W. D., and L. E. Webster. 1976. Breeding status of the Mountain Plover. *Condor* 78: 265-67.

Greenwood, J. G. 1983. Post-nuptial primary moult in Dunlin *Calidris alpina. Ibis* 125: 223-28.

Grieve, A. 1987. Hudsonian Godwit: New to the Western Palearctic. *Brit. Birds* 80: 466-73.

Grinnell, J. 1921. Concerning the status of the supposed two races of the Long-billed Curlew. *Condor* 23: 21-27.

Groves, S. 1978. Age-related differences in Ruddy Turnstone foraging and aggressive behavior. *Auk* 95: 95-103.

—. 1984. Chick growth, sibling rivalry, and chick production in American Black Oystercatchers. *Auk* 101: 525-31.

Guiguet, C. J. 1955. *The Birds of British Columbia.* Vol. 3, *The Shorebirds.* Handbook No. 8, Brit. Columbia Prov. Mus., Victoria.

Gullion, G. W. 1951. Birds of the southern Willamette Valley, Oregon. *Condor* 53: 129-49.

Hagar, J. A. 1966. Nesting of the Hudsonian Godwit at Churchill, Manitoba. *Living Bird* 5: 5-43.

—. 1983. The flight of the godwit. *Sanctuary* 22: 3-6.

Haig, S. M., and L. W. Oring. 1985. Distribution and status of the Piping Plover throughout the annual cycle. *J. Field Ornith.* 56: 334-45.

—. 1988a. Mate, site, and territory fidelity in Piping Plovers. *Auk* 105: 268-77.

—. 1988b. Distribution and dispersal in the Piping Plover. *Auk* 105: 630-38.

Hale, W. G. 1980. *Waders.* The New Naturalist, Collins, London.

Hamilton, R. B. 1975. Comparative Behavior of the American Avocet and the Black-necked Stilt (Recurvirostridae). *Ornith. Monogr.* 17, Am. Ornithol. Union.

Hamilton, W. J. 1959. Aggressive behavior in migrant Pectoral Sandpipers. *Condor* 61: 161-79.

Hammond, N., and M. Everett. 1980. *Birds of Britain and Europe.* Pan Books, Ltd., London.

Handel, C. M., and C. P. Dau. 1988. Seasonal occurrence of migrant Whimbrels and Bristle-thighed Curlews on the Yukon-Kuskokwim Delta, Alaska. *Condor* 90: 782-90.

Haney, J. C., and A. E. Stone. 1988. Littoral foraging by Red Phalaropes during spring in the northern Bering Sea. *Condor* 90: 723-26.

Hanna, G. D. 1921. The Pribilof Sandpiper. *Condor* 23: 50-57.

Harrington, B. A. 1982. Morphometric variation and habitat use of Semipalmated Sandpipers during a migratory stopover. *J. Field Ornith.* 53: 258-62.

—. 1983. The migration of the Red Knot. *Oceanus* 26: 44-48.

Harrington, B. A., and S. Groves. 1977. Aggression in foraging migrant Semipalmated Sandpipers. *Wilson Bull.* 89: 336-38.

Harrington, B. A., J. M. Hagan, and L. E. Leddy. 1988. Site fidelity and survival differences between two groups of New World Red Knots (*Calidris canutus*). *Auk* 105: 439-45.

Harrington, B. A., and L. E. Leddy. 1982. Are wader flocks random groupings? A knotty problem. *Wader Study Group Bull.* 36: 20-21.

Harrington, B. A., F. J. Leeuwenberg, S. Lara Resende, R. McNeil, B. T. Thomas, J. S. Grear, and E. F. Martinez. 1991. Migration and mass change of White-rumped Sandpipers in North and South America. *Wilson Bull.* 103: 621-636.

Harrington, B. A., and R. I. G. Morrison. 1979. Semipalmated Sandpiper migration in North America. *Stud. Avian Biol.* 2: 83-100.

Harrington, B. A., and A. L. Taylor. 1982. Methods for sex identification and estimation of wing area in Semipalmated Sandpipers. *J. Field Ornithol.* 53: 174-77.

Harris, P. R. 1979. The winter feeding of the Turnstone in North Wales. *Bird Study* 26: 259-66.

Harrison, C. 1978. *A Field Guide to the Nests, Eggs and Nestlings of North American Birds.* Collins, Glasgow.

Harrison, P. 1985. *Seabirds: An Identification Guide.* Croom Helm, London.

Hartwick, E. B. 1974. Breeding ecology of the Black Oystercatcher *Haematopus bachmani. Syesis* 7: 83-92.

—. 1976. Foraging strategy of the Black Oystercatcher *Haematopus bachmani. Canad. J. Zool.* 54: 142-55.

—. 1978a. Some observations on foraging by Black Oystercatchers *Haematopus bachmani. Syesis* 11: 55-60.

—. 1978b. The use of feeding areas outside of the territory of breeding Black Oystercatchers. *Wilson Bull.* 90: 650-52.

Hartwick, E. B., and W. Blaylock. 1979. Winter ecology of a Black Oystercatcher population. *Stud. Avian Biol.* 2: 207-15.

Hatler, D. F., R. W. Campbell, and A. Dorst. 1978. Birds of Pacific Rim National Park. *Brit. Columbia Prov. Mus., Occ. Pap.* No. 20.

Hayman, P., J. Marchant, and T. Prater. 1986. *Shorebirds: An Identification Guide to the Waders of the World.* Houghton Mifflin Co., Boston.

Hays, H. 1972. Polyandry in the Spotted Sandpiper. *Living Bird* 11: 43-57.

Henshaw, H. W. 1910. Migration of the Pacific Plover to and from the Hawaiian Islands. *Auk* 27: 245-62.

Herman, S. G., and J. B. Bulger. 1981. *The Distribution and Abundance of Shorebirds During the 1981 Spring Migration at Grays Harbor, Washington.* U. S. Army Corps of Engineers, Seattle District.

Herman, S. G., J. B. Bulger, and J. B. Buchanan. 1988. Snowy Plover in southeastern Oregon and western Nevada. *J. Field Ornith.* 59: 13-21.

Higgins, K. F., and L. M. Kirsch. 1975. Some aspects of the breeding biology of the Upland Sandpiper in North Dakota. *Wilson Bull.* 87: 96-102.

Higgins, K. F., L. M. Kirsch, M. R. Ryan, and R. B. Renken. 1979. Some ecological aspects of Marbled Godwits and Willets in North Dakota. *Prairie Nat.* 11: 114-18.

Hildén, O. 1975. Breeding system of Temminck's Stint *Calidris temminckii. Ornis Fenn.* 52: 117-46.

—. 1978. Population dynamics in Temminck's Stint *Calidris temminckii. Oikos* 30: 17-28.

—. 1979a. The timing of arrival and departure of the Spotted Redshank *Tringa erythropus* in Finland. *Ornis Fenn.* 56: 18-23.

—. 1979b. Territoriality and site tenacity of Temminck's Stint *Calidris temminckii*. *Ornis Fenn*. 56: 56-74.

Hildén, O., and S. Vuolanto. 1973. Breeding biology of the Red-necked Phalarope *Phalaropus lobatus* in Finland. *Ornis Fenn*. 49: 57-85.

Hindwood, K. A., and E. S. Hoskin. 1954. The waders of Sydney (County of Cumberland), New South Wales. *Emu* 54: 217-55.

Hobson, W. 1972. The breeding biology of the Knot. *Proc. Western Found. Vert. Zool*. 2: 5-26.

Hockey, P. A. R. 1983. A bibliography of world oystercatcher literature. *Wader Study Group Bull*. 37: 25-28; Bull. 38: 23-28.

Hogan-Warburg, A. J. 1966. Social behaviour of the Ruff, *Philomachus pugnax* (L.). *Ardea* 54: 109-229.

Hoge, G., and W. Hoge. 1980. *Birds of Ocean Shores*. Privately published.

Höglund, J., and A. Lundberg. 1989. Plumage color correlates with body size in the Ruff (*Philomachus pugnax*). *Auk* 106: 336-38.

Höhn, E. O. 1957. Observations on display and other forms of behavior of certain Arctic birds. *Auk* 74: 203-14.

—. 1967. Observations on the breeding biology of Wilson's Phalarope (*Steganopus tricolor*) in central Alberta. *Auk* 84: 220-44.

—. 1968. Some observations on the breeding of Northern Phalaropes at Scammon Bay, Alaska. *Auk* 85: 316-17.

—. 1969. The phalarope. *Sci. Am*. 220: 104-11.

—. 1971. Observations on the breeding behavior of Grey and Red-necked Phalaropes. *Ibis* 113: 335-48.

Höhn, E. O., and J. R. Barron. 1963. The food of Wilson's Phalarope during the breeding season. *Canad. J. Zool*. 41: 1171-73.

Holland, P. K., J. E. Robson, and D. W. Yalden. 1982. The breeding biology of the Common Sandpiper (*Actitis hypoleucos*) in the Peak District. *Bird Study* 29: 99-110.

Holmes, P. F. 1939. Some oceanic records and notes on the winter distribution of phalaropes. *Ibis* (14) 3: 329-42.

Holmes, R. T. 1966a. Breeding ecology and annual cycle adaptations of the Red-backed Sandpiper (*Calidris alpina*) in northern Alaska. *Condor* 68: 3-46.

—. 1966b. Molt cycle of the Red-backed Sandpiper (*Calidris alpina*) in western North America. *Auk* 83: 517-33.

—. 1966c. Feeding ecology of the Red-backed Sandpiper (*Calidris alpina*) in arctic Alaska. *Ecology* 47: 32-45.

—. 1970. Differences in population density, territoriality, and food supply of Dunlin on arctic and subarctic tundra. *Symp. Brit. Ecol. Soc*. 10: 303-19.

—. 1971a. Latitudinal differences in the breeding and molt schedules of Alaskan Red-backed Sandpipers (*Calidris alpina*). *Condor* 73: 93-99.

—. 1971b. Density, habitat, and the mating system of the Western Sandpiper (*Calidris mauri*). *Oecologia* 7: 191-208.

—. 1972. Ecological factors influencing the breeding season schedule of Western Sandpipers (*Calidris mauri*) in subarctic Alaska. *Am. Midland Nat*. 87: 472-91.

—. 1973. Social behaviour of breeding Western Sandpipers *Calidris mauri*. *Ibis* 115: 107-23.

Holmes, R. T., and F. A. Pitelka. 1962. Behavior and taxonomic position of the White-rumped Sandpiper. *Proc. 12th Alaskan Sci. Conf*.: 19-20.

—. 1964. Breeding behavior and taxonomic relationships of the Curlew Sandpiper. *Auk* 81: 362-79.

—. 1968. Food overlap among coexisting sandpipers of northern Alaskan tundra. *Syst. Zool*. 17: 305-18.

Hosking, E., and W. G. Hale. 1983. *Eric Hosking's Waders*. Pelham Books, London.

Howe, M. A. 1974. Observations on the terrestrial wing displays of breeding Willets. *Wilson Bull.* 86: 286-88.

—. 1975a. Social interactions in flocks of courting Wilson's Phalaropes (*Phalaropus tricolor*). *Condor* 77: 24-33.

—. 1975b. Behavioral aspects of the pair bond in Wilson's Phalarope. *Wilson Bull.* 87: 248-70.

—. 1982. Social organization in a nesting population of Eastern Willets (*Catoptrophorus semipalmatus*). *Auk* 99: 88-102.

Hudson, G. E., and C. F. Yocom. 1954. A distributional list of the birds of southeastern Washington. *Res. Stud. State Coll. of Washington* 22: 1-56.

Hunn, E. S. 1982. *Birding in Seattle and King County*. Seattle Audubon Society, Seattle.

Hussell, D. J. T., and G. W. Page. 1976. Observations on the breeding biology of Black-bellied Plovers on Devon Island, N. W. T., Canada. *Wilson Bull.* 88: 632-53.

Ivey, G., and C. Baars. 1990. A second Semipalmated Plover nest in Oregon. *Oregon Birds* 16: 207-208.

Ivey, G. L., K. A. Fothergill, and K. L. Yates-Mills. 1988. A Semipalmated Plover nest in Oregon. *Western Birds* 19: 35-36.

Jackson, A. C. 1918. Notes on the relation between moult and migration as observed in some waders. *Brit. Birds* 11: 197-203.

Jehl, J. R., Jr. 1963. An investigation of fall migrating dowitchers in New Jersey. *Wilson Bull.* 75: 250-61.

—. 1968. The systematic position of the Surfbird *Aphriza virgata*. *Condor* 70: 206-10.

—. 1973. Breeding biology and systematic relationships of the Stilt Sandpiper. *Wilson Bull.* 85: 115-47.

—. 1979. The autumnal migration of Baird's Sandpiper. *Stud. Avian Biol.* 2: 55-68.

—. 1981. Mono Lake: A vital way station for the Wilson's Phalarope. *Nat. Geog.* 160: 520-25.

—. 1985. Hybridization and evolution of oystercatchers on the Pacific coast of Baja California. In Neotropical Ornithology, ed. P. A. Buckley, M. S. Foster, E. S. Morton, R. S. Ridgely, and F. G. Buckley. *Ornith. Monogr.* No. 16.

—. 1986. Biology of Red-necked Phalaropes (*Phalaropus lobatus*) at the western edge of the Great Basin in fall migration. *Great Basin Nat.* 46: 185-97.

Jehl, J. R., Jr., and B. G. Murray, Jr. 1986. The evolution of normal and reverse sexual size dimorphism in shorebirds and other birds. *Current Ornith.* 3: 1-86.

Jewett, S. G., W. P. Taylor, W. T. Shaw, and J. W. Aldrich. 1953. *Birds of Washington State*. Univ. of Washington Press, Seattle.

Johns, J. E. 1969. Field studies of Wilson's Phalarope. *Auk* 86: 660-70.

Johnsgard, P. A. 1981. *The Plovers, Sandpipers, and Snipes of the World*. Univ. of Nebraska Press, Lincoln.

Johnson, C. M., and G. A. Baldassarre. 1988. Aspects of the wintering ecology of Piping Plovers in coastal Alabama. *Wilson Bull.* 100: 214-23.

Johnson, O. W. 1973. Reproductive condition and other features of shorebirds resident at Eniwetok Atoll during the boreal summer. *Condor* 75: 336-43.

—. 1977. Plumage and molt in shorebirds summering at Enewetak Atoll. *Auk* 94: 222-30.

—. 1979. Biology of shorebirds summering on Enewetak Atoll. *Stud. Avian Biol.* 2: 193-205.

—. 1985. Timing of primary molt in first-year golden-plovers and some evolutionary implications. *Wilson Bull.* 97: 237-39.

Johnson, O. W., and P. M. Johnson. 1983. Plumage-molt-age relationships in "over-summering" and migratory Lesser Golden-Plovers. *Condor* 85: 406-19.

Johnson, O. W., P. Johnson, and P. Bruner. 1981. Wintering behaviour and site faithfulness of American Golden Plovers *Pluvialis dominica fulva* in Hawaii. *Wader Study Group Bull.* 31: 44.

Johnson, O. W., M. L. Morton, P. L. Bruner, and P. M. Johnson. 1989. Fat cyclicity, predicted migratory flight ranges, and features of wintering behavior in Pacific Golden-Plovers. *Condor* 91: 156-77.

Johnson, O. W., and R. M. Nakamura. 1981. The use of roofs by American Golden Plovers *Pluvialis dominica fulva* wintering on Oahu, Hawaiian Islands. *Wader Study Group Bull.* 31: 45-46.

Johnson, T. B., and R. B. Spicer. 1981. Mountain Plovers on the New Mexico–Arizona border. *Cont. Birdlife* 2: 69-73.

Johnston, D. W., and R. W. McFarlane. 1967. Migration and bioenergetics of flight in the Pacific Golden Plover. *Condor* 69:156-68.

Jones, R. E. 1975. Food of Turnstones in the Wash. *Brit. Birds* 68: 339-41.

Kagarise, C. M. 1979. Breeding biology of Wilson's Phalarope in North Dakota. *Bird-Banding* 50: 12-22.

Kaiser, G. W., and M. Gillingham. 1981. Some notes on seasonal fluctuations in the weight of Dunlin *Calidris alpina* on the Fraser River delta, British Columbia. *Wader Study Group Bull.* 31: 46-48.

Kålås, J. A. 1986. Incubation schedules in different parental care systems in the Dotterel *Charadrius morinellus*. *Ardea* 74: 185-90.

Kålås, J. A., and I. Byrkjedal. 1984. Breeding chronology and mating system of the Eurasian Dotterel (*Charadrius morinellus*). *Auk* 101: 838-47.

Kaufman, K. 1990. *A Field Guide to Advanced Birding*. Houghton Mifflin Co., Boston.

Kautesk, B. M., R. E. Scott, D. S. Aldcroft, and J. Ireland. 1983. Temminck's Stint at Vancouver, British Columbia. *Am. Birds* 37: 347-49.

Keast, J. A. 1949. Field notes on the Grey-tailed Tattler. *Rec. Aust. Mus.* 22: 207-11.

Keith, S., and J. Gooders. 1980. *Collins Bird Guide*. Collins, London.

Kelly, P. R., and H. R. Cogswell. 1979. Movements and habitat use by wintering populations of Willets and Marbled Godwits. *Stud. Avian Biol.* 2: 69-82.

Kenyon, K. W. 1949. Observations on behavior and populations of Oystercatchers in Lower California. *Condor* 51: 193-99.

Kessel, B., and D. D. Gibson. 1978. Status and distribution of Alaska birds. *Stud. Avian Biol.* No. 1.

Kieser, J. A. 1983. Jizz of Spotted Sandpiper. *Brit. Birds* 76: 313-14.

Kieser, J. A., and G. A. Kieser. 1982. Field identification of common waders: Marsh Sandpiper and Greenshank. *Bokmakierie* 34: 63-66.

Kinsky, F. C., and J. C. Yaldwyn. 1981. The bird fauna of Niue Island, south-west Pacific, with special notes on the White-tailed Tropic Bird and Golden Plover. *Nat. Mus. New Zealand, Misc. Ser.* 2: 1-49.

Kirsch, L. M., and K. F. Higgins. 1976. Upland Sandpiper nesting and management in North Dakota. *Wildlife Soc. Bull.* 4: 16-20.

Kistchinski, A. A. 1975. Breeding biology and behaviour of the Grey Phalarope *Phalaropus fulicarius* in east Siberia. *Ibis* 117: 285-301.

Kitson, A. R. 1978. Identification of Long-toed Stint, Pintail Snipe and Asiatic Dowitcher. *Brit. Birds* 71: 558-62.

Knowles, C. J., C. J. Stoner, and S. P. Gieb. 1982. Selective use of black-tailed prairie dog towns by Mountain Plovers. *Condor* 84: 71-74.

Koopman, K. 1986. Primary moult and weight changes of Ruffs in The Netherlands in relation to migration. *Ardea* 74: 69-77.

Krohn, W. B. 1971. Some patterns of Woodcock activities on Maine summer fields. *Wilson Bull.* 83: 396-407.

Kull, R. C., Jr. 1978. Color selection of nesting material by Killdeer. *Auk* 94: 602-4.

Kus, B. E., P. Ashman, G. W. Page, and L. E. Stenzel. 1984. Age-related mortality in a wintering population of Dunlin. *Auk* 101: 69-73.

Labutin, Y. V., V. V. Leonovitch, and B. N. Veprintsev. 1982. The Little Curlew, *Numenius minutus*, in Siberia. *Ibis* 124: 302-19.

Lane, B. A. 1987. *Shorebirds in Australia*. Nelson Publishers, Melbourne.

Lank, D. 1979. Dispersal and predation rates of wing-tagged Semipalmated Sandpipers *Calidris pusilla* and an evaluation of the technique. *Wader Study Group Bull*. 27: 41-46.

Lank, D. B. 1989. Why fly by night? Inferences from tidally-induced migratory departures of sandpipers. *J. Field Ornithol*. 60: 154-61.

Lank, D. B., and C. M. Smith. 1987. Conditional lekking in Ruff (*Philomachus pugnax*). *Behav. Ecol. & Sociobiol*. 20: 137-45.

Lauro, B., and J. Burger. 1989. Nest-site selection of American Oystercatchers (*Haematopus palliatus*) in salt marshes. *Auk* 106: 185-92.

Legg, K. 1954. Nesting and feeding of the Black Oystercatcher near Monterey, California. *Condor* 56: 359-69.

Lehman, P., and J. L. Dunn. 1985. A little-known species reaches North America. *Am. Birds* 39: 247-50.

Lenington, S. 1980. Bi-parental care in Killdeer: an adaptive hypothesis. *Wilson Bull*. 92: 8-20.

Lenington, S., and T. Mace. 1975. Mate fidelity and nesting site tenacity in the Killdeer. *Auk* 92: 149-51.

Lenna, P. 1969. Short-billed and Long-billed Dowitchers in the Point Reyes, California area. *Point Reyes Bird Obs. Bull*. 13: 2-10.

Lessels, C. M. 1984. The mating system of Kentish Plovers *Charadrius alexandrinus*. *Ibis* 126: 474-83.

Lethaby, N., and J. Gilligan. 1991. An occurrence of the Great Knot in Oregon. *Oregon Birds* 17: 35-37.

Lewis, M. G., and F. A. Sharpe. 1987. *Birding in the San Juan Islands*. Mountaineers, Seattle.

Lifjeld, J. T. 1984. Prey selection in relation to body size and bill length of five species of waders feeding in the same habitat. *Ornis Scand*. 15: 217-26.

Lincoln, F. C., and S. R. Peterson. 1979. Migration of Birds. *Fish & Wildlife Serv. Circ*. 16, U. S. Dept. of Interior.

Lind, H. 1961. Studies on the behaviour of the Black-tailed Godwit. *Medd. Naturfredningsrådets Reservatudvalg* 66, Copenhagen.

Littlefield, C. C. 1990. *Birds of Malheur National Wildlife Refuge, Oregon*. Oregon State University Press, Corvallis.

Littlefield, C. C., and J. E. Cornely. 1984. Fall migration of birds at Malheur National Wildlife Refuge, Oregon. *Western Birds* 15: 15-22.

Littlefield, C. C., and E. L. McLaury. 1973. Bird arrival dates on Malheur National Wildlife Refuge, Oregon. *Western Birds* 4: 83-88.

Loftin, H. 1962. A study of boreal shorebirds summering on Apalachee Bay, Florida. *Bird-Banding* 33: 21-42.

L'Hyver, M-A., and E. H. Miller. 1991. Geographic and local variation in nesting phenology and clutch size of the Black Oystercatcher. *Condor* 93: 892-903.

McCaffery, B. J., T. A. Sordahl, and P. Zahler. 1984. Behavioral ecology of the Mountain Plover in northeastern Colorado. *Wader Study Group Bull*. 40: 18-21.

MacDonald, S. D., and D. F. Parmelee. 1962. Feeding behaviour of the Turnstone in arctic Canada. *Brit. Birds* 55: 241-44.

Mace, T. R. 1978. Killdeer breeding densities. *Wilson Bull*. 90: 442-43.

McGill, A. R. 1960. The Little Whimbrel. *Emu* 60: 89-94.

McKee, J. 1982. The winter feeding of Turnstones and Purple Sandpipers in Strathclyde. *Bird Study* 29: 213-16.

MacLean, S. F., Jr., and R. T. Holmes. 1971. Bill lengths, wintering areas, and taxonomy of North American Dunlins, *Calidris alpina*. *Auk* 88: 893-901.

Madge, S. C. 1977. Field identification of Spotted Sandpipers. *Brit. Birds* 70: 346-48.

Mahoney, S. A., and J. R. Jehl, Jr. 1985. Adaptations of migratory shorebirds to highly saline and alkaline lakes: Wilson's Phalarope and American Avocet. *Condor* 87: 520-27.

Mallory, E. P. 1982. Territoriality of Whimbrels *Numenius phaeopus hudsonicus* wintering in Panama. *Wader Study Group Bull.* 34: 37-39.

Mallory, E. P., and D. C. Schneider. 1979. Agonistic behavior in Short-billed Dowitchers feeding on a patchy resource. *Wilson Bull.* 91: 271-78.

Marchant, J. H. 1986. Identification, habits and status of Great Knot. *Brit. Birds* 79: 123-35.

Marks, J. S., R. L. Redmond, P. Hendricks, R. B. Clapp, and R. E. Gill, Jr. 1990. Notes on longevity and flightlessness in Bristle-thighed Curlews. *Auk* 107: 779-81.

Maron, J. L., and J. P. Myers. 1984. A description and evaluation of two techniques for sexing wintering Sanderlings. *J. Field Ornithol.* 55: 336-42.

—. 1985. Seasonal changes in feeding success, activity patterns, and weights of nonbreeding Sanderlings (*Calidris alba*). *Auk* 102: 580-86.

Marsh, C. P. 1986. Rocky intertidal community organization: The impact of avian predators on mussel recruitment. *Ecology* 67: 771-86.

Marshall, D. B. 1989. A review of the status of the Snowy Plover in Oregon. *Oregon Birds* 15: 57-76.

Martin, F. W. 1964. Woodcock sex and age determination from wings. *J. Wildlife Manage.* 28: 287-98.

Mason, C. F., and S. M. Macdonald. 1976. Aspects of the breeding biology of the Snipe. *Bird Study* 23: 33-38.

Matthiessen, P. 1973. *The Wind Birds*. Viking Press, New York.

Mayfield, H. F. 1973. Black-bellied Plover incubation and hatching. *Wilson Bull.* 85: 82-85.

—. 1978a. Red Phalaropes breeding on Bathurst Island. *Living Bird* 17: 7-39.

—. 1978b. Undependable breeding conditions in the Red Phalarope. *Auk* 95: 590-92.

Meinertzhagen, R. 1925. The distribution of the phalaropes. *Ibis* (12) 1: 325-44.

Melville, D. S. 1981. Spring measurements, weights and plumage status of *Calidris ruficollis* and *C. ferruginea* in Hong Kong. *Wader Study Group Bull.* 33: 18-21.

Metcalfe, N. B. 1986. Variation in winter flocking associations and dispersion patterns in the Turnstone *Arenaria interpres*. *J. Zool. Lond.* (A) 209: 385-403.

Metcalfe, N. B., and R. W. Furness. 1985. Survival, winter population stability and site fidelity in the Turnstone *Arenaria interpres*. *Bird Study* 32: 207-14.

—. 1987. Aggression in shorebirds in relation to flock density and composition. *Ibis* 129: 553-63.

Middlemiss, E. 1961. Biological aspects of *Calidris minuta* while wintering in south-west Cape. *Ostrich* 32: 107-21.

Miller, E. H. 1979a. Flight display of Least Sandpipers. *Wader Study Group Bull.* 26: 44-45.

—. 1979b. Functions of display flights by males of the Least Sandpiper, *Calidris minutilla* (Vieill.), on Sable Island, Nova Scotia. *Canad. J. Zool.* 57: 876-93.

—. 1983a. Habitat and breeding cycle of the Least Sandpiper (*Calidris minutilla*) on Sable Island, Nova Scotia. *Canad. J. Zool.* 61: 2880-98.

—. 1983b. Structure of display flights in the Least Sandpiper. *Condor* 85: 220-42.

—. 1983c. The structure of aerial displays in three species of Calidridinae (Scolopacidae). *Auk* 100: 440-51.

—. 1984. Communication in breeding shorebirds. In *Shorebirds: Breeding Behavior and Populations, Behavior of Marine Animals,* Vol. 6, ed. J. Burger and B. L. Olla. Plenum Press, New York.

—. 1985. Parental behavior in the Least Sandpiper (*Calidris minutilla*). *Canad. J. Zool.* 63: 1593-1601.

—. 1986. Components of variation in nuptial calls of the Least Sandpiper (*Calidris minutilla;* Aves, Scolopacidae). *Syst. Zool.* 35: 400-413.

Miller, E. H., W. W. H. Gunn, and R. E. Harris. 1983. Geographic variation in the aerial song of the Short-billed Dowitcher. *Canad. J. Zool.* 61: 2191-98.

Miller, E. H., W. W. H. Gunn, and S. F. MacLean, Jr. 1987. Breeding vocalizations of the Surfbird. *Condor* 89: 406-12.

Miller, E. H., W. W. H. Gunn, J. P. Myers, and B. N. Veprintsev. 1984. Species-distinctiveness of Long-billed Dowitcher song (Aves: Scolopacidae). *Proc. Biol. Soc. Wash.* 97: 804-11.

Miller, E. H., W. W. H. Gunn, and B. N. Veprintsev. 1988. Breeding vocalizations of Baird's Sandpiper *Calidris bairdii* and related species, with remarks on phylogeny and adaptation. *Ornis Scand.* 19:257-67.

Miller, J. R., and J. T. Miller. 1948. Nesting of the Spotted Sandpiper at Detroit, Michigan. *Auk* 65: 558-67.

Moore, R. T. 1912. The Least Sandpiper during the nesting season in the Magdalen Islands. *Auk* 29: 210-23.

Morrier, A., and R. McNeil. 1991. Time-activity budget of Wilson's and Semipalmated Plovers in a tropical environment. *Wilson Bull.* 103: 598-620.

Mundahl, J. T. 1982. Role specialization in the parental and territorial behavior of the Killdeer. *Wilson Bull.* 94: 515-30.

Munro, J. A., and I. McT. Cowan. 1947. A Review of the Bird Fauna of British Columbia. *Special Publ. No. 2, Dept. Educ., Brit. Columbia Prov. Mus.*, Victoria.

Murie, O. J. 1924. Nesting records of the Wandering Tattler and Surf-bird in Alaska. *Auk* 41: 231-37.

Murray, B. G., Jr. 1983. Notes on the breeding biology of Wilson's Phalarope. *Wilson Bull.* 95: 472-75.

Myers, J. P. 1979. Leks, sex, and Buff-breasted Sandpipers. *Am. Birds* 33: 823-25.

—. 1980a. Sanderlings *Calidris alba* at Bodega Bay: facts, inferences and shameless speculations. *Wader Study Group Bull.* 30: 26-32.

—. 1980b. Territoriality and flocking by Buff-breasted Sandpipers: variations in non-breeding dispersion. *Condor* 82: 241-50.

—. 1981. A test of three hypotheses for latitudinal segregation of the sexes in wintering birds. *Canad. J. Zool.* 59: 1527-34.

—. 1982. The promiscuous Pectoral Sandpiper. *Am. Birds* 36: 119-22.

—. 1983a. Conservation of migrating shorebirds: Staging areas, geographic bottlenecks and regional movements. *Am. Birds* 37: 23-25.

—. 1983b. Space, time and the pattern of individual associations in a group-living species; Sanderlings have no friends. *Behav. Ecol. Sociobiol.* 12: 129-34.

Myers, J. P., P. G. Connors, and F. A. Pitelka. 1979. Territory size in wintering Sanderlings: The effect of prey abundance and intruder density. *Auk* 96: 551-61.

—. 1981. Optimal territory size and the Sanderling: Compromises in a variable environment. In *Foraging Behavior: Ecological, Ethological and Psychological Approaches.* ed. A. C. Kamil and T. D. Sargent. Garland STM Press, New York.

Myers, J. P., O. Hildén, and P. Tomkovich. 1982. Exotic *Calidris* species of the Siberian tundra. *Ornis Fenn.* 59: 175-82.

Myers, J. P., and L. P. Myers. 1979. Shorebirds of coastal Buenos Aires Province, Argentina. *Ibis* 121: 186-200.

Myers, J. P., M. Sallaberry A., E. Ortiz, G. Castro, L. M. Gordon, J. L. Maron, C. T. Schick, E. Tabilo, P. Antas, and T. Below. 1990. Migration routes of New World Sanderlings (*Calidris alba*). *Auk* 107: 172-80.

Myers, J. P., C. T. Schick, and C. J. Hohenberger. 1984. Notes on the 1983 distribution of Sanderlings along the United States' Pacific coast. *Wader Study Group Bull.* 40: 22-26.

Myers, J. P., S. L. Williams, and F. A. Pitelka. 1980. An experimental analysis of prey availability for Sanderlings (Aves: Scolopacidae) feeding on sandy beach crustaceans. *Canad. J. Zool.* 58: 1564-74.

National Geographic Society. 1987. *Field Guide to the Birds of North America.* Washington, D. C.

Navarro, R. A., C. R. Velásquez, and R. P. Schlatter. 1989. Diet of the Surfbird in southern Chile. *Wilson Bull*. 101: 137-41.

Nehls, H. 1989. A review of the status and distribution of dowitchers in Oregon. *Oregon Birds* 15: 97-102.

Nethersole-Thompson, D. 1973. *The Dotterel*. William Collins Sons & Co., Glasgow.

Nethersole-Thompson, D., and M. Nethersole-Thompson. 1979. *Greenshanks*. Buteo Books, Vermillion, South Dakota.

—. 1986. *Waders: Their Breeding, Haunts and Watchers*. T. & A. D. Poyser, Calton, England.

Nettleship, D. N. 1973. Breeding ecology of Turnstones *Arenaria interpres* at Hazen Camp, Ellesmere Island, N. W. T. *Ibis* 115: 202-17.

—. 1974. The breeding of the Knot *Calidris canutus* at Hazen Camp, Ellesmere Island, N. W. T. *Polarforschung* 44: 8-26.

Neufeldt, I., A. V. Krechmar, and A. I. Ivanov. 1961. Studies of less familiar birds 110. Grey-rumped Sandpiper. *Brit. Birds* 54: 30-33.

Newlon, M. C., and T. H. Kent. 1980. Speciation of dowitchers in Iowa. *Iowa Bird Life* 50: 59-68.

Nicholls, J. L., and G. A. Baldassarre. 1990a. Winter distribution of Piping Plovers along the Atlantic and Gulf coasts of the United States. *Wilson Bull*. 102: 400-412.

—. 1990b. Habitat associations of Piping Plovers wintering in the United States. *Wilson Bull*. 102: 581-90.

Nickell, W. P. 1943. Observations on the nesting of the Killdeer. *Wilson Bull*. 55: 23-28.

Nielsen, B. P. 1975. Affinities of *Eudromias morinellus* (L.) to the genus *Charadrius* (L.). *Ornis Scand*. 6: 65-82.

Niemi, G. J., and T. E. Davis. 1979. Notes on the nesting ecology of the Piping Plover. *Loon* 51: 74-79.

Nisbet, I. C. T. 1961. Dowitchers in Great Britain and Ireland. *Brit. Birds* 54: 343-57.

Nol, E. 1989. Food supply and reproductive performance of the American Oystercatcher in Virginia. *Condor* 91: 429-35.

Nol, E., and A. Lambert. 1984. Comparison of Killdeers, *Charadrius vociferus,* breeding in mainland and peninsular sites in southern Ontario. *Canad. Field-Nat*. 98: 7-11.

Norton, D. W. 1972. Incubation schedules of four species of *Calidris* sandpipers at Barrow, Alaska. *Condor* 74: 164-76.

Norton, D. W., and U. N. Safriel. 1971. Homing by nesting Semipalmated Sandpipers displaced from Barrow, Alaska. *Bird-Banding* 42: 295-97.

Norton-Griffiths, M. 1967. Some ecological aspects of the feeding behaviour of the Oystercatcher, *Haematopus ostralegus,* on the edible mussel, *Mytilus edulis. Ibis* 109: 412-24.

Oddie, W. E. 1980. Leg colour and calls of Spotted Sandpiper. *Brit. Birds* 73: 185-86.

Oddie, W. E., and B. A. E. Marr. 1981. Identification of Semipalmated Sandpipers and Little Stints in autumn. *Brit. Birds* 74: 396-98.

Oring, L. W. 1964. Displays of the Buff-breasted Sandpiper at Norman, Oklahoma. *Auk* 81: 83-86.

—. 1968. Vocalizations of the Green and Solitary Sandpipers. *Wilson Bull*. 80: 395-420.

—. 1973. Solitary Sandpiper early reproductive behavior. *Auk* 90: 652-63.

Oring, L. W., and M. L. Knudson. 1972. Monogamy and polyandry in the Spotted Sandpiper. *Living Bird* 11: 59-73.

Oring, L. W., and D. B. Lank. 1982. Sexual selection, arrival times, philopatry and site fidelity in the polyandrous Spotted Sandpiper. *Behav. Ecol. Sociobiol*. 10: 185-91.

Oring, L. W., D. B. Lank, and S. J. Maxson. 1983. Population studies of the polyandrous Spotted Sandpiper. *Auk* 100: 272-85.

Oring, L. W., and S. J. Maxson. 1978. Instances of simultaneous polyandry by a Spotted Sandpiper *Actitis macularia. Ibis* 120: 349-53.

Ouellet, H., R. McNeil, and J. Burton. 1973. The Western Sandpiper in Quebec and the Maritime

Provinces, Canada. *Canad. Field-Nat.* 87: 291-300.

Owen, R. B., Jr., and W. B. Krohn. 1973. Molt patterns and weight changes of the American Woodcock. *Wilson Bull.* 85: 31-41.

Page, G. W. 1974a. Age, sex, molt and migration of Dunlins at Bolinas Lagoon. *Western Birds* 5: 1-12.

—. 1974b. Molt of wintering Least Sandpipers. *Bird-Banding* 45: 93-105.

Page, G. W., F. C. Bidstrup, R. J. Ramer, and L. E. Stenzel. 1986. Distribution of wintering Snowy Plovers in California and adjacent states. *Western Birds* 17: 145-70.

Page, G. W., and M. Bradstreet. 1968. Size and composition of a fall population of Least and Semipalmated Sandpipers at Long Point, Ontario. *Ontario Bird-Banding* 4: 80-88.

Page, G. W., and B. Fearis. 1971. Sexing Western Sandpipers by bill length. *Bird-Banding* 42: 297-98.

Page, G. W., B. Fearis, and R. M. Jurek. 1972. Age and sex composition of Western Sandpipers on Bolinas Lagoon. *Calif. Birds* 3: 79-86.

Page, G. W., and A. L. A. Middleton. 1972. Fat deposition during autumn migration in the Semipalmated Sandpiper. *Bird-Banding* 43: 85-96.

Page, G. W., and A. Salvadori. 1969. Weight changes of Semipalmated and Least Sandpipers pausing during autumn migration. *Ontario Bird-Banding* 5: 52-58.

Page, G. W., and L. E. Stenzel. 1981. The breeding status of the Snowy Plover in California. *Western Birds* 12: 1-40.

Page, G. W., L. E. Stenzel, and C. A. Ribic. 1985. Nest site selection and clutch predation in the Snowy Plover. *Auk* 102: 347-53.

Page, G. W., L. E. Stenzel, W. D. Shuford, and C. R. Bruce. 1991. Distribution and abundance of the Snowy Plover on its western North American breeding grounds. *J. Field Ornithol.* 62: 245-255.

Page, G. W., L. E. Stenzel, D. W. Winkler, and C. W. Swarth. 1983. Spacing out on Mono Lake: Breeding success, nest density, and predation in the Snowy Plover. *Auk* 100: 13-24.

Paget-Wilkes, A. H. 1922. On the breeding-habits of the Turnstone as observed in Spitsbergen. *Brit. Birds* 15: 172-79.

Parkes, K. C., A. Poole, and H. Lapham. 1971. The Ruddy Turnstone as an egg predator. *Wilson Bull.* 83: 306-8.

Parmelee, D. F. 1970. Breeding behavior of the Sanderling in the Canadian high Arctic. *Living Bird* 9: 97-146.

Parmelee, D. F., D. W. Greiner, and W. D. Graul. 1968. Summer schedule and breeding biology of the White-rumped Sandpiper in the central Canadian Arctic. *Wilson Bull.* 80: 5-29.

Parmelee, D. F., and R. B. Payne. 1973. On multiple broods and the breeding strategy of arctic Sanderlings. *Ibis* 115: 218-26.

Paton, D. C., and B. J. Wykes. 1978. Re-appraisal of moult of Red-necked Stints in southern Australia. *Emu* 78: 54-60.

Paton, D. C., B. J. Wykes, and P. Dann. 1982. Moult of juvenile Curlew Sandpipers in southern Australia. *Emu* 82: 54-56.

Patten, M. A., and B. E. Daniels. 1991. First record of the Long-toed Stint in California. *Western Birds* 22: 131-138.

Paulson, D. 1979. Dotterel at Ocean Shores, Washington. *Cont. Birdlife* 1: 109-11.

Paulson, D. R. 1983. Fledging dates and southward migration of juveniles of some *Calidris* sandpipers. *Condor* 85: 99-101.

—. 1986. Identification of juvenile tattlers, and a Gray-tailed Tattler record from Washington. *Western Birds* 17: 33-36.

—. 1990. Sandpiper-like feeding in Black-bellied Plovers. *Condor* 92: 245.

Paulson, D. R., and W. J. Erckmann. 1985. Buff-breasted Sandpipers nesting in association with Black-bellied Plovers. *Condor* 87: 429-30.

Pearson, D. J. 1981. The wintering and moult of Ruffs *Philomachus pugnax* in the Kenyan rift valley. *Ibis* 123: 158-82.

—. 1984. The moult of the Little Stint *Calidris minuta* in the Kenyan rift valley. *Ibis* 126: 1-15.

Pearson, D. J., J. H. Phillips, and G. C. Backhurst. 1970. Weights of some Palearctic waders wintering in Kenya. *Ibis* 112: 199-208.

Peterson, R. T. 1990. *A Field Guide to Western Birds*. Houghton Mifflin Co., Boston.

Pettingill, O. S., Jr. 1936. The American Woodcock *Philohela minor* (Gmelin). *Mem. Boston Soc. Nat. Hist.* 9: 169-391.

Phillips, A. R. 1975. Semipalmated Sandpiper: Identification, migrations, summer and winter ranges. *Am. Birds* 29: 799-806.

Phillips, R. E. 1972. Sexual and agonistic behaviour in the Killdeer (*Charadrius vociferus*). *Anim. Behav.* 20: 1-9.

Pickett, P. E., S. J. Maxson, and L. W. Oring. 1988. Interspecific interactions of Spotted Sandpipers. *Wilson Bull.* 100: 297-302.

Pickwell, G. 1925. The nesting of the Killdeer. *Auk* 42: 485-96.

Pienkowski, M. W. 1982. Diet and energy intake of Grey and Ringed plovers, *Pluvialis squatarola* and *Charadrius hiaticula,* in the non-breeding season. *J. Zool., Lond.* 197: 511-49.

—. 1983a. Changes in the foraging pattern of plovers in relation to environmental factors. *Anim. Behav.* 31: 244-64.

—. 1983b. The effects of environmental conditions on feeding rates and prey-selection of shore plovers. *Ornis Scand.* 14: 227-38.

—. 1984. Breeding biology and population dynamics of Ringed Plovers *Charadrius hiaticula* in Britain and Greenland: Nest-predation as a possible factor limiting distribution and timing of breeding. *J. Zool., Lond.* 202: 83-114.

Pienkowski, M. W., and G. H. Green. 1976. Breeding biology of Sanderlings in north-east Greenland. *Brit. Birds* 69: 165-77.

Piersma, T. 1986. Eastern Curlews *Numenius madagascariensis* feeding on *Macrophthalmus* and other ocypodid crabs in the Nakdong Estuary, South Korea. *Emu* 86: 155-60.

Pitelka, F. A. 1943. Territoriality, display, and certain ecological relations of the American Woodcock. *Wilson Bull.* 55: 88-114.

—. 1950. Geographic variation and the species problem in the shorebird genus *Limnodromus*. *Univ. Calif. Publ. Zool.* 50: 1-108.

—. 1959. Numbers, breeding schedule and territoriality in Pectoral Sandpipers of northern Alaska. *Condor* 61: 233-64.

Portenko, L. A. 1933. Some new materials adding to the knowledge of breeding ranges and life history of eastern Knot, *Calidris tenuirostris* (Horsf.). *Arctica* 1: 75-98.

Potts, W. K. 1984. The chorus-line hypothesis of manoeuvre coordination in avian flocks. *Nature* 309: 344-45.

Prater, A. J. 1972. The ecology of Morecambe Bay III. The food and feeding habits of the Knot (*Calidris canutus* L.) in Morecambe Bay. *J. Appl. Ecol.* 9: 179-94.

—. 1982. Identification of Ruff. *Dutch Birding* 4: 8-14.

Prater, A. J., and J. H. Marchant. 1975. Primary moult of *Tringa brevipes* and *Tringa incana*. *Bull. Brit. Ornithol. Club* 95: 120-22.

Prater, A. J., J. H. Marchant, and J. Vuorinen. 1977. Guide to the Identification and Ageing of Holarctic Waders. *BTO Guide* 17. Brit. Trust for Ornithol., Tring.

Prevett, J. P., and J. F. Barr. 1976. Lek behavior of the Buff-breasted Sandpiper. *Wilson Bull.* 88: 500-503.

Pringle, J. D. 1987. *The Shorebirds of Australia*. Angus & Robertson, North Ryde, Australia.

Pruett-Jones, S. G. 1988. Lekking versus solitary display: temporal variations in dispersion in the Buff-breasted Sandpiper. *Anim. Behav.* 36: 1740-52.

Pulliainen, E. 1970. On the breeding biology of the Dotterel *Charadrius morinellus*. *Ornis Fenn*. 47: 69-73.

Purdue, J. R. 1976a. Thermal environment of the nest and related parental behavior in Snowy Plovers, *Charadrius alexandrinus*. *Condor* 78: 180-85.

—. 1976b. Adaptations of the Snowy Plover of the Great Salt Plains, Oklahoma. *Southw. Nat*. 21: 347-57.

Purdue, J. R., and H. Haines. 1977. Salt water tolerance and water turnover in the Snowy Plover. *Auk* 94: 248-55.

Puttick, G. M. 1978. The diet of the Curlew Sandpiper at Langebaan Lagoon, South Africa. *Ostrich* 49: 158-67.

—. 1979. Foraging behaviour and activity budgets of Curlew Sandpipers. *Ardea* 67: 111-22.

—. 1981. Sex-related differences in foraging behaviour of Curlew Sandpipers. *Ornis Scand*. 12: 13-17.

Pym, A. 1982. Identification of Lesser Golden Plover and status in Britain and Ireland. *Brit. Birds* 75: 112-24.

Rabe, D. L., H. H. Prince, and D. L. Beaver. 1983. Feeding-site selection and foraging strategies of American Woodcock. *Auk* 100: 711-16.

Rands, M. R. W., and J. P. Barkham. 1981. Factors controlling within-flock feeding densities in three species of wading bird. *Ardea* 12: 28-36.

Reader's Digest. 1986. *Reader's Digest Complete Book of Australian Birds*. Reader's Digest, Sydney.

Redmond, R. L., and D. A. Jenni. 1982. Natal philopatry and breeding area fidelity of Long-billed Curlew (*Numenius americanus*): Patterns and evolutionary consequences. *Behav. Ecol. Sociobiol*. 10: 277-79.

—. 1986. Population ecology of the Long-billed Curlew (*Numenius americanus*) in western Idaho. *Auk* 103: 755-67.

Reynolds, J. D. 1987. Mating system and nesting biology of the Red-necked Phalarope *Phalaropus lobatus*: What constrains polyandry? *Ibis* 129: 225-42.

Reynolds, J. D., M. A. Colwell, and F. Cooke. 1986. Sexual selection and spring arrival times of Red-necked and Wilson's Phalaropes. *Behav. Ecol. Sociobiol*. 18: 303-10.

Reynolds, J. D., and F. Cooke. 1988. The influence of mating systems on philopatry: A test with polyandrous Red-necked Phalaropes. *Anim. Behav*. 36: 1788-95.

Ridgway, R. 1919. The Birds of North and Middle America. *U. S. Nat. Mus. Bull*. 50, Part VIII.

Ridley, M. W. 1980. The breeding behaviour and feeding ecology of Grey Phalaropes *Phalaropus fulicarius* in Svalbard. *Ibis* 122: 210-26.

Roberson, D. 1980. *Rare Birds of the West Coast*. Woodcock Publications, Pacific Grove, California.

Robertson, H. A., and M. D. Dennison. 1979. Feeding and roosting behaviour of some waders at Farewell Spit. *Notornis* 26: 73-88.

Rohwer, S., D. F. Martin, and G. G. Benson. 1979. Breeding of the Black-necked Stilt in Washington. *Murrelet* 60: 67-71.

Rowan, W. 1932. The status of the dowitchers with a description of a new subspecies from Alberta and Manitoba. *Auk* 49: 14-35.

Rowlett, R. A. 1980. Little Stint (*Calidris minuta*) in Delaware. *Am. Birds* 94: 850-51.

Royal, L. A. 1939. Feeding habits of the Sanderling at Copalis Beach, Washington. *Murrelet* 20: 25.

Ruiz, G. M., P. G. Connors, S. E. Griffin, and F. A. Pitelka. 1989. Structure of a wintering Dunlin population. *Condor* 91: 562-70.

Russell, R. P., Jr. 1983. The Piping Plover in the Great Lakes region. *Am. Birds* 37: 951-55.

Ryan, M. R., and R. B. Renken. 1987. Habitat use by breeding Willets in the northern Great Plains. *Wilson Bull*. 99: 175-89.

Ryan, M. R., R. B. Renken, and J. J. Dinsmore. 1984. Marbled Godwit habitat selection in the northern prairie region. *J. Wildlife Manage*. 48: 1206-18.

Sadler, D. A. R., and W. J. Maher. 1976. Notes on the Long-billed Curlew in Saskatchewan. *Auk* 93: 382-84.

Sauer, E. G. F. 1962. Ethology and ecology of Golden Plovers on St. Lawrence Island, Bering Sea. *Psychol. Forsch.* 26: 399-470.

Sauppe, B., B. A. MacDonald, and D. M. Mark. 1978. First Canadian and third North American record of the Spoon-billed Sandpiper (*Eurynorhynchus pygmeus*). *Am. Birds* 32: 1062-64.

Schamel, D., and D. Tracy. 1977. Polyandry, replacement clutches, and site tenacity in the Red Phalarope (*Phalaropus fulicarius*) at Barrow, Alaska. *Bird-Banding* 48: 314-25.

—. 1987. Latitudinal trends in breeding Red Phalaropes. *J. Field Ornithol.* 58: 126-34.

—. 1988. Are yearlings distinguishable from older Red-necked Phalaropes? *J. Field Ornithol.* 59: 235-38.

—. 1991. Breeding site fidelity and natal philopatry in the sex role-reversed Red and Red-necked Phalaropes. *J. Field Ornithol.* 62: 390-398.

Schmitt, M. B., and P. J. Whitehouse. 1976. Moult and mensural data of Ruff on the Witwatersrand. *Ostrich* 47: 179-90.

Senner, S. 1976. The food habits of migrating Dunlins (*Calidris alpina*) and Western Sandpipers (*Calidris mauri*) in the Copper River Delta, Alaska. *Quart. Rep. Alaska Coop. Wildlife Res. Unit* 27: 2-9.

Senner, S. E., and E. F. Martinez. 1982. A review of Western Sandpiper migration in interior North America. *Southw. Nat.* 27: 149-59.

Senner, S. E., and P. G. Mickelson. 1979. Fall foods of Common Snipe on the Copper River Delta, Alaska. *Canad. Field-Nat.* 93: 171-72.

Senner, S. E., G. C. West, and D. W. Norton. 1981. The spring migration of Western Sandpipers and Dunlins in southcentral Alaska: numbers, timing and sex ratios. *J. Field Ornithol.* 52: 271-84.

Sheldon, W. G. 1967. *The Book of the American Woodcock.* Univ. Massachusetts Press, Amherst.

Shepard, J. M. 1976. Factors influencing female choice in the lek mating system of the Ruff. *Living Bird* 14: 87-111.

Siegfried, W. R., and B. D. J. Batt. 1972. Wilson's Phalaropes forming feeding association with Shovelers. *Auk* 89: 667-68.

Silliman, J., G. S. Mills, and S. Alden. 1977. Effect of flock size on foraging activity in wintering Sanderlings. *Wilson Bull.* 89: 434-38.

Simmons, K. E. L. 1961a. Foot-movements in plovers and other birds. *Brit. Birds* 54: 34-39.

—. 1961b. Further observations on foot-movements in plovers and other birds. *Brit. Birds* 54: 418-22.

Sinclair, J. C., and H. Nicholls. 1976. Red-necked Stint identification. *Bokmakierie* 28: 59-60.

Sinclair, J. C., and G. H. Nicholls. 1980. Winter identification of Greater and Lesser Sand Plovers. *Brit. Birds* 73: 206-13.

Skaar, D., D. Flath, and L. S. Thompson. 1985. P. D. Skaar's Montana Bird Distribution. *Montana Acad. Sci. Monogr.* No. 3.

Skeel, M. A. 1978. Vocalizations of the Whimbrel on its breeding grounds. *Condor* 80: 194-202.

—. 1982. Sex determination of adult Whimbrels. *J. Field Ornithol.* 53: 414-16.

—. 1983. Nesting success, density, philopatry, and nest-site selection of the Whimbrel (*Numenius phaeopus*) in different habitats. *Canad. J. Zool.* 61: 218-25.

Sloanaker, J. L. 1925. Notes from Spokane. *Condor* 27: 73-74.

Smith, N. G. 1969. Polymorphism in Ringed Plovers. *Ibis* 111: 177-88.

Smith, P. C., and P. R. Evans. 1973. Studies of shorebirds at Lindisfarne, Northumberland. I. Feeding ecology and behaviour of the Bar-tailed Godwit. *Wildfowl* 24: 135-39.

Smith, R. W., and J. S. Barclay. 1978. Evidence of westward changes in the range of the American Woodcock. *Am. Birds* 32: 1122-27.

Soikkeli, M. 1967. Breeding cycle and population dynamics in the Dunlin (*Calidris alpina*). *Ann. Zool. Fenn.* 4: 158-98.

—. 1970a. Dispersal of Dunlin in relation to sites of birth and breeding. *Ornis Fenn.* 47: 1-9.

—. 1970b. Mortality and reproductive rates in a Finnish population of Dunlin *Calidris alpina*. *Ornis Fenn.* 47: 149-58.

Sordahl, T. A. 1979. Vocalizations and behavior of the Willet. *Wilson Bull.* 91: 551-74.

—. 1982. Antipredator behavior of American Avocet and Black-necked Stilt chicks. *J. Field Ornithol.* 53: 315-25.

—. 1984. Observations on breeding site fidelity and pair formation in American Avocets and Black-necked Stilts. *N. Am. Bird Bander* 9: 8-11.

—. 1990. Sexual differences in antipredator behavior of breeding American Avocets and Black-necked Stilts. *Condor* 92: 530-32.

Southern, H. N., and W. A. S. Lewis. 1937. The breeding behaviour of Temminck's Stint. *Brit. Birds* 31: 314-21.

Spaans, A. L. 1976. Molt of flight and tail feathers of the Least Sandpiper in Surinam, South America. *Bird-Banding* 47: 359-64.

—. 1980. Biometrics and moult of Sanderlings *Calidris alba* during the autumn in Suriname. *Wader Study Group Bull.* 28: 33-35.

Stenzel, L. E., H. R. Huber, and G. W. Page. 1976. Feeding behavior and diet of the Long-billed Curlew and Willet. *Wilson Bull.* 88: 314-32.

Stephens, D. A., and J. E. Stephens. 1987. An American Oystercatcher in Idaho. *Western Birds* 18: 215-16.

Stern, M. A., and G. A. Rosenberg. 1985. Occurrence of a breeding Upland Sandpiper at Sycan Marsh, Oregon. *Murrelet* 66: 34-35.

Stevenson, H. M. 1975. Identification of difficult birds: III. Semipalmated and Western Sandpipers. *Florida Field Nat.* 3: 39-44. (Also published in *Birding* 11: 84-88.)

Stinson, C. 1977. The spatial distribution of wintering Black-bellied Plovers. *Wilson Bull.* 89: 470-72.

Stout, G., ed. 1967. *The Shorebirds of North America*. Viking Press, New York.

Strauch, J. G., Jr. 1967. Spring migration of Dunlin in interior western Oregon. *Condor* 69: 210-12.

Strauch, J. G., Jr., and L. G. Abele. 1979. Feeding ecology of three species of plovers wintering on the Bay of Panama, Central America. *Stud. Avian Biol.* 2: 217-30.

Sugden, J. W. 1933. Range restriction of the Long-billed Curlew. *Condor* 35: 3-9.

Summers, R. W., L. G. Underhill, M. Waltner, and D. A. Whitelaw. 1987. Population, biometrics and movements of the Sanderling *Calidris alba* in southern Africa. *Ostrich* 58: 24-39.

Sutton, G. M. 1967. Behaviour of the Buff-breasted Sandpiper at the nest. *Arctic* 20: 3-7.

—. 1968. Sexual dimorphism in the Hudsonian Godwit. *Wilson Bull.* 80: 251-52.

—. 1981. On aerial and ground displays of the world's snipes. *Wilson Bull.* 93: 457-77.

Sutton, G. M., and D. F. Parmelee. 1955. Breeding of the Semipalmated Plover on Baffin Island. *Bird-Banding* 26: 137-47.

Swarth, H. S. 1935. Systematic status of some Northwestern birds. *Condor* 37: 199-204.

Taverner, J. H. 1982. Feeding behaviour of Spotted Redshank flocks. *Brit. Birds* 75: 333-34.

Taylor, D. M., and C. H. Trost. 1987. The status of rare birds in Idaho. *Murrelet* 68: 69-93.

Taylor, K. 1984. *A Birders Guide to Vancouver Island*. Privately published.

Taylor, R. C. 1980. Migration of the Ringed Plover *Charadrius hiaticula*. *Ornis Scand.* 11: 30-42.

Terres, J. K. 1980. *The Audubon Society Encyclopedia of North American Birds*. Alfred A. Knopf, New York.

Thomas, D. G., and A. J. Dartnall. 1971a. Ecological aspects of the feeding behaviour of two calidritine sandpipers wintering in south-eastern Tasmania. *Emu* 71: 20-26.

—. 1971b. Moult of the Red-necked Stint. *Emu* 71: 49-53.

—. 1971c. Moult of the Curlew Sandpiper in relation to its annual cycle. *Emu* 71: 153-58.

Thompson, M. C. 1974. Migratory patterns of Ruddy Turnstones in the central Pacific region. *Living Bird* 12: 5-23.

Tinbergen, N. 1935. Field observations of east Greenland birds. 1. The behaviour of the Red-necked Phalarope (*Phalaropus lobatus*) in spring. *Ardea* 24: 1-42.

Tomkins, I. R. 1965. The Willets of Georgia and South Carolina. *Wilson Bull.* 77: 151-67.

Townshend, D. J. 1981. The importance of field feeding to the survival of wintering male and female Curlews (*Numenius arquata*) on the Tees Estuary. In *Feeding and Survival Strategies of Estuarine Organisms,* ed. N. V. Jones and W. J. Wolff. Plenum Press, New York.

Townshend, D. J., P. J. Dugan, and M. W. Pienkowski. 1984. The unsociable plover—use of intertidal areas by Grey Plovers. In *Coastal Waders and Wildfowl in Winter,* ed. P. R. Evans, J. D. Goss-Custard and W. G. Hale. Cambridge Univ. Press, Cambridge.

Tree, A. J. 1974. Ageing and sexing the Little Stint. *Safring News* 3: 31-33.

—. 1979. Biology of the Greenshank in southern Africa. *Ostrich* 50: 240-51.

Tree, A. J., and J. A. Kieser. 1982. Field separation of Lesser Yellowlegs and Wood Sandpiper. *Honeyguide* 110: 40-41.

Tuck, L. M. 1972. The snipes: A study of the genus *Capella*. *Canad. Wildlife Serv., Monogr. Ser.* No. 5: 1-428.

Udvardy, M. D. F. 1977. *The Audubon Society Field Guide to North American Birds*. Western Region. Alfred A. Knopf, New York.

Urban, E. K., C. H. Fry, and S. Keith. 1986. *The Birds of Africa*. Vol. II. Academic Press, London.

van Rhijn, J. G. 1973. Behavioural dimorphism in male Ruffs, *Philomachus pugnax* (L.). *Behaviour* 47: 153-229.

—. 1983. On the maintenance and origin of alternative strategies in the Ruff *Philomachus pugnax*. *Ibis* 125: 482-98.

—. 1985. A scenario for the evolution of social organization in Ruffs *Philomachus pugnax* and other Charadriiform species. *Ardea* 73: 25-37.

—. 1991. *The Ruff*. T. & A. D. Poyser, London.

Vaughan, R. 1980. *Plovers*. Terence Dalton Ltd., Suffolk.

Vaurie, C. 1964. Systematic notes on Palearctic birds. No. 53. Charadriidae: The genera *Charadrius* and *Pluvialis*. *Am. Mus. Novitates* No. 2177.

Veit, R. R. 1988. Identification of the Salton Sea Rufous-necked Sandpiper. *Western Birds* 19: 165-69.

Veit, R. R., and L. Jonsson. 1984. Field identification of smaller sandpipers within the genus *Calidris*. *Am. Birds* 38: 853-76.

Vinicombe, K. 1983. Identification pitfalls and assessment problems. 4 Buff-breasted Sandpiper *Tryngites subruficollis*. *Brit. Birds* 76: 203-06.

Vogt, W. 1938. Preliminary notes on the behavior and ecology of the Eastern Willet. *Proc. Linn. Soc. N. Y.* 49: 8-41.

Vuolanto, S. 1968. On the breeding biology of the Turnstone (*Arenaria interpres*) at Norrskär, Gulf of Bothnia. *Ornis Fenn.* 45: 19-24.

Wahl, T. R. 1975. Seabirds in Washington's offshore zone. *Western Birds* 6: 117-34.

Wallace, D. I. M. 1968. Dowitcher identification: A brief review. *Brit. Birds* 61: 366-72.

—. 1970. Identification of Spotted Sandpipers out of breeding plumage. *Brit. Birds* 63: 168-73.

—. 1974. Field identification of small species in the genus *Calidris*. *Brit. Birds* 67: 1-17.

—. 1979. Review of British records of Semipalmated Sandpipers and claimed Red-necked Stints. *Brit. Birds* 72: 264-74.

Wallis, C. A., and C. R. Wershler. 1981. Status and breeding of Mountain Plover (*Charadrius montanus*) in Canada. *Canad. Field-Nat.* 95: 133-36.

Ward, S. D., and D. J. Bullock. 1988. The winter feeding ecology of the Black-tailed Godwit—a preliminary study. *Wader Study Group Bull.* 53: 11-15.

Warriner, J. S., J. C. Warriner, G. W. Page, and L. E. Stenzel. 1986. Mating system and reproductive success of a small population of polygamous Snowy Plovers. *Wilson Bull.* 98: 15-37.

Webb, B. E., and J. A. Conry. 1979. A Sharp-tailed Sandpiper in Colorado, with notes on plumage and behavior. *Western Birds* 10: 86-91.

Weber, J. W. 1981. Status of the Semipalmated Sandpiper in Washington and northern Idaho. *Cont. Birdlife* 2: 150-53.

—. 1985. First specimen record of the Short-billed Dowitcher from eastern Washington; subspecific identification of Idaho specimens. *Murrelet* 66: 31-34.

Weber, J. W., and E. J. Larrison. 1977. *Birds of Southeastern Washington*. Univ. Press of Idaho, Moscow.

Weber, W. C., R. W. Phillips, and J. A. Roach. 1976. Second breeding record of the Semipalmated Plover at Vancouver, British Columbia. *Canad. Field-Nat.* 90: 55.

Webster, J. D. 1941a. The breeding of the Black Oyster-catcher. *Wilson Bull.* 53: 141-56.

—. 1941b. Feeding habits of the Black Oyster-catcher. *Condor* 43: 175-80.

—. 1942. Notes on the growth and plumages of the Black Oyster-catcher. *Condor* 44: 205-11.

Weeden, R. B. 1959. A new breeding record of the Wandering Tattler in Alaska. *Auk* 76: 230-32.

—. 1965. Further notes on Wandering Tattlers in central Alaska. *Condor* 67: 87-89.

Wetmore, A. 1925. Food of American Phalaropes, Avocets, and Stilts. *U. S. Dept. Agric. Bull.* 1359.

Weydemeyer, W. 1973. The spring migration patterns at Fortine, Montana. *Condor* 75: 400-413.

White, R. P. 1988. Wintering grounds and migration patterns of the Upland Sandpiper. *Am. Birds* 42: 1247-53.

Whitfield, D. P. 1986. Plumage variability and territoriality in breeding Turnstone *Arenaria interpres:* status signalling or individual recognition? *Anim. Behav.* 34: 1471-82.

—. 1988. The social significance of plumage variability in wintering Turnstone *Arenaria interpres. Anim. Behav.* 36: 408-15.

Widrig, R. S. 1979. *The Shorebirds of Leadbetter Point*. Privately published.

—. 1983. A Bristle-thighed Curlew at Leadbetter Point, Washington. *Western Birds* 14: 203-4.

Wiens, T. P., and F. J. Cuthbert. 1988. Nest-site tenacity and mate retention of the Piping Plover. *Wilson Bull.* 100: 545-53.

Wilcox, L. 1980. Observations on the life history of Willets on Long Island, New York. *Wilson Bull.* 92: 253-58.

Wild Bird Society of Japan. 1982. *A Field Guide to the Birds of Japan*. Wild Bird Society of Japan, Tokyo.

Wilds, C. 1982. Separating the yellowlegs. *Birding* 14: 172-78.

Wilds, C., and M. Newlon. 1983. The identification of dowitchers. *Birding* 15: 151-66.

Williamson, F. S. L., and M. A. Smith. 1964. The distribution and breeding status of the Hudsonian Godwit in Alaska. *Condor* 66: 41-50.

Williamson, K. 1946. Field-notes on the breeding-biology of the Whimbrel. *North Western Nat.* 21: 167-84.

—. 1950. The distraction behaviour of the Faeroe Snipe. *Ibis* 92: 66-74.

Wilson, E. M., and B. R. Harriman. 1989. First record of the Terek Sandpiper in California. *Western Birds* 20: 63-69.

Wilson, H. 1989. Short-billed Dowitcher migration at Bottle Beach, Grays Harbor, Washington, during spring 1989. *Washington Birds* 1: 37-39.

Wilson, H. W., Jr. 1990. Relationships between prey abundance and foraging site selection by Semipalmated Sandpipers on a Bay of Fundy mudflat. *J. Field Ornith.* 61: 9-19.

Wilson-Jacobs, R., and G. L. Dorsey. 1985. Snowy Plover use of Coos Bay North Spit, Oregon. *Murrelet* 66: 75-81.

Wilson-Jacobs, R., and E. C. Meslow. 1984. Distribution, abundance, and nesting characteristics of Snowy Plovers on the Oregon coast. *Northwest Sci.* 58: 40-48.

Wood, A. G. 1986. Diurnal and nocturnal territoriality in the Grey Plover at Teesmouth, as revealed by radio telemetry. *J. Field Ornith.* 57: 213-21.

Wood, A. G. 1987. Discriminating the sex of Sanderling *Calidris alba*: some results and their implications. *Bird Study* 34: 200-04.

Zeillemaker, C. F., M. S. Eltzroth, and J. E. Hamernick. 1985. First North American record of the Black-winged Stilt. *Am. Birds* 39: 241.

Index

The primary account for each species is indicated in bold face; other pages refer to material in the introduction and significant comparisons among species.